HANCOCK HOUSE

Encyclopedia of the Lories

HANCOCK HOUSE

Encyclopedia of the Lories

by Rosemary Low

ISBN 0-88839-413-6

Cataloging in Publication Data
Low, Rosemary.
  Hancock House Encyclopedia of the Lories

  Includes bibliographical references and index.
  ISBN 0-88839-413-6

1. Lories 2. Psittacines 3. Brush-tongued parrots 4. Aviculture I. Title.
SF473.L57L68 1997   636.6'865   C97-910519-6

Printed in Hong Kong—ColorCraft

Editing: Dave Bender and Nancy Miller
Production: Lorna Brown, Nancy Miller and Ingrid Luters
Cover Painting: Red-collared lorikeets by Rachel Lewis

Published simultaneously in Canada and the United States by

HANCOCK HOUSE PUBLISHERS LTD.
19313 Zero Ave., Surrey, B.C. V4P 1M7
(604) 538-1114 Fax (604) 538-2262

HANCOCK HOUSE PUBLISHERS
1431 Harrison Ave., Blaine, WA 98230-5005
(604) 538-1114 Fax (604) 538-2262
Web Site: www. hancockhouse.com *email:* sales@hancockhouse.com

Contents

Dedication

To John Courtney whose knowledge and understanding of little and musk lorikeets in the wild inspired me to write a second book on lories.

Foreword

I am delighted to write the foreword for this important book on lories, not only because I share the author's enthusiasm for these charming birds, but also her concerns about their conservation and welfare. Rosemary Low is uniquely qualified to write this monograph. She is the author of hundreds of articles and numerous books, including the standard reference *Parrots, their care and breeding.* Rosemary has a wealth of personal experience as former curator of two major parrot collections, Loro Parque and Palmitos Park, and continues to work with lories at home. She has also traveled widely, observing and studying lories in the wild.

Quintessentially tropical, lories are brightly colored, vibrant, highly engaging parrots whose avicultural popularity has, in the recent past, resulted in large numbers entering international trade. It is difficult to be certain of the numbers involved; figures from quotas, capture, transport and CITES permits are often in conflict with each other. Sadly, some lories, destined to be lost in living room cages, will have been purchased by pet-keepers ignorant of the biology of these specialized pollen and nectar feeders. Others acquired by well-meaning aviculturists and bred with some success no longer grace their aviaries simply because breeding them became commercially unprofitable. Zoos displaying lories must also consider the sustainability of their stock. Currently there is great diversity with a large number of species held but, with a few notable exceptions, little emphasis on managed breeding programs. Lories can make excellent educational exhibits and it would be sad to deny future generations the opportunity to enjoy these attractive parrots. There are other good reasons to maintain lories in captivity. Knowledge obtained of their husbandry may one day be applied to their advantage should captive breeding be required as part of a conservation program. Information on basic nesting biology is difficult to obtain from wild lories; observations from captive breedings fill important gaps in our knowledge. All too often such information is not documented.

In this book Rosemary Low has ensured that her original observations and those of many of her colleagues are made available for the benefit of others. This excellent book fulfills its title by being encyclopedic in scope, addressing the natural history of lories and the conservation issues associated with habitat destruction and trade. It is in the detailed species accounts that the author's comprehensive knowledge and total immersion in her subject shows most clearly. Her observations and experiences, and those of other aviculturists, are presented in such detail that no one with access to this book can plead ignorance of their lories' biology, behavior or particular requirements. *The Hancock House Encyclopedia of the Lories* can be highly recommended. It should not only sit on every parrot enthusiast's book-shelf, but should be frequently off the shelf and regularly consulted.

ROGER WILKINSON, B.Sc., Ph.D.
Curator of Birds, Chester Zoo
Chair, European Parrot Taxon Advisory Group

Preface

There are 53 species of lories and lorikeets, small to medium-sized parrots from the Pacific region. They are arguably the most colorful and beautiful birds in the world—although pheasant enthusiasts might not agree. While some of the birds of paradise are spectacular, females are drab and a number of species are black. For sheer color the lories are unrivalled. Ever since 1896, when Mivart's *A Monograph of the Lories or Brush-tongued Parrots* figured for the first time most of the lory species which were then known to science, these birds have attracted attention. No other volume covered this group of parrots until my *Lories and Lorikeets* was published in 1977. It was a treatise for the aviculturist. That decade saw the commencement of large-scale importations of a number of species of lories and lorikeets but, until then, little had been published to guide the keeper of these birds. A lot remained to be learned (as it does today) regarding their care. My book was the first (by nearly a decade) of several on the subject. But no one had ever tried to bring together available information on aviculture, natural history and conservation of the Loriinae. This idea matured in my mind over a long period.

During that time lories and lorikeets were an important part of my life. From the time I kept my first pair, in 1971, to the present day, I have been extremely fortunate in my contact with these wonderful birds. After the decade of the 1970s, the most exciting era for the lory enthusiast, I made the huge leap from private aviculturist to curator of the world's largest (at that time) collection of parrots—Loro Parque in Tenerife, Canary Islands. Here was assembled more species of lories (indeed, of Amazons, of conures, of cockatoos, etc.) than anywhere in the world. Although the number of lory species in the collection is even larger today than the 26 species and 11 subspecies present then, I had in my care more lories than most aviculturists could hope to keep in several lifetimes. I was equally fortunate when I left Loro Parque for Palmitos Park on the neighboring island of Gran Canaria; a large proportion of the parrot collection there comprised lories, although of fewer species. I was a "hands-on" curator. During my nearly nine years with those

collections, I ringed all the chicks, inspected the nestboxes and hand-reared many of the young. I was constantly observing and learning. Alas, when you work with large numbers of birds, you cannot devote enough time to each individual. Now that I am once again a private aviculturist with a relatively small number of birds, I can do just that—learn to know each bird and its special needs—which, to me, is the most interesting part of bird keeping. Although I have had more than 200 species of parrots under my care and supervision over the years, and throughout that time have maintained my own small personal collection (mainly lories and Amazons), it is the lories which grace my aviaries almost exclusively today. Their color, behavior and personalities are a source of never-ending joy.

In his preface to *Understanding the Bird of Prey*, Nick Fox wrote: "My main reason for writing this book has been to teach myself. Whenever I have dug into a subject, I've written it up as a way of summarizing my thoughts. Sometimes I had to leave an aspect for a few years until I've had time to do more research or investigation into it, or until the subject itself has matured. For example, it is only in the last few years that domestic breeding has become routine."

Those words perfectly summarize how I feel about this book. Much of my knowledge crystallized in its writing, and I learned a lot more by reviewing the literature and through conversations and correspondence with other lory enthusiasts. More importantly, however, in this book all aspects of lories and lorikeets have been brought together under one cover. Now the ornithologist as well as the aviculturist has a source of reference for the accumulated data on this fascinating group of birds.

Everyone is aware of what is happening to our world as a result of human greed and over-population. Very few of the millions of species of fauna and flora which inhabit the earth have remained untouched for the worse. As yet, we are only on the brink of discovering how the influence of man has affected the status of birds and animals worldwide. This book could not have been written even a decade ago because, except in Australia, there had been virtually no field studies of lories and lorikeets. Little was known of the status of these birds on Indonesian and Pacific islands, for example, or in New Guinea. Our knowledge is still small but increasing at an extraordinary rate. A few years hence this book will need to be revised and updated. But our limited knowledge even today shows that many species of lories and lorikeets are declining in the wild for human-related reasons—destruction of habitat, introduction of rats and trapping for trade, for example. The latter should be stopped, but, alas, people in too few countries have been properly educated about the usually unnecessary and often cruel trade in wild-caught birds. Deforestation continues, or is commencing, in all countries where lories are found. A few species can adapt to altered habitats, but many species will face such problems as fragmented habitats and insufficient forested areas for foraging. Virtually all of these birds will suffer from the loss of nesting sites. Logging is already having an effect in Australia, where suitable cavities for nesting form only in very mature trees. In that country, lorikeets are popular garden inhabitants. Unfortunately, misguided feeding by well-meaning people is causing the deaths of probably thousands of lorikeets. It is for this reason that a special section on lorikeets in Australian gardens has been included.

Australia was just one of the countries which I visited when researching this book. It was the first country in which I saw lories in the wild. I was captivated. I returned to Australia on two more occasions and also visited Fiji, the Cook Islands and New Guinea. Observing lories and lorikeets in the wild was essential to my understanding of them. To return again to New Guinea and to visit other areas where lories occur is my aim. I need to know more about the way they live. During my travels my camera was always in my hand. Most of the photographs in this book are my own; the majority are from a vast collection which form a record of the birds with which I have worked, and their young. I prefer to photograph captive birds in their usual surroundings because they look and behave in a more natural way. The photographs may lack technical perfection but they show the birds as and where they are, sometimes as part of a family, one of a flock, or while performing some form of behavior which cannot be recorded in a photography cage.

Many people, aviculturists and ornithologists, have provided information, and I thank them all. Acknowledgments are listed separately. Undoubtedly my greatest inspiration was John Courtney, a remarkable Australian who has studied the lorikeets in his area for more than forty years. This book is dedicated to him. All the opinions expressed are my own, based on what I have seen, but do not necessarily apply in all situations. Any errors are also my own, since I personally typed every word on to disk! My goal was to make current knowledge of every aspect of lories and lorikeets easily accessible, and available to all in book form. My hope is that this book will improve the welfare of the lories and lorikeets kept in aviculture or as pets and will contribute to our understanding of these birds in the wild and, in the long-term, to their conservation.

Acknowledgments

A worldwide network of contacts, avicultural and ornithological, provided information at my request, or copies of published scientific papers. Without their invaluable help the book could not have been written.

Australia   Phil Bender, Alan Dear, Carl Clifton, W. Deuchar, Phil Digney, Gordon Dosser, Mr. Dowling, Roy Dunn, Janet Flinn, Dr. A. Gallagher, Kate Gorringe-Smith (RAOU), Gould League (Victoria), Barry Hutchins, Graeme Hyde (Avicultural Society of Australia), Andrew Isles, Shannon Johnson, David Judd, Monica Kargulewicz, Joseph Kenning, Rachel Lewis, Graham and Glenys Matthews, Harald Mexsenaar, Peter Odekerken, J. Pearson, William Peckover, Len Robinson, Tony Rose, Mark Shephard, Nigel Steele-Boyce (*Australian Birdkeeper*), J. E. Wallace

Belgium   Steffen Patzwahl

Canada   John Tabak

Canary Islands   Roger Sweeney (Loro Parque), David Waugh (Loro Parque)

Denmark   Leif Rasmussen, Peter Them

France   Vincent Guionnet, Roland Seitre

Germany   Thomas Arndt, D. Brockhoff, Armin Brockner, Gerhard Hofmann, Martina Muller (Vogelpark Walsrode), Dr. Robert Peters, Ruediger Neff, Nekton-Produkte, Paul Tiskens

Indonesia   Paul Jepson (BirdLife International), Victor Mason

Japan   H. Shimura

Netherlands   Gert van Dooren, R. D. Flierman, Peter Holsheimer, Jos Hubers, Herman Kremer, H. van Keulen, Anton Spenkelink

New Guinea   Phil Gregory, Mr. and Mrs. B. Love, I. Richardson

New Zealand   P. Eglinton, K. Muller, Fred Rix, Kerry Jane Wilson

Pacific   Claudia Matavalea, Gerald McCormack, Dr. Dieter Rinke, Dick Watling

Philippines   Antonio de Dios

Singapore   Lawrence Kuah, Janice Buay and Imelda Santos (Jurong BirdPark), Patrick Tay

South Africa   Fred Barnicoat, Rex Duke, Arthur Kemp, Dr. Chris Kingsley, Luch Luzzatto, Antonie Meiring, Jan-Hendrik van Rooyen, David Russell, Gavin Zietsman

Sweden   H. Andersson

Switzerland   Dr. R. Burkard, Lars Lepperhof, H. Walser

U.K.   N. Brickle, Trevor Buckell, Peter Clear, Nigel Collar (BirdLife International), Dulcie Cooke, Carol Day, David Field, Jim Hayward, D. Holyoak, Simon Joshua, Adrian Long (BirdLife International), Jonathan Newman, Steve Peacock, Cathy Peak, Tony Pittman, Michael Poulsen, Jon Riley, Craig Robson, Tom Tew (CITES, JNCC), W. Timmis, Rosie Trevelyan, Dr. Roger Wilkinson (Chester Zoo), Rosemary Wiseman, Val Wittcke

U.S.A.   Bruce Beehler, Gwendolyn Campbell (San Diego Zoo), Sharon Casmier, Linda Coates (Library, San Diego Zoo), Roland Cristo, Dr. Dominique Homberger, Jay Kapac, Cyndi Kuehler, Ed Lewins (San Diego Zoo), Richard Parks, David Rimlinger (San Diego Zoo), Dick Schroeder, David Steadman

Thanks are due to S. Sayers of St. Martin's Press for permission to quote from *The Human Nature of Birds* and to University of Texas Press for permission to quote from *A Naturalist in New Guinea*.

Special thanks are due to John Courtney, Dr. Stacey Gelis and Jan van Oosten who read various parts of the manuscript and made valuable suggestions.

I am indebted to the following who supplied illustrations for publication: Thomas Arndt, Bruce Beehler, Armin Brockner, Thomas Brosset, Gwendolyn Campbell, Sharon Casmier, Peter Clear, John Courtney, Carol Day, Alan Dear, the Gould League of Victoria, David Field, R. H. Grantham, Charles Hibbert, Gerhard Hofmann, Jos Hubers, David Judd, J. Kenning, H. van Keulen, Judith Kunzle,

Rachel Lewis, Claudia Matavalea, Ron and Val Moat, Martina Muller and N. Neumann (Vogelpark Walsrode), Ruediger Neff, Jan van Oosten, Richard Parks, William Peckover, Imelda Santos, Roland Seitre, Syd and Jack Smith, Roger Sweeney, Patrick Tay, Rosie Trevelyan and Dick Watling.

All other photographs are by the author, with the exception of the ultramarine lory.

Finally, my thanks go to Klaus Paulmann of Palmitos Park and Wolfgang Kiessling of Loro Parque who provided me with the opportunity to work with so many lory species.

About the Book

This book has been arranged to provide rapid access to information. Part 1 covers many topics pertaining to the natural history and aviculture of lories and lorikeets. These are presented alphabetically by subject. Part 2 comprises detailed accounts for each of the 53 species. Information is presented under set headings. At the beginning of each account the species is identified by its scientific name, followed by the name of the author who described the first specimen to science, and the date of publication of this description. Synonyms of common names follow, with native name, where known. These may vary according to locality (in New Guinea there are 500 languages), thus in some cases the name of the tribe precedes the name. Under the headings of Description and Immature bird, details of plumage and the color of bill, legs and soft parts are given. These descriptions have been made from live birds in many instances, or by reference to published descriptions, or a combination of both. Length is taken mainly from published sources and from live birds in some instances. Weight has been obtained by weighing live captive birds where possible or from published sources in other instances. All these features vary to a slight degree in individual birds. Notes on Range and Status reflect what is known at the present time. In many cases, this may prove to be incomplete or even inaccurate. Few field studies have been carried out on lories and lorikeets, except in Australia. The information in the species accounts is based on personal experience, reviewing avicultural and ornithological literature up to 1996, and also originates from many contacts worldwide who made available unpublished records or anecdotal information. The source of such information is indicated by pers. comm. (personal communication) or verbal comm. The information under Chick weight table is derived from captive birds. Note that when these have been hand-reared from an early age, the weights for the first few days are much lower than those of parent-reared chicks.

In Part 1 and Part 2 cross references in capital letters refer to other relevant entries in Part 1 (e.g., see Feeding Behavior). Many islands on which lories occur are little known to the general reader. Brief descriptions of these islands and their localities appear in the Gazetteer which comprises Part 4. The list of References Cited gives the sources of the published information referred to and, in many cases, allows the reader to obtain further information.

Taxonomic order

I debated long and hard before diverging from the established taxonomic order. This should surely reflect relatedness of the genera, insofar as is possible, or approximately. It is for this reason that I have placed *Lorius* at the beginning. It is closest to *Eos* and *Chalcopsitta*; its former placement between genera of small lorikeets with which it has little in common seemed quite illogical.

Part 1

Alphabetical Listing of Topics

Art, Lories in

By coincidence, rainbow lorikeets were the subjects of two historical landmarks in avian art: the oft-regarded probable first-published illustration of an Australian bird and the first known painting of an Australian bird. It is easy to imagine how this colorful species would have captured the attention of artists who had never seen their like. In 1776 Brown's *New Illustrations of Zoology* was published; the rainbow was featured under the name "blue-bellied parrot." In 1772, the Welsh artist Moses Griffith painted this species—almost certainly from life, judging by the pose and expression. The subject was possibly the bird taken alive on Cook's first voyage. It became the pet of the Tahitian priest Tupaia and finally came to rest as a stuffed specimen in a private London museum. Willis (1991) suggests that the lorikeet may have survived the voyage and was seen by Griffith, who was the servant and traveling companion of Thomas Pennant, who made numerous important contributions to zoology. Griffith might have had the opportunity to see the lorikeet when he and Pennant visited Joseph Banks in September, 1771. Banks accompanied Cook on his epic voyage.

The "discovery" of this painting is another fascinating story. Rick Willis, from the Institute of Education, University of Melbourne, was studying at the Natural History Museum in London in 1987. One lunchtime, while out window-shopping, he was attracted by antique bird prints in an expensive antique shop. His attention was drawn by a gouache (water color) on vellum of *A Green Parrot*. But it actually depicted a rainbow lorikeet—and was dated 1772. Here, before his eyes, was the first known painting of an Australian bird! He bought it! Describing it as "one of the great finds in Australian zoological art," he suggested that "an estimate of $100,000 is probably fairly realistic" regarding its monetary value (Willis 1991).

The first great artist to portray various lory species was the renowned John Gould. Fifteen of his enchanting plates in *The Birds of New Guinea and the adjacent Papuan Islands*, volume 5, depicted lories. Many of these would have been drawn from skins. They must have brought to the attention of countless people the beauty of this group of parrots. The volumes were published between 1875 and 1888.

During the late 1920s and 1930s, *The Avicultural Magazine* (journal of the Avicultural Society) published numerous plates by the talented artist Roland Green. A glance at any of these shows that the birds were drawn from life. Green was one of the first artists, in my opinion, to paint parrots in a truly realistic way that captured the essence of the species. See, for example, the Duyvenbode's lory with golden under wing coverts exposed, in the May 1931 issue.

Today, lorikeets have attracted the attention of several of a wave of young Australian artists who capture on canvas the beauty of their native wildlife. Outstanding among these is Rachel Lewis, whose work enhances the cover of this book. Her almost photographic style shows not only the beauty of the Australian lorikeets but exposes her own passion for these birds.

In another area, that of philately, lories are worthy of mention. These birds are featured not only on the stamps of the countries of origin, but also those of states where they do not occur. In 1996 the Solomon Islands issued a set which featured each of the six lory species which occur there. To date they are unique; all other lories and lorikeets featured on stamps are part of a mixed set. Examples are the beautifully portrayed purple-bellied lory and green-naped lorikeet, K1.50 value. These appear on the Papua New Guinea Research and Conservation Foundation stamps issued in 1993. Collecting lory stamps makes an interesting side hobby for the lory keeper. See also **Conservation** for information on limited edition prints of lories.

Aviaries and Cages

There are at least two different concepts in accommodating lories and other parrots. One method involves housing the birds in a way which is pleasing to the eye, such as the garden aviary or zoo aviary. The other is the farming concept in which as many birds as possible are maintained in a usually limited space for breeding purposes with no regard to the appearance of the cages; indeed, these cages often offend the eye. Perhaps the difference lies in whether birds are being kept for pleasure or for profit. However, appearance is not the most important factor; ease of cleaning, and comfort and safety of the birds are the first priorities, along with size.

Minimum cage sizes

Lories are exceptionally active and playful birds. They do not sit still for much of the day, like an Amazon parrot, for example. Therefore they need more space than other parrots of comparable size. I believe that a breeding pair of a small species up to the size of a perfect lorikeet should be kept in a cage at least 1.8 m (6 ft.) long; other species need a minimum length of 3 m (10 ft.). In European and South African aviculture these requirements are met by the majority of breeders. In the U.S.A. they are not. Those planning accommodation for lories should use the suggested lengths as an absolute minimum. If they cannot be met, the breeder would do better to keep lovebirds or other smaller or less active species; alternatively, keep half the number of lories in larger cages. It grieves me to see, for example, a pair of Musschenbroek's lorikeets kept in a cage 61 cm^2 (2 sq. ft.) and to be told, on inquiring, that in the seven years since purchase they had not bred. It mystifies me why people keep an exquisite bird like a Stella's lorikeet, whose beauty can be best appreciated only when it is in flight, in breeding cages 1.2 m to 2 m (4 ft. to 6 ft.) in length. In the long term, constructing larger aviaries is not much more, or any more, expensive than making

small ones. The occupants will be slimmer and fitter and happier, therefore production will be maintained at a good level over a long period. In small cages, initial production may be satisfactory but over the years it will decline and the quality of young produced will be inferior to those bred in larger aviaries. Disease is more likely to spread in small cages which are crowded together.

A number of South African aviculturists have very impressive lory aviaries, notable for their size and cleanliness. As an example, Gavin Zietsman has between 40 and 50 aviaries for breeding pairs, each one being about 4 m (13 ft.) long and 1.5 m (5 ft.) wide. The aviaries are well designed, with a flat metal tray (to hold food dishes) which slides out for easy cleaning. Overhead sprinklers are used in hot weather. The lory breeding aviaries of veterinarian Dr. Chris Kingsley were even larger; each one was a generous 6 m (20 ft.) long. My impression was that this is the perfect length for medium-sized and large lories. There was sand on the floor (which would not be practical in a wet climate). Only the top half of the aviaries were double-wired; no problems resulted.

Aviary construction

Aviary construction is a complex subject. The reader is referred to books such as *Build your own Aviary* by D. Pearce, published by Blandford/Cassell (London), and *Building an Aviary* by C. Naether, published by TFH in the U.S.A. A few facts should be known about aviaries for lories. The first is that they must be constructed in such a way that they are easy to keep clean. The second is that there should be double wiring between each aviary, or at least at perch level. Some lories are aggressive and if they can reach neighboring birds they will not hesitate to bite their feet or beaks.

Suspended cages

These days many people use suspended cages for lories, which prevents injuries by aggressive birds in neighboring enclosures. Unfortunately, this usually results in small cages. In cages only 1.8 m (6 ft.) long the larger lories seldom even try to fly and there is a danger that they will become overweight. Although the term "suspended cages" is in popular use, few such cages are suspended from, for example, metal beams. Most welded mesh cages are placed on a metal framework approximately 1 m (3 ft.) high. There is a metal surround to the door, but the cage is otherwise made entirely of welded mesh. It is practical to make these cages 3 m (10 ft.) long or even longer. As already mentioned, this is the minimum humane size for the larger lories. The cages should be placed in a position where pressure washing them is possible. If water pressure is too low to permit this, a removable wire grid should be made in the floor of the cage at the center. It should measure approximately 50 cm (20 in.) square. This allows the attendant to stand up inside the cage to reach the corners for cleaning, to replace perches and to catch birds. It is noticeable that perches in suspended cages are seldom changed until they fall down. This is partly due to difficulty of access. The problem can be solved by making a small door in the mesh next to the main perch. Alternatively, a novel solution I saw in California was to use two long metal rods, each with a dip at the end. The perches were scooped out of the holders using the rods and a dexterous handler could replace the perches in the same way. Catching birds in suspended aviaries is achieved either with a long-handled net, or by a slot in the center of the cage which allows a divider to be inserted. The latter method is possible only if there is sufficient space between each cage.

Because all-wire cages are relatively light-weight, an entire row can be moved to another location. This is advantageous if cages are located on grass, instead of concrete, to prevent a build-up of bacteria on the ground below. But for ease of cleaning, the area below the cages should be constructed of concrete. It is not necessary to have solid dividers between each breeding cage in a row of cages. Unlike gray parrots and Amazons, for example, lories do not need seclusion; they tend to be stimulated by the sight of the courtship of their neighbors.

Wire mesh/welded mesh

The mesh suggested is 16 ga., or the heavier 14 ga. for the larger species; ½ in. square mesh is recommended as it keeps out all vermin except young mice. The "Easierect" mesh, 2 in. x ½ in., is popular because no framework is needed, thus aviary construction is quicker and easier. Mesh size of 1 in. (2.5 cm) square is a danger to the smaller lory species as they could trap their heads in it.

Various aviary designs

An example of well-designed lory aviaries is the Lori Atrium at Vogelpark Walsrode, in Germany, with 35 enclosures. These aviaries have two sections. In the heated indoor part, which the public cannot view, the floor is concrete, and the walls are of stone. The feeding tray is set into the door; dishes are removed through a small door in the main entry door. An observation panel in the door allows viewing of lories without entering. Nestlogs are placed in this area. It is serviced from a passage behind, along which the food cart is pushed by a keeper. In the outer section, part of the floor is of concrete and the other part is covered in pebbles (small stones). Sides and roof are constructed of welded mesh. A sprinkler is used every morning (Patzwahl 1991). These aviaries have an attractive appearance and are very suitable for exhibiting lories to the public.

The private birdkeeper will usually have less to spend on his or her aviaries. If the construction is to last and be serviceable it will not be cheap; economy at the construction stage usually results in more expenditure later. These days there is no benefit in using wood for aviary framework. Wood is expensive, difficult to maintain and keep clean, and easily destroyed by the birds. Anodized aluminum is expensive but has proved to be the perfect material with which to construct lory aviaries. It is light, easy to keep clean and always looks new. In the Netherlands, Jos Hubers uses black, anodized aluminum with black rivets. Special plastic corners and T-pieces join the framework together. When I wanted to have cages made from aluminum framework, the cage maker told me he could not make them as he had no way of attaching the welded mesh to the aluminum frame. He suggested the use of steel angle instead, to which I reluctantly agreed; it made the cages very heavy. As they are all on frames with castors, however, they were reasonably easy to move around. Later I discovered a cage maker with a special tool which enabled him to attach welded mesh to aluminum angle. One cage was ordered as a comparison and proved superior in every way—and no more expensive. For aviary framework I would recommend this type of construction. It will prove to be very long lasting. The angle used in my aviaries is 3/4 in. x 3/4 in. and the welded mesh is 1 in. x ½ in., 16 ga. It is strong, light and indestructible. Double wiring between sections is achieved by angling the wire outward at top, bottom and sides for 3 cm (1¼ in.). This is preferable to leaving a space between the aviaries into which there is no access for cleaning. A V-section of wire 3 cm (1¼ in.) wide is used to separate the two sections.

My 5 m (16 ft.) outdoor flights are placed on two courses of bricks on concrete footings. The floor surface is of small (approx. 2 cm long) stones, through which grass and dandelions grow; these are nibbled by the birds, and soften the otherwise harsh appearance of the floor. Framework and floor are easily cleaned using a pressure washer. The indoor section consists of a building lined with melamine in which there are 2 m (6 ft.) long cages on castors which line up with the entrance to the outdoor flight. These can be wheeled outside or pressure washed *in situ*. An entrance/exit flap 20 cm (8 in.) square is closed to shut the birds in at night. It rests on a 7.5 cm (3 in.) bracket. On the first night, the birds needed only a few minutes of persuasion to enter. After the second night they entered readily when required. Inside, the floor is concrete and slopes to a drain and soakaway in the center.

Newspaper is placed beneath the cages and renewed daily. A window on one side has glass reinforced with wire mesh. Lights are on a dimmer for winter use, reducing the lighting gradually over a half-hour period.

Four smaller aviaries measure 2.1 m (7 ft.) long and 91 cm (3 ft.) wide. They form a unit with two aviaries on each side of a 1.2 m x 2 m (4 ft. x 6 ft.) service area. The rear portion of each aviary is open-fronted, but otherwise enclosed. The central division is of aluminum 86 cm (34 in.) long. The sides are made of tongue and groove wood, painted green, and the roof is made of galvanized corrugated sheet. It has a 7.5 cm (3 in.) slope from back to front to allow rain to run into a gutter along the front. The central part of the roof is of corrugated PVC sheeting attached by screws to wooden spars. In winter, the front of the unit is enclosed as protection against the weather, by placing heavy plastic panels on wooden framework into position. These are held in place by wing nuts.

The front and dividing sections of the aviary (the back is against the wall of the other block) are double wired. This prevents injury due to aggression from the occupants of the adjoining aviary or from predators such as cats and owls. The floor is of concrete which slopes to a drain and soakaway in the

service area. All the aviaries are protected by an electronic alarm system.

Except in a tropical or subtropical climate, the aviary should have an inside shelter with service passage. The feeding is carried out from this area. By feeding in the covered shelter, contamination of food by wild birds is avoided, and in cold weather nectar can be prevented from freezing by heating the indoor section with an infrared lamp in each shelter. The birds can be shut inside the shelter at night to keep them safe; this is also an advantage with the larger, noisier lories whose early morning calls might disturb close neighbors.

Where temperatures are consistently low in winter a different style of building might be adopted. In Washington State, Sharon Casmier chose a greenhouse type which can be opened up during the summer months. The building is 12 m (40 ft.) long and 4.6 m (14 ft.) wide. The roof is constructed from clear, corrugated fiberglass sheeting which provides natural lighting, even on cloudy days. Back and sides are constructed from T1-11 siding. Large fiberglass doors run the length of the building. Each door is 3 m (10 ft.) wide and 2 m (6 ft.) high, mounted on a track that allows them to be opened easily on sunny days. The building can be heated by a forced-air oil furnace during long cold spells. The cages are raised off the ground on concrete blocks to provide sufficient space for cleaning beneath them. The floor is covered with 17 cm (7 in.) of sand and gravel. Quick cleaning is achieved by raking the gravel and hosing the cages and floors at the same time (Casmier and Wisti-Peterson 1995).

Careful thought should be given to the floor surface in an aviary. It can be made of concrete, in which case it should slope toward a drain placed in each aviary. Cheaper alternatives are paving slabs or a 15 cm (6 in.) layer of 2.5 cm (1 in.) pebbles. Both slabs and pebbles can be hosed clean. They are easier to lay than concrete.

The aviary roof is best made of welded mesh with a covered section of corrugated galvanized sheet and/or corrugated PVC to protect the occupants from the weather. In hot climates, a false roof a few inches above the real one will help to reduce the temperature inside the aviary.

Planted aviaries

Outside Australia, few aviculturists keep lories in planted aviaries. This is a pity—because it is ideal from the birds' point of view. Lories love to be in foliage. With earth floors, however, precautions must be taken to keep out mice and other vermin. Half-inch square welded mesh should be sunk several inches under the ground before the aviary is constructed. The sides of the welded mesh should rise up to meet the aviary framework. Lories are essentially arboreal species; in the wild they spend their entire lives among foliage. In aviaries their behavior is more natural and pleasing when their aviary is planted; they must experience a feeling of well-being which is totally missing in a cage or bare aviary. Another advantage is that it reduces aggression in species such as *Vini* lories because the less dominant bird can quickly disappear into the foliage at times of stress. Lories are naturally playful and enjoy swinging on thin branches. I recall seeing a pair of Tahiti blue lories at San Diego Zoo swinging and tumbling together on a chewed up palm frond inches from the floor—and thinking that this is how all lories should enjoy aviary life.

At San Diego Zoo all the lory species on exhibit (blue-crowned, solitary, red-flanked, Stella's and cardinals) are housed in aviaries about 2.4 m (8 ft.) square, densely planted with palms, staghorns and shrubs. The *Vini* species are quite destructive to most plants but they do not damage Australian tree fern (*Cyathea cooperi*), holly fern (*Cyrtomium falcatum*) or papaya (*Carica papaya*). At times it can be difficult to view the birds, but usually they are visible—and what a joy it is to see them moving through the plants, especially when the misters are turned on. Then they bathe in the foliage. Two species which could not be kept in planted aviaries there were iris and Musschenbroek's; they proved too destructive.

At Umgeni River Bird Park in Natal, South Africa, the lories are kept in a U-shaped aviary complex which is quite heavily planted. The plants, including *Agapanthus*, extend inside and outside all the aviaries. The effect is most pleasing and very little damage is done to the growing shrubs by the lories.

In England, Rosemary Wiseman keeps small lorikeets in planted aviaries. Even the iris lorikeets had not destroyed the *Ficus* trees, which were self-planted. They had grown from the soaked dried figs fed to the lorikeets! However, I suspect that when the *Ficus* trees reach more than 1.2 m (4 ft.) high,

the lorikeets will be more destructive. In the aviary containing a family group of red-flanked lorikeets, a hybrid conifer (*Cupressocyparis leylandii*) and an elder bush (*Sambucus* sp.) were growing with minimal damage. The lorikeets did eat the flowers of the elder. Another useful plant, introduced from the Himalayas, is Indian balsam (*Impatiens glandulifera*), which spreads quickly and produces an attractive pink flower. It grows up to 2 m (6 ft.) tall and is found, in the U.K., growing near riverbanks and wet waste places.

Dick Schroeder of California has bred both Meyer's and Stella's lorikeets (at different times) in a planted softbill aviary 32.5 m^2 (350 sq. ft.) in area and 3 m (10 ft.) high (Schroeder 1995b). Wilson (1992) preferred to house his musks in aviaries at least 4 m (12 ft.) long, 1 m wide and 2.4 m (8 ft.) high. He commented that "ideally a planted aviary would suit Musks beautifully, although some softer species of bush may not survive. Hardy, robust bushes and small trees can even be grown in the sheltered area with care and frequent watering, and Musks love to play in water soaked shrubs." He noticed that on very hot days his lorikeets would hang on the wire closer to the ground, but those in planted aviaries would stay higher up, in the bushes, taking advantage of the coolness that the foliage offered.

Aviaries can be planted with Russian vine (*Polygonum baldschuanicum*), elder (*Sambucus* spp.), honeysuckle (*Lonicera* spp.), *Ficus* and passion flower (*Passiflora* spp.). In warmer climates, hibiscus is suitable. In Australia, native shrubs can be used, including those which lorikeets feed on in the wild, such as *Banksia*, *Callistemon* and *Grevillea*. However, unless the aviary is large or the plants are very well established before the lorikeets are let loose, they may do so much damage that the shrubs will not survive. Obviously, the larger the aviary, the less damage lories are likely to inflict upon foliage. I once kept a pair of Tahiti blue lories in a tiny planted house which measured 1.2 m (4 ft.) square and 1.8 m (6 ft.) high. It was thermostatically controlled at 18°C (65°F) and a sprinkler system was installed for the benefit of birds and plants. The lories delighted in bathing in a *Kentia* palm when the sprinkler was turned on. In this small space they were quite destructive; they totally destroyed a passion flower vine and severely damaged a *Ficus*. It was necessary to rotate plants in tubs to allow them to recover.

In 1978 when I visited Joe and Margie Longo of Kent, Washington, they told me that their lory aviaries were not purposely planted. The wild plants in the aviaries died down during the fall and winter, but grew back in the spring. Their growth was so rapid that even the larger species such as the Duyvenbode's and black-caps did not destroy them. By late spring the aviaries were at least half-covered with green plants.

Bacterial Infections

Bacterial infections are a major cause of death of captive lories. Species of bacteria most likely to be responsible are those of the genera *Salmonella* and *Klebsiella*, and also *Escherichia coli*. However, *E. coli* may be part of the normal intestinal flora and is not necessarily pathogenic. The main reasons why death from gram-negative bacterial infections is common in lories is because they are fed a liquid diet which often has a high content of simple sugars. In warm temperatures this is a perfect medium for the growth of bacteria. This is why the liquid food should be changed twice daily—or even three times in temperatures over 35°C (95°F). Lory cages and aviaries need cleaning more frequently than those of other parrot species. If this task is neglected, harmful bacteria will proliferate.

In my experience, salmonellosis is the most serious disease from the viewpoint of the lory keeper in a warm climate. Annually it will make itself known in summer. Because affected birds seldom show abnormal behavior until a few hours before death, treatment usually comes too late. If birds are not inspected several times daily, nothing abnormal will be noticed and a bird could be picked up dead. Although at least one avian pathologist of note believes that lories which carry salmonellosis do not exist (i.e., exposed birds are either noncarriers or die), this was not my experience. I recall the sad case of a beautiful male red-flanked lorikeet. Two of his females and most of his young had died. Repeated tests gave the result that he was shedding *Salmonella*. The reluctant decision was made to euthanize him.

Poor breeding results, and deaths of adults and chicks, should always be investigated by an avian veterinarian. There is a high possibility that gram-negative bacteria are responsible. The cause could be poor nest hygiene. It is not advisable to accept chicks for hand-rearing from other collections because of the disease risk. However, a decade or more ago a breeder asked me to hand-rear the chicks from his pair of blue-streaked lories, whose chicks invariably died at a few days old. He brought me a newly hatched chick; it, too, died soon after. The postmortem report read as follows:

> This chick showed a very pale, mottled liver and some congestion of the lungs. Culture from lungs, liver and intestine all produced a heavy, pure growth of a haemolytic *Streptococcus* (apparently of Group C, otherwise known as *Streptococcus zooepidemicus*). There is little doubt that the origin of the bacteria is the nest environment. Contamination of the chicks just after hatching, or possibly of the eggs prior to hatching, is the cause of the problem.

I impressed on the owner the need to clean and disinfect the nestbox. In the next clutch both eggs hatched. I cannot recall whether I received them just prior to hatching or as newly hatched chicks. This time both chicks were successfully reared.

It must be realized that bacteria are found naturally in most living organisms, internally and externally. The type depends upon the species. The normal intestinal flora of parrots is made up predominantly of a group known as gram positive, such as lactobacillus, staphylococcus and streptococcus. These bacteria usually take several weeks to colonize the tracts of newly hatched chicks and are passed to the chick by its parents or other organisms with which it comes into contact. The natural intestinal bacteria (or flora) aid in the digestion of food which the bird itself cannot digest, and protect against the colonization of harmful bacteria such as those mentioned above, and also *Pseudomonas*. It must be remembered that treating bacterial infections with antibiotics and antifungal preparations also results in the destruction of the "friendly" microorganisms. In other words, their use weakens the bird's immune system. This is why they should never be given on a regular or prophylactic basis or given at a greater strength than that recommended by a veterinarian.

Probiotics are said to counteract the harmful effects of antibiotics by replacing the natural flora destroyed during treatment. In recent years a number of different kinds have been marketed for birds. They vary in their efficacy. Frankly, some appear to be useless. However, it is much easier to administer them to lories than to other parrots as they can be added to the nectar at the recommended dosage.

The opinion of Ted Zompa, a microbiologist and lory breeder, is worth quoting. "Of all the parrots encountered during my experience in aviculture, I find lories and lorikeets the most resistant to infection. In fact, many of the bacterial diseases seen in amazons and other parrots are rarely seen in these birds." He stated that *E. coli* and *Salmonella* are the organisms most often responsible for bacterial infections in lories. *E. coli* can be cultured from virtually any surface that comes into contact with humans:

> Contamination from *E. coli*, and to a lesser extent *Salmonella*, usually results from transfer by a human to birds, thus thorough hand washing is essential to protect your birds from these organisms. Additionally, cleanliness of cages and food dishes (particularly with *Salmonella*) is essential.

> For treatment and prophylaxis of bacterial infections I use the human antibiotic Floxin. This drug is related to the antibiotics Cipro (human) and Baytril (veterinary). It is a member of the class of antibiotics known as quinolones which inhibit the replication of bacterial DNA. They are excreted by the kidney and only minimally metabolized in the liver. This means that they have a relatively wide margin of dosing safety. In other words, a mild overdosage is unlikely to cause any harm, nor will extended treatment lengths. All three of these drugs are very effective for the treatment of gram-negative infections, particularly *E. coli* and *Salmonella*. While this drug is not recommended for children, I have on occasion used it for chicks without any adverse effects (Zompa 1995).

Baytril (enrofloxacin), made by Bayer, is very widely used by veterinarians worldwide for treating bacterial infections.

A final point regarding *Salmonella*. This organism is often found in uncooked chicken. On no account should a chopping board used to cut fruit for lories also be used to cut up raw chicken. A separate chopping board should be kept for bird food.

Banding, see **Ringing**

Bathing

Few birds are more enthusiastic bathers than lories. This habit, indulged in daily or as often as possible, is essential to keep their plumage in good condition. As they feed on nectar and pollen, their feathers could easily become soiled. Lories tend to do everything with great energy and enthusiasm, and this is certainly true where bathing is concerned. In the wild, varied lorikeets have been observed bathing by dashing straight at the water and hitting it with their wings and bodies (Tarr, in Forshaw 1981). They then flew to a nearby tree where they preened themselves.

An unexpected bath site was that recorded by Morrison (1987). She observed a flock of musks in a large yellow box in Wattle Park, Burwood, Melbourne. A group of them congregated around a fork in the tree, about 3 m (10 ft.) above ground. She and other observers saw birds appearing out of the fork looking very bedraggled. Their places were taken by two more who disappeared, headfirst, inside. When they reappeared with wet plumage she realized that the musks were bathing in the fork. It was full of rain water! "Two by two the lorikeets headed in to come out totally saturated from their beaks to their tails. Soon the air was filled with spray as the bathing and preening went on at a frantic pace until all the birds had bathed several times."

Lories will also bathe in wet foliage, after or during rain. Jan van Oosten has observed this in the following species: cardinal, Massena's and yellow-bibbed in the Solomon Islands and green-naped, purple-bellied and red-flanked in New Guinea. Sometimes they hang upside down with wings extended and tails fanned, until they are soaking wet. Massena's lorikeets have been seen bathing among the needles of *Casuarina* trees (van Oosten pers. comm. 1996).

Lories kept in planted aviaries will bathe in wet foliage. It is unusual for captive lories to show reluctance to bathe, but should this occur a fresh branch with leaves which have been sprayed with water will probably encourage them to take a bath. Most lories will bathe in a container of standing water. This should be shallow and considerably larger than their body size. They have a habit of laying sideways in the water and vigorously beating their wings. The water goes everywhere; the surroundings will be soaked! Bathing is a contagious activity. Once one bird starts, most of those within sight will follow suit. In some zoos which have colonies of lories and provide a large bathing pool for them, the sight of a group of bathing lories will attract a large crowd of bystanders! The colorful soaking-wet lories provide hilarious entertainment. One friend, whom I have no reason to disbelieve, told me of a lory aviary in a bird park where a small chute, about 46 cm (18 in.) long, was placed above the bathing pool. Apparently the lories would follow each other down into the water, like children at a water park!

Captive birds will even break a thin layer of ice on water in an outdoor aviary, to take a bath. My dusky lories regularly did this. Some crave a bath, no matter how cold the day—and it never did the duskies any harm! Make sure the container is large enough. My iris will bathe in a hook-on water dish only 12 cm (5 in.) long and 6 cm (2½ in.) deep in which their body neatly fits—but they prefer a larger dish in which they can really splash about. Some of my lories will even bathe late at night and go to roost with the plumage still damp. But roosting side by side, they dry quickly. This would not be recommended for a single bird.

Lories are unhappy if deprived of water and captive birds face a greater likelihood of becoming ill. Fungus and bacteria can thrive on the skin and feathers; regular bathing prevents this. If a bath is not available, lories desperate for moisture on their plumage will even bathe in nectar containers. If this is allowed to happen on a regular basis, their health will suffer.

Beak Color

Observant aviculturists will have noticed that the beak color in lories may vary in individuals of the same species. There are four basic reasons. The first possible reason is the case of an immature bird, especially in the many species with orange beaks; young birds of these species have brown beaks until they leave the nest. It is generally at the age of three or four months that adult beak color is acquired. Secondly, birds of orange-beaked species which have a very pale, washed out orange beak ("light-beak syndrome") are sick; liver disease is a possibility. The so-called light-beak syndrome is seen in *Charmosyna* species, notably Stella's and

Josephine's. Blood profiles of some of these birds have revealed a low level of haemoglobin; affected birds always seem to be hungry. It has been suggested that a little iron should be administered to such birds. However, this is not recommended as it would be easy to kill them with an overdose. Another suggestion is that the liver is damaged, therefore amino acids cannot be metabolized. This sounds feasible. It has been recommended that glucose be added to the nectar of affected birds. It is worth recording that the female of a very prolific breeding pair of Josephine's lorikeets which bred in my care over a period of five years, had a light beak and a swollen liver for most of this period. She was still alive, in this condition, at the time of writing.

An incident with another female Josephine's lorikeet produced a startling example of how quickly the beak color can change if the liver is threatened. The morning after being placed in a new cage, a pair of Josephine's lorikeets appeared sick. (See **Poisoning, Heavy Metal**.) The cause was ingestion of zinc from galvanized wire of their new cage. By the following day the upper part of the female's upper mandible had turned white. She received treatment and two days later her beak color had returned to normal. On this day she appeared to be fully recovered. The speed of the change in the beak color and that of its return to normal was remarkable.

Where the beak color of every bird in a collection is slightly duller than normal, the reason is probably a dietary deficiency. According to Jan van Oosten (pers. comm. 1996), this occurs due to lack of Vitamin A. Within 60 to 90 days of adding liquidized raw carrot, beet or Spirulina to the nectar, the deep orange beak color will be restored, if no liver or kidney problems exist.

It should be noted that in some *Charmosyna* (e.g., Stella's), *Trichoglossus* and *Eos* species, the beak darkens when the bird comes into breeding condition. Kathy Peak noted that both mandibles of her two pet male Swainson's (rainbow) lorikeets flushed much darker red, as opposed to their normal orange color, "when they came into mating condition. At this time they have been seen to copulate with toys, perches and even my arm, given the chance!" (Peak pers. comm. 1995).

In immature birds of *Trichoglossus* species (including varied lorikeets), a difference in beak color of young in the same nest has been noted. It is possible that there is a sexual correlation. For example, in one nest of varieds, the male had the beak black on fledging and the female had a pinkish beak. Breeders might keep records of beak colors on fledging of birds which they intend to keep or to have sexed at an early age.

Beginners, Species for

The most readily available and least expensive lory species are in this category; initially they were imported in large numbers and bred freely. They are more tolerant of mismanagement than, for example, *Charmosyna* and *Chalcopsitta* species. The lorikeets recommended most highly for beginners are green-naped and Goldie's or, in Australia, the rainbow. All are popular, easily located, beautiful and prolific. Other suitable species are violet-necked (*Eos squamata*) and Meyer's and, in Australia, scaly-breasted.

It is most unwise to start with the rarer species, and unfair to the birds. Beginners may be misled into believing that they can keep any lory primarily on a dry diet. While some species will survive without nectar for a period, the rarer ones (except iris and Musschenbroek's) would die. If one bird of a pair of a rare species dies, it would be difficult to replace without good avicultural contacts, which beginners do not have. The rarer species need a greater degree of expertise and an improved quality of accommodation.

Behavior, Agonistic (Combative)

In the wild

Agonistic behavior is sometimes seen in lories in the wild. It is mainly territorial, observed while lories are feeding or protecting their nest site, and can be directed toward their own or other species. As an example, Phil Gregory once watched two black-capped lories of the race *somu* at Ok Menga. They were fighting so heatedly that they fell from their perch, high in a tree, and tumbled to the ground, squawking noisily and still locked together. Then they recovered and flew off (Gregory pers. comm. 1996). In a study of the feeding behavior of scaly-breasted lorikeets, Tim Hamley classified agonistic

behavior under "squabbling" and "fighting," based on the outcome of the encounter. Fighting behavior consisted of one bird chasing another and directing pecks at it. Often the bird's wings remained folded and the slight patch of red feathers on the lesser wing coverts appeared to be more evident. This behavior appears to be highly ritualized; in more than 20 encounters no attempted pecks made contact. "Squabbling," although seemingly composed of the same movements, differed in that both birds pecked at each other and although they might move from branch to branch, or even tree to tree, the encounter did not end with one of the birds leaving the area. Both types of agonistic behavior were accompanied by a call in which the elements seemed more rapid and of a higher pitch than normal. The only interspecific agonistic behavior observed occurred with noisy miners (*Manorina melanocephala*). Here again, squabbling and fighting could be differentiated on the part of the lorikeet, the difference being that the miners never retaliated. They either moved away or left the area (Hamley 1977).

In captivity

Agonistic encounters often lead to death because the subordinate bird cannot flee. There is little doubt, however, that some lories work in pairs with the deliberate aim of killing another bird, showing a malicious kind of behavior which is rarely seen in other parrots. This is one reason why few lory species are suitable for mixed collections. Some species, such as the chattering, are notorious killers; by now this fact is probably so well recorded that they are no longer kept in circumstances where this can arise. What is not well known is that the smallest and most innocent-looking species can be equally guilty. A well-known aviculturist with enormous planted aviaries related a shocking story. He found a touraco dead with injuries which had obviously been inflicted by another bird but he was at a loss to know which one. Some time later he heard a crested jay shrieking in terror. On hurrying to the aviary he found it was being attacked by a pair of red-flanked lorikeets (*Charmosyna placentis*); they were standing on its back and pecking at its eyes. The unfortunate jay died of its injuries. It is strange indeed that they should have attacked species so much larger than themselves. I have seen this species living and breeding in harmony with finches in a small aviary. Stella's lorikeets had been kept in the same aviary before the red-flanked were introduced and had never caused any problems. However, I believe the potential for agonistic behavior exists in this species. Roland Cristo told me how a couple of young Stella's kept in a cage with two young Duyvenbode's deliberately kept the latter away from the food, so that the two species had to be separated. Even young lorikeets are potentially dangerous. I once housed together two newly independent iris lorikeets with two young Meyer's. One of the latter had most of its upper mandible ripped off by the iris. (The Meyer's survived for years, although the beak never grew.) This type of incident is fairly unusual, but breeders should be alert to the possibility.

A single lory can act with deadly aggression toward another bird, as for example, when a potential mate is introduced into its aviary. (This is a serious mistake. All parrots should be introduced in neutral territory.) My pet black-capped lory can react with unbelievable speed when what he considers to be a foreign object is introduced into his cage. On a couple of occasions when I made the mistake of trying to sponge clean the perch (instead of removing it for cleaning), his reaction was so fast and malicious that I knew if that sponge had been another bird, it would have been killed instantly. And yet he is normally sweet-tempered and affectionate.

Birdrooms

A birdroom can be located inside a house or other building, or it can be a separate, free-standing structure. A room in the house needs to be tiled or lined with a material which is easily cleaned, such as shiny white wallboard or modern bathroom wall materials. An indoor birdroom is usually suitable only for a small number of birds. Cleaning will become an overwhelming task—unless there is a drain in the floor. An outdoor birdroom may be a permanent building, custom-built, a modified garage or even a modified trailer. Ease of cleaning is of primary consideration; therefore, in a custom-built room, a sealed concrete floor, drain and soakaway should be incorporated into the design.

In a trailer, Robin Stockton and Samuel Tucker of Pennsylvania used fiber-glass shower bottoms 91 cm (3 ft.) square, as the cage floor. A piece of 7.5 cm (3 in.) diameter PVC pipe extended from the drain of the shower bottom through the floor of the

trailer to discharge on the ground below. The pipe was curved, therefore always contained water, preventing anything from entering. This unusual method proved very successful (Tucker 1994).

In a birdroom all corners and joints should be sealed with silicone to make it waterproof. For preference, glass used in windows should be of the type reinforced with fine wire, thus assisting security and preventing escapes due to glass breakage. If the cages lead to outdoor flights there must be a means of closing the entrance/exit hole (see under **Aviaries and Cages**). This ensures that the birds can be confined inside or outside during cleaning, and prevents heat loss during cold weather. A good method is a flap which drops down to form a landing platform.

Other important considerations are electric power points (see **Electricity**) and ventilation. Professional advice should be sought on both. Good ventilation is essential; poor ventilation presents a serious health risk because viruses, bacteria and fungal spores will stay in the atmosphere. An ionizer will help to remove them but is not an adequate substitute for a good ventilation system. A poorly ventilated room is not a pleasant atmosphere in which to work or to live.

Bleeding (as result of injury), see **Nails, Overgrown**

Breeding (Captive)

Generally speaking, lories are not difficult to breed in a captive environment provided that certain requirements are met (see also **Aviaries and Cages, Diets** and **Nestboxes**). They are sexually mature at between one and four years (see **Reproductive Span**). If male and female are introduced when young, they are usually compatible; difficulties may occur when trying to pair older birds, especially if one of them was previously paired to a bird with which it had formed a very strong bond. Nest mates of the opposite sex should not be kept together for too long as they may form a very strong bond which makes it difficult to pair them later. This is especially true of *Charmosyna* lorikeets. Lories are extremely affectionate—more so than many parrot species—and a compatible pair spend much of the day in close physical contact. This is the case regardless of their sex; a strong bond can form be-

tween two males or two females. If a breeder discovered that he had two birds of the same sex
together, he or she would make a big mistake in
believing that a third bird of the opposite sex could
be introduced into the aviary in the hope that it
would chose its partner from one of the existing two
birds. In the case of the medium-and large-sized
lories, it would almost certainly be killed within
minutes, unless the birds were very young. This
would also apply with some of the smaller species.
Neither do lories accept other birds in the aviary
when they are nesting, although there are some rare
exceptions (see **Colony Aviaries**). However, lories
can exist in trios of two females and one male. This
seems to be possible only when the females or all
three have grown up together. It is only likely to
occur when several young birds are kept together up
to breeding age. It is a pointless exercise as it would
be surprising if the male fertilized both females.
Both Jorgen Bosch and I kept together a trio of
Goldie's lorikeets. Both females laid at the same
time but only one (always the same one) produced
fertile eggs (see under **Goldie's Lorikeet**). This
also happened in a trio of black lories. The three
birds shared a great rapport and the unmated female
helped to feed the young bird after it left the nest
(and possibly before).

Breeding pairs

Of course, the normal way of accommodating
breeding pairs is one pair per aviary (or cage).
Young, unrelated birds should be acquired as breeding stock. Because many people are impatient, they
start with adult birds. As breeding pairs are seldom
available, this usually means buying single adult
birds and trying to pair them up. Sometimes the
male and female are instantly compatible and breed
soon after purchase, especially if they have been
kept alone and are craving the company of their own
kind. Often, however, it may be several years before
two adult birds start to reproduce. It is often difficult
to know what triggers the desire to breed. A friend
had a compatible pair of Stella's lorikeets; she kept
them for seven years, during which time they made
absolutely no attempt to breed. The other lorikeets
in her aviaries all bred well. Finally she decided that
her conditions apparently did not suit them and she
gave them to a well-known bird garden. Within
weeks they were on eggs!

Failure to nest

It is very rare for a compatible pair of lories to make
no attempt to breed over the long term. I have seen
lories breeding in all kinds of circumstances—some
of which were far from ideal. Undoubtedly, some
species are easier to breed than others. Generally
speaking, those which are common and numerous
in the wild nest more readily in captivity than those
which occur in small numbers. As with many other
parrots, this is partly because they are not habitat
specialists; they take a wider variety of foods and
are more adaptable.

If lories make no attempt to breed, the most
likely reasons are as follows:

1. Male and female are incompatible (see
   Incompatibility).
2. Poor management: diet or accommodation are at fault. An apparently small point
   such as the nestbox sited in the wrong
   position could be to blame.
3. Bacterial infection (possibly low-grade)
   or other infection (see **Bacterial Infections**).
4. The "pair" was incorrectly sexed.
5. One or both birds are too young.
6. Stress, including a disturbance in the immediate environment such as the presence
   of noisy or aggressive birds.

Most people who start to breed lories are fascinated by their color and beauty; they wish to acquire
as many species as possible. In due course they will
realize the wisdom of acquiring two or three unrelated pairs of a few species, rather than one pair each
of a greater number of species. Little progress is
made with a species over the long term by a breeder
who has only one pair. This is especially so with the
rarer lories. It is an enormous advantage and of
benefit to aviculture in general to be able to offer for
sale unrelated young pairs. Another mistake which
is made by even very experienced breeders is not to
retain young birds for future breeding purposes. I
recall being shown a 12-year-old male of a certain
species which was bred from an extremely prolific
pair. This pair had bred for 20 years, rearing two or
three clutches a year. But the breeder had kept back
only a couple of youngsters and when one of them
died, he was unable to find a mate for the other one
in the entire United States. If he, and a few other
breeders, had kept two or three pairs of the species,

the situation would not have arisen. The problem occurred because the birds had been too widely dispersed. They had been so prolific he could not envisage the day when he would not be able to obtain more from other breeders, if he so desired. But the truth is, without a long-term breeding plan which involves several breeders, any species of exotic bird not present in large numbers can quickly be lost from aviculture.

Forming new pairs

Another possible problem with prolific species is that very successful pairs can dominate the gene pool in the country in which they are kept. Marc Valentine, the well-known American geneticist, breeds only yellow-backed lories. To prevent one successful pair from dominating the gene pool, he has broken up pairs at the end of the breeding season. In his experience, by the start of the next breeding season they have formed a pair bond. One male has produced young with three females (Valentine verbal comm. 1996).

Hand-reared birds for breeding

When lory breeding commenced in earnest, in the 1970s, it was, of course, during the days of mass importations, and virtually all the birds available were wild-caught. Now, two decades later, most of the lories available are captive-bred—or domestically bred, to use the American term—although most species are a long way from being truly domesticated. Comparatively few species have been bred for more than four or five generations. Many of the lories available are hand-reared. It is often debated whether hand-reared lories are suitable for breeding. The answer is that it depends on the way in which the bird has been reared. If it has had no contact with its own species during the first year or so of its life (as is often the case with birds reared for the pet trade), it may be totally useless for breeding. It has never learned the normal behavior and vocalizations of its own kind. In addition, some species imprint very easily on humans so that this preference may be irreversible—or take years to reverse. Species most susceptible are the *Chalcopsitta* species and dusky lories. However, almost any species can become imprinted if the bird has been kept as a pet with a very close relationship with one human.

If lories are hand-reared with the intention of using them as breeding stock, and precautions taken to ensure that imprinting does not occur, they should be just as suitable for breeding as are parent-raised birds. Jay Kapac of California noted that his most productive pairs were hand-reared; they were calmer and quicker to reproduce. However, tame birds can be very aggressive toward the person caring for them when they have eggs or young. At this time, some never miss an opportunity to bite! This also applies to some parent-raised birds but is more likely to be a problem in those which have been hand-reared.

Keeping records

One aspect to which the breeder must pay attention is keeping records and identifying individual birds; without identification, most aspects of record keeping have little use. A chart should be kept for each pair showing details of all clutches laid and their outcome, together with identification numbers of the young reared. Ringing (banding) is the most usual method of marking individual birds for identification purposes. See **Ringing**.

Specialization

Few private breeders choose to specialize in one species or genus. Among the exceptions are Leann Collins and David Sefton who chose the red lory. They pinpointed some of the advantages: "By specializing in a species, you are able to match up compatible pairs. Three pairs just laid for the first time this year. Previously, they were set up for two and a half years with other mates. They had been with their new mates for less than six months. Species specialization allows us this flexibility. Also, you learn about behavior, color patterns and many species-specific characteristics that would never be possible if you weren't able to observe so many of one species together" (Collins and Sefton 1994).

Studbooks

Since the late 1980s, zoo organizations and specialist avicultural societies have started to keep studbooks for the rarer species of birds and animals. Those who are interested in certain types of parrots, or one particular species, be it a lory or lories, cockatoos or Amazons, form groups. They decide which species are most in need of attention (rare in aviculture and/or rare in the wild) and appoint someone to keep records of all the birds of a certain species whose owners wish to participate. Hatchings, deaths and changes of ownership are recorded, together with surplus birds and those wanted. This is known as a studbook. There are now studbooks for a number of lory species. They indicate upward or downward trends in numbers being reared, thus pinpointing the species which need most attention. Studbooks enable those who participate to find birds of that species and to dispose of their young to serious breeders looking for the same species. The information is not available to nonmembers. All breeders of the rarer species have a responsibility to participate, in order to enhance the future of that species in aviculture. The names of studbook keepers are available from lory societies and, in the U.K., from Dr. Roger Wilkinson (Chester Zoo, Upton, Chester CH2 1LH), the chairman of the EEP Parrot TAG.

Breeding Problems

During the breeding season problems can occur, as with all livestock. Some of the most common are mentioned here.

Broken eggs

Not uncommon with young pairs, it is important to stop this as soon as possible before it becomes a habit. When shell remnants soon after laying indicate what is happening, it is more likely to be the male who is to blame. As soon as the next egg is laid it should be replaced by a plastic or otherwise artificial egg. The egg is then placed under another female or in an incubator. If the aim is to have parent-raised young, the pipping egg can be returned to the nest. In many cases, the male will then stop trying to destroy eggs. A nestbox with a baffle also helps to prevent egg-breaking. A baffle is a shelf placed above the area where the incubating bird sits; it prevents a bird which is entering the nest from dropping straight down on to the eggs. Many lory breeders use this type of nest. If these actions do not prove successful, as a last resort, the male can be removed from the aviary just before the egg is due to be laid. In most lory species imminent egg-laying is indicated by a pronounced bulge near the female's ventral region. At least removal will pinpoint whether the male or female is responsible for egg-breaking. In some parrots, eggs are broken because the female has produced thin-shelled eggs due to a calcium deficiency. This is very rare in lories and would indicate a totally inadequate diet (or possibly a very old female).

Males not incubating in species in which incubation should be shared, usually indicates the male is too young.

Failure to feed newly hatched chicks

This is usually due to inexperience. If an older lory chick is available, perhaps one of four or five days, the small chick should be removed for (hopefully) temporary hand-feeding and the older chick placed in the nest. Its more aggressive demands for food usually result in it being fed. It should be checked after three or four hours to discover whether this has occurred. If so, the younger chick can be placed in the nest next day and the older chick removed.

Chicks dying in the nest before two weeks of age

The most common cause is due to the parents ceasing to brood the young during cold weather. If chicks are found dead with full crops, this could be the reason. Alternatively, a bacterial or *Candida* infection might be suspected. Nests containing chicks should always be inspected at first light (in pairs used to nest inspection) during cold weather. Chilled—and even nearly-dead—chicks often can be revived using heat. The temperature should be gradually increased over a period of a few minutes.

Chicks plucked in the nest

This is probably the most common problem with which lory breeders have to contend. When the feathers start to appear at between three and four weeks (or even the down at an earlier stage), they are removed by one or both of the parents. They usually start with the feathers or down on nape and wings, and progress to the underparts. The degree of plucking varies; it may not be very serious or it may result in all the chick's contour feathers being removed. In the worst cases chicks will have to be removed for hand-rearing. In hot climates and in suspended cages, young can be allowed to fledge into the cage provided that their flight feathers are present and they are able to fly. They should feather up normally within about six weeks of being removed from their parents. However, if there is any chance of them catching cold, they should be removed for hand-rearing at an earlier stage. Can feather plucking be prevented? It is less likely to occur if the nest is kept dry; if necessary, the nest litter should be changed daily. One can also try placing strips of bark or small twigs in the nest to divert the plucker from removing the chicks' feathers. This has been known to be successful. Budgerigar breeders have solved the problem by placing the nestbox on the cage floor with the lid open. The light interior prevents the adult birds from spending much time in the nest. This might be tried with lories but one would need to check frequently to ensure that the chicks were being fed. (See also **Feather Plucking**.)

Wet nests

Soiling the nest interior by adults occurs with some pairs. This is especially the case with small *Charmosyna* species. I believe it is due to the fact that in the wild they often nest in epiphytes, not in tree holes, and the excreta would drain out of the bottom of the nest. Jay Kapac of California solved this problem in a pair of dusky lories by replacing the nestbox with a smaller one measuring only 18 cm (7 in.) square and 25 cm (10 in.) high. His theory was that if the nest is so small that the bird's tail touches the back, it would exit the nest rather than soil it. I have tried this with a pair of *placentis* which soiled their nest but it was not successful. The best remedy is to drill holes in the bottom of the nest; also have an identical nestbox ready to change over. Young chicks could die from chilling in a wet nest.

Parental aggression after young leave nest

As soon as this is noted remove the aggressive parent until the young are independent. The other parent will feed the young—except in the case of a female who has already laid the eggs of the next clutch. If this happens the young must be removed. Hand-feeding is seldom necessary as most lories learn to drink nectar within a couple of days of leaving the nest. Obviously they should be watched closely after removal to ensure that they are independent.

Aggression directed at the female

Lories are extremely excitable birds. During the breeding season some males become highly excitable and territorial. This is quite rare but the outcome can be fatal. A female red lory was found with

foot injuries believed to have been caused by trapping her nail in the wire. She was removed from the cage for treatment. During the period that the male was alone he became very aggressive toward the keeper and would strike, drawing blood. A few days later the female was reintroduced to the male. Several days after that she was found dead in the nestbox. Necropsy showed massive trauma to the throat and neck area (Casmier and Wisti-Peterson 1995). At Palmitos Park, Gran Canaria, a pair of red lories were nesting. They had not raised young for several years. When the chick was about two weeks old the male became very aggressive. On opening the nest inspection door to check the chick, the male flew in at great speed and bit me on the hand, drawing blood—totally abnormal behavior. He continued to be very aggressive. A few days later I found the female with her wing trapped in the mesh of the cage floor; this was something that had never happened before. When I found her with a leg injury two days later I realized that the male had been to blame on both occasions; on the first occasion he must have attacked her and in her panic she had caught her wing in the mesh. The female was removed to the clinic for treatment and the chick was removed for hand-rearing. After about a week the female was returned. The male's aggressiveness had subsided. It might have been stimulated by his desire to protect the nest but was misdirected at the female because she was the only available target.

Pairs aggressive to humans

This problem must be dealt with before the bird learns it has the upper hand! Sometimes merely showing the bird a catching net will cause it to back off. However, showing the net too often just serves to increase the bird's aggressive tendencies. The problem may occur because the cage is too small and the bird considers its nest to be threatened, or perhaps because the nestbox is too near the service hatch or door, making each use of the door an invasion of its protected nesting territory. Switching cages, or changing the position of the nestbox within the cage, may solve the problem. I use a small piece of cardboard which I slide between the welded mesh and the nest entrance to trap an aggressive female in the nestbox while I am changing the food, or outside the nestbox while checking its contents. This is simple but effective! (I always use a thin card in case I forget to remove it. The birds can then bite their way in or out.)

Breeding Season

In their natural habitat, lories breed only after the rainy season, due to their dependence on blossoming trees. They require some pollen, with its higher protein level than nectar, on which to rear their young. They also need a reliable food source in the area. Out of the breeding season many species are nomadic and can travel over wide areas to find food. During the breeding season they are restricted to foraging in the immediate locality of the nest site. Therefore, nesting must occur in an area where many food trees are in blossom.

For the New Guinea species, only sparse records of nests and young are available, but these suggest that breeding can occur at almost any time of the year, according to local conditions. In Australia, the same applies to rainbow and scaly-breasted lorikeets, for example. The start and duration of the breeding season is affected by weather conditions.

In captivity, some lories are seasonal nesters, but many—in fact the majority—nest throughout the year, ceasing only during and after the molt. In my experience, the dusky lory is the most seasonal, starting to lay in April or May and not usually laying after July. This applied both in the U.K. and in the Canary Islands. Other species are seasonal in regard to the start of nesting. For example, my Musschenbroek's and iris are most likely to lay in December or January and may or may not continue after a second clutch.

A brooder must be thermostatically controlled and easy to clean.

Brooder

When lory chicks must be hand-reared, the use of a brooder is essential until they no longer require a heated environment. One can purchase brooders which have been constructed for parrot chicks (refer to an avicultural magazine) or make one's own. Important features of a brooder are easy and accurate temperature control, ease of cleaning and the ability to view the chicks without opening the brooder and disturbing them, thus losing heat. A small aquarium can be adapted by making a wooden top into which is fitted two red light bulbs and a thermostat (or a manual light dimmer). The advantage of an aquarium is that it is very easy to clean; this is important because of the liquid droppings of lory chicks. A box constructed with melamine (a surface which is easy to wipe clean) with the front made of Perspex, is another alternative. It provides a greater feeling of security.

Small chicks can be kept in plastic containers. Older chicks can walk freely on a base of small-size welded mesh under which newspaper is placed. A very small mesh should be used to prevent a chick's foot becoming wedged in the mesh. Welded mesh provides excellent foot exercise and helps to prevent disorders such as splayed legs which can result from keeping a chick on a smooth surface. The heat source in a brooder should be either at one side or above the floor—not under the floor unless it is a heating pad which cannot produce a very high temperature. An overheated chick could die within minutes.

The brooder should contain a small spill-proof jar of water with the top covered with welded mesh to provide some humidity. A digital thermometer with a probe can be placed outside the brooder to prevent soiling; only the probe extends inside. The thermometer should be periodically checked against another one to ensure accuracy.

Call, Juvenile Food-Begging

A vocalization known as the juvenile food-begging call is present in the chicks of most species of birds having altricial young (Courtney 1996). These calls invariably consist of a single, simple, repeated note, thus differing from the complex sounds of most birds in their other vocal communications. The begging call is species specific; it varies so much in pitch, tone, timing and general construction between the different natural groups of birds that random duplication of this sound between unrelated species would seem to be rare or unlikely. However, there is an undeniable similarity in the calls of closely related species and genera. This suggests that the begging call is slow to change, apparently because it is unlikely to be influenced by the external evolutionary pressures that more quickly shape morphological characters. This information is from the work of John Courtney who started to record the food-begging calls of Australian parrots in 1963. From 1980, I provided him with tape recordings of the juvenile food-begging calls of nine lory species from seven genera (unfortunately not including *Charmosyna*). In addition, he has studied the food-begging calls of all Australian lorikeets except the purple-crowned.

John Courtney concluded (Courtney in press) that the food-begging call divides the lorikeets into two categories, those which utter a hissing note and those which utter a tremulous trill. The hissing note, "which sounds like a simple, repeated, somewhat high-pitched hiss, is shown to be complex and has a very distinctive shape when analyzed electronically by audiograph. The somewhat sharp-sounding, high-intensity note uttered when young are hungry and being fed, begins with a structureless hiss abruptly giving way to a momentary high-energy, high-pitched "blip" which then initiates a structured hiss for the remainder of the call. In lower intensity begging, as is typical of much calling, often only the structured hiss is given."

The food-begging call of the *Vini* lories is described by Courtney (in press) as "a brief repeated trill instead of a hiss." The differences between the two types of calls are easily discernible to the human ear; thus I can state with confidence that the four *Charmosyna* species with which I am familiar (*papou, josefinae, placentis* and *pulchella*) all belong in the same category as *Vini* lories. I would expect this to apply to all the members of the genus *Charmosyna*, with the possible exception of *multistriata* which in some respects is not a typical member of the genus. I would also expect it to apply to *Oreopsittacus* which is behaviorally closely allied to *Charmosyna*.

The chicks of *Vini* and *Charmosyna* species with which I am familiar are similar in appearance. When very young they have a distinctive head shape. Ruediger Neff supplied me with excellent photographs of chicks of *Oreopsittacus* which shows that they are indistinguishable from those of *Vini* and *Charmosyna*.

John Courtney believes that the extraordinarily uniform food-begging call of the lorikeet species that hiss suggests that all arose relatively recently from common ancestral stock, possibly in a staggered "explosion" of speciation as they spread widely throughout the Australian region, adapting to the proliferation of flowering trees over the last few million years.

Based on juvenile food-begging call, appearance of chicks and behavior of adult birds, I believe that the lories and lorikeets form two distinctive groups: *Vini, Charmosyna* and *Oreopsittacus*, which are perhaps the oldest species, and the other genera which perhaps evolved later.

Candidiasis

Candidiasis is a fungal infection caused by the yeast cells of *Candida albicans* and other species of *Candida*. Stress and vitamin A deficiency, also the prolonged use of antibiotics, predispose lories to this infection. When most of these birds available to aviculturists were wild-caught, i.e., in the 1970s and early 1980s, candidiasis was common. Prior to export, lories were often crowded together and were usually fed inadequately, perhaps for weeks or months. Fortunately, candidiasis is no longer common, due to improved standards of care and nutrition and the decrease in wild-caught birds. However, lories are susceptible because they are offered liquid food. *Candida* cells can be found anywhere on a bird but are most likely to occur near mucous membranes, such as the mouth. Lories kept under unhygienic conditions are at risk from candidiasis, as are those exposed to liquid food which is not changed often enough during periods of high temperature. In warm and humid conditions *Candida* cells multiply very quickly.

Also at risk are lories which have suffered an injury in or near the mouth. Here the *Candida* fungus will grow in the form of whitish or yellowish plaque, starting as a white spot. The lesions could spread to the area under the tongue and hard plaque could form at the corners of the mouth or under the upper mandible. Untreated, the lesions could spread down the oesophagus into the crop. This can cause vomiting and the infection is then very difficult to treat. It is negligent to allow the growth of *Candida* to proceed so far; the lesions cause much discomfort and the chance of recovery is greatly reduced.

If in doubt regarding the diagnosis of candidiasis, consult a vet. It is a matter of seconds for him or her to swab the mouth and make a culture, or culture the feces. The culture plate is then stained and if *Candida* cells are present they can be observed through a microscope. It should be mentioned that fecal analysis of lories and other parrots often reveals the presence of *Candida* cells. However, unless microscopic examination of the cultured feces shows large numbers of *Candida* per field, or that the *Candida* are actively dividing, there is usually no need to treat the bird. Low levels of *Candida* are not harmful until the bird is stressed; then the cells increase very rapidly.

Treatment and prevention

If the fungus is already well developed as a plaque in the mouth (sometimes described as having a cheesy appearance), and is still growing, bleeding will occur if an attempt is made to remove it. In this case ointment or gel should be applied directly on the affected area. If the fungus is hard and dead, gentle removal with a moistened Q-tip (cotton swab) may be successful. If the fungus is still growing, treatment twice daily is recommended. The affected part should be gently cleaned with a moistened sterile Q-tip, then Nystatin ointment (Nystan-Squibb) or Daktarin gel (Janssen) should be applied. (Note that in the U.K. and Australia and perhaps elsewhere, Daktarin can be bought over the counter whereas one needs a vet's, or a doctor's, prescription to obtain Nystatin.) Ensure that the oral gel is used and not the external cream used to treat tinea. To treat systemic candidiasis, Nystatin suspension should be added to the nectar twice daily. A small quantity of nectar should be given so that the whole dose is taken; then the nectar dish can be refilled.

Chicks can be treated by giving Nystatin at the recommended dose three times daily. This is most effective when the crop is empty, so that the Nystatin comes into contact with the lining of the crop. In chicks there is a danger of *Candida* cells lining the crop. Parent-reared chicks suffering from candidiasis (often as a result of the fungus growing in nest litter in humid climates) can be treated in the nest, parents permitting, twice daily. Nystatin diluted slightly in warm nectar can be given directly into the mouth of infected young before they are fully feathered. After this stage it is more difficult. Duration of treatment in young birds and adults is usually at least five days. Note that Nystatin should not be given to chicks as a prophylactic (preventative) measure because it will upset the balance of the normal intestinal flora. Young fed on a diet containing adequate levels of vitamin A and kept under stress-free and hygienic conditions do not normally suffer from candidiasis—although they are more susceptible than adults. Long-term antibiotic treatment makes all parrots more susceptible. For this reason, antibiotic treatment should be accompanied by increased levels of vitamin A. With lories this is easily achieved by adding a multivitamin preparation to the nectar.

Two red-spotted lorikeets which developed candidiasis after being under extreme pressure from their partners, did not respond to treatment with Nystatin. The condition was cured with the use of Daktarin (Day pers. comm. 1995). The drug Ketoconazole (Nizoral) has also been used to cure candidiasis in lories but I believe that it should be used with extreme caution and only when other methods of treatment have failed. Great care should be taken not to overdose it. The more recently developed Itraconazole appears to be less toxic to the liver than Ketoconazole but is more difficult to use as it is available only in capsule form. It can be given orally or in the food. The dose is 10 mg per kg of body weight, twice daily for 10 to 14 days.

It is advisable to inspect the mouth and the corners of the beak of all newly purchased lories and any which appear to be slightly or greatly below normal health status. A lory which has a small part of the mandible missing may have suffered from a *Candida* infection which rotted part of the beak. Also look with suspicion on any lory which has liquid food matted on its head feathers or one which is shaking its head, then vomiting.

It is generally stated that candidiasis is an infectious disease (e.g., Kaal 1991). In the 1970s one member of more than one pair of newly imported lories was infected, necessitating long-term treatment. But the birds' partners did not catch candidiasis. I believe that unless a bird is susceptible for the reasons already stated, candidiasis is not infectious.

It is worth noting the advice of Alicia McWatters regarding treating birds with candidiasis:

> These birds are repeatedly placed on an antifungal for yeast and, if bacteria is present, an antibacterial medicine. Then we risk the possible side effects of these drugs. A weak immune system is the underlying problem that needs to be dealt with. There are yeast-fighting foods and supplements that can often inhibit yeast growth and bring about normal health (gut flora) in your birds without the use of antifungals.
>
> ...In our experience, antifungals have not made an improvement, while a combination of the following products have completely cured *C. albicans* for our birds.

She lists the following: Kaprycidin-A, also called coconut extract or Caprylic acid; available from a natural foods store. It results in increased efficiency of the immune system. If used in a time-release form, it is said to pass through the intestinal tract, killing yeast with no side effects. She also uses foods high in vitamins C, A and E, plus the mineral selenium; these work as antioxidants. It is stated that vitamin C strengthens the immune system and prevents *C. albicans* and viral and bacterial infections. Also mentioned are "friendly bacteria," such as those found in live yoghurt, buttermilk and cultured milk products; the probiotic *Lactobacillus acidophilus* (not milk-free strains); and odorless garlic extract such as Kyolic (McWatters 1996).

Casuarina

The casuarina or Australian pine (*Casuarina equisetifolia*) is an evergreen, not a true pine, with long, drooping green needles. It has been introduced into many tropical and subtropical countries throughout the world. In New Guinea and Pacific islands, it is a favorite food source of lorikeets, such as those near Port Moresby (at Taurama Barracks) in Papua New Guinea. During the wet season in December, 1964, green-naped lorikeets (presumably *micropteryx*) started to frequent a 550 m (600 yd.) long row of planted *Casuarina equisetifolia*, 15 m (50 ft.) high. At first these parties consisted of 6 to 10 birds but numbers gradually built up to at least 150 birds which were present all day, making a

considerable amount of noise. They could be seen at any time, tearing open casuarina fruits and consuming the seeds inside. As the fruiting ceased, the birds turned their attention to the dried seed cases. By late March, 1965, only empty seed cases could be found. They then fed on the seed pods of ornamental *Cassia* until this source was also depleted. Despite a moderate flowering of *Eucalyptus papuana* in January, only one group of lorikeets was ever seen to feed on these blossoms. In September, 1965, Major Bell reported that the lorikeets had returned, to tear open the green fruits of the *Casuarinas*. They showed no interest in the sugared water put out for them at a couple of nearby homes (Bell 1966).

In the Solomon Islands, Jan van Oosten has observed cardinal lories and Massena's and red-flanked lorikeets feeding on the unripe seeds of casuarina trees (pers. comm. 1996).

In Australia, in eastern New South Wales, rainbow lorikeets were seen feeding on the seed of *Casuarina* trees (Lepschi 1993). Again, in eastern New South Wales, John Courtney recorded this event in his diary for October 22, 1964:

> I saw a number of little lorikeets feeding on sap (?) exuding from some bare patches on sides of trunks of river oaks (*Casuarina*). There were four separate "sap patches" within a radius of 30 yards. The lorikeets mostly fed in pairs, sometimes singly, and others waiting would be repulsed until the originals left. Many little lorikeets were "shuttling" back and forth between the river oak trees, and a yellow box tree in heavy blossom several hundred yards away. Sometimes at least two of the sap patches were occupied at the same time. A number of musk lorikeets were present within this 30-yard radius, but were not observed to feed on the sap patches. The musks continually climbed about in the drooping leaves of the river oaks and appeared to be gathering something with their tongues. Sometimes they came within feet of the Little Lorikeets feeding at the sap patches, but never flew over to join them.

In the immediate vicinity of the breeding station of Palmitos Park, Gran Canaria, in the mountains above the park, are a number of *Casuarina equisetifolia* trees. The wild canaries feed avidly on the seeds, perhaps in preference to almost any other seed. In size and shape it is not unlike canary seed. One day I was surprised to see two free-flying green-naped lorikeets from the park feeding in a large *Casuarina* outside the food preparation kitchen. How they knew that it was fruiting was a mystery. I never saw them there again.

The common name of this species of *Casuarina* varies according to locality. McCormack and Kunzle (1993) state that *Vini kuhlii* feeds on flowers of Pacific ironwood (*Casuarina equisetifolia*) "and appeared to extract the seeds from the fruit." This encouraged me to offer small branches of fruiting *equisetifolia* to a number of species of lories. The fruits look like tiny cones and, when full grown, measure about 1.6 cm x 1.4 cm (about 3/5 in. x 1/2 in.). The seed kernels are about the size of canary seed with a "wing" that makes them appear larger. Each little kernel contains about 40 seeds. Results were as follows: an entry in more than one column indicates that different pairs of the same species reacted differently.

Comparative Interest in *Casuarina*

Species	No interest	Some interest	Everything destroyed and/or eaten	Seeds eaten
C. a. insignis		x		
C. duyvenbodei		x		
C. scintillata		x	x	x
T. h. haematodus			x	x
T. euteles			x	x
T. iris			x	few seeds only
E. bornea				x
L. lory			x	x
C. placentis				x
C. pulchella	x			
C. papou goliathina		x		x
C. josefinae		x		x
N. musschenbroekii				x

The different results from pairs of the same species is partly explained by the more adventurous nature of hand-reared birds. However, after it was offered a few times, nearly all pairs greeted its appearance with great enthusiasm and excitement. In many cases, there was no trace of it in the cage within a few hours—just some debris on the floor below!

Lory keepers who live in areas where *Casuarina equisetifolia* grows should offer fruiting branches regularly. It has been introduced into various countries and is common in, for example, California and southern Florida. Planting this tree to provide an endless source of enjoyment for their lories should be the aim of all lory keepers in appropriate climates.

Chicks, Foot and Leg Problems

Attention should be paid to the toes and feet of chicks; most problems can be corrected if measures are taken early enough.

Toes pointing forward

This problem was evident in a 12-day-old black-capped lory which had to be removed for hand-rearing at about 7 days. Additional calcium was added to the food, but at 18 days there was no improvement. A false floor of 1/4 in. square welded mesh (of the kind used to reinforce glass) was placed on the floor of the brooder. This had the desired effect of making the chick use its feet more often. Three days later all the toes were in the correct position. I prefer to place chicks which are being hand-reared on a face cloth until they are about two weeks old (or longer for very small species) and then on small welded mesh, of a size in which the leg cannot become trapped.

Swollen toes

A retarded blue-streaked lory chick was kept on a bed of tissues, instead of on a wire base. By 33 days its toes were badly swollen. I suspected that this was due to the length of time its feet were in contact with tissues, which probably contain harmful chemicals. For this reason, chicks being hand-reared should be kept only on very soft tissues and not after the age of two weeks. I gave an antibiotic, Synulox palatable drops (Beecham Animal Foods) in the rearing food twice daily and placed the chick on a different surface. Four or five days later the infection was cured.

Greenstick fractures

When a parent-reared red-flanked lorikeet was about 11 days old I noticed one leg jutting out at an unnatural angle. It looked as though it had suffered a greenstick fracture. Because this is such a small species there was nothing I could do until it was old enough to ring at 17 days—and it then weighed only 13 g. I ringed both feet and tied the legs with ribbon, so that they were normally spaced. Within a couple of weeks the leg appeared normal. The ribbon was not removed until the young lory left the nest—when there was no sign of the earlier injury.

Chicks, Killing of

In captivity lories sometimes kill their chicks. I believe that there are two types of infanticide: that of newly hatched chicks, and deaths of other chicks, usually before they are about to feather. It seems that some lories, as may happen in other parrot species, kill their first chick as soon as it hatches, perhaps due to inexperience. The chick is found dead soon after hatching, badly mutilated or as a scarcely recognizable mangled little corpse. Or it, and perhaps the second egg, may disappear without trace, almost certainly bitten into tiny pieces.

Then there is infanticide which is probably the result of disturbance. For example, a black-capped lory (*Lorius lory erythrothorax*) laid her very first clutch; later, in the next cage, another pair of *erythrothorax* had a pipping egg. Previously, all chicks from this pair had died young. I wanted to foster their chick so I gave the pipping egg to the female who had not had chicks previously. The chick hatched and the pair proved to be excellent parents. Two weeks later her own first egg hatched in her nest; the second egg was in an incubator. If I had taken her egg when it was pipping I would have needed to hand-rear two chicks. That season I had to keep numbers of chicks hand-reared to a minimum and decided, unwisely as it proved, to remove the older chick and leave the newly hatched chick to the care of the parents. By the next morning it had been killed, so I ended up having to hand-rear two chicks anyway. My timing was at fault. My removal of the older chick was probably equivalent to the nest failing, and therefore anything left in the nest would be destroyed. The foster chick had been well cared for. Had I not taken it away, it would have been interesting to see if the female would have fed both it and the younger chick.

The killing of nestlings by the parents can in many cases be traced directly to an obvious disturbance. An instance of this type was recorded by Kuah (1993) in respect to purple-naped lories (*Lorius domicellus*): "In some cases, pairs can turn on and kill each other but, in the majority of cases, they form strong pair bonds, often shrieking and screaming, if their mate is removed or when checking the nest. For these reasons, I do not attempt to look into their nest boxes, once the eggs have hatched because chicks have been literally torn to pieces after I inspected them."

Herein lies the problem. If lories are not accustomed to regular nest inspection, they resent the intrusion and may immediately kill their young. I inspect the nest of an incubating female at least twice weekly and, when chicks hatch, inspection is carried out daily except in the case of a few extremely aggressive/protective pairs in which inspection could lead to problems. However, this is discovered during the incubation period—not after the chicks hatch. In normal circumstances, if inspection is carried out in a sensible manner, by speaking quietly to the bird, then knocking gently on the box, no harm will ensue, in my experience. However, the person inspecting the nestbox should be known to the birds, and familiar with their behavior.

CITES

The **C**onvention on **I**nternational **T**rade in **E**ndangered **S**pecies of Wild Fauna and Flora is an international treaty which seeks to protect species affected by significant trade, and to regulate or prohibit trade in those species endangered by trade.

Clutch Size and Laying Interval

Two eggs form the normal clutch of most non-Australian lory species. Perfect lorikeets usually lay three eggs, and the Australian lories (except the subspecies of *Trichoglossus haematodus*) lay 3 to 5 eggs. The Australian species which are most reliant on pollen for rearing their young need to take advantage of this food source while it is abundant, perhaps only for a short period, thus they rear more young in a clutch. Species whose food is less seasonal and less specialized can breed more than once annually, thus their clutches are smaller. This is, I believe, the explanation for the varying clutch sizes.

In aviaries, there are exceptions to two-egg clutches in non-Australian species. Sometimes the explanation is clear. For example, if the first egg is laid from the perch and therefore broken (as sometimes happens with inexperienced females—or if a female is frightened or disturbed just prior to egg-laying), two more eggs may be laid in the nest. An aviculturist could take advantage of this by removing the first egg as soon as it is laid, then placing it in an incubator or with another pair. It is then likely that two more eggs will be laid. However, I would recommend this procedure only in a special case, where it is important to breed as many young as possible in a short time from a certain pair.

It is rare for three eggs to be laid in other circumstances. From my own experience I can recall only one instance, that of a fairy lorikeet (*Charmosyna p. pulchella*) which has been breeding for five years, to date, but has produced only one such clutch. All three hatched but the youngest chick had to be hand-reared because it was not growing well. In Sweden, a pair of red-flanked lorikeets (*Charmosyna placentis*) belonging to Hans Andersson of Kungsbacka, laid three eggs in her second-ever clutch. All hatched and all three young were parent-reared. They were four months old when this incident was reported to me (Andersson pers. comm. 1994). Mr. Andersson believed that this was the only pair of *placentis* in Sweden at that time. Mother and daughter red-spotted lorikeets belonging to Carol Day regularly produce three eggs in a clutch. It is of interest that the mother's mother also did so. Thus it seems that this ability can be inherited.

When fostering eggs, care must be taken with the timing because it might affect the number of eggs laid by the female. I once gave a female Meyer's lorikeet an iris egg on the day after she laid her first egg. She did not lay a second egg; but on another occasion when the same practice was carried out, she did lay a second egg.

The interval between each egg in the clutch is normally either two or three days. For example, in the little lorikeet, Harald Mexsenaar recorded intervals of two days between three eggs and three days between five eggs (Mexsenaar pers. comm. 1996). In the wild, intervals of two days have been recorded for the little lorikeet (Boyd 1987). Atypically, the interval can be four days, as I have noted with the green-naped lorikeet, the iris lorikeet and the black-capped lory on rare occasions. For example, in the black-cap the laying interval is usually three days, but could be two or four days. To be certain of the egg interval it is necessary to check the nestbox twice daily—which is not normal avicultural practice; a female does not necessarily lay at the same time of day.

The interval between the "end" of one clutch (i.e., when the female has destroyed eggs which were not viable, when chicks have been removed for hand-rearing, or when fledged young have been removed) and the laying of the next, has also been noted. It is usually about three weeks, but has varied between 16 and 28 days. See also **Eggs**.

Colony Aviaries

A colony aviary of lories looks wonderful—but can it work? Is it practical and is it wise? The answer to the last question is usually "No." Of course there are exceptions, but due to the nature of lories, most are especially unsuited to being kept in this way. Instant death for the unfortunate co-inhabitants would normally be the result of trying to keep other birds with *Lorius* lories. Smaller species such as Goldie's and Meyer's have been kept on the colony system but deaths through fighting or, at the least, poor breeding results, will follow. Of course much depends on the size of the aviary and how the birds are introduced, but the main long-term successes seem to be with green-naped and Swainson's lorikeets (*Trichoglossus h. haematodus* and *T. h. moluccanus*). Stan Sindel had a colony of eight or ten red-collared lorikeets (*T. h. rubritorquis*), not all pairs, housed in an aviary 15 m (50 ft.) long, 3 m (10 ft.) wide and 2.4 m (8 ft.) high. They lived amicably with several parrakeets and some large softbills and bred constantly (Sindel 1987).

He offered some very sound advice: "Over the years I have had several such colonies and have seen numerous others. All have worked well until something upsets this balance. Never add to or remove from a successful colony, except the young. Never try to relocate the colony or refurnish the aviary and pray you never lose a key bird in the group. The whole success usually depends on everything being the same."

Dr. R. Burkard mentioned studies of aviary groups of Swainson's and red-collared lorikeets which indicated that only one pair of young from the group could be reared successfully each time. One scientist who observed such a group found that there were distinct rank orders which are established through fighting, biting or pursuing. Once the order was established, the colony lived peacefully. He described this as a hierarchic group.

Another category he defined as the partial hierarchic group. The implementation of the hierarchy is less violent, and more than one pair will breed in a confined area. Dr. Burkard's aviary containing six pairs of Goldie's lorikeets was an example. They lived in an indoor flight measuring 3.5 m^2 (37 sq. ft.) with an outdoor flight of 4.5 m^2 (48 sq. ft.). Up to three pairs hatched young at the same time, but on occasions young had to be hand-reared due to parental neglect. These young could not be integrated into the flock when independent (Burkard 1983).

However, Bosch (1993) recommended that adult Goldie's not be kept in a colony, as young birds would not live long. "Single, weak or inexperienced individuals will at some time be attacked and will be, because the attacker is not a skilled slayer and the victim cannot escape, slowly and agonizingly killed."

At the 1979 convention of the American Federation of Aviculture, Dr. Ray Jerome spoke on lories. He discussed colony breeding. He mentioned a female dusky which had helped to rear the young of a pair of Swainson's in a colony! Paolo Bertagnolio, a respected Italian aviculturist, promoted the idea of keeping certain lorikeet species on the colony system. Of lories in general he wrote:

Perhaps the lack of distractions, more than possible dietary deficiencies, lead many pairs to breed satisfactorily for a limited number of seasons only, after which we very often find a gradual deterioration of parental care, with ever increasing plucking of the young and, in the ultimate stage, killing of the chicks as soon as they hatch. Eating of the eggs is a less common occurrence and the culprit is more likely the male. Often when he is caged in the shelter a few days after laying, the female proves able to rear by herself without inconveniences.

The fact that Lories and Lorikeets often do better in Zoos than in private aviaries confirms that, intelligent and temperamental birds as they are, they need to fill their time and the best distraction of all is to keep them in flocks. This can be done if very large flights only are used, because each pair requires a small territory for itself, absence of which will invariably lead to mortal fights.

Apart from the fact that the cost of space per pair is lower, colony breeding has a number of other advantages. First of all there will be a spontaneous choice of the partner, a fact that anyone who has spent hours trying to make up his mind whether he has a true pair or not will

appreciate. Also the group's stimulus will start them off earlier and with a better result not restricted to a limited number of years. The need to defend a small territory around their nest will keep the pairs occupied, rendering more natural and strong the link with their chicks that truly become something to protect and not to pluck, taste or bite in order to discharge aggressiveness...

A number of pairs of Swainson's Lorikeets (*T. h. moluccanus*) that had stopped producing, limiting themselves to kill or to let die newly hatched chicks, started to rear again when mixed together in a very large aviary, with at disposal only nectar, sunflower, fruits and green food but no bread and milk, biscuits, mealworms, insectile mixtures or other sophisticated additives (Bertagnolio 1978).

I agree regarding the stimulus to breeding which a flock provides—however, in my opinion, it would be most unwise to try this with any species but *Trichoglossus*, and preferably *T. haematodus*. At the time of writing, Bertagnolio was still experimenting with the size and shape of communal flights but suggested that an area of 5 m^2 (53 sq. ft.) per pair, and a height of 2 m (6 ft. 6 in.) to 2.5 m (8 ft.), was sufficient. The design suggested was roughly square with a number of partial subdivisions in order to render the internal space more complicated and articulated. This allows dominated birds to put themselves out of reach of possible aggressors ranking higher on the hierarchic scale. This is especially important for young birds that might be killed by dominant pairs as soon as they leave the nest.

Regarding the construction of a colony aviary, Bertagnolio suggested the use of cement poles for the perimeter. The roof would be supported by well-tightened, galvanized 5 mm wire, anchored into the ground by cement blocks.

In addition to providing a stimulus, there are a few other reasons why it might be desired to keep lories in a colony. The first is as an attractive display at a zoo, and the second is in the case of young birds which are being retained for breeding. Thirdly, for the sheer pleasure of watching the social behavior and the color of a group of lories. It opens up new vistas for the aviculturist and reveals aspects of behavior which cannot be observed with a single pair. In all cases, several species might be kept together. In my experience this can be successful if the following points are strictly adhered to:

1. Place all the birds in the aviary at the same time.
2. Ensure that all are in excellent health and can fly strongly.
3. Remove any bird which appears unwell or cannot fly strongly. Also remove aggressive trouble-makers.
4. On no account try to reintroduce birds which have been removed. They will probably be killed.
5. Unless the lories are all young of the same season, no new birds must be added.
6. Never provide nestboxes. These will be claimed and defended by dominant pairs, resulting in fighting.
7. No species should be in the minority; in other words, if there are eight birds each of two species and only two of a third species, the third species is likely to be attacked.

Provided that no nestboxes are available, young birds can be kept together until they are sexually mature. At Palmitos Park, young black-caps and rajah lories, for example, lived together in a colony with other lory species (a total of about 30 birds) until they were four years old (at the time of writing). During this period (before the birds were sexually mature) one black-cap was killed and two more were injured. Later on, copulation was frequently observed. Once this occurs, the risk is greater and it is advisable to replace the entire colony with young birds.

A major disadvantage of keeping a number of birds of one species together occurs when a bird is observed to be slightly unwell. After entering the aviary, it is usually difficult to identify individuals.

Consciousness

The Concise Oxford Dictionary of Current English defines the word "conscious" as "aware, knowing (of fact, of external circumstances); with mental faculties awake (of actions, emotions, etc.)."

"Consciousness" is the word that comes to mind when I consider how birds, including parrots and lories, act on a daily basis and in certain situations. Their mental faculties are much more finely tuned than the average person realizes. Their actions are at times instinctive or unconscious—just as ours are—but they are capable of actions and feel emotions which are far beyond the realms of the instinctive.

It has apparently taken man thousands of years to realize this; the idea is only now gradually being accepted. Anthropomorphism, which seems to have come to mean attributing humanlike thoughts and actions to animals, was ridiculed even 30 years ago, thus the concept that animals and birds lack ideas and emotions is accepted by many people in the Western world. It is not accepted by those who work closely with animals and spend much time observing them. The evidence is overwhelming that parrots, for example, know the whole range of emotions that are common to humans—love, dislike, fear, jealousy, etc.

I had to learn this by observation from childhood, since it was not what I would have been taught. At first, as a young teenager, I questioned my own interpretation of what I saw, since my views were definitely "anthropomorphic." I kept them to myself, not expecting others to understand. Fortunately, today one can talk about the emotions which parrots, for example, experience. Worldwide, there are millions of people who keep parrots in their own homes who have no reason to doubt, from their own observations, that many of these birds are highly intelligent, even manipulative and sometimes even exhibiting a pronounced sense of humor.

In the U.S.A. there are a number of parrot behaviorists who teach parrot owners how to discipline their birds, based on their knowledge of the birds' intelligence and emotions. Everyone who keeps parrots should be aware of the fact that interpreting their actions in an "anthropomorphic" manner will assist them in understanding their birds, and thereby create a happier world for them.

A letter from one of the many aviculturists who contributed to this book was typical of that awareness. Writing about her red-spotted lorikeets, Carol Day stated: "I 'stole' another female's newly hatched chick to give to a pair who had failed several times. The male was getting really frantic when another pair of eggs was overdue. He hardly leaves the nest when he is sitting. They adored that baby from the moment they saw it and went into the nest very carefully. They have since reared their own chick."

In a book which should be compulsory reading

for anyone whose lives are touched by birds, *The Human Nature of Birds*, Theodore Xenophon Barber wrote:

> Since avian intelligence and awareness is a factual, demonstrable conception, it cannot be squashed and is bound to prevail. How quickly humanity's conception of reality changes will depend to a significant degree on how profoundly and personally readers of this book understand the sensitivity, awareness, and intelligence of birds and how effectively the readers communicate this understanding to others. The forthcoming revolution in human thought will be led by men and women who are no longer intimidated by the taboo against perceiving birds as conscious individuals; who devote much time and effort befriending birds in the wild and raising birds freely in their open homes...; who guide birds to bond with them and learn their language; and who transmit to others (via lectures, demonstrations, television shows, news articles, and group projects) their perception of the humanlike characteristics of birds. The avian revolution will be complete when the new generation accepts as natural that people and birds can understand each other and relate to each other not only as equals but also as friends.

Conservation

What are the main threats facing lories in their natural habitat? They include loss of that habitat and loss of nest trees, and over-trapping for trade. (Trade is now better controlled. See **Trade**.) The main threat to the tiny *Vini* lories is the introduction of ship rats to the islands they inhabit. These aggressive vermin have already been responsible for wiping out populations on some islands and could cause the total extinction of more than one species unless certain measures are taken within the next few years.

Overall, the most serious threat is habitat destruction. By the early 1990s the use of satellites to photograph the earth's surface indicated that about 20,000 km^2 (7,700 sq. mi.) of tropical forest are destroyed every year. Although this figure was smaller than expected, the effect is greater because the fragmentation of forest which occurs leaves large areas which are unsuitable for many of the species they formerly contained. The total impact is said to be equivalent to the destruction of 45,000 km^2 (17,370 sq. mi.) per year. It is well known that thousands of species of animals and plants have already been lost without ever being scientifically described. The situation will continue to become more and more serious. The reserves and national parks which already exist are unlikely to be able to support all the species found within their boundaries when they were set up. This is especially the case if they are encroached upon (as has already happened), or if surrounding areas of forest are increasingly fragmented or destroyed. In ecological terms the reserves become the equivalent of islands. The threats to lories vary in different regions.

Australia

Courtney (pers. comm. 1995) points out: "The future looks very bleak for the lorikeets, a group highly specialized in their nest requirements. Most other species of parrots use holes of quite variable size and are easy to accommodate." This was the conclusion he reached after a lifetime of watching little and musk lorikeets in the eastern part of the Inverell District, northern New South Wales. It seems likely that wherever several lory species occur in the same area, each species has quite specialized nesting requirements or, more accurately, nest entrance requirements. This is because the larger species apply extreme selection pressure on the smaller ones, forcing them to use the smallest holes into which they can squeeze. This would ensure that their nests are not taken over by larger species. At Bonshaw, on the Queensland-New South Wales border, Courtney has observed little, musk, scaly-breasted and rainbow lorikeets nesting in huge river red gums along the river flat. He noted how they all "have to wriggle a little to get in! I have no doubt that all have their special-sized holes graduating from the little to the rainbow."

This has had serious implications for the little lorikeet, which has undergone a dramatic crash in its breeding population during a very short period of time due to lack of nest sites. It seems likely that this selection pressure is nonexistent or less significant for species which inhabit islands or areas where there are no other hole-nesting birds of similar size to themselves.

Trade in wild-caught birds is a factor which is often quoted in connection with declining populations of parrots. How serious is this threat where lories are concerned? In Australia, the catching of native birds is forbidden and aviary-bred lorikeets of most species are available at a low price. If there is any illegal catching of wild birds this would be so small-scale that its impact would be nil.

Indonesia (including Irian Jaya) and Papua New Guinea

Throughout Indonesia, including Irian Jaya, lories were captured and traded in large numbers. In 1995, the export of many species was banned but the domestic trade still involves large numbers. Quotas have been set, but in most cases these are exceeded (see **Trade**). Species endemic to smaller islands have suffered substantial population declines. However, in most cases these are the result of the combined pressures of destruction of habitat and trapping. Most of the lories from Indonesia and Papua New Guinea were almost unknown in aviculture until the areas they inhabit were opened up by logging. This applied mostly to Indonesian islands off the coast of Irian Jaya. Fortunately, 70 percent of New Guinea is still clothed in dense forest. The complex mountain ranges that stretch for nearly 2,000 km (1,240 mi.) have helped to protect the forest; even today there are few roads and in many localities removing timber is difficult and expensive. All this could change—and will almost certainly do so when the wealth of New Guinea's oil, natural gas, gold and copper are exploited. However, this country is one of the few worldwide which still has the opportunity to conserve large areas of rainforest. The next few decades will reveal whether this important opportunity is seized.

Bruce Beehler, the ornithologist who has devoted so much of his life to the avifauna of this country, wrote:

> Some 75% of New Guinea's original forests remain undisturbed. This is as large a percentage as for any humid, equatorial region. However, this expanse of forest is small by tropical standards; it is dwarfed by those of the continental masses of Africa or South America. The entirety of New Guinea's prime lowland forests could be clearfelled in a blink of the evolutionary time scale, a single human lifetime.
>
> Although New Guinea still supports vast tracts of original forest, it is never too early to make efforts to set aside representative areas of habitat that support the natural riches of the island. The Indonesian government has spon-

sored the development of a remarkable system of natural reserves in Irian Jaya, created with the aid of World Wildlife Fund and the International Union for the Conservation of Nature. Yet Indonesia is a very populous and a very poor country, and its bureaucracy is not noted for its ability to resist the pressures of development and short-term economic incentive. Will the vast reserves of Irian, today relatively pristine by virtue of their isolation and economic insignificance, remain inviolate decades from now? What can be done to insure their long-term preservation?

> In Papua New Guinea, the problem is quite different. Virtually all of the lands are traditionally owned, and the village elders...have direct say over the preservation or exploitation of these lands. In general, this has a conservative effect. It is more difficult for large timber and mining companies to gain easy access to land for exploitation. But this is not to say that it is impossible. And what match is a village elder against some clever businessman, offering visions of cash, cars and luxury? Fortunately, the Papua New Guinea government has been protective of village interests in such schemes, yet it increasingly faces the need to reduce foreign debt. Of late, there is evidence that the timber concerns have begun to gain an upper hand.

> The national park and reserve system in Papua New Guinea is in its infancy, mainly because of the tribal land tenure system. Certainly the national government should actively pursue some means of accommodating the needs of its local landowners, while attempting to set aside significant blocks of land for internationally recognized habitat reserves. At present, a system of locally administered wildlife management areas is being established, but this is not adequate, considering the future pressures from a growing world hunger for raw materials. (Beehler, 1991).

As mentioned by Bruce Beehler, Worldwide Fund for Nature (formerly World Wildlife Fund) has a special conservation program in Irian Jaya. It commenced in the early 1980s and has sought to identify the important conservation areas, lobby for their protection, assist the government agencies to manage the areas, work on the ground with local people to effect boundary protection and sustainable patterns of exploitation, and to conduct biological surveys.

Once felling starts on a large scale, it will be the lowland forests which go first, taking with them lowland lories which are primarily forest dwellers, such as *Chalcopsitta* and *Lorius* species. Only a few lories are so adaptable that they can survive in most kinds of habitat, including agricultural land and city streets with flowering trees. They include the green-naped lorikeet and its subspecies, and the dusky lory. On some islands the red-flanked lorikeet would also survive. The highland species would be the least threatened because of the expense and difficulty of extracting timber in the mountains.

During the 1980s, field work, and studies which also investigated the effects of trade on parrots and other birds, commenced in Indonesia. This resulted in recommendations which have reduced the number of birds exported to the more responsible consumer countries, such as those in Europe. However, this may merely have diverted the trade to other countries (such as Japan) which have no restrictions concerning species which can be imported. Thus, trade, especially between countries which find it difficult to implement CITES (Indonesia) and those which are not members of CITES (Japan), may continue to threaten the existence of some species.

The first lory species to be placed on Appendix 1 of CITES following an investigation into trade was the red and blue lory (*Eos histrio*). The University of York expeditions to Sangihe and the Talaud Islands in 1995 and 1996 were partly financed by The International Loriinae Society (in the U.S.A.). Jan van Oosten raised funds by commissioning paintings of various lory species, producing limited edition prints and selling the originals. Administrative costs were borne by himself and 100 percent of the proceeds were donated to lory conservation. Up to early 1996, U.S.$12,000 had been donated, including $1,300 to the University of York's Sangihe-Talaud Expedition. Prints issued to date are of the following lories: Tahiti blue, black-capped, red and blue, and ornate lorikeet. (At the time of going to press, prints cost U.S.$45 each, plus $12 for post and packing and were available from the International Loriinae Society, Sharon Casmier, P.O. Box 850, Sumner, WA 98390, U.S.A.)

Education

The key to saving all wildlife lies in educating local people regarding its importance, especially that of endemic species. It is the younger generation which will hold in its hands the power to preserve the habitats which have survived. If we can instill in them a love of nature and all that it holds, there is hope. We can hope that man's selfish disregard of the consequences of actions such as deforestation for dollars and short-term gain, will cease. Schools must teach the need to conserve; this can only be done by opening the eyes of children at an early age to the natural beauty which surrounds them. Birds, especially colorful parrots such as lories, catch the imagination of the young. Children delight in painting colorful birds. At a school on Tonga, a young boy's painting of a blue-crowned lory (native name "Henga") was chosen for inclusion in a calendar. Using such appealing subjects, it is not difficult for teachers to elaborate on the theme of the intrinsic value of birds to mankind.

The World Parrot Trust

The organization which is doing more than any other, worldwide, to educate and inform regarding the plight of endangered and threatened parrots is The World Parrot Trust. One-third of all parrot species, that is, more than 100, are in danger of extinction. The Trust finances and supports conservation projects, creates links between conservation and aviculture and promotes high standards in the keeping of parrots. By 1996, it had members in 52 countries worldwide and branches in nine countries. Four times a year it publishes *PsittaScene*, a color magazine with the latest news of parrot conservation, and articles on aviculture. In order to advance its work, the Trust needs many more members. Further information can be obtained from The World Parrot Trust, 4 Glanmor House, Hayle, Cornwall TR27 4HY, UK, fax (44) 1736 756438 or, in the U.S.A., World Parrot Trust, P.O. Box 341141, Memphis, Tennessee 38184.

Crop Pests

The hornbills (Bucerotidae) comprise a family of large fruit-eating birds from Asia, Indonesia and Africa; in common with the Loriinae, much of the population of most hornbill species is nomadic. Thus they are able to locate isolated trees, just as the fruit is ripening. They are constantly sampling resources over a wide area and must have a well-developed information system (Leighton 1986). The same must be true of lorikeets. This information system enables large numbers to utilize a good food source, be it natural or cultivated. In Australia, lorikeets sometimes feed on cultivated crops, cereal and fruit. This is not a recent problem. In 1938, Sydney Porter recorded his concern at the many factors adversely affecting parrot populations in Australia and, at the same time, described the slaughter of many parrots:

> One person I met stated that he recently saw and counted over five thousand various Parrots, including many Leadbeater's Cockatoos, at one drinking place which had been poisoned. I was told of fruit growers who sprayed their fruit trees with poison to kill the Lorikeets and in one large orchard four dray loads of Musky Lorikeets were carted away. Yet I was refused permission by the authorities to take away one single pair of these birds (Porter 1938).

In 1995, the Australian press reported the outrage of fruit growers in northern Victoria whose crops were being attacked by musk lorikeets. They retaliated by killing the birds. For three years, a farmer had been campaigning the Department of Conservation to take action against these birds. He stated that if nothing was done the industry would press for the mass slaughter of the lorikeets, which were said to devour and damage $2–3 million worth of fruit in a season. One newspaper report stated that the lorikeets wreak havoc on 10 percent of the 300 orchards in the area each year. The problem was getting worse, as lorikeets attacked a wider range of fruit in an increasing number of orchards. It had reached the stage where many growers were shooting the birds without bothering to obtain a license.

At about the same time as these newspaper reports were published, I received a letter from a lady in Mornington, coastal Victoria (south of Melbourne). She had lived there for 19 years and, in 1995, had witnessed something which had never happened before. She wrote: "In March we experienced an invasion of hundreds of rainbow lorikeets. They have been coming in each day and eating the fruit on the trees—first nectarines, then cherry plums, apples and figs. Every one has been devoured! They have now moved on as there is no more fruit left on the trees. Is there a shortage of natural food this year?"

My appeal for further information resulted in a reply from Ian Dowling of McCrae. He wrote:

> I live on the Mornington Peninsula, about 10 to 12 km south of your Mornington correspondent. Twenty years ago lorikeets (musk and rainbow) were seen only in the northern sections of the peninsula. Today they are found in all the residential areas in large numbers. Residential development has exploded throughout the coastal zone, resulting in the introduction of nonindigenous trees and has changed the birdlife in the area. Places which five years ago never sighted rosellas, now regard them as common.
>
> The shooting of lorikeets has occurred in the main fruit-growing areas 100 km to 120 km north of Melbourne. I can see that it will not be long before the cherry orchards here on the Mornington Peninsula will want to do the same. The establishment of vineyards is a rapidly growing activity and I wonder how long it will take the rosellas and lorikeets to find the grapes (Dowling pers. comm. 1995).

Barry Hutchins of Blair Athol observed fruit crop damage by musk lorikeets during January 1995. In the early part of the month he saw small flocks at Inglewood, in the Mount Lofty Ranges of South Australia. The numbers seemed to be unusual for that time of year and the birds appeared almost agitated. When he returned two weeks later he was amazed by the increased numbers. Flocks of 30 or more were observed during most of the day. By mid-February numbers had doubled. Other fruit-growing areas, such as Gumeracha, Lower Hermitage and Chain of Ponds, were also invaded by musks. Mr. Hutchins recorded the following:

> The target fruit was pears, all varieties. With the co-operation of one fruit grower I watched flocks numbering from 20 to 200 birds arrive about 9:00 A.M. in his orchard of 8,000 pear trees. They would congregate in sections of the orchard and feast on the pears, mainly green to three-quarters ripe, and leave only the stalk and core. This feeding would continue until 12:00 mid-day when most of the birds would fly into tall eucalypts and rest until about 3:30 P.M. Then feeding would occur again until approximately 7:00 P.M. On the property under observation from mid-January until April the damage to pears was 50% loss, with some trees up to 75%. Fruit growers in the target area said that it was the largest invasion by musk

lorikeets in memory and estimated the numbers in tens of thousands.

At the same time the following year, by the beginning of February only small numbers of musks were present. They were feeding in flowering eucalypts, which strengthened the belief that in the previous year feeding in orchards was due to poor flowering of the eucalypts (Hutchins pers. comm. 1996).

I sympathize with the fruit growers whose crops are invaded in this way. Some retaliate by shooting lorikeets but this is not a humane or desirable answer to the problem. Surely fruit growers should try to come to terms with it in a civilized way. During nearly a decade of living in the Canary Islands, I admired the way that growers of papaya and some other fruit crops would neatly net over the whole area which contained the fruit trees. The mesh had a rigid appearance so may have been plastic-covered. While initially this is an expensive exercise, it would soon pay for itself in terms of increased yield.

Some farmers, at least, are more tolerant of lorikeets feeding on grain crops. In May, 1969, a study was carried out on lorikeets frequenting sorghum crops in the Giru district of north Queensland. Flocks consisting of approximately 1,000 rainbow lorikeets and 500 scaly-breasted lorikeets invaded a 500-acre farm at Horseshoe Lagoon. The surrounding region was open forest-woodland of a variety of tree and grass species, with occasional large areas of open water. These were comparatively dry following poor wet season rains during the previous January-March. These lorikeet flocks were the largest seen for many years. The birds fed on the sorghum throughout the daylight hours, particularly during the late afternoon.

On May 19–20, 1969, 51 rainbow lorikeets and 24 scaly-breasted lorikeets were collected from flocks feeding at both the mature and immature (soft dough stage) sorghum plants, and from resting flocks nearby. Most had inactive gonads. Analysis of stomach contents revealed sorghum seeds in all the rainbow lorikeets examined. Grain particles (endosperm) occurred on the bill, face and cheeks, throat, breast and wings of almost as many birds. Some small pieces of bark were present in many stomachs. The only other contents were occasional hard seeds, notably of lantana and prickly pear. The largest number of sorghum seeds present in one stomach was nearly 500, with most of the material being endosperm. On mature sorghum, the lorikeets ate at the base of the heads. Only one mature seed was found, and this was undigested, as were the other hard seeds in the stomach. Both species took, and were otherwise liable to damage significantly, only immature grain.

The lorikeets appeared one week before harvesting, although immature seeds had been available for much longer. Reasonably large amounts of grain were eaten and still more seeds were wasted. However, less than 1 percent of the crop was made unavailable for harvesting. Attempts at reducing local lorikeet flocks in sorghum fields result in as much damage and economic loss by trampling, plus the costs of ammunition and poisons, as from the birds feeding on the crops. Effective shooting was not possible because of the diverse routes and small flocks, and destruction of the lorikeets' nearby roosting trees would have involved much tree-felling. In any event, most bird damage was carried out over a longer period by the smaller, quieter and less distinctive flocks of resident birds, such as grass finches. Control measures other than incidental frightening of lorikeets in the course of other farm work were unwarranted. Early harvesting was unnecessary since only immature grain was damaged to any extent, and earlier or later planting was impractical. Moreover, the problem is neither perennial nor predictable.

It was concluded that "since invasions took place during a period of severe drought and since all birds were in extremely good condition with more than one-half of the number of each species having large deposits of subcutaneous fat, the crop undoubtedly provided relief from starvation and thus assisted at negligible cost in the conservation of major populations of these peculiar and decorative species" (Lavery and Blackman 1970).

Diets (Captive Birds)

On the subject of lories in aviculture, undoubtedly more has been written about feeding them than on any other aspect. Most of the published information is anecdotal rather than scientific and may leave the reader asking more questions than it answers. Before formulating a diet for any animal it is necessary to know on what it feeds in the wild. In the case of lories this depends on the species, location and season. Principal items of the diet are nectar, pollen and small seeds, also flowers, buds, fruits, insects and, in agricultural areas, cereal and fruit crops (see **Crop Pests**).

The tongue (see **Tongue**) of a lory is specially adapted to allow it to gather pollen and nectar. The crop is less developed because it is not used to store food and the gizzard is less developed than in seed-eating birds because lories eat smaller amounts of fibrous material. The intestinal tract is shortened because most items eaten are rapidly digested. As the moisture content of their diet is usually high and the nutrient content low, lories must consume much food to satisfy their energy requirements. The passage of food through the digestive tract is rapid, partly because of the high moisture content. Whereas most parrots in the wild have two main feeding periods, morning and late afternoon, and their crops take probably two or three hours or more to empty, lories may feed throughout the day in appropriate weather conditions. They have high energy requirements.

One of the most common mistakes in feeding lories is to make the food too liquid. In early accounts of lory keeping dating back to the later years of the nineteenth century (see the **Red and Blue Lory**), it was evident that newly captured lories did not live long, and usually died of fits or cramps. This was also happening in the 1970s when many people were keeping lories for the first time and feeding was still based on guesswork. According to Peter Holsheimer, an avian research nutritionist: "High contents of lactose or galactose in the diet cause ataxy; falling from the perch, rolling over on the floor, followed by a spontaneous recovery" (Holsheimer 1996). In the 1970s there were no commercial lory foods and most homemade diets relied quite heavily on milk-based foods. For about 17 years my own birds were fed on a milk-based baby cereal with the addition (except in hot weather) of tinned condensed milk, also malt extract and either glucose or honey. On this ill-advised diet they seemed to thrive and they bred, although breeding results and longevity might have improved on a different diet. Probably everyone who has kept lories over several years has seen one suffer from ataxy. While it may be due to a faulty diet, I am not convinced that milk products are always to blame. After all, ataxy is often seen by Australian veterinarians in wild lorikeets which have been picked up

in distress. While this is probably usually the result of lorikeets feeding in gardens on honey and water, milk is not used in these artificial diets. In the U.K. most home-made lory foods (called "nectar") are based on baby foods which contain milk. If this is the only ingredient of the "nectar" which contains milk the quantity which the birds receive is probably not excessive but I would not recommend the addition of any other milk products. See **Nectar** for detailed information.

Commercially formulated nectar powders may claim to be a complete diet. Nutritionally this may be true. But it would be wrong to feed lories on nothing else. This is partly because they enjoy variety in their food (and no intelligent animal should be expected to live on a single item) and partly because they also need some form of roughage. This is not found in a powdered food to which only water is added. Most lories and lorikeets enjoy fruits and vegetables and some also eat seed. Lories need food items other than dry powder and liquid nectar on a daily basis.

Fruits

The following are suitable. **Apple**: golden delicious are preferred by most lories, in a slightly soft, not crisp, condition. Of course, other types will be eaten. In the 1970s, some wild-caught lories refused to sample apple. Yet, perversely, they enjoyed eating cooking apples that fell on to the roof of the aviary from the tree above! **Pear**: conference are generally preferred; must be ripe, not hard. **Grapes**: all kinds eaten. **Oranges**: some lory keepers report that their birds do not like oranges. This has not been my experience, but the oranges must be sweet. Tangerines and satsumas are also relished. **Pomegranate** (*Punica granatum*): the number one favorite of all lories, in my experience; also excellent nutritionally. The vitamin B2 content is high, also that of vitamin C and the valuable mineral manganese. Pomegranates should not be offered until the skin (and therefore the contents) is ruby-red. The fruit is best cut into pieces, leaving the skin intact. (Wear gloves when cutting this fruit as the juice stain looks unpleasant and is difficult to remove.) Unfortunately, the season of availability is only about three months and pomegranates cannot be preserved frozen. **Papaya**: an excellent food but not all lories will eat it. **Guava**: a great favorite of most lories; simply cut in half. **Cactus fruit** (prickly pear,

Opuntia species): probably the second most favored fruit and relished by all lories. In some countries these fruits are cultivated for human consumption; they are sold and exported. In others, such as Australia, the plant is a pest. Armed with a pair of leather gloves, this crop is free for the picking! Lories generally prefer the type which are red in the center, rather than those which are yellow. The fruits can be offered sliced in half down the middle. The spines do not cause any discomfort to the birds. (Wear gloves and use tongs when cutting the fruits; if the spines penetrate the skin they cause unpleasant sores.) **Banana**: a nutritious and palatable food which is readily accepted by most of the larger lories, especially *Lorius* species. Fed to newly caught birds by trappers and dealers, it is one of the few foods which wild-caught birds are familiar with when imported into a foreign country. It is usually refused by the smaller species. Banana should be fed when ripe (just soft) but not hard or overripe. Its acceptance depends on being offered in the correct state of ripeness. **Melon**: I have no experience in feeding this but one breeder reported that underripe, less sweet melons were preferred. Melons, especially cantaloupe, must be well washed as the skin may be eaten. **Plums**: some lories relish ripe plums; others ignore them. **Strawberries**: I have never been able to persuade lories to eat them. **Loquats**: readily eaten by some species. **Mangoes**: little interest shown, in my experience. One breeder reported an allergic reaction in some parrots. **Dried figs**: iris and Musschenbroek's lorikeets relish them, dried or soaked; they eat only the seeds and leave the flesh. They should be fed sparingly as the iron content is very high.

Vegetables

Carrot: raw carrot is enjoyed by most of my lories. If it is refused, it can be liquidized and added to the nectar. Its high vitamin A content makes it a very valuable food. **Fresh corn**: cut into small sections about 2.5 cm (1 in.) long, it is relished by most lories. Some larger species eat it by tethering it to the perch with one foot. **Chickweed** (*Stellaria media*): an excellent food which most lories enjoy. Gather it only from a safe source free of contamination—preferably one's own premises. It is easy to grow in damp, shady areas—throughout the year in frost-free locations. To ensure a supply in winter, it

can be rooted in potting compost, in containers with drainage holes.

Sprouted foods

Fred Bauer from California promotes the use of sprouted foods because "lories fed low sugar, live diets, will have a much greater continuance of energy output and will not show hyper or starvation symptoms without food before them all day." He devised an excellent sprouted food for his lories. He uses only human-grade grains, seeds and legumes, all hulled, except buckwheat (which must be whole in order to germinate). The hulled sprouts were evenly coated with various supplements such as Spirulina, trace mineral marine earth, six herbs and spices and six beneficial microbes. To prevent souring or fungal growth, a citrus bioflavonoid stabilizer (500 ppm) was used. This extract, originally found in grapefruit seed, prevents pathogen proliferation in sprouted foods. Fred Bauer emphasized: "Success with any sprouting method is only as good as your water. No matter how 'clean' they are after soaking, if they are rinsed with contaminated water, there will be trouble." If the water is not sufficiently pure, he suggests that the sprouted food is not rinsed; the stabilizer has an inoffensive taste. Because grains and other sprouted items are alive, they still possess natural defense mechanisms and will not become sour as rapidly as other foods in high temperatures and humidity (Bauer 1995).

Fred Bauer related a very interesting happening. He sold a pair of green-naped lorikeets and, three years later, was asked to sell the numerous offspring produced by the pair. When he went to collect them, he noticed a bowl of sprouted food in the flight of the breeding pair. The owners were surprised when he commented on this. Then the truth emerged. They had gained the impression that this was the only food to be offered to this species! The lorikeets had thrived, were in exceptional condition and had a content manner. Obviously this diet had suited this species very well. The green-naped is one of the omnivorous lorikeets. It is unlikely that the true nectar feeders, such as the *Charmosyna* species, would have survived if fed in this way.

Cooked foods

Some parrot keepers who have mixed collections offer their lories food prepared for other species, in addition to nectar. For example, a pair of blue-streaked lories received a mixture of boiled pinto beans, cooked rice, cooked mixed vegetables, cooked sweet potatoes and fresh vegetables such as broccoli, carrots, beets, occasionally spinach, white cheese, hard-boiled eggs, drained tuna fish in spring water, plus two or three fruits. Dog kibble and bird pellets were ground up and added to this mixture! (McGregor 1991).

Seed

The myth has grown that lorikeets should not be offered seed. This results from the early attempts at keeping lories when many birds were offered only seed. Not surprisingly, they were short-lived. Some lories in the wild feed quite extensively on small, mainly unripe, seeds at a certain time of the year. Yes, for some species, mainly *Trichloglossus*, seed is a natural component of the diet (see **Casuarina**). The most suitable seeds for captive birds are dry or sprouted spray millet and soaked or sprouted sunflower seed. It is interesting to note that the German ornithologist Heinrich Bregulla observed Mt. Apo lorikeets on Mindanao eating wild-grown sunflower seed! (Burkard 1983). The only species which need a large proportion of seed in their diet are iris and Musschenbroek's lorikeets.

Flowers

Flowers provide great enjoyment for lorikeets. Fortunately, natural food trees of lorikeets in Australia, such as eucalypts and bottlebrush (*Callistemon*), have been introduced into many countries for ornamental purposes. Branches can be cut in the late afternoon and kept in water overnight. When they are offered to the lories next morning, all the pollen will be available—not depleted by wild birds. This idea was given to me by Fred Barnicoat of South Africa, who uses clothes pegs (spring type) to clip the blossoms to a perch. Cultivated garden flowers can also be offered, such as fuchsias, honeysuckle, wallflowers, begonias and almost any flowers which contain nectar. The flowers of the common garden weed dandelion (*Taraxacum*) are also suitable. When first offered to birds which are not used to receiving flower heads, they may be very cautious.

Guide to foods favored by captive birds

Note that this is a very general indication by genus and that species and individuals may have marked preferences which are not reflected in the table below.

Captive Lory Food Preferences

** indicates main dietary items
* indicates also eaten

	Nectar	Fruits	Vegetables & sprouted foods	Green-food	Seed	Flowers
Chalcopsitta	**	**	*			*
Eos	**	**	*	*	*	*
Pseudeos	**	**			*	*
Trichoglossus (except iris & Mount Apo)	**	**	*	*	*	*
Iris & Mount Apo	*	**	**	*	**	
Glossopsitta	**	*				*
Neopsittacus	*	*	*	*	**	*
Lorius	**	**	*			*
Vini	**	*				*
Charmosyna	**	**		*		*
Oreopsittacus	**	*		**		*

See also **Casuarina, Dry Diets, Feeding Behavior, Grit, Insects as Food, Nectar, Protein**.

Disease, see **Bacterial Infection, Candidiasis, Flagellate Infection, Gout, Malnutrition, PBFD, Worms**.

Display

Most of the ritualized components of the display of lorikeets are not unique to them. Behavior such as head bobbing, lateral swaying, wing fluttering and, of course, courtship feeding, occur in other, unrelated parrots. An exaggerated swaying walk is a pronounced feature of the display of some lories, such as *Trichoglossus* and *Eos* species. *Lorius* species and some *Trichoglossus* (e.g. iris), when on the ground, jump quite high, up and down, or in a small circle, in a very excited manner. Probably the only element of display which is unique is that sometimes described as "hiss-ups" (Serpell 1981). The bird will suddenly extend itself vertically upward while simultaneously sticking out the tongue and uttering a hissing sound.

The courtship and precopulatory display of lories is energetic and exaggerated. It is often designed to show off some colorful part of the plumage, especially the underwing coverts. In closely related species the pattern or color of the underwing coverts may be strikingly different, such as the solid red underwing coverts of the rajah lory (*Chalcopsitta atra insignis*); in the closely related Duyvenbode's the underwing coverts are brilliant yellow. Displaying birds slowly flap their wings, and I have even seen a Duyvenbode's revolve around the perch with open wings.

A striking example of brilliant coloration under the wings is Musschenbroek's, in which this area is scarlet. It seems that many actions are designed to emphasize this. When a bird lands on a perch it holds the wings open for a fraction of a second. During copulation the wings are held open, unlike most other lories, which mate with the wings closed.

Species which do not have brightly colored underwing coverts, such as Meyer's, iris and perfect, do not feature any wing movements in their display, but make exaggerated prancing steps and head movements. Other members of the same genus (*Trichoglossus*), such as subspecies of *haematodus*, use these movements in addition to a slow flapping of the wings to reveal the brightly colored underwing coverts.

Lorius species such as the black-cap also briefly hold the wings open in display. In this species the pattern of the coverts varies in the subspecies. The mainly red members of the genus *Charmosyna* are sexually dimorphic and brilliantly colored. In Josephine's and Stella's the rump is brilliant yellow in the female. In display, the rump feathers stand out, away from the body. Also in Stella's, the streaky blue feathers of the crown stand up from the head. A displaying male weaves his head and body in snaky movements.

It is important for breeders to realize that two lories seen displaying to each other and attempting copulation are not necessarily male and female. Likewise, in an aviary containing several species they are not necessarily of the same species, even where the same species is present. In short, a lory might display to, preen and form a pair bond with any other lory—or even with any other parrot of any size. I know of a very strange alliance—that of a male perfect lorikeet and a male yellow-fronted Amazon. The two birds live in the same cage and are devoted to each other.

I once watched the behavior of two female perfect lorikeets, kept together for want of males. One of them took the male role, prancing along the perch, following the other female, with head bowed. They traversed the perch, up and down, a few times. Then the bird in the male's role tried twice, unsuccessfully, to mount the other female. During this

performance she was hissing and bowing in the usual *Trichoglossus* manner.

During copulation and in some cases during the precopulatory display, all lories make a ticking or clicking sound, or a rhythmic squeaking in the case of *Lorius* species. The male mounts the female with both feet on her back; he usually grasps her flight feathers on one wing with a foot. Copulation usually lasts about two minutes. It occurs several times daily for one to two weeks before the female lays.

Peter Odekerken noted: "*Trichoglossus* species have rather amusing and elaborate displays. Wing whirring often occurs as either a threat or dominance form of display to females and other males lower in the pecking order. It seldom occurs that birds get hurt. Confrontations usually result in the less-dominant bird retreating in haste, however I have seen birds roll on their backs in a submissive posture to pacify the dominant bird when feeding on the ground" (Odekerken 1993a). The displays of *Trichoglossus* species were the subject of several research projects, carried out by James Serpell. His detailed work serves as a baseline for anyone intent on making a scientific study (see Serpell 1979, 1981, 1982).

Drinking

In the wild, lories drink fresh water wherever they can find it. This is usually water trapped in foliage, or they might climb down to a stream. Varied lorikeets have been seen to descend a branch overhanging a stream or climb onto an exposed twig on a partly submerged log (Forshaw 1981). The tiny little palm lorikeet has been observed drinking water from leaves (Bregulla 1992).

John Courtney recorded an interesting passage in his diary for October 5, 1967.

2:20 P.M. I saw a pair of little lorikeets fly into a small river oak on the Swanbrook Creek. One flew down and had a long drink from a small rock pool, 1 ft. away from the flowing creek. It dipped and raised its bill a number of times and when swallowing rapidly moved its mandibles, as if rapidly 'brushing' nectar from a blossom. It then flew up into the tree, and its mate descended to the same rock pool, and drank in similar fashion. They then departed together. The rock pool was fresh and clear, and was fed from the main stream by a crack in the rock.

On October 8, 1968 another passage regarding lorikeets drinking was entered in John Courtney's diary: "11:30 A.M. I saw two musk lorikeets fly down and sit on stones at the edge of Swanbrook Creek. They bent down and drunk from the creek. They then flew up and went straight into their nest. I approached and could hear the hissing begging call of young in the nest."

H. E. Tarr described the drinking method of the varied lorikeet. "The bird gradually climbs down a drooping branch to the water and then hangs with the head down and bent backwards while the liquid is swallowed" (Tarr 1963).

In the Central Highlands of New Guinea I saw a small group of whiskered lorikeets fly down to a stream which was little more than a trickle at the side of the road. A grassy bank rose up from one side of the stream, which was clear but full of stones and grit. Joseph, the ornithologist resident in the area, told me that he had often seen these lorikeets drinking in similar locations.

By observing lories in aviaries one can see that the method of drinking differs slightly. Bosch (1993) noted of Goldie's lorikeets: "When drinking nectar only the tongue is dipped in, never the whole beak...the feeding area therefore stays much cleaner than with *Charmosyna* species which, after every sip, shake their beaks and fling drops across the whole area."

Most lories drink by dipping the tip of the beak into the liquid, but I have seen black-caps dip almost the whole beak, then gulp the liquid down. *Charmosyna* species flick the head after a few sips and wipe around the beak with the tongue after almost every sip.

Clean drinking water is extremely important to lories; unlike many parrots, they will not drink dirty water. Some keepers believe that because they drink nectar, drinking water is not important. This is not true. In fact, with the trend of using dry foods, fresh water is even more important. In spacious aviaries, it is beneficial to offer a water container for drinking near the food, and a larger container for bathing in the flight area. I use stainless steel dishes which hang by three chains (attached to a dog clip) from the aviary roof.

Cleanliness of water dishes is very important. Lories are prone to bacterial infections, and contaminated water is one of the most likely sources.

Frankly, when visiting breeders' aviaries, I am appalled at how little importance many of them attach to providing water in clean containers. In a large collection there is not usually time to change or clean all the water dishes daily—or weekly, so the owner will tell you. If this is the case, he has more birds than he can care for properly. Clean water is vitally important. Automatic watering systems are increasingly used in large collections. A pipe flows through all aviaries and delivers water automatically. The system should be designed so that a disinfectant can be flushed through it on a regular basis. Automatic watering is an excellent labor-saving device; however, it does not mean that the cleanliness of the water containers can be totally ignored!

Dry Diets

The idea of feeding dry foods to lories is not as new as some people might believe. I mentioned it as long ago as 1977 in *Lories and Lorikeets*, and Stan Sindel in Australia had then been using it for some years. However, at that time I suggested that the more omnivorous species such as the iris would benefit from this addition to the diet. That it should be used to totally replace nectar was, and still is, unthinkable to me. Unfortunately, in the U.S.A. the formulation of dry foods for lories was latched on to by breeders and pet shops to promote these birds as pets. What was not made clear, however, was that not all lories will accept dry foods. Some would—and did—starve to death rather than eat them. Dry foods are an excellent supplement to the diet of many species, but all lories (except some iris and Musschenbroek's) need to have nectar available for at least two-thirds of the daylight hours.

Two different types of dry diets are commercially available. One is a powder, available from several companies; this food is readily eaten by many lories. The other is an extruded product (pellet), produced by Pretty Bird, called Lory Select. It contains 16 percent protein, 6 percent fat, 1.5 percent fiber and 12 percent moisture. It has the appearance of tiny red and green grains. My experience of this food, which I was eager to test, was that my lories and lorikeets refused to sample it. The species range from omnivorous to mainly nectar-feeding.

Presumably not all breeders have had this experience. The companies who produce dry diets state that they should be supplemented with fruits and nectar. One company promoting their dry diet stated: "Some nectars become contaminated during preparation or may even contain contaminated ingredients. Water supplies can also be contaminated and can cause many problems to the aviculturist. Dry foods which are extruded are unlikely to be contaminated due to high temperatures used during production" (Clubb 1996). It should be pointed out that lories fed mainly on a dry diet need to drink large quantities of water, so if the water is contaminated the use of dry diets will not prevent water-borne bacteria from causing disease.

I continue to be concerned about promoting dry foods unless it is made very clear that not all lories can survive unless nectar forms the main part of the diet. Susan Clubb does make this clear by stating: "Smaller Lories such as *Charmosyna*, *Glossopsitta* and *Oreopsittacus*, are more difficult to convert to a dry diet, and some of these birds are dependant upon nectar." It should be made quite clear that these species are impossible to convert to a diet of extruded foods. One breeder described using the Pretty Bird product and stated that the lories "play with their pellets by floating them in their water bowls before consuming them." When hand-reared dusky lories were about five weeks old they were offered fresh fruit sprinkled with the extruded food and a warm bean and vegetable mixture. "If they are going to the pet trade we recommend they maintain a diet without nectar to avoid mess" (Jones 1995). (Dusky lories are one of the omnivorous species which readily accept a bean and vegetable mixture.)

Some lory keepers invent a dry diet—and it does not have to be suitable to be accepted! Where birds are concerned, the proof of the pudding is not in the eating but in the quality of the young reared over a period. The excellent breeding results of Harald Mexsenaar in Queensland with his little and purple-crowned lorikeets are recounted in the species section when discussing these species. His dry mix contains the following:

2 cups of Heinz Rice Cereal

2 cups of a proprietary egg and biscuit food

1 cup of glucose powder

1 teaspoon multivitamins

¼ cup of skimmed milk powder

2 cups of rice flour

Similar dry mixes are widely used in Australia, an adaptation of Stan Sindel's original recipe. This consisted of:

2 cups of Heinz Rice Cereal

2 cups of rice flour

2 cups of canary egg and biscuit food

1 cup of glucose powder

1 teaspoon of multivitamin and mineral powder

1 dessertspoon (1½ teaspoons or ½ tablespoon) of pollen is optional (Sindel 1987).

One of the most successful lory breeders in Germany (he has had exceptional results with whiskered lorikeets) is Ruediger Neff. His dry diet and nectar solution are based on the same foods (Neff 1993), as follows:

Neff Diets

Ingredients	Nectar Solution	Dry Diet
	To 1 liter of lukewarm water add	to make 1 kilo:
Bee pollen	30 g (non-milled)	430 g (milled)
Oat cereal	65 g	360 g
Honey	50 g	–
Glucose	–	140 g
Brewer's yeast (specially prepared)	5 g	50 g
Calcium and minerals	0.5 g	20 g
Multivitamins	3 drops	–

In the U.K., Pete Clear made up his own dry diet, which was used for Musschenbroek's, emerald and Goldie's lorikeets and black lories. It contained Readybrek oat cereal, Vitafood, glucose, ground rice, Horlicks, demarara sugar, sunflower kernels and SA37 (protein and vitamin supplement). The sugar and sunflower kernels were reduced to a powder in a coffee grinder. This food was very well accepted (Clear pers. comm. 1995).

I carried out a simple experiment to test the palatability of a dry food offered to 15 species of lories. I mixed together Milupa baby cereal, wheat germ cereal, fructose (natural, pure sugar from fruits) and a small amount of Nekton MSA (minerals, trace elements, Vitamin D_3 and amino acids). It tasted sweet and pleasant. The birds to which it was offered were housed in pairs (plus a few single birds and one cage of three), the inhabitants of two rows of cages. Species were as follows: 4 pairs and 1 single black lory (mainly rajahs), 2 pairs of Duyvenbode's, 3 pairs of yellow-streaks, 1 pair of red, 2 pairs of duskies, 2 pairs of green-naped, a cage containing a single yellow-streaked and a single green-naped, 2 female perfects, 3 pairs of Goldie's,

6 pairs and 1 single iris, 1 pair and 1 single Stella's, 2 pairs of Josephine's, 1 pair and 1 single red-flanked, 5 pairs of black-caps, a single yellow-bibbed, and 3 pairs and a single Musschenbroek's.

On the first occasion the food was offered when the birds were not hungry, that is, about one hour after receiving the second nectar feed of the day. A very small amount of finely chopped pear or apple was placed on top of the dry mixture to encourage them to sample it. Nearly all the lories investigated it at once and ate the fruit. The dry mixture was then eaten at once by a pair of Josephine's (the other pair was incubating), the single young Stella's and the single yellow-bibbed. It is worth mentioning that these are all very tame captive-bred birds. The iris are also tame but showed little enthusiasm. The dishes were removed five hours later. Only the two pairs of Josephine's, the single Stella's, the yellow-bibbed and the three young black-caps had eaten it all. A significant amount had been consumed by the young pair of green-naped (also hand-reared), the single iris, a pair of black-caps, the pair of Stella's and a pair of Musschenbroek's. Most of the others had tasted it but were obviously not impressed. It was of interest to find how readily the Josephine's and Stella's took it, and how it was virtually ignored by the *Chalcopsitta* species. On the following day, all food was withdrawn from the same birds at 11:00 A.M. At 11:45 A.M. it was replaced by the same dry mixture. The results were almost the same as on the previous day except that the single black lory, a pair of yellow-streaks and a pair of iris sampled a little. A pair of rajahs (*insignis*) and a pair of Duyvenbode's were waiting for food to feed their young but did not touch the dry food. At 2:00 P.M. the latter was withdrawn and replaced with the usual nectar; every bird fed immediately.

The next stage of the experiment was to repeat the process substituting a commercial dry lory diet. The results were exactly the same. It was eaten with great enthusiasm by the Josephine's lorikeets while they were rearing young and henceforth became part of their diet at this time. When they did not have chicks they showed no interest in it.

In my opinion, dry diets are of great value for certain species which readily accept them. Their use means that during hot weather, or when for some reason no attendant can be present twice a day, food which will not spoil is always available. It is also useful as part of the diet for a pet bird of the more omnivorous species (such as dusky and green-naped). It reduces the quantity of liquid droppings and therefore makes these birds more acceptable as pets. It cannot be overemphasized that clean, fresh water must always be available for birds fed on dry foods. Ideally, there should be a small water pot next to the container of dry food and a larger water container for bathing in another location, away from the food.

Eggs

Lory eggs are white, like all parrot eggs. However, some of those of Charmosyna species with which I have worked (*josefinae*, *papou*, *placentis* and *pulchella*) alter if they are fertile. They become grayish with small marks, almost like pencil lines. The shape of lory eggs varies according to species and individual birds. Eggs in the same clutch may differ in shape and, of course, in size. I have noticed a tendency for one of my iris to lay eggs which are pyriform (almost round)—unusual among lories. Most lay elliptical (oblong oval) or ovate (oval) eggs.

In the text for each species some recorded egg sizes are given, where known. However, as the size variation in eggs laid by the same species is great, these figures are not too significant. I would suggest that the sizes of at least ten eggs laid by three different females would be needed to obtain representative data.

The ratio between body size and egg weight in birds varies and is partly dictated by whether the species is nidicolous, as are the parrots (young remain in the nest after hatching) or nidifugous (are feathered and able to walk almost as soon as they hatch). Species in the latter category lay larger eggs. Also, it seems that the more aerial species lay eggs which are relatively lighter for their body size, as heavy eggs would hamper flight. Other factors influence egg size, such as availability of food prior to the period of egg formation.

Lory eggs weigh from 1.7 g in the tiny Wilhelmina's lorikeet up to 10 g in the largest *Lorius* species. Ratio of new laid egg weight to body weight varies from 1:10 in the smallest lorikeets to about 1:22 in the largest *Lorius*. Some examples are as follows:

Lory Body and Egg Weight

Species	Body weight*	Egg weight (new laid)
Lorius lory	220 g	10 g
Eos cyanogenia	170 g	9 g
Eos reticulata	160 g	9 g
Trichoglossus h. mitchellii	100 g	5.5-6 g
Vini solitarius	75 g	5.3 g
Trichoglossus iris	65 g	5 g
Charmosyna rubronotata	30 g	2.5 g
Oreopsittacus arfaki major	22 g	2.5 g
Charmosyna wilhelminae	20 g	1.7-1.8 g

* based on weight of adult female

See also **Clutch Size** and **Laying Interval**, **Incubation**

Electricity

In indoor and outdoor birdrooms electricity is usually essential. Tungsten, fluorescent or Vita-Lite can be used. Fluorescent often provides a harsh light and should not be left on all day unless absolutely essential. It can prove stressful to lories housed below it. Artificial light sources which mimic natural sunlight (full-spectrum lights) are recommended. They produce ultraviolet radiation which permits birds to synthesize vitamin D3. This is most important for those which do not have access to sunshine. Broad- or wide-spectrum lights do not have this advantage.

On a hot day, lights could result in an unwelcome increase in the temperature. Tungsten lights can be wired to a dimmer which gradually lowers the lights (either when required or on a timer, at a preset hour). This is necessary because if the lights are suddenly extinguished when it is dark, a pair with eggs or young could be outside the nest and afraid to enter in the dark. When planning a birdroom, include several more power points than are believed to be necessary. They should be positioned throughout the birdroom, enabling one to place a heat lamp over or by the cage of a sick bird. This is often preferable to removing it to a hospital cage.

The following birdroom equipment results in a need for power points: brooder, dimmer, fan, heat lamp, incubator, ionizer, hospital cage, humidifier (or dehumidifier), pressure washer and ventilators, plus an electric drill for maintenance tasks.

Environmental Enrichment

Lories are active and inquisitive birds. Kept in suspended cages, which provide no outlet for their playful and often amusing behavior, they soon become bored. Anyone who keeps them in such cages must try to alleviate this boredom. The best and most natural way is the regular provision of leafy branches or those with buds, or even buds and blossoms. In Australia, eucalyptus is the perfect choice. Most nonpoisonous trees are suitable, especially fruit trees, poplar, willow, eucalyptus and mulberry. One only has to witness the enjoyment they derive from such branches; their pleasure is quickly transferred to the observer, making the effort well worthwhile. Ideally, fresh branches should be provided daily—but this is seldom possible. Another source of enjoyment is a brightly colored flower, such as fuchsia and hibiscus. However, if these are not offered on a regular basis, they may cause alarm when first introduced!

It is not only pet birds that need "toys" in their cage. Two owners of a large lory collection wrote: "It is important to understand that lories are easily bored and very excitable. Therefore, breeding birds should be provided with large flights and outside distractions so that they can release pent up energy. Toys such as rings, chains, and ropes are very appropriate and lories enjoy them tremendously" (Casmier and Wisti-Peterson 1995).

Rope swings and thick rope hung from the aviary roof are excellent forms of entertainment. One must ensure that the correct kind of rope is used and that there are no strands which might unravel. These could trap a bird by the leg or by the neck. An inexpensive and readily available toy is a child's skipping rope with wooden handles (unvarnished). Simply knot the rope three or four times, and attach a dog clip through the top loop and clip to the cage or aviary roof. My own birds spend hours of enjoyment climbing, swinging and biting on skipping ropes.

Another simple toy is made by attaching a length of chain to the end of a piece of dowel. Attach a dog clip to the other end of the chain and hang with the dowel furthest from the top. Make two identical toys and hang them from the roof about 7 cm (3 in.) apart. Young lories will use them like swings, or several will spin around on them at the same time.

A shallow, concrete pool set into the aviary floor, or even a bird bath suspended by chains from the roof, is a worthwhile addition to a zoo or private aviary. The lories' bathing antics will be enjoyed by all onlookers. Bathing is a contagious activity and lories in a colony will all indulge simultaneously! An excellent hanging bath can be made by buying a stainless steel dish and drilling three holes in the rim at equal distances. Put a large key ring through each hole and add a chain. Clip the three chains together at the top by means of a dog clip, then clip the chain to the roof. The rings and chains can be purchased in hardware stores. I hang these containers in my aviary flights as baths; the drinking water is inside the shelter where the food is located. I also make swinging perches, using natural branches and the rings and chains previously described in this paragraph. Choose a branch with a natural bump near what will be the end of each perch. The chain can be looped round the inner side of the bump and secured with a key ring.

Many lory breeders in warm climates install a misting system above their aviaries. The lories delight in bathing in the fine spray. If leafy branches are placed in the aviary below the spray, they will rain-bathe energetically and excitedly, sustaining this activity for minutes. This keeps them happy and healthy; their plumage will be literally shining as a result of regular and prolonged contact with water.

Ethics (Avicultural)

High standards of self-imposed ethics on the part of aviculturists are essential if lory breeding is to survive long-term. Important aspects are considered below.

1. All lories should be kept, fed and housed in a manner appropriate to the species. If the correct conditions cannot be provided, especially where hygiene and cage size is concerned, lories should not be kept. They are intelligent, very active, playful and affectionate birds and should not be housed under conditions suitable for a budgerigar or canary. Both those species have been domesticated for countless generations. Lories and lorikeets need space in which to fly!

2. Species should not be hybridized under **any** circumstances. Gene pools for some lory species are rapidly diminishing; avoiding the contamination that results from crossbreeding is essential. For the more common species there is no excuse; with a little effort, the correct partner can usually be found. In the case where this is impossible, a lory should have a partner of the same sex, and of a species similar in size. Lories need company, unless tame and very imprinted, and two birds of the same sex can be content together. Zoos which maintain mixed-species lory exhibits should never allow hybridization to occur within the colony. The zoo community should set an example in this respect. Mutations will always be popular, but hybridizing in order to introduce a mutation into another species is deplorable.

3. Young birds produced for the pet trade should be sold only to responsible people, and accompanied by written instructions (which can be passed on) detailing their proper care. One breeder in California, Bobbie Meyers, gives away an inexpensive published book on lories with every lory she sells as a pet; this gesture is highly commended, and might well be emulated by other breeders.

5. Unweaned young should be sold only to experienced hand-feeders who are breeders; they should not go directly into the pet trade.

6. Aviculturists must be honest regarding the birds they sell to other breeders. They must disclose undesirable traits, such as an aggressive male, or a history of ill health.

7. Breeders must be responsible enough to voluntarily participate in stud books for the rarer species. Only through cooperation will these species be preserved in aviculture.

Exotic Bird Registration (Australia)

As from October 2, 1996, all owners of exotic bird species in Australia need to register under the National Exotic Bird Registration Scheme. Certain exotic parrot species are exempt from this scheme. None are lories or lorikeets. Therefore, all owners of exotic lory species who have not already registered their birds should contact National Exotic Bird

Registration Scheme, GPO Box 1443, Canberra, ACT 2601.

Extinctions

It seems likely that most South Pacific islands of sufficient size and suitable habitat, and not currently occupied by lories, were formerly inhabited by one or more species. During this century, the rapid reduction of the ranges of some species has highlighted the fact that many islands are now bereft of lories. Members of the genus *Vini* have suffered the most extirpations in recent years. For example, the Tahiti blue lory probably became extinct on Tahiti and Moorea during the early years of the twentieth century. It was formerly recorded from more than 20 islands, but is now extinct on most of them. The ultramarine lory has recently become extinct on Nuku Hiva and is now feared extinct on Ua Pou. Only relict populations exist of these two exquisite birds. The introduction of ship rats to their native islands is almost certainly to blame. Bones aged at least 700 years indicate that Kuhl's lory was once widespread in the southern Cook Islands; it has long been extinct there. It was too beautiful to survive and was probably trapped to extinction for its feathers.

At least two *Vini* species have become extinct and are known only from sub-fossils. Perhaps more await discovery...

See *Vini sinotai, Vini vidivici*.

Feather Plucking

Lories, like all parrots, may pluck themselves or bite their feathers in times of stress or ill health. An example of feather biting due to stress was a tame black-capped lory placed in a large aviary with many other young lories and kept there for 2 years. As soon as he was removed and kept on his own he ceased to bite his tail and flight feathers and never did so again. A hand-reared bird, he preferred human companionship to that of his own kind. Feather plucking, where the whole feather is pulled out, usually commences with the removal of the breast feathers. In serious cases the feathers may be removed from other areas. Some lories will pluck the head feathers from their mates, a bad habit which appears to stem from overenthusiastic preening.

This could lead to permanent loss of feathers in that area, due to destruction of the feather follicles.

Unfortunately, there is no ready cure for feather plucking. One has to search for the cause, which is seldom easy to find. On the other hand, some females pluck their breast feathers when breeding, and not at other times, and the condition never becomes serious. Feather plucking is common in the red lory—but why this species is so susceptible is unknown.

Plucking of chicks in the nest is a not uncommon problem in lories. Many pluck the down from their chicks; one need not worry about this except in cold weather. Plucking of the contour feathers is more serious, especially if the majority of feathers is removed. Again, red lories are particularly prone to develop this habit. Young will have to be taken for hand-rearing if most of the body feathers are removed—or if the flight feathers are plucked and the birds are housed in an aviary. In a suspended cage in a warm climate, badly plucked chicks might survive, but it is wiser to remove them. If this is impossible, a nectar container should be placed where the young bird can reach it easily. As soon as it is seen to feed itself it should be removed to a cage on its own. The feathers will start to grow as soon as plucking ceases; it should be fully feathered in about six or seven weeks.

Obviously it is very important to stop lories from plucking their young. An effort to do so should be made immediately when plucking starts. It is essential to keep the nest dry. Holes can be drilled in the bottom of the nestbox and the nest litter changed daily, or at least every other day. Alternatively, a nestbox with a welded mesh base can be utilized. If two layers of welded mesh are used, the wood shavings can be placed directly on top of the mesh. This allows the liquid feces to drain out of the box, thus the shavings will not need to be changed so frequently. In a large collection, where changing nest litter is very time-consuming, this method can be used to advantage. At Loro Parque, Tenerife, where the idea originated, it has proved to be successful (Sweeney pers. comm. 1995). Another idea is to place strips of bark in the nestbox every three or four days. The theory is that chewing it up will distract the parents from plucking their young. This method solved the problem for M. Gammond, whose perfect lorikeets were habitual pluckers. An-

51

other idea, which perhaps originated among budgerigar breeders, and would probably be best suited to a suspended cage, is to place the nestbox on the cage floor with the top open. (The use of a nestbox which opens from the top is not recommended under normal circumstances.) This procedure stopped a pair of Rosenberg's lorikeets from denuding their young (Hubers 1994).

In the wild, lories must work hard to find enough pollen and nectar to rear their young. They do not have time to spend long periods in the nest. In the artificial conditions which exist in captivity, they spend hours in the nest and excessive preening of the young leads to plucking. Self-plucking is not always psychological in origin. It can indicate an underlying disease problem. If this is suspected, a vet should be consulted.

Feeding Behavior

In the aviary

In most lories, the instinct to feed as a flock is very strong. When one places a number of young birds in an aviary after removing them from their parents, the younger ones quickly find food by copying the others. In any case, lories are so quick to investigate everything, that unless the food is placed on the floor, it will be found and sampled within minutes. It should never be placed on the ground, due to the increased risk of contamination, not only from the feces of the birds themselves, but also from insects or even vermin. Additionally, in a very small aviary (but not in a suspended cage) some lories do not like to descend to the floor. Another important consideration is that feeding from the ground represents an unnatural behavior to most lories.

If food is not presented in the right way (even favorite items of food) it may be ignored. When a group of young lories (previously housed in suspended cages with their respective parents) was placed in a new aviary, a dish of pears on the food shelf next to the nectar container was ignored. When I wedged the pears into the welded mesh in the same place, the lories immediately came up and ate them. This is not an isolated case. It has repeated itself on other occasions with different groups of young birds.

Feeding methods of different species vary, according to their main food source. Unlike many parrots, most species do not easily hold food in the foot. The fact that iris and Musschenbroek's hold nearly all solid items in this manner suggests that the main components of their diet in the wild must be seeds, insect larvae or some other items which they need to manipulate with their powerful bills. The primarily nectar- and fruit-eating species are unable to do this. Some, such as black-capped and yellow-streaked lories, will hold food in the foot on occasion, but seldom prolong the attempt because they are not adept at it. They will, however, anchor food by the foot (a half banana in the skin, for example), then tear at it with the beak. In my opinion, this indicates that such species eat some fairly solid items in their natural habitat.

In the wild

When a lorikeet is feeding on small flowers, it is difficult for the observer to ascertain whether it is eating pollen or nectar or both. Each flower is enclosed within the beak so that anthers and nectaries could be touched by the tongue simultaneously. Freshly opened flowers offer pollen and sometimes nectar, whereas older flowers offer only nectar (Hopper 1980). Hopper wanted to determine what purple-crowned lorikeets were feeding upon, so he studied them utilizing a species of eucalyptus (*Eucalyptus occidentalis*) with large flowers requiring the lorikeets to assume different feeding positions for pollen and for nectar. This was necessitated by the fact that the anthers and the nectaries are well separated in this species. This eucalyptus flowers in autumn and in winter, and is endemic in southern Western Australia. It can grow to 20 m (66 ft.) in height with erect, spreading branches and a flat-topped canopy. A mass-flowering species, the entire canopy appears creamy yellow when in full bloom. Each inflorescence has up to seven elongate pendulous buds. The stamens attain a length of about 15 mm. Pollen can be gleaned from anthers only when the bud is mature and in young flowers whose stigmata have not yet become viscid (sticky).

On the morning of August 20, from 10:50 A.M. to 11:05 A.M., lorikeets were the most numerous birds feeding on the nectar, in a pure woodland of *E. occidentalis*. There were up to five birds per tree. Two purple-crowned lorikeets low in the canopy were each observed for five minutes. Both took pollen from mature buds and nectar from open flowers. Hopper recorded:

They approached buds in which the elongate operculum had dehisced from the base but still remained attached to the stamens and removed the operculum with the beak, thus uncovering the tightly packed cylinder of freshly dehisced anthers. With their heads at right angles to the stamen cylinder, they could enclose the entire bundle of anthers in their beak and nibble at them to remove pollen. No attempts to take nectar from the base of these buds were observed. On open flowers, however, the Lorikeets thrust their bills into the central cup for nectar and ignored the splayed array of anthers. They fed on all open flowers and mature buds within reach from a given perch and spent about as much time taking pollen as taking nectar. Only a few seconds were needed to forage on each flower or bud. The Lorikeets worked their way through dense clusters of flowers on a given branch, moving as little as possible (usually only a few decimeters) between clusters.

Stephen Hopper made another feeding study of purple-crowned lorikeets which suggested that this species can harvest pollen and nectar from small-flowered *Eucalyptus* at the rate of one flower every one to three seconds. This is five to ten times faster than the rate of nectar harvesting estimated by Churchill and Christensen (1970) as being necessary to satisfy basal energy requirements. If the feeding rate observed by Hopper was maintained throughout the day, it would take a lorikeet only 2 1/2 hours to harvest pollen and nectar from 3,000 flowers, leaving ample time for other activities, such as preening, flying and social interactions (Hopper and Burbridge 1979). Their findings cast considerable doubt on the claim that nectar is at best a supplementary item of diet of these lorikeets and that pollen is the staple item. It would seem that either pollen or nectar could be used as the principal food, because each alone can be harvested at a rate more than adequate to fulfill daily nutritional needs.

What is the preferred feeding height of lories in foliage? This depends on the species and the type of habitat. Brooker, Braithwaite and Estbergs (1990) studied two lorikeet species in Australia and tabled their foraging height according to season and habitat. No birds were observed below 2 m (6 ft. 6 in.).

Foraging Habits

| Species | Habitat | Season | Foraging height in meters | | | |
| | | | 2-3 | 4-7 | 8-14 | above 14 |
			Percentage of observations			
Red-collared	MF	DW	5	5	15	75
	OFW	D	10	43	38	10
		W	21	36	43	
Varied	OFW	D		31	31	38
		W		20	45	35

MF = monsoon forest
OFW = open forest and woodland
D = dry season
W = wet season

Feeding scaly-breasted lorikeets were closely observed at Toowong (Brisbane) for eight days from March 5, 1977. This was followed by four days of observation from a roof to give an unobscured view of the birds. Two different feeding methods were observed on umbrella trees (*Brassia actinophylla*). The first consisted of a slower, more deliberate action. The beak was held wide open and the tongue was thrust deep into the flower. This method of feeding was employed almost exclusively during the first few days of observation; small grains, presumed to be pollen, could be seen on the lorikeets' tongues. The average duration of this method was between one and two seconds per flower. The second method of feeding was much more rapid (two to three flowers per second). It consisted of placing the upper surface of the tongue in a bent position on the flower. This method was most common after the fourth day and coincided with an increase in the number of bees around the flower stems, and the presence with what was assumed to be nectar clearly visible on the surface of the flower (Hamley 1977). See also **Honeyeaters**.

Feral Population

Feral populations of parrots, notably ringneck parrakeets (*Psittacula krameri*) and Quaker parrakeets (*Myiopsitta monachus*), have become established in many countries. There does not appear to be any instances of such populations of lorikeets living in a different country, although some have colonized islands on which they did not originally occur (e.g., Tahiti blue lory on Aitutaki). In Australia, however, the rainbow lorikeet (*Trichoglossus haematodus moluccanus*) has populated an area which is 3,200 km (2,000 mi.) from its natural range. It has been found in Perth since at least 1968. In 1988 it occurred within a triangle between Mosman Park, Scarborough and South Perth. It has been suggested that the lorikeets originated from Rottnest Island in 1960. The board manager there requested permission to release rainbow lorikeets and cockatiels; his request was refused. Nevertheless, unidentified green parrots, budgerigars and cockatiels were observed on feeding platforms or in the settlement area there. It was eventually reported 13 months later that six or seven rainbow lorikeets had escaped from a cage which overturned in a gale. From there they are believed to have colonized the Perth area (Coyle 1988).

In Coolgardie, Western Australia, 500 km (310 mi.) east of Perth, a pair of rainbow lorikeets was observed. Presumably aviary escapees, they were squabbling with a pair of Port Lincoln parrakeets (*Barnardius zonarius*) over a hollow in a salmon gum. They fed on the flowers of that species in a nearby tree. Twenty days later, on July 6, a pair of rainbow lorikeets was observed to fly from the same hollow. Several days later, the pair was captured and taken into captivity; two eggs were removed from the hollow. The Port Lincolns then established themselves in the same hollow. This suggests that rainbow lorikeets are aggressive nesters with the potential to displace locally indigenous species, even those considerably larger than themselves (Chapman and Hazeldon 1994).

Field Studies

Few lory species have been studied in depth in their natural habitat. A little more is known about Australian species than those from other areas, many of which are seldom visited by ornithologists. Our knowledge of the natural history of lories is, generally speaking, sketchy. We know only the most basic facts of their lifestyle. In a number of species, nests have yet to be found.

New Guinea species, for example, have been little studied. Why is it that most ornithologists visiting New Guinea (a paradise for parrot lovers) become totally seduced by and concentrate their studies on the birds of paradise? Yes, they are behaviorally fascinating and many are gorgeously colored, but so are lories! Joseph Forshaw (1989) mentions a group of purple-bellied lories (*Lorius hypoinochrous*) on Normanby Island. "Their noise, continual movements and striking red plumage coloration were most distracting to field workers trying to observe displays of Goldie's Birds of Paradise (*Paradisaea decora*) in trees above." What a pity the field workers did not consider that

lories are also worthy of study! As a group, the ecology of birds of paradise is likely better known!

Bruce Beehler, the well-known American ornithologist, who has devoted nearly a lifetime to studying the birds of New Guinea, recorded how little is understood of the avifauna there.

> August 1986 was a month of rain, mist and westerly winds. August 1987 was a month of sun and cloudless sky. All of eastern New Guinea was caught in a deep drought, and gardens were wilting, villagers hungry. In 1986 the three most common resident birds at Lake Omha [eastern Papua New Guinea, near English Peaks] were New Guinea Thornbill, Crested Berrypecker, and MacGregor's Bird of Paradise. In 1987 the three most common species were Red-collared Myzomela, Plum-faced Lorikeet [*Oreopsittacus arfaki*], and Sooty Melidectes. The seasonally resident bird community changed drastically from one August to the next. This sort of change in community dynamics has not previously been exhibited in New Guinea. But it is obvious that major events are happening on an irregular schedule, and predictability is probably not a significant feature of some Papuan habitats. It just so happens this was easier to document in an impoverished, high-altitude environment, where the changes were not masked by environmental complexity (Beehler 1991).

Few sponsored or other studies have been made in which a lory is the central species. To date, only *Vini* lories and the endangered red and blue (*Eos histrio*) have been the subject of such research. In several studies concerned with parrots in Indonesia, including those which are traded, lories have featured prominently. Unfortunately, the main thrust of research into the natural history of lories in the future will arise only as a result of the threatened or endangered status of certain forms, especially those from islands.

Flagellate Infections

Flagellate microbes are single-celled organisms and can reside in the gastrointestinal tract of parrots. They include *Trichomonas gallinae*, *Hexamita* and *Giardia* species. They do not require an intermediate host and are transmitted through direct contact or through ingestion of contaminated water or food (Ritchie, Harrison and Harrison 1994). According to Hubers (1995b), symptoms are crop inflammation and an inclination to vomit, although only a few droplets of nectar may appear at the tip of the beak.

In the early phase the bird does not lose weight or show other signs of illness. Hubers tried unsuccessfully for two days to empty the crop of an affected bird; he was not successful until he gave 2 ml of physiological saline (9 g salt to 1 liter of water) into the crop using a crop needle. The crop was massaged. Then the lory was given 1 ml of nectar containing ronidazole. Within one day the crop had emptied and the bird made a full recovery.

Hubers states that examination of crop samples does not always reveal the presence of flagellates. A crop smear and a microscope with a magnification of 100x to 400x usually reveals their presence. Because flagellates die quickly outside of their normal environment, diagnosis should be carried out within an hour of making the smear. Treatment consists of ronidazole (also known as Ronitrol) 2.5 percent. It is not advisable to use metronidazole (Emtryl) as it has no effect in low doses and is highly toxic if overdosed. The ronidazole (in powder form) should be mixed with a small quantity of nectar. Consult a veterinarian for the correct dosage. The treatment should be given for ten days. Birds which have suffered from a previous infection should be treated twice a year as flagellates may survive in the folds of the crop wall and eventually cause another infection. Hubers states that flagellate infections are more widespread in lories than is generally believed. Such an infection was responsible for the death of all of a group of iris lorikeets imported into the Netherlands. Hubers treats all his birds twice annually as a precautionary measure. In South Africa flagellate infections in lories are common. The treatment used there is Tricho-Plus, added to the nectar twice daily for five days.

Fledging

This term is usually defined to mean when a young bird has acquired all its feathers. As related to lories and other parrots, however, it usually infers the time they actually spend in the nest before entering the outside world. This information is difficult to acquire in lories in the wild, partly because nests are not usually easily accessible, and partly because few people have studied them. In captive lories, the age at which young leave the nest is well documented. However, it can be affected by environmental conditions and the food on which they were reared. On a poor diet lories, which normally spend 70 days in the nest,

could fledge 20 days late. In the coverage of the various species, fledging information is given for each species under the heading of **Young in nest**. Normal fledging periods are summarized below.

Lory Fledging Periods

Chalcopsitta:	70 to 82 days
Eos:	70 to 80 days
Pseudeos:	70 days
Trichoglossus:	60 to 70 days for larger species; 55 to 65 days for small species
Varied lorikeet:	37 to 41 days
Glossopsitta:	about 40 to 52 days
Neopsittacus:	47 to 60 days
Lorius:	70 days
Vini:	about 60 days
Charmosyna:	50 to 60 days, larger species; 45 to 50 days, some small species
Oreopsittacus:	about 38 to 40 days

The age at which a young lory leaves the nest for the first time is often unknown; this is because it appears very briefly, perhaps for a few minutes only, and then returns.

Folklore

It is not surprising that such colorful and conspicuous birds as lories should be the subject of folklore. Heinrich Bregulla (1992) recounts a delightful tale about Massena's lorikeet which comes from Pongkil in the Erromango group of islands of Vanuatu. It tells how the flying foxes (bats) there came to be black. One day the flying fox and Massena's lorikeet were sitting together talking. The flying fox suggested to the lorikeet that they should make themselves beautiful. The lorikeet agreed and asked the bat to paint her first, which it did, giving her many lovely colors. The lorikeet was pleased and asked the bat to go to sleep while she painted it. The lorikeet then painted it entirely black and flew away. When the bat awoke it was horrified at its appearance and went away and hid itself. I find this story appealing because it seems to reflect the sense of humor and mischief which is so often apparent in lorikeets.

Foods (captive birds), see **Diets (main entry), Dry Diets, Nectar, Protein**

Fostering Chicks

Lories are generally most obliging foster parents and will rear young of perhaps any other lory species, provided that the fostering is undertaken with care. Indeed, they have even been known to rear chicks of totally unrelated species. Problems are unlikely to arise if eggs are placed under an incubating bird and hatch under the foster parents. Alternatively, a chick up to one week old can be placed in a nest which already contains one young chick. Lories will feed chicks which are different in size and appearance to their own, even when they have a chick of their own in the nest. The following are examples from my own experiences. Rajah lories reared a green-naped, also, on a different occasion, a black-capped placed in the nest on hatching, when they had a two-day-old chick of their own. A pair of iris reared a yellow-streaked until it was as large as they were and it was then removed for hand-rearing. A pair of iris also reared a Musschenbroek's. A pair of yellow-streaked lories reared two yellow-bibbed lories from hatching to independence—on exhibit at Loro Parque. And my own yellow-streaked reared a red lory.

At Rotterdam Zoo, Goldie's lorikeets reared musks, and at San Diego Zoo, Goldie's reared a Tahiti blue lory and the latter reared a Goldie's. (It was feared to entrust the first chick of the blue lory to its parents—but this fear was proved to be unjustified.) Some pairs will even rear an additional chick to bring the total to three. Jay Kapac's blue-streaked were given a newly hatched dusky when one of their own hatched. The second blue-streaked hatched and the pair reared all three until they were removed for hand-rearing at ten days (Kapac pers. comm. 1996). He also overcame a problem with a pair of hand-reared dusky lories which would not feed their own chicks by fostering to them a five-day-old chick from another pair. Presumably the more vigorous cries for food from the slightly older chick stimulated their feeding response. They fed it and thenceforth fed their own young (Kapac 1989). Paul Mespelt had a lone dusky lory who was laying eggs. He gave her three fertile eggs—two Edwards' and one green-naped. She hatched and reared all three chicks, presumably until removal for hand-rearing.

The most amazing fostering experiment involved an adult Duyvenbode's, reputedly a female. Dr. A. J. Wright, a South African aviculturist, kept it in his birdroom, awaiting the arrival of a male. He was hand-rearing an ornate lorikeet and noticed that the Duyvenbode's "became agitated" when he fed it. When he showed the chick to her, she came to the side of the cage to investigate. He moved the chick close to the cage and she started to feed it through the bars! The chick was introduced to the Duyvenbode's three times daily to feed it and was then left with her permanently. The ornate was successfully reared (Wright 1981). A word of warning! I would not recommend trying to duplicate this experience. I believe that in most circumstances the chick would be killed, especially if the single bird was an aggressive male. In this case the Duyvenbode's had probably reared young previously, and the sight of the chick being fed stimulated her strong maternal instinct, partly because she was deprived of the company of her own kind. Dr. Wright also recorded that a pair of scaly-breasted lorikeets hatched and reared first, two ornate lorikeets, and later on, two Buru red lories.

Lories will feed chicks of species much larger than themselves, but these must eventually be removed for hand-rearing. Under no circumstances should small species be at risk from filling the crops of chicks larger than themselves. The chicks must be removed before the foster parents become overworked. For example, a pair of iris hatched one green-naped lorikeet, and reared the other from the day it hatched, until the foster chicks were removed at the age of 42 and 43 days.

A most interesting happening was related to me by Imelda Santos, curator of birds at Jurong Bird-Park, Singapore. In 1994 a green-naped lorikeet egg was pipping in the incubator. As no other small chicks were being hand-reared at the time, she wanted to place the chick with foster parents, but no other lories had eggs. She decided to take a chance and placed it in the nest of a pair of red and blue lories (*Eos histrio*). Not only did they accept it and rear it, they reared a purple-capped lory (*Lorius domicellus*) at the same time. I believe that the chances of an egg or a chick being accepted by a pair without either are remote—unless the female was preparing to lay. It would not be advisable to attempt this with eggs of rarer species!

As the food soliciting calls of lory chicks are different from those of other parrots, does this mean that lories would not feed other parrot chicks? This might depend partly on whether they have previously reared lory chicks. Armin Brockner fostered two eggs of blue-throated conures (*Pyrrhura cruentata*) to a pair of Stella's lorikeets, replacing their infertile eggs. The Stella's fed the *cruentata* chicks until they were 14 days old and removed for hand-rearing (Brockner pers. comm. 1995). As *Pyrrhura* chicks are difficult to hand-rear from the egg, the Stella's had performed a very useful service!

Do lories reared by other species suffer a disadvantage? Do they learn behavior that is foreign to their own species which renders them unsuitable for breeding? I doubt that this would occur if they were introduced to their own species as soon as they were independent. They are more likely to be at a disadvantage if hand-reared without the companionship of other lories, in my opinion.

Free-flying

The most joyful way to keep lories is undoubtedly at liberty. Their quality of life will be far superior to that of those which are kept in cage or aviary. It has to be said, however, that they will not live as long. There are too many hazards. Some lories do make excellent liberty birds but few keepers can provide the right conditions. These include a property of several acres which is free of cats, raptors and other predators which are quick enough to catch lories. Generally speaking, dusky lories and green-naped lorikeets are the best species. *Lorius* are totally unsuitable. I have no experience of *Charmosyna* species at liberty but, in California, Fred Bauer kept a pair of Stella's for over a year; then they disappeared, probably taken by a hawk.

At Loro Parque in Tenerife, several species were kept at liberty during the time I was curator. A flock of male duskies, seven or eight birds, were the star attraction. I have written often of how they delighted in flying fast and low in a squadron, dividing into two groups at the last moment and flying on both sides of a surprised visitor—at ear level. It was obvious that they enjoyed this game! Zoo or bird park personnel who decide to keep lories at liberty should be aware that young, hand-reared birds could be too tame, or become nippy, or be in danger of being stolen. Lories for release should first be caged in the area. When they are released the food should be placed near the cage. It could be dangerous to place it inside the cage as they could be trapped there by a predator. Their food should be offered fresh at least twice a day in a site to which they can safely return. It is not advisable to release three birds if both sexes are present as the third bird

could be driven away by the pair. I have no experience of releasing a single mature pair but suspect that if no nestbox was available the birds would stray in search of a nest site.

On no account should lories be released because the owner no longer wants them. An avicultural society or pet shop could assist with the placement of the unwanted birds if this circumstance arises.

As I once recorded:

> To see lories at their best one needs to see them in flight. I well recall walking in the mountains above Palmitos Park, Gran Canaria, one day, when two brightly colored parrots flashed past beneath me. They were Green-naped Lorikeets. A small group of this species is at liberty in the park. I often see Quaker Parrakeets and various *Aratinga* Conures flying in the mountains (on their way to visit their relatives in our breeding center) but for sheer colour and aerial maneuverability the lorikeets are the clear winners (Low 1992b).

Some of the green-naped lorikeets which I hand-reared at Palmitos Park for release were so tame that I am told they allowed visitors to touch them. They could sometimes be seen sitting near the turnstile at the entrance, greeting visitors. However, at least one loss was believed to be due to theft. But the uninitiated who tried to catch them would receive an unpleasant surprise. Few parrots give a bite more painful (for its size) than a lory!

Jay Kapac in California also mentioned this fact in regard to dusky lories which were kept in small cages:

> In the morning when the feeding crock is removed for cleaning and feeding I fasten the door of the cage open. The duskies then fly off to exercise. Some of them flutter in the surrounding shrubbery while others fly high in big circles landing on neighboring trees. I have experimented with carrying them further down the street from my house just to watch them return, and they all do. In fact to this day not one bird has been lost. In fact last winter two birds were hit by migrating hawks but survived without ill effects. They probably bit their captor. And we all know how Lories can bite (Kapac 1985).

Some of the young hatched in these small cages (only 75 cm x 41 cm x 31 cm) did not fly well. Most of them only flew around for an hour or so before becoming hungry. Probably the secret lies in releasing them before the morning feed. Also, both members of a pair should not be released on the same day.

As long ago as 1884, W. T. Greene (*Parrots in Captivity*, vol. 1, p. 43) mentioned letting a Swainson's (rainbow) lorikeet fly free to feed on the blossom of gorse. In the same volume, Rev. F. G. Dutton recorded how he let his pair loose in the garden for one summer:

> They might sometimes fly nearly half a mile away, to a covert out of sight of home, but always about five they would be found on the cornice of the house, where they roosted, if they did not feel sufficiently hungry to come to their cage. It was very pretty to see them fly home. They were like living jewels as their bright scarlet bodies flashed through the air. They did not always come home together; sometimes one would be back twenty minutes earlier than the other, but at five the cornice of the house was pretty sure to hold them.

Perhaps the largest flock of free-flying lories in recent years is that in Jutland, Denmark, near the town of Kolding. Since 1991, garden nursery owner and parrot breeder Jan Hedegaard has released his lories. In 1996, 12 species were free-flying, ranging in size from Goldie's and Meyer's to Duyvenbode's and yellow-streaked. The other species were *Trichoglossus* and *Eos*, and dusky lories. The latter were the most aggressive, and headed the pecking order. Most return to roost in their own aviaries at night and when they are breeding. The small species are released only in spring and summer, but the others fly free all the year. Most do not wander far, but the larger ones fly up to 2 km (just over one mile). Mr. Hedegaard's neighbor reported that apple production has increased in his garden since the lories have been flying free! Previously the old trees produced a few very small apples but since the lories have thinned out the blossoms, the trees produced excellent large apples. However, the lories probably ate these!

Some countries or states have legislation forbidding the release of nonindigenous species. This should be consulted before considering releasing lories or any other birds.

Fructose, see **Nectar**

Gout, Articular

This condition is not common in lories (captive), but it can occur as the result of an incorrect diet. If gout is chronic (long-standing), it is doubtful whether a bird would respond to treatment. Excessive sugar in the diet affects the kidneys. The lumps around the feet that can occur are deposits of uric acid crystals which are deposited in the feet (articular gout) or around organs in the body (visceral gout). Sometimes white deposits as large as 4 mm can be seen beneath the skin and could be mistaken for abscesses. Uric acid is synthesized by the liver mainly (also in the kidneys) and transported in the blood stream to the kidneys, where it is then secreted in the feces. Much remains to be learned about this disease. Toxins from spoiled food and lead poisoning have also been implicated in kidney damage and gout. Uric acid deposits surrounding the internal organs is a not uncommon necropsy finding in lories. Suggested causes are excessive protein, calcium and vitamin D_3 in the diet, lack of vitamin A, dehydration and stress. A combination of these factors may create this condition.

Current veterinary advice (MacWhirter 1994) is that birds with gout or renal disease should be put on diets which lower the workload of the kidneys, i.e., lower protein, calcium, phosphorus, magnesium, sodium and vitamin D_3. Adequate levels of vitamin A and increased vitamin B are necessary (to compensate for losses in the urine). A veterinarian or avian nutritionist should be consulted on this matter. Treatment with the drug allopurinol and supportive fluid therapy may be instituted by the attending veterinarian but success with treating this condition is limited (Gelis pers. comm. 1996).

Grieving

In compatible pairs of lories the pair bond is very strong and the devotion between male and female is a joy to observe. In other lories (as in many human couples) the bond is not so strong and, given the opportunity, they do not remain faithful to their partner. One cannot generalize that parrots pair for life, any more than one can say that of human beings. However, in devoted pairs the bond is so strong that when one has died, the other has been known to die of grief. I do not invariably accept such stories without knowing the circumstances because there is a possibility that the first bird died of disease and that the second bird succumbed for the same reason. However, an instance of the male of a long-established breeding pair dying after the death of the female convinced me that lories can indeed die of grief. The male was distraught; at that time his species was rare in aviculture and I could not obtain another. He lost interest in everything after his female died and he died soon after.

Patricia King had two black-winged lories which were kept together for some years before it was discovered that they were both females. A male was obtained and paired with one of the females. The other was so distraught at the loss of its companion that it pined and died (Low 1986a).

Lories are such sociable and affectionate birds that one which is suddenly deprived of its partner should be placed in stimulating surroundings to prevent it from grieving. If a replacement for its mate can be found this should be carefully introduced by caging the two birds side by side. If this is not possible, placing the bird in an aviary with a number of young lories may help to lift its spirits.

Grit

Should one feed grit to lories? I believe that it can be offered to seed-eating species, although most lories ignore it. It is of interest that small stones were found in the gizzard of a green-naped lorikeet in the wild (Coates 1985) and gravel was found in the gizzard of a Ponapé lorikeet.

Hand-rearing

There are several reasons for hand-rearing lories. The most usual are (1) removing chicks from the nest, just before the feathers erupt, of a pair who pluck their young; (2) taking chicks at an early age, when they hatch outdoors in a cold climate; (3) in the U.S.A., producing tame young birds for the pet market.

As hand-rearing is not always convenient, or where parent-raised young are preferred, fostering may be an option in the case of chicks from pairs who pluck their young (see **Fostering Chicks**). Lories of species which make suitable pets, such as the dusky, do not have to be hand-raised for this purpose. Most young birds will become tame a couple of months after being removed from their parents, if kept in close association with people. If the young are handled daily in the nest, from just before they start to feather, most will be unafraid when they leave the nest and very easy to tame. However, in the U.S.A., prospective pet owners expect to be offered hand-raised birds as pets. Breeders should know that parent-raised young can become tame because it is inhumane to invariably remove eggs or chicks and never allow a pair the opportunity to rear their young.

Lory chicks are easier to hand-rear than those of any other parrots. They start to feed on their own at an earlier stage in their development and many wean with great rapidity. Unless they have been handled regularly, lory chicks will be difficult to feed if removed from the nest when they are partly feathered. Weight loss may occur for the first two days because they are nervous. If for some reason they must be removed just before they would have left the nest, they should be encouraged to feed themselves, and will usually learn to do so very quickly. The best age to remove chicks is at about the time the eyes open, two or three weeks, depending upon the species.

Newly hatched chicks need to be fed small quantities of very liquid food every hour to every hour and a half between about 6:30 A.M. and 11:00 P.M. for the first two days. If they are of a small species which weighs 3 g or less on hatching, a 3:00 A.M. feed is advisable for the first three nights. One can use Lactated-Ringers solution or (in the U.S.A.) Pedialyte to dilute the food for the first two days as this is fortified with electrolytes. These are minerals which become charged with negative or positive ions when dissolved in water. Many of these ions are important for normal cell function in the kidneys, heart muscle and nerve tissues, for example. Lactated-Ringers can be obtained from a veterinarian or from a medical supplier. For the first two feeds it is suggested that only Lactated-Ringers is given; thereafter, the food can be diluted with it for the first two or three days.

Formulas

Several companies produce hand-feeding formulas for parrots; some of these have proved suitable for rearing lory chicks. Probably only one produces a food for hand-feeding lories specifically, or to offer to pairs of lories rearing young. The company is Aves Product from the Netherlands and the product is Loristart (widely available in Europe). Alternatively, lories can be hand-reared using a home-made formula. A few examples are given below by regions, because of the availability of items mentioned.

Australia

Rachel Lewis uses a mixture of 2 kg of rice flour, 1½ kg of dextrose (glucose), ½ kg of egg and biscuit mix, ¼ kg of rice cereal baby food, ½ kg full cream powdered milk, 250 g of high protein baby cereal and 200 g of brewer's yeast. To this mixture, which is blended with water, canned apples and mixed vegetables are added about once daily. Many lorikeets have been reared over a period of six years on this mixture.

Canary Islands

During 1995 and 1996, Pretty Bird hand-feeding formula (19 percent protein and 8 percent fat) was used at Loro Parque for lories and lorikeets, with excellent results. When I worked at Loro Parque (1987-89), commercial foods were not available. I used Nestle's 7 Cereal baby cereal, baby food in jars (mainly vegetable), papaya and a small quantity of pollen grains (in jars), all blended with water. I preferred the mixture I used at Palmitos Park which consisted of Milupa baby cereal and NektonLori blended with water, with the occasional addition of papaya. Wheat germ cereal was added for chicks over about two weeks of age.

U.K.

For many years Patricia King of Cornwall used a mixture which consisted of equal parts of her nectar (equal parts of honey, Complan and Horlicks) and a mixture of Cow and Gate Fruit Delight baby food, some natural yoghurt, one teaspoonful of Farlene baby cereal and half a teaspoonful of wheat germ cereal.

U.S.A.

Dick Schroeder uses Lory Life nectar, mixed 50 percent with water, with the addition of a small amount of apple sauce (Schroeder 1991). Jay Kapac, also in California, has used Pretty Bird (8 percent fat) since 1990. A little more water than recommended is added for the lories, plus a little fructose. Once a day a different food is given, one

which he makes up and cooks. It includes sunflower kernels and oatmeal. Also once a day, the juice of a fresh orange is squeezed into the food. Roland Cristo uses only Roudybush No. 3 hand-rearing formula.

These diets are given only as examples which have proved successful over a long-term. Dozens of different items have been used for the purpose of hand-feeding lories, usually with good success.

Consistency of food

For the first three days the food should be thin, no thicker than milk. If food stays in the crop for more than 1 3/4 hours, it is too thick in my opinion, and difficult to digest. The crop capacity is very small initially (for example, a newly hatched 4 g perfect lorikeet was fed 0.2 ml). Its capacity is influenced by how much food is fed, as this stretches the crop. Unlike most who are hand-feeding, I do not fill the crop to its maximum capacity. Most do, and therefore may feed chicks only about four times daily. Generally speaking, the formula fed to lories is thinner than that fed to most other parrots, therefore it does not stay in the crop as long. As the chicks grow, it can be thickened slightly.

Frequency of feeding

Chicks being fed from the egg need liquid (electrolytes or thin food) about every 1½ hours for the first day, from about 6:30 A.M. until about 11:00 P.M. From 2 days old, chicks need to be fed every 2 hours during the day. From the age of about 7 days I feed every 2½ hours. The period can be extended to about every three hours when appropriate for the chicks in question. They can be offered the warm rearing food in a small, shallow container, held in front of the beak, at an early age and most will soon learn to feed themselves. The weaning procedure is described under Weaning. As an example, taken at random, I initially fed two rajah lories 7 or 8 times daily; by the ages of 4 and 8 days they were on 6 feeds, by 7 and 11 days they were on 5 feeds, and by 20 and 24 days they received 4 feeds. Obviously, feeding schedules vary according to people's circumstances. The important factor is that the schedule should be maintained and not varied from day to day. In California, Jay Kapac feeds all newly hatched chicks 6 times daily for 1 or 2 days, then 5 times daily between 6:00 A.M. and 10 P.M.. After 2 weeks chicks receive only 3 feeds daily (Kapac pers. comm. 1995). For the method I use, this would not be enough, but of course much depends on the food given and on the crop capacity of the chick. Also in California, Bobbie Meyer removes her chicks when they are between 3 and 4 weeks old. Initially they receive 4 feeds daily. By the time they are moved to a small wire cage they receive 3 feeds daily.

Feeding utensils

A spoon with the sides bent inward is the only method I normally use for lory chicks. A small spoon can be used for even newly hatched chicks of the smallest species; it should be warmed by placing it in the heated food but its temperature should be tested on the back of the hand before giving the first spoonful. If either the spoon or the food is too hot (or too cool) the chick will shake its head and prove reluctant to feed. A very small plastic spoon might be preferred for very young chicks. For young which must be removed at a later age when they are more difficult to feed, a 5 ml syringe is useful for the first few days. In California, the use of 2½ oz. paper or plastic cups is popular among lory breeders. John Vanderhoof explains why:

> ...the cups are completely disposable, thus affording a much more hygienic atmosphere for the youngsters. Using a separate cup for each brooder box containing 3-4 babies eliminates transfer of contamination. If plastic is used, they may be soaked in disinfectant and then reused. Also, these cups can be squeezed into a narrow spout to accommodate mouths of different sizes. The cup is tilted slightly so the formula flows slowly into the baby's mouth...
>
> As this is being written, Mary is feeding 63 Lories, ranging in age from 2 weeks to 8 weeks—at which time the weaning process is nearing completion. This number stays fairly constant from approximately March through August. From August through February the numbers decline to 4-15 on any given day. It takes 55-60 minutes to feed 60-66 babies. This is actual feeding time... By using the cup, Mary has cut the overall time of feeding and maintenance by approximately one-half from previous years when using other techniques (Vanderhoof 1994).

Brooder temperature

Newly hatched chicks need to be kept in a brooder

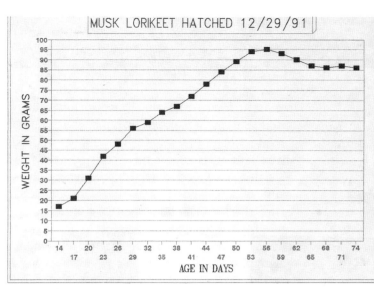

This graph, showing weight gains of a musk lorikeet, is typical of the growth pattern expected of chicks that have been left with the parents for two weeks.
Graph: Sharon Casmier

at which the temperature is thermostatically controlled. For the first week or so the temperature is the same as that at which the egg was incubated—about 37°C (98.5°F). It is reduced very gradually, depending on circumstances. Obviously several chicks in one container will not need as much heat as a single chick. After chicks are three weeks old they can be removed to a brooder with a heat source such as a heat pad. (See **Brooder**.) Except in a hot climate, they will need some heat until they are almost feathered, that is, until they are approximately six weeks old. Lory chicks can cause heart-stopping moments for those who are not aware that they commonly sleep in strange positions—on their side, on their back or with their legs stretched out. After the age of about two weeks, they need to be kept on a surface which they can grip, either very small welded mesh or even, for the small species, a plastic colander with a meshlike surface. I use the type which fits into a plastic holder; the holder is lined with paper to absorb the feces which fall through the mesh. Cleaning this is easier than cleaning welded mesh.

More detailed information on hand-rearing is given in *Hand-rearing Parrots and other Birds* by Rosemary Low, published by Blandford/Cassell (U.K.) and distributed by Sterling Publishing Co. in the U.S.A.

Chick weights

Throughout the text, under the species headings, one or more examples of weights of chicks being reared are given. These should be consulted as a guide only. Weights for any species can vary according to food used, dilution, ambient temperature, health and sex of the chick and, if hand-reared, the expertise of the feeder and the number of feeds given per day. However, most of these factors will be stable in any collection, thus weighing chicks daily and keeping records, is a valuable indication of whether a young bird is growing at the normal rate. If it is not, action is called for. This might be fecal analysis, an improved diet or removal for hand-rearing. It should be noted that chicks hand-reared from the egg or from a very early age, grow very slowly for the first week or so. Weight gains are minimal.

Future of hand-reared young

If required for breeding purposes, young should be reared, weaned and allowed to socialize with other lories, never kept on their own, and handled minimally. They should be kept with other young birds of their own species from the earliest time possible. If they are to be sold as pets, they should be weaned and confident before leaving for their new home. Selling an unweaned youngster often results in its death. There are too many pitfalls for the inexperienced.

Behavior of hand-reared lories

Many hand-reared young lories display adolescent behavior—such as food begging—for months. In all aspects of behavior, parent-reared birds appear much more mature and confident. Most hand-reared lories imprint readily on the rearer unless care is taken to prevent this, and immature behavior is apparent longer than in many other parrots.

Hardiness

Lory keepers in Europe and other sometimes inhospitable climates, will want to know about the tolerance of lorikeets to low temperatures. This depends basically on three factors: the species, how they are housed and how they are fed. Most species show no discomfort when the temperature drops to 5°C (40°F) and some will happily play in snow or even break ice-covered water dishes to take a bath. However, low temperatures are not the only factor involved. Lories and most birds except small species can tolerate cold on its own quite well but damp cold, such as is common in some of the major

lory-keeping countries—England, the Netherlands and Germany—can be very harmful to birds in outdoor aviaries. Draughts can be literally lethal. I have read a couple of accounts by breeders who found lories literally frozen to the perches in the indoor part of the aviary, almost certainly because they were in a draught. No bird should have to tolerate this kind of management.

The following points must be borne in mind.

1. The small *Charmosyna* and *Vini* species cannot tolerate low temperatures or cold damp weather. Fairy lorikeets, for example, look uncomfortable when the temperate falls below 10°C (50°F). If not kept in a warm climate, they should be taken into an indoor birdroom when the weather becomes cooler, or kept in an aviary which has an enclosed and heated shelter, preferably with lighting to extend the feeding hours in winter. The nestbox should be placed in the shelter. The shelter must be draught-proof and frost-proof.

2. Although small *Charmosyna* species such as red-flanked, red-spotted and fairy will survive in a temperature of 13°C (55°F), breeding results will improve dramatically when they are maintained at 18–21°C (65–70°F). An infrared lamp can be used to maintain this temperature when the temperature in other areas of the room is lower than this.

3. All lories kept in temperate climates should have an aviary with an enclosed shelter or, if it is in a very sheltered location and the species is large, an open-fronted shelter. The nestbox must always be available for roosting purposes and placed in the shelter.

4. Their tolerance to cold will depend partly on the food supply. If the temperature is so low that the nectar freezes, it must be renewed at first light and an hour before dusk so that they roost with full crops of fresh, unspoiled food. This is extremely important. No one should keep lories outdoors in winter unless they ensure that they have fresh food at least twice daily. If this is not possible, the lories should be housed indoors during cold weather.

Historical

The evolution of parrots is a subject which has long held the interest of George Smith (author, aviculturist and former veterinarian). He believes that most parrots have evolved comparatively recently (in geological terms)—less than 35 million years ago, and that lories evolved less than 10 million years ago. "Lories show, from the poor differentiation of genera and the multiplicity of subspecific forms, all the evidence of a very rapid, post-glacial, evolution" (Smith 1988).

In modern times, little is known about the relationship of lories and man until the eighteenth century. Mivart (1896) wrote of the ornate lorikeet: "This bird shares with *Charmosyna papuensis*, and probably with *Lorius lory*, the distinction of being one of the longest known species of the whole family of Loriidae, it having been figured by Seba in 1734." In fact, as shown by the remarkable avicultural historian Josef Lindholm III, the ornate lorikeet was kept in captivity more than 8 centuries earlier. Known as the "Five-colored Parrot," at least 7 ornate lorikeets were present in China during the Tang Dynasty (618-906 A.D.). Huizong, Emperor of China, painted an ornate lorikeet that lived in his capital city of Kaifeng, more than 2,500 miles from its island home. He depicted it some time before 1126. The Emperor wrote an inscription, part of which can be translated as: "Tame and lovely, it flies back and forth at will in its specially built enclosure" (Lindholm 1995). His painting has resided in Boston's Museum of Fine Arts since 1933. The bird it depicts is unmistakably an ornate.

It was not until the eighteenth century that a more scientific interest was taken in birds and animals and their taxonomy. The first lory species known to science were named by Linnaeus in 1758. They were representatives of three genera—the red lory (*Eos bornea*), the black-capped lory (*Lorius lory lory*) and, of course, the ornate lorikeet. Most lory species had been named by the end of the nineteenth century, but a few were not officially known to science until the twentieth century. The last two species to be named were the white-naped (*Lorius albidinuchus*) in 1924 and the blue-fronted or Buru lorikeet (*Charmosyna toxopei*) in 1930.

Note that in this book, following the scientific name in the heading of each species, appears a name and a date. This represents the first person to describe it to science, and the date of the publication of that description.

See also **Art, Lories in**.

Honey, see **Nectar**

Honeyeaters, Association with

Honeyeaters form the largest family of birds in the New Guinea region; different genera approximate to the sunbirds, warblers, flycatchers and jays found elsewhere. Some are specialized nectar-feeders (Beehler, Pratt and Zimmerman 1986). They share broadly the same geographical range as lories, being widely distributed throughout the Australasian region. However, honeyeaters occur in New Zealand; lories do not. They do not occur in continental Asia, and neither do lories.

Lories are often seen feeding in the same trees as honeyeaters of various genera (such as *Melidectes*, *Melilestes*, *Meliphaga* and *Myzomela*). This is obviously because both groups of birds use the same food source. In some cases there is a strong relationship between the two groups. Diamond (1972) recounts J. W. Terborgh's observations at *Ficus* trees at Karimui and Miarosa in the Eastern Highlands of New Guinea. The fruits of these *Ficus* were about 8 cm (3 in.) long and 4 cm (1½ in.) in diameter and "protected by a thick, woody pericarp which precluded any direct assault on the soft pulp inside. Underneath the trees were lying fallen fruits with neatly cut holes about 1 cm in diameter at the blossom ends. Observations of the feeding birds showed that these holes were being made by *Trichoglossus haematodus* and another parrot, *Psittaculirostris desmarestii*, which tossed aside chips with a shake of their heads and gained access to the pulp within five minutes. Not even these parrots, though, could penetrate the side walls of the pericarp, and they always had to abandon the fruit with most of the pulp remaining inside because of their short bills."

However, the honeyeaters were able to insert their long delicate bills which were useless for opening the fruit, and they ate most of the pulp. Initially, when most of the fruits were unopened, 99 percent of bird-usage in these *Ficus* trees was by *Trichoglossus* and *Psittaculirostris* (fig parrots) and 1 percent by honeyeaters. But when most of the fruits had been opened and the parrots had obtained as much pulp as their bills could reach, honeyeaters accounted for up to 93 percent of the usage, and parrots for only 7 percent. This demonstrates the useful role which lories (in this case the green-naped, *T. h. intermedius*) play in opening fruits, making available to other species the pulp, and thereby assisting in seed dispersal. When watching red-collared myzomelas (*Myzomela rosenbergii*), a most beautiful scarlet and black honeyeater, feeding in the same tree as Stella's lorikeets, I sometimes saw a scarlet flash in the dense foliage. For a few seconds, until more of the bird was visible, I was uncertain which I was looking at! Red is an unusual color for a honeyeater, thus it is interesting that it should occur in the same place as a scarlet lorikeet.

How tolerant are lorikeets toward honeyeaters when the latter are attempting to share their food source? Joseph Forshaw watched a single scaly-breasted lorikeet from the kitchen window of his holiday home. It fed on the brilliant red flowers of a *Callistemon* bush. So abundant were the blossoms that it apparently remained in the same bush from dawn to dusk, day after day. Aggression toward other nectar-feeding species seemed to be triggered by their size. Other lorikeets and large honeyeaters, such as wattlebirds and friarbirds, were driven off immediately. Most smaller honeyeaters were ignored. A male scarlet myzomela (*M. sanguinolenta*), the smallest species, was unmolested as it fed there most of the day (Anon 1989).

In Fiji, an observer noted that collared lories commonly displaced feeding honeyeaters. He saw a lory nip the wing of a wattled honeyeater (*Foulehaio carunculata*), causing it to flee (Holyoak 1979).

Bruce (1973) noted how the presence of red wattlebirds (*Anthochaera carunculata*) altered the feeding behavior of rainbow and scaly-breasted lorikeets. He watched them feeding on eucalypts at Lindfield, New South Wales. When only rainbow lorikeets were present, they fed over the whole of the grove of eucalypts, as did the red wattlebirds. When the three species were together, however, the lorikeets fed in the outer foliage of a few trees while the wattlebirds moved through the canopies. The wattlebirds and lorikeets kept to these parts of the trees and whenever one approached the other, the wattlebird would chase the lorikeet. At this time, the availability of nectar would limit numbers of feeding birds, thus aggressiveness would be expected;

but this occurred only while feeding and not in an attempt to remove competitors from the area. The temporary feeding territories adopted by the birds were a possible consequence of the wattlebirds' failure to chase the lorikeets away. Aggressiveness on the part of the wattlebirds changed the feeding behavior of all three species present. Although the lorikeets could easily have moved from one tree to another, they remained within the outer foliage of the tree into which they were chased.

Husbandry, see **Aviaries and Cages, Bathing, Birdrooms, Breeding, Diets, Electricity, Ethics, Feather Plucking, Hardiness, Mites, Molt, Nails, Perches, Quality of Life**

Hybrids

Lories have evolved relatively recently and are therefore all quite closely related. Thus most or all species of approximately similar size could hybridize. This physical aspect may, in fact, be the only barrier between hybridization. Hybrids occasionally occur in the wild, especially the rainbow x scaly-breasted (*Trichoglossus h. moluccanus* x *T. chlorolepidotus*). In December 1994, at Ashmore on Queensland's Gold Coast, I saw in a feeding flock of rainbow lorikeets, a hybrid of this origin. It was larger than a scaly-breast with a light blue head, but otherwise resembled a scaly-breast except for the red flecking on the breast. Some rainbow x scaly-breast hybrids have much more yellow on the breast than in pure scaly-breasts. Hybrids between the same two species can differ in appearance, possibly depending on the sex of the parents. Two types of hybrids between rainbow and scaly-breasted lorikeets have been described. My guess would be that the type described above has a scaly-breasted female parent. This type is commonly seen in the wild. It seems much less likely that a male scaly-breast would mate with a female rainbow in the wild, although there is one record which might have been the result of such a liaison. It has been described from captive birds (Leggett and Woodall 1987). The plumage generally resembles the rainbow but the blue coloration on the abdomen is either reduced or missing. There is also variation in the coloration of head, breast and abdomen. Hybrids between these two species are intermediate in size.

It seems that hybrids are not necessarily the

result of a permanent liaison. At Currumbin I saw a displaying male scaly-breast about to mount a female of his own species displaced briefly—but not successfully—by a rainbow lorikeet. The latter species being larger, and with a more dominant personality, perhaps quite often seizes the opportunity to mate with scaly-breast females.

I believe that it is irresponsible of aviculturists to purposely produce hybrids—for whatever reason, including developing color mutations. The argument that hybrids can be sold as pets is invalid because the seller cannot control their future and the fact that they might eventually come into the hands of an unsuspecting breeder. This could easily happen if the hybrid bore a strong resemblance to one parent. The breeder might be unaware that it was not pure-bred. Lorikeet hybrids are known to be fertile and this probably applies to most crosses. Because lories readily befriend other species, given the opportunity, the production of hybrids is very easy. A few irresponsible breeders producing hybrids could do untold damage to the gene pool of captive birds, especially of the rarer species.

Lory hybrids quite often look to the uninitiated as though they might be good species, i.e., their coloration is even and convincing although not bearing a resemblance to any known species. This convincing appearance led to the naming of several hybrids as good species! (See **Hybrids, Species Named from.**)

Intergeneric crosses

These are not rare—because all lories are so closely related. Probably the most common involve members of the genera *Trichoglossus* and *Eos*, such as the rainbow x red. This hybrid is green above, with the crown and nape red and most of the rest of the head blue; breast red, abdomen dark blue and flanks green; underside of tail yellow. Shape is more like the rainbow but size is slightly larger. Another recorded cross is a male rainbow x female chattering. It resembled the female in size and build; crown feathers were purple with the tip golden; nape feathers, purple with a green tip; back, wings and pointed tail, green; breast, red; and thighs and underparts green (Hartley 1912). The two young were alike in plumage. A second example is black x Edward's (head, neck and tail black; underside of tail, yellow; patch of red on ear coverts, dark blue chest, underside of wings, black flecked with red; rest of wings

and body green; size intermediate (Von Sorghenfrei 1978)).

Probably the most talked-about hybrid involving a lorikeet was the reputed cross between a male rainbow (Swainson's) and a female king parrot (*Alisterus scapularis*) which was bred in Australia in 1980 (see **Nomenclature**). Many people refused to believe that such a cross was possible. A number of prominent aviculturists believed that it was. Furthermore, photographs of the hybrid showed that it carried characteristics of both species. George Smith commented: "Over the past two or three years I have come to accept, for various anatomical and behavior reasons, that seemingly quite diverse Parrots may not have taken as long to evolve their differences as once I thought" (Smith 1985). In other words, he believed that such a cross was feasible.

Hybrids, Species Named from

In the nineteenth century it was not unusual for a new bird species to be named from a single type specimen, or two specimens, even though their origin was unknown. In this way several "species" and subspecies of lories which do not exist were described.

Verreaux's lory (*Trichoglossus verreauxius*)

In 1854, Bonaparte named Verreaux's lory. A plate of the only two specimens known is depicted in Mivart (1896). Mivart stated that Count Salvadori had arrived at the conclusion that this lorikeet was a hybrid between a rainbow and a musk lorikeet. Salvadori was correct. In 1992 I photographed such hybrids (although I believe that the musk was the male parent) in Australia. They were identical to the plate in Mivart.

Chalcopsitta atra spectabilis

One subspecies of the black lory was named *Chalcopsitta atra spectabilis* by van Oort. It is known only from the unique type specimen, collected at Mamberok Peninsula, northwestern New Guinea. Mayr (in Peters 1937) suggested that this bird was a hybrid between a rajah (*insignis*) and a yellow-streaked lory (*scintillata*) or an intermediate race nearer to *insignis*. It is surely a hybrid! See the photograph of the type specimen above. Beehler, Pratt and Zimmerman (1986) state that easternmostbirds (i.e., *insignis*) may intergrade with *scintillata*—but I find this very difficult to envisage.

The head coloration of this bird resembled that of *insignis*; the hindneck and mantle were streaked with greenish yellow and the back and wings were greenish brown. The bend of the wing was described as violet, and the breast was dark violet marked with red and streaked with yellow. (The black areas of *insignis* are revealed as purplish violet when seen in sunlight.) The rump was bright blue. Abdomen and upper and under tail coverts were dark green and the under wing coverts, red. The tail was olive green above and olive yellow below, marked with red at the base.

Although I am totally opposed to hybrid breeding, I did try to prove the origin of *spectabilis*. I placed a male *scintillata* with a female *insignis*. The birds steadfastly ignored each other for three months; as the *insignis* seemed a little afraid of the male, I returned her to her rightful mate. Both were adult birds. This experiment would have more chance of success using immature birds.

Blue-thighed lory (*Lorius tibialis*)

This bird was named by Sclater in 1871 from a single specimen. A female, it was purchased in Calcutta by the British importer Jamrach about 1867 and presented to London Zoo. After examining the type skin, Forshaw suggested that it was probably an aberrant specimen of the purple-naped lory (*Lorius domicellus*). Nevertheless, he included it as a species in all three editions of *Parrots of the World*. (This is regrettable because authors relating the number of lory species in existence always count it!) I disagree with Forshaw, especially as it seems unlikely that the black cap would be lost. I always suspected that the bird was a hybrid between a *garrulus* and a *domicellus*. Lories have been traded between islands of the Moluccas for decades. It is quite feasible that an escaped bird of one of these two species joined a wild flock of the other species, paired with a wild bird and produced this hybrid. I found it hard to believe that the mutation of a gene or genes would produce a bird more or less intermediate in appearance between the two species. Confirmation of my hypothesis came when I was re-reading a booklet on the yellow-backed lory, published in four languages by the late Marie Louise Wenner, who gave me a copy in 1979. She wrote: "In 1973 in the Tampa Busch Garden Zoo in Flor-

ida, there were just such hybrids. The father is a *Domicella Domicella* (Purple-capped Lory) and the mother a Chattering Lory. The young have no purple cap but rather a small yellow band across the crop and their thighs are blue with dark green feathers" (Wenner 1979). She mentions the incorrect coloration of the cere and the skin surrounding the eye in Cooper's illustration in Forshaw. Both are actually gray (as depicted in the plate in Mivart).

Importance of Lories

In their natural habitat, lories are very important pollinators because, unlike insects, they can fly in all weather. Plants such as *Banksia* release only small amounts of nectar at a time, thus encouraging them to fly from plant to plant. This increases the chances of pollination. Unfortunately, in some areas of Australia, *Banksias* are dying due to the die-back fungus which was accidentally introduced on lemon trees.

Purple-crowned lorikeets feed on the flowers of *Eucalyptus buprestium*, which are small, with receptacles less than 5 mm in diameter. This is equally true of other *Eucalyptus* species on which these lorikeets feed: *accedens, baxteri, cornuta, diversicolor, fasciculosa, leucoxylon, marginata, odorata, salmonophloia, sargentii* and *wandoo*. The harvesting behavior, as observed by Hopper and Burbridge, would probably lead to effective pollination of small-flowered eucalypts because the stigma of a flower would be brushed frequently by the pollen-bearing tongue as it moved round the ring of stamens (Hopper and Burbridge 1979). These lorikeets may be more efficient pollinators of small-flowered eucalypts than honeyeaters. The latter, when probing for nectar, are not compelled to orientate their beaks in a direction which ensures that the stigma will be brushed by pollen-bearing surfaces. It is worth noting that lorikeets are unable to pollinate some introduced tree species. Observation of musk and rainbow lorikeets showed that they were unable to pollinate coral trees (*Erythrina x sykesii*) because their tongues were apparently too short to reach the open end of the corolla (Wood 1992).

In New Guinea, Beehler (1991) found that so-called bat pollinated rhododendrons were pollinated by bats at night and by lorikeets in the daytime. In Western Australia, Hopper (1980) observed purple-crowned lorikeets feeding on *Eucalyptus occiden-*

talis (see **Feeding Behavior**). He concluded that they were important pollen vectors of this species in the Stirling Range area. They were the most numerous nectivorous birds. While foraging, their bodies came into contact with stamens and stigmata. Several birds were interrupted while feeding by red wattlebirds (*Anthochaera carunculata*)—a honeyeater about 35 cm (14 in.) long—which chased them aggressively from tree to tree. This may have resulted in cross pollination. Lories perform another important service. They provide access to fruits for other birds, by opening them (see under **Honeyeaters**).

Feathers

No creature should be judged by its usefulness to man—since that is not its purpose—but in Australia and New Guinea lories are prized for their bright feathers. David Attenborough described how, in Australia, aborigines used "brilliant orange lory feathers" (presumably from *T. h. rubritorquis*) to decorate a giant didgerido (trumpet) that was elaborately painted with goanna symbols and used in a special ceremony. At each end it was ringed with a band of lory feathers (Attenborough 1981).

Throughout New Guinea the men of most tribes adorn themselves with feathers, plants and even bird's beaks and animals' teeth, in an attempt to look impressive. Along with the feathers of birds of paradise and Pesquet's parrot (*Psittrichas fulgidus*), lory feathers and whole skins are the most sought after adornments. The Huli native (illustrated on page 97) from the Southern Highlands of New Guinea was a highly important member of his tribe. As befitted his status as the medicine man, his headdress was probably the most valuable and the most carefully preserved. It included the tail feathers of King of Saxony's bird of paradise (*Pteridophora dalberti*), the breast plate of a superb bird of paradise (*Lophorina superba*), four cockatoo tail feathers and the fur of the cuscus. But it was the lorikeet feathers—yellow, green and red—which provided the color. Partial skins of Josephine's lorikeets and feathers from Musschenbroek's featured prominently. Important but lesser members of the tribe also use the feathers of these lorikeets in their headdresses. The headdresses are carefully tended and it seems unlikely that large numbers of lorikeets lose their lives for the purpose of decoration. Nevertheless, they are an important part of native culture. At least in that area,

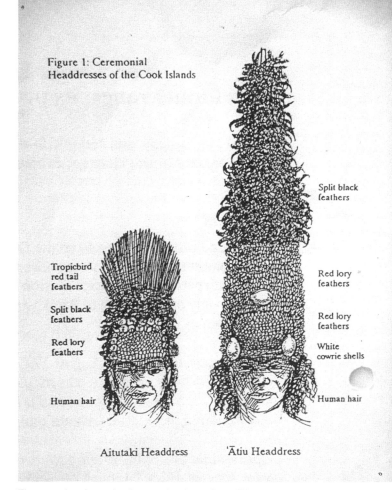

Figure 1: Ceremonial Headdresses of the Cook Islands

Split black feathers

Red lory feathers

Red lory feathers

White cowrie shells

Human hair

Tropicbird red tail feathers

Split black feathers

Red lory feathers

Human hair

Aitutaki Headdress   'Ātiu Headdress

Thousands of scarlet breast feathers from Rimatara or Kuhl's lory were contained in the ceremonial headdress from the Cook Islands. The species is extinct there.
Drawing: Judith Kunzle (Cook Islands Natural Heritage Project)

headdresses are not made entirely of lorikeet feathers. However, a photograph in a New Guinea newspaper depicted a headdress which was made entirely of tail feathers from either small *Charmosyna* or *Neopsittacus* species. For this adornment dozens of birds must have died.

Near the towns, where the inhabitants wear westernized clothes (a strange assortment of colorful garments), feathers might still be worn in the hair; in the case of one man visiting a local market, the entire skin of a Josephine's lorikeet was used. Mounted on a stick, it was thrust into the dense frizzy hair which is typical of many Papuan races.

The skins of fairy lorikeets, so small and colorful, are also highly valued. Mayr and Rand (1937), described "a native near Mafulu with one of these skins worn in his ear, by having the head of the skin thrust through a hole in the lobe of his ear. The tail feathers are prized by the Kuni people, a mountain tribe, for incorporation in their head dress."

In the Cook Islands, the scarlet breast feathers

of the Rimatara or Kuhl's lory (now extinct there) added the color to the ceremonial headdresses. Human hair, cowrie shells and black feathers also formed part of this intricate adornment. The headdress from the island of Atui was taller and must have contained thousands of lory breast feathers. Perhaps this excessive demand from a tiny bird led to its local extinction.

Another *Vini* lory might have met the same fate—but it was protected by law and is thriving. "Valued like gold" is how the red feathers of the exquisite "*Kula*" have been described. *Kula* means red or riches in the Fijian language and is the native name of the collared lory (*Vini solitarius*). The Fijian islands have been inhabited for 3,000 years, and for much of this period trade has been carried out between Fiji and other Pacific islands. Fiji possessed one "commodity" that was valued so highly that it was coveted by the inhabitants of Samoa, 1,000 km (620 mi.) distant, and by the Tongans. The Samoans are renowned for their finely woven mats and carpets, made for houses and for ceremonial occasions. They use strips of leaves from the pandanus palms. These mats were decorated with the red feathers of the collared lory. Dr. P. Bahr recorded how the Samoans made an "annual pilgrimage" to Fiji in order to shoot collared lories and red shining parrots (*Prosopeia tabuensis splendens*) for their red breast feathers. The feathers became symbolic of wealth and social standing and were obtained at great expense from Fiji; they were used like currency. Some live birds must also have been traded because escaped collared lories were formerly seen in Samoa, although they never became established there. Trade in feathers and in live birds in Fiji was officially prohibited at the beginning of the twentieth century. Now the mats are edged with wool—red, of course.

Food

In some areas of New Guinea where cultivation of crops is impossible, the people struggle to survive. Even lories and other small birds are worth killing for food. Beehler (1991) described how the Yali people, who have no livestock except a few pigs, eat birds, forest rats, tree kangaroos and wallabies. One of the natives showed Beehler a blind he had built high in a tree in anticipation of lorikeets visiting a Schefflera. When it flowered he would perch in the tree for several days, shooting any bird

which came to take the nectar. He commented that the Yali people knew the habits of the birds and the seasons of the plants. With this information they could crop the forest's available protein in an efficient manner.

The fate of a pair of red-flanked lorikeets on Lihir Island, New Ireland Province, was recounted by Peter Odekerken. He had been watching them excavate a nest chamber in an epiphyte. His interest must have attracted the attention of some young children who captured them at night as they roosted in their new home (Odekerken 1995b). To them they were not objects of beauty—just food.

Trade

Almost throughout their range, except in Australia and the Pacific, lories are caught for trade, providing a small but regular source of income for local people. The damage caused by trade is emphasized throughout this book. (See **Trade**.)

Pleasure

In Australia, lorikeets give pleasure to thousands, even those who live in cities, whose backyards and gardens attract nectar-loving birds. They can observe these colorful, playful and often conspicuous parrots at close quarters as they feed on shrubs loaded with nectar or pollen. Lorikeets contribute to an understanding and appreciation of nature which city-dwellers in particular might not otherwise have acquired.

Incompatibility

Aviculturists who are unsuccessful in breeding from certain pairs should note that the problem may be incompatibility. Lories usually nest quite readily when mature. If a male and female ignore each other, it may be because one of them was previously mated to another bird and its bond to that bird remains strong. If it was, and the previous mate is still in the collection, it is unlikely to accept a new partner as long as its old one is within sight or earshot.

Some species are more difficult to pair up than others. Most *Charmosyna* lorikeets readily accept the partner they are offered. The larger species in such genera as *Lorius* and *Chalcopsitta* are more difficult to pair up. Placing birds together when young is recommended. Sometimes a male and female may initially show no interest in each other but after some months will accept each other and even breed. In other cases male and female continue to ignore each other; nothing will alter their dislike. An example among my own birds was a five-year-old iris lorikeet who had never laid. I changed the male and just three weeks later she laid her very first egg! It was not fertile, but a fertile egg was produced in the second clutch. The interesting fact about this bird is that she had been totally incompatible with several males; the stress had resulted in her mutilating her flight feathers. She had only two remaining in each wing. But by the time she laid her third clutch with the compatible male (after about five months) most of her flight feathers had grown back and were intact.

The most difficult dilemma to deal with is a male and female who are constantly fighting—fighting so viciously that they may actually fall to the ground locked together. I have had this happen with hand-reared Musschenbroek's lorikeets (see page 397). In such a case there is a real risk that one bird will kill the other, and the two must be separated for their own safety. In my opinion, the problem is more likely to occur with hand-reared birds, many of which are aggressive and totally fearless when adult. When fighting is observed, one must try to understand what has caused it. Another pair of Musschenbroek's (wild-caught male and parent-reared female) were observed fighting first thing every morning for several days, when previously they had always been compatible. It was not difficult for me to understand why this was happening. The female had laid but showed little interest in her eggs. The male was trying to force her back into the nestbox to incubate them. Fortunately, after a few days this behavior abated. With the other pair that was fighting, I had to remove the male after six days. However, the female laid four days later and the eggs were infertile. The pair had been alternately mating and fighting. I would not describe this as true incompatibility as the female readily accepted the male. They were fighting for some reason which I did not understand. If a pair is truly incompatible, fertile eggs will not be produced and the only solution is to find both birds new partners.

Extreme caution should be exercised when trying to pair up ex-pet males, especially the larger species such as yellow-backs and black-caps. There is a danger that the male will instantly kill the female

as an intruder. This is because he is "bonded" to his human partner. Some such birds should be considered as unpairable and maintained as pets. If a pet bird receives plenty of attention, is very tame and has been alone for some years, the fact should be considered that not only is he a potential killer, but that he may be happier on his own, any way!

Incubation

In most lory species incubation is carried out only by the female. In the genera *Vini* and *Charmosyna* the task is shared by male and female. One might incubate for several hours, and then the other take a turn; or the male might incubate for most of the day and perhaps the female at night. More rarely, both birds will incubate an egg each!

Some females start to incubate as soon as the first egg is laid; some wait until the second egg appears. In *Charmosyna* species it is common for the eggs to lay unincubated in the nest for several days. They may even be covered over! This happens in other species more rarely, although in two pairs of *Trichoglossus haematodus* of different subspecies, I have known the eggs to be ignored for several days before incubation commenced. This means that both eggs will hatch on the same day.

Some lorikeets are nervous incubators, the female leaving the nest at the smallest disturbance. However, they are usually quick to return. On more than one occasion, breeders have recorded never finding the breeding pair in the nestbox during the incubation period. This is quite possible if there is a concealed approach to the aviary, as the female could leave as soon as footsteps are heard, and be sitting unconcernedly on the perch in the presence of the owner. This is why regular nestbox inspection is important. One pair of Duyvenbode's lories took their secrecy to extremes. Their owner told me that the first time he knew that they had a youngster was when it was sitting on the perch in the aviary! (One cannot help wondering about someone who did not inspect a nestbox during a period of over four months!)

Candling eggs

Fertility in lory eggs can be ascertained after four or five days, depending on the efficiency of the candling device. When I used an adapted flashlight,

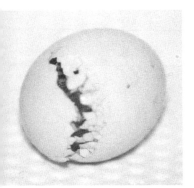

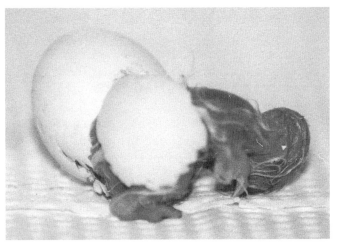

Forsten's lorikeet hatching at Loro Parque on March 28, 1988, photographed at 9:04 P.M., 9:23 P.M. and 9:31 P.M.

hatched after 22½ days. But under a female the incubation period is usually 23 to 25 or even 26 days. In Stella's lorikeets, where incubation is normally delayed, the two eggs in a clutch may hatch as long as 31 and 29 days after being laid, although the actual incubation period is probably only 24 or 25 days. The incubation period should be calculated from the laying of the last egg to the hatching of the last egg, not counting the day on which the egg was laid.

Generally, incubation periods (in days) are as follows:

Lory Incubation Periods

Chalcopsitta	24 or 25
Eos	24 or 25
Pseudeos	24 or 25
Trichoglossus	23 or 24, less often 22 and 25
Glossopsitta	20 to 22, less often 19, 23 and 24
Neopsittacus	23 or 24
Lorius	24 to 26
Charmosyna	23 to 25
Vini	25
Oreopsittacus	21 or 22

Artificial incubation

In an avicultural emergency it may be necessary to place lory eggs in an incubator. If other birds in the collection are incubating, it is often preferable to place the eggs under another parrot, even if there is a size discrepancy. For example, I placed a fertile egg of a Stella's lorikeet under an Amazon for two weeks before moving the egg to an incubator. It hatched. If an incubator must be used, I prefer to hand-turn eggs of small species, at least six times daily. Incubation temperature should be 37°C (approx. 98.5°F). The humidity should be in the region of 30 to 40 percent, increased when external pipping starts. Eggs should then be placed in a hatcher.

Insects as Food

There are numerous reports of lories eating insects and their larvae. One can assume this practice is more widespread than in the species for which it has been recorded. It seems likely that animal protein would be utilized most when chicks were being reared, as is common with other birds. In the days when many birds were taken for museum specimens, crop contents would be recorded. North (1911) often found the remains of beetles in the stomachs of rainbow lorikeets in the Sydney area.

fertility could not be determined until an egg had been incubated for five days. With a more powerful light source (such as a commercially produced candler containing a halogen lamp), fertility is hinted at after three days and determined with certainty at four days. When candling an infertile egg, only the outline of the yolk is visible. In a fertile egg the faint tracery of veins is apparent, plus a small black spot. By about six days, in a species with an incubation period of 24 days, one can see a circle of veins with a few lines crossing each other from a central point. By about eleven days the egg looks dark, except at both ends, and it is then more difficult to see the veins.

Incubation periods

The incubation period in the same species can vary by three days or more, depending on the habits of the incubating bird and environmental factors such as temperature. For example, I placed a fresh egg of a Swainson's (rainbow) lorikeet in an incubator; it

In some birds the stomachs were crammed with them. Grubs and caterpillars have been found in the crops of various lorikeets from Australia and elsewhere and, in New Guinea, a flock of 50 dusky lories were seen feeding on the pupae of teak moths (*Hyblaea puera*). Crop contents of a Stella's lorikeet included three kinds of insects and the larvae of one of these species (Forshaw 1989). Smaller fly larvae are sometimes assumed to be ingested with fruits, but I suspect that many lories seek out such food. Small caterpillars and psyllid lerps were found in the crop of a Musschenbroek's lorikeet collected in the Western Highlands.

When I offered wax moth (*Galleria mellonella*) larvae to yellow-streaked, iris and Meyer's lorikeets the larvae were ignored. But the Musschenbroek's (wild-caught and captive-bred), who had never seen them before either, reacted immediately. They picked them out of the container, held one at a time in the foot, and ate them frantically. The method was either to run the larvae rapidly through the bill, squeezing out the contents, or to nibble from one end to the other. Their satisfaction was very evident. They continually returned to the container looking for more—which I provided for the sheer satisfaction of watching their enjoyment. I then started to question why Musschenbroek's have such powerful beaks and came to the conclusion that it is to enable them to remove wood-boring grubs from bark. The way they deal with branches or twigs offered for gnawing seems to confirm this. Whereas other lories lose interest in these when they are no longer soft and green, I do not remove those given to the Musschenbroek's. They like to gnaw at the hard wood and eventually it is completely destroyed.

Some aviculturists offer mealworms to their lories. For example, Regler (1994) provided them two or three times a week for his Goldie's lorikeets (15 per pair). When chicks were being reared they received mealworms on a daily basis. (He had kept and bred Goldie's for 15 years at the time of writing.) Rucker (1980) considered that drone larvae together with the comb was an ideal food for his blue-streaked lories. He commented that unfortunately, this byproduct of beekeeping was available only from April to June. The pair readily took maggots during the rearing period and mealworms if maggots were not available. Great care must be exercised when feeding maggots to ensure that they come from a clean source—or they could prove toxic. One keeper of Musschenbroek's lorikeets apparently saw one of them eating a slug in the aviary. This is not to be recommended, but it does indicate that this species has a need for live food.

The strangest and most amusing account of the liking that lories have for live insects must be that recorded by S. Dillon Ripley in *Trail of the Money Bird*. While he was in Sainkedoek, in western New Guinea, he saw that the natives used a pale piece of cassowary bone to search for the dark lice in their hair:

> The insect usually stuck to the bone, which was then withdrawn from the hair and passed back to the mouth with a swift motion. The woman was faring better than the man, for he was having competition. On her head was sitting a small pet blue and red lory which grabbed for a louse every time the man killed one. I had often wondered why Wilfred, the pet lory I bought in Sarong, used to love to perch on my head and explore all over, preening my hair with delicate motions of his bill. Now I realized life must be really empty for a lory brought into American surroundings from such fertile fields as those of a good head of Papuan hair (Ripley 1947).

Island Species, Captive Breeding of, see **Small Island Endemics, Captive Breeding of**

Lifestyles

The gaudily-colored lories might seem to epitomize the inhabitants of tropical south seas islands. And yet fewer than half of them live such a lifestyle. In Australia, musk and little lorikeets meet the winter snows, and in New Guinea nine species inhabit the often cold and always damp highlands.

Mountain species

Several species, such as Stella's and whiskered, are found right up to the tree line. They are true mountain species and do not occur in the lowlands. The same is true of the little emerald lorikeet. None of these species normally occurs below about 1,500 m (4,900 ft.). Where these lorikeets live, the temperature fluctuates between about 8°C (47°F) and 20°C (68°F). It is damp and cloudy and frequently overcast. The sun, when it appears, easily burns in the thin air. In some areas over 3,000 m (10,000 ft.), frost falls on 30 days of the year, or more. At these altitudes lorikeet densities are comparatively lower than at lower altitudes. Large flocks, such as those of the dusky and green-naped lorikeets which are found lower in the highlands, do not occur. Because there are fewer of them, this probably means that many of the lorikeets living at higher altitudes are able to find a hole in which to roost at night—either a hole in the base of an epiphyte or in a tree. Where lorikeets occur in hundreds and gather at dusk to roost, there are not enough tree cavities to accommodate them. They therefore roost in huge congregations in certain trees which are close together, as there is protection in numbers.

Migration

It is easy to see flocks of lorikeets in Australia; now more than ever before they are found in the vicinity of human habitation, attracted by immense numbers of flowering tropical trees on which they can feed. While some flocks are nomadic, others are resident. The nomadic birds move around during much of the year, in search of pollen and nectar. Almost nothing is known about their migrations. Even less is known for the New Guinea species. In December, 1936, the naturalist S. Dillon Ripley set sail for New Guinea. In the vicinity of the Tamrau mountains, in the Vogelkop, Irian Jaya, he had an extraordinary experience.

> Clouds were all about, racing down over the flanks of the mountains. The wind was strong. At one moment the whole range would be revealed. Then at another, half of it would be blotted out. As we looked across the stretch of clouds toward the mountains there came faintly down the wind a seething sort of noise. It sounded like a great concentration of starlings along the cornices of buildings in a city. At first we heard only the noise. Then I realized that high over the valley at about our level there was a dark patch moving along, changing its shape all the time like a piece of paper blown by the wind. I ran into the house and got out my binoculars. When I came back the spot was still there, moving in a haphazard way. It was an enormous flock of lories, those small red, blue and green parakeets so prized by the Papuans. There must have been well over a thousand birds in the flock, possibly more. To me it was an amazing sight. Lories occasionally associate together in small groups around flowering trees, but a vast aggregation like this is unheard of. From time to time during the late afternoon I again heard that sound of far-away screams and

whistles and, looking out, saw similar flocks, all passing in the same direction from west to east.

> What this type of migration can mean is anyone's guess. I think it is possible to hazard the opinion that where there are vast areas of jungle stretching for many thousands of square miles, birds which are gregarious to begin with may tend to flock together in ever larger and larger groups. This great movement of birds which I saw in the Tamrau might have been simply a food migration on an enormous scale (Ripley 1947).

In Ripley's day, New Guinea's vast rainforests were virtually untouched. Now they have been disturbed by logging and mining, yet high in the mountains huge tracts of forest remain little touched by man's hand. So, do large-scale migrations as observed by Ripley still occur? None are reported in the ornithological literature—but that may be because such areas, if they exist, are seldom visited by ornithologists. Or perhaps immense numbers no longer occur and migration is on a smaller scale. Where many lorikeets congregate it may reflect scarcity of food in other areas. This probably means that very large tracts of forest are necessary to maintain large numbers of lorikeets in New Guinea, to provide the requisite diversity of food sources. This suggests that if, in the future, the forests there are fragmented, lorikeet numbers will decline very rapidly.

Town dwellers

As already mentioned, lorikeets live alongside man in busy towns and suburbs. This happens not only in Australia, but on a number of islands in the Pacific. On Guadalcanal, the most important island of the Solomons, the capital city of Honiara is an excellent location to observe lorikeets. Massena's can be seen and heard all day long. Jan van Oosten has observed them annually, from 1992 to 1996, always in April. At dawn they fly half a mile from their roost in a tall tree behind the prison. They head for the beach to feed on the *Casuarina* trees which run the whole length of the town, about 2.4 km (1½ mi.) long. During the heat of the day, from about 11:30 A.M. to 2:00 P.M., they rest on the east side of the town, near a large, open-air market. They are invariably found in a large tree with big leaves. Fifty to 100 lorikeets can be seen, resting or playing. Some of them are extremely playful and indulge in very active games, which include chases. They feed again in the afternoon, then begin working their way back to their roosting tree. By 5:30 P.M. most are settled for the night. However, by 1996 their numbers had increased to the point where a number flew south to a roosting tree elsewhere. The gradual increase in numbers coincided with the end of a four-year drought (van Oosten pers. comm. 1996).

Small island endemics

On small islands, large flocks of lorikeets could not survive; there would not be enough food to support them. The *Vini* lories epitomize the species from small islands. They are usually seen in pairs or family groups (up to 4 birds) or perhaps up to eight birds when a couple of families congregate to feed. Their numbers are limited by the size of the island. This is not the case, however, where birds are known to move from island to island. *Vini australis* either does so or used to do so and was reported in large flocks. Island endemics have few predators except introduced rats which have exterminated entire populations. Coconut palms are an important food source; they feed on the pollen of the inflorescences. The islands are hot, thus the birds' main feeding sessions occur during the early morning and late afternoon. This is in contrast to the mountain species which feed throughout the day in their cloudy environment. On these small islands, the sun is strong and the light is intense. Presumably this is why *Vini* lories have such small eyes in comparison with other lories.

It is possible that some lories inhabiting very small islands which are isolated by a considerable distance from other land masses were blown there in a hurricane. But it is also known from subfossil evidence (see **Extinct Species**) that lories once occurred on almost all the Polynesian islands. However they arrived, they probably had small founder populations and may be highly inbred (see **Small Island Endemics**).

Light Beak Syndrome, see **Beak Color**

Liver, Diseases Affecting the

As in all captive parrots, lories often suffer from diseased livers. There are many causes, so it is important that the underlying one is identified in each case, so that appropriate treatment can be implemented. The most likely reasons are an infection, ingestion of toxins or a faulty diet over a long period. Other causes may include cancers and metabolic diseases (e.g., diabetes, visceral gout and fatty liver disease). Affected birds may show many of the nonspecific signs associated with any sick bird—loss of appetite, weight loss, diarrhea, polyuria (excess urine), lethargy, etc. The presence of green, discolored droppings is highly suggestive of liver disease as the green color indicates the presence of the bile pigment biliverdin. Be aware, however, that a bird which is not eating may also pass green droppings.

The first signs of disease affecting the liver can sometimes be seen in very young chicks. A pair of green-naped lorikeets occasionally hatched chicks which would die at the age of a few days. The orange tinge to the skin was evident from the day of hatching. On at least one occasion postmortem examination revealed a diseased liver. In 1980, a pair of Goldie's lorikeets in my care hatched two chicks. One thrived; the down color of the other had changed from the normal white to yellow by the time it was 12 days old. It grew slowly and its weight was lower than normal. It was sick and nearly died soon after leaving the nest and eventually died at a few months old. Although a postmortem examination was not carried out, it seems likely that a diseased liver would have been found as part of the disease process.

Changes in plumage coloration in adult birds and down coloration in chicks are common in affected parrots of any species. Even today, many breeders do not relate this to disease but believe that they have produced a rare and perhaps valuable mutation. One breeder informed me: "I have a very unusual mutation of dusky lory—and I believe that it is a genetic fault since they usually die before weaning. The top of the head is normal and the underparts are a little light, but the rest of the body is orange and the wings are speckled with black, with the flights yellow. I suppose it will survive as it is weaned. It is not a strong bird, but stronger than the others, possibly due to much more pigment and less orange than the others."

A vet should be consulted at the first indication of liver damage. If an infection is responsible, treatment might be successful. Identification and elimination of the cause, where possible, may help to prevent further problems from occurring.

A lory with a diseased liver usually shows a lighter beak color, unless the normal color is black. (See **Beak Color**.)

Longevity, see **Reproductive Span**

Malnutrition

A captive lory suffering from malnutrition is easily identified. Species with red in the plumage have dull colored, even pinkish, plumage, often patchy in appearance. They have yellow feathers or patches of yellow in areas which should be red or green. In species in which the beak is red or deep orange, the beak color is lighter. The plumage lacks the natural glossy appearance.

A red lory which had been fed mainly on apple sauce was described by Casmier (1989). This unfortunate bird had five owners in the two years after she sold it as a pet. "His feather condition was appalling. He had no tail feathers, his wing feathers were frayed and badly clipped, he was bare of feathers around his cere, eyes and his thighs and legs were bare. His beak was pale yellow blending into white at the tip." She bought the bird back and fed him only on Nekton-Lori. Upon being given the first dish he licked it dry and needed many refills daily. In less than four weeks he started to grow new feathers and his beak started to turn orange (in a striped pattern). Sharon Casmier wrote: "I feel he has suffered enough in the role of a pet at the hands of man and deserves a healthy, happy life." She gave him a mate.

What can breeders do to try to prevent similar treatment of pet birds? I would suggest that when a young bird is sold, a diet leaflet is given to the buyer. The request should be made that if ever the bird is sold, the leaflet or the breeder's telephone number should be given to the new owner. The fact that one lory had that many owners in a short time demonstrates how lightly many people take on the responsibility of a pet bird. This one was lucky; it was finally rescued by its breeder. If you know someone who has a lory that is not receiving an adequate diet, please try to inform them regarding the proper care of their bird.

Mimicry, see **Pets**

Mites

Mites (ectoparasites) are rarely, if ever, seen on captive birds. Red mite (*Dermanyssus gallinae*) can cause problems during hot weather. They multiply rapidly, hide in nestboxes or at the ends of perches and suck the blood from birds, but do not live in their plumage.

Chicks are the most likely victims and could die of anemia if rapid action is not taken. During hot weather chicks in the nest should be examined for the presence of minute, moving red dots (red mite). Birds and aviary must be sprayed with a suitable insecticide specially formulated for birds, the nestbox must be changed and, in severe cases, the chicks must be removed for hand-rearing (after ridding them of the mites).

In the wild any bird species may carry mites. They are unlikely to cause much harm to a healthy bird; in a debilitated one they could result in its death. Filippich and Domrow (1985) recorded the effect of an infestation of harpyrhynchid mites in a scaly-breasted lorikeet captured in southeastern Queensland. Numerous white lumps between 0.5 and 1.5 cm long were noted to cause ruffling of the feathers of its wings and body; it was otherwise in good condition. The lumps were confined to the skin. On cross-section they consisted of shapeless white material interspersed with remnants of feather shafts. Histopathological examination revealed the lumps to be a papillomatous reaction to mites contained in multiple spaces lined by epidermis. Each cyst contained large numbers of mites including eggs and debris. Dense accumulations of bacteria, indicative of secondary infection, were present in some cysts, close to the outer margins of the lesion. The mites were identified as *Harpyrhynchus rosellacinus*. Mites from a musk lorikeet caught in South Australia in 1977 in the South Australian Museum were identified as the same species. The specimen label indicates that the mites were found in a large fibromalike tumor.

Molt

Most lories fly fast and high and many are nomadic; thus they move over long distances in a fairly short time. If their powers of flight were impaired, they would not survive long. Thus in lories the annual molt is not a short, fairly rapid occurrence, as in many birds (some of which are even rendered flightless for a while) but is more protracted.

According to Stresemann and Stresemann (in Diamond and Lecroy 1979), the molt in lories is regular, beginning with the sixth primary and moving from there in both directions. *Trichoglossus haematodus* follows this pattern. Eleven of 18 specimens examined were in some stage of molt. Five of these had the sixth primary missing; in five others the molt had proceeded farther but apparently centered around primary number six. On no occasion was more than one primary missing in a single wing. The left and right wings on one bird might be at different stages of molt.

Young lories start to molt into adult plumage at between about 5½ and 10 months. The smaller species start to molt at a younger age. The head and breast feathers are usually the first to be molted but this gradual molt covers a period of several months. Probably not all the feathers are molted the first time. Species which have long tails leave the nest with the tail very short. In Stella's, for example, the tail is about 6 cm (2 in.) long on fledging, 17 cm (7 in.) at 3 months and 20 cm (8 in.) at 7 months. When a fully mature bird loses a tail feather—which measures about 27 cm (11 in.)—it takes approximately 14 weeks for the tail to grow out to its full length.

Nails, Overgrown

Overgrown nails are a common problem in the larger lories. Their nails tend to grow in a close curve and are a danger when overgrown, especially to birds kept in cages made of welded mesh. Many lories like to move by jumping along the floor; if the nails are too long they could become entangled in the welded mesh and the bird's attempt to pull free will result in pulled or broken nails—and bleeding. Nails could also become trapped in welded mesh when the bird is climbing around the aviary. A regular check should be made of the feet of all lories; overgrown nails should be carefully cut with nail clippers while the bird is held securely in a towel. With the larger species this is usually a task for a couple of people—one holding, the other cutting. It must be realized that the vein reaches almost to the end of the nail; it can be seen in good light. Always cut below the vein; bleeding, possibly serious, will occur if the nail is cut above the vein. If bleeding

does occur, use special remedies such as Quik-stop or a cauterizer or, if nothing else is at hand, hold the nail in flour until the bleeding stops. One of these remedies should be kept ready in case of need. In serious cases of loss of blood, a veterinarian must inject the bird with vitamin K, to assist clotting of blood. Subcutaneous fluids may also be necessary to replace the fluids lost. Both these treatments should be considered urgent; more fluids are essential if bleeding continues. The bird will soon be on the floor of the cage in a distressed condition.

According to one avian vet, lories are more susceptible to heavy bleeding than other parrots. He has found it wise to inject some lories which are to undergo surgery with vitamin K half an hour before the operation. This includes surgical sexing.

Failure to trim nails, should this procedure become necessary, could cost a lory its life—or the owner a hefty vet bill. One lady, whose companion was a nine-month-old lory (species not stated), suffered an embarrassing visit to the vet with her bird trapped by its nails in the metal screen of the speaker of a portable stereo. (She had recently cut two of the lory's nails but had not continued because he objected.) She was charged $50 for the anesthetic needed to disentangle the bird, $35 for treatment and $7 for nail clipping. This left her wishing she had found time to take her lory to the pet shop which charged only $5 for nail-clipping (Zawila 1992).

Nectar

Nectar is the name given to the liquid food offered to lories which is the basic diet in captivity of most species. Until the mid-1980s lory breeders had to improvise and make up their own food. Since that time a number of commercial foods have been marketed. Some are produced by bird-food manufacturers who employ nutritionists, and have carefully formulated and tested their products. Others have been offered by breeders who have decided to market their own "mixtures," not all with the help of an avian nutritionist. Some of these are neither nutritionally balanced nor palatable. Birds fed on one such food had dull plumage which lacked the glossy texture of the feathers found on healthy lories. Foods produced by companies which have their own nutritionists give the best results. Choose a product which has either been on the market for some years or which has been produced by a large animal food company which conducts lengthy trials on its products. After many years of making my own nectar, I would not return to this method of feeding, simply because unless you are an avian nutritionist you are unlikely to produce a food which is perfectly balanced nutritionally.

Simon Joshua summed up the qualities to look for in a nectar food as "a tried and tested product which is of the same composition for every batch, contains all the essential nutritional requirements for a wide range of species in the right proportions, caters for breeding and non-breeding birds, is economical to use, easy to feed and, above all, one that Lories and Lorikeets like" (Joshua 1993b).

There are other factors to be considered. The product or mixture should not separate when mixed with water; if part of it settles on the bottom of the container, the smaller lories will not drink the thick food at the bottom and could suffer from malnutrition. The keeping quality once mixed should also be considered. Years ago I used condensed milk in the mixture but this had to be omitted in warm weather. The final factor to consider is that the food should be free of harmful bacteria.

Since returning to the U.K. to live in 1995, I have used Lorinectar made by Avesproduct BV in the Netherlands (P.O. Box 671, 7400 AR Deventer). It contains fructose and glucose as found in flower nectar, pollen, protein of animal and plant origin (14.2 percent crude protein), 2.8 percent crude fat, 1 percent calcium and all 14 essential vitamins and 12 minerals in a balanced ratio. Energy value is 14.4 MJ/kg. This product is widely used by lory breeders in Europe, many of whom (myself included) add more water than the instructions recommend. Aves Product's nutritionist, Peter Holsheimer, commented as follows on this: "I have no objection to the diet containing more water than the recommended 1 part powder to 4 parts water, measuring 20% dry matter and 80% water. For *Charmosyna* species I recommended a greater dilution, say 1:5, as it seems that the birds prefer this solution. The consistency of the droppings is also a point to be considered. When the droppings are firmer, and the birds stay in good health, I think this should be considered a safe way of feeding lories. Thicker food does not result in kidney and liver damage" (Holsheimer pers. comm. 1996). Small *Charmosyna* species do not like thick mixtures. In my

experience, this mixture does settle slightly, thus as the amount in the container decreases, the mixture becomes thicker. Even the large species will leave the thick mixture at the bottom. This is mentioned because some lory keepers might believe there is no need to offer more while some remains in the container. However, as the food should be offered fresh twice daily, this is not a problem.

Avesproduct states that young lories have different dietary requirements than adults; the company therefore produces Loristart for pairs feeding chicks, and for hand-rearing. (However, excellent chick growth has been attained when the parents are fed Lorinectar as the basis of the diet.) Loristart contains 17.1 percent crude protein, 1.6 percent crude fat, 1 percent calcium and all the required vitamins and minerals. Energy value is 14.0 MJ/kg.

Another well established product is Nekton-Lory, made by Nekton Produkte (7530 Pforzheim, Germany). It contains dextrose, fructose, soybean meal, ground rusks, bee pollen and all the necessary vitamins and minerals. Crude protein content is 21.9 percent. The protein is combined to take into account the rapid digestion. The 18 different free amino acids which it contains allows the dietary protein to be rapidly and fully absorbed.

The same company produces Nektar-Plus for small lorikeets. They state that whiskered and red-flanked have been kept and bred solely on this food with excellent results. It contains dextrose, fructose, sucrose, bee pollen, soy protein isolate, plus necessary vitamins and minerals.

There are several other commercial nectar foods available which I have not used. They all make the preparation of the basic food for lories simple; water only is added to the powder. It is a complete food, and food supplements such as vitamins and minerals must not be added. These days there is no basis in the statement that lories are too difficult to feed. In fact, feeding them is easier than feeding any other parrot. Manufacturers recommend that the nectar mixture is fed fresh twice daily to prevent spoilage. However, Nekton has produced a disinfectant called Desi-Plus which, if added to the nectar mixture, results in the need to change it only once daily. Desi-Plus is a disinfectant suitable for all food dishes and ideal for stainless steel.

Some lory keepers are unwilling to change to commercially prepared foods on the grounds that they are more expensive. Whether or not this is so depends on the ingredients of the homemade food. However, foods formulated by an avian nutritionist promote excellent health in lories. Bearing in mind that so many die as a result of an incorrect diet, correctly formulated foods pay for themselves over the long term in increased longevity and productivity. On the subject of homemade lory foods, Schroeder (1991) commented: "Many do it yourself diets are done for the convenience of the keeper rather than the welfare of the birds. Canned 'fruit nectar' and apple sauce are not a balanced lory diet. neither are parrot mix or pancake flour. If bird keepers don't want to cater for the special needs of their birds, maybe they should take up gardening [instead]." It must be pointed out, however, that some homemade foods are quite complicated and time-consuming to produce. They are probably used to lower cost or out of habit, rather than for convenience.

Some keepers might argue that it is impossible to produce a food suitable for all species in all seasons—and this is probably correct. However, nectar does not comprise 100 percent of the diet and the other items fed will help to cater for the special needs. Perhaps we should also be considering factors such as ambient temperature. In low temperatures energy requirements are greater. Glucose and honey will quickly furnish the demand for energy.

If lories are acquired unexpectedly and a visit to a supermarket is called for, water can be mixed with two parts of a baby cereal such as Milupa (a fruit flavor is recommended) and one part glucose—or better still, equal parts of glucose and fructose. If the cupboard must be raided for something suitable in an emergency, tinned fruit cocktail will suffice until the following morning. Below are listed some of the items which have been used in homemade nectar mixtures.

Fructose (natural fruit sugars) can be obtained from a health food store; it is more expensive than glucose. It is thought to be less disruptive to blood sugar levels in humans than white sugar and the same may apply to birds. It does not dissolve as easily as glucose.

White sugar should be offered only in small amounts. An excess of sugar in the diet is very harmful to lories and can cause diabetes.

Honey is said to have healing and medicinal properties and even to be a natural antibiotic. For many years I used a small quantity daily for my lories. The main problem is that in summer it can attract bees and wasps to an aviary. Because honey is nectar which has already been digested by bees, it is easier for the kidneys to process honey than any other sugar. (In view of the high incidence of kidney disease in lories, this point is worth bearing in mind.) Certainly in humans honey is rapidly assimilated, quickly furnishing the demand for energy. It is nonirritating to the lining of the human (and avian?) digestive tract. When buying honey, note the color. Generally speaking, the darker it is, the more vitamins and minerals it contains. Those present will depend on the type of honey; one of good quality will contain iron, copper, sodium, manganese, calcium, magnesium, phosphorous and various vitamins of the B complex. Its composition will depend on the flowers upon which the bees fed, and on the season. Much poor quality honey is offered for sale in the supermarkets, produced by feeding bees on a sugar solution placed near the hive; if in doubt, buy from a health food store.

Pollen is a very important food source of lories. Analysis of pollen grains shows great variation in composition, presumably dependent on the plant species and locality from which it was collected. Protein levels (dry weight) vary from 7 percent to 40 percent and carbohydrates (sugars and starch, dry weight) from 13 percent to 39 percent (Brice, Dahl and Grau 1991). Pollen contains up to 16 different vitamins, up to 18 enzymes and coenzymes, up to 16 minerals and up to 18 amino acids, plus many other substances. Some breeders who make up their own food add commercially available pollen to it; most commercial nectar mixtures contain pollen; only a small quantity is needed. However, the dried pollen one can buy in jars may have lost important enzymes in the processing. One lory breeder used fresh pollen obtained by a beekeeping friend. Traps were made to brush and collect pollen from bees as they entered the hive. Alternatively, one can offer to lories flower-heads and blossoms which are rich in pollen. Holland (1994) suggested the use of dandelion (*Taraxacum officinale*) and red clover (*Trifolium pratense*).

Brice, Dahl and Grau (1989, 1991) carried out feeding trials with pollen, the results of which were unexpected and not easy to comprehend. Two lorikeet subspecies were used in the trials, green-naped (*Trichoglossus h. haematodus*) and Swainson's (rainbow) lorikeets (*T. h. moluccanus*, confusingly called Moluccan lorikeets by the researchers). Nestlings of the latter race, and adult and nestling cockatiels, were also used in the trial, which was designed to give information about pollen digestion. The adult lorikeets digested very small amounts of the pollen eaten from *Eucalyptus*, between 3.9 percent and 7.0 percent, yet the nestlings absorbed between 19.5 percent and 32.4 percent. Strangely, adult cockatiels absorbed from 13.6 percent to 22.7 percent, and nestling cockatiels between 33 percent and 46 percent. The significance of this experiment is not clear. The results are given here only because they appear to be the only trials carried out on captive birds to try to ascertain how much pollen ingested is actually digested.

Milk and milk products, including all forms of milk powder and condensed milk, should be avoided. According to the nutritionists, birds cannot digest milk products well and this results in diarrhea. This was unknown in the 1970s. The fact is that almost all home-made nectar recipes include milk powder, either on its own or as a component of baby cereal, or both.

Cereals, particularly baby cereals, wheat germ cereal and rice flour are commonly used.

Homemade nectar, a few examples are given of mixtures which have been used successfully by breeders over a period of several years or more. They have very few ingredients in common! Recipes are listed by country due to the availability of products mentioned.

Australia

Peter Odekerken's nectar contains 1 part Farex baby cereal, 1 part blended plain sweet biscuit (Nice or Marie) and ½ part wheat germ cereal. These are mixed and stored in an airtight container. When required, one cup is mixed with 1 liter of water, 1 dessertspoon of honey, 3 dessertspoons of tinned mixed vegetables (blended) and 1 dessertspoon Nestle's malt powder (Odekerken 1993b). (Note: 1 dessertspoon = 1½ teaspoons or ½ tablespoon)

Rachel Lewis uses 2 kg of rice flour, 2 kg of dextrose, 1 kg of full cream powdered milk [the

necessity for this might be questioned], 500 g of high protein baby cereal and 400 g of brewer's yeast. Warm water is added to form a mixture of the consistency of milk, then a small can of baby food (usually mixed vegetables) is added (Lewis pers. comm. 1996).

David Judd used 4 parts of his own mixture with 1 part honey and 1 part (wet) Wamparoo Lory Food. His own mixture contained one packet of Heinz High Protein baby cereal (19 percent protein, 5.5 percent fat, 58 percent carbohydrate), 1 cup of brown sugar, ½ cup skimmed milk powder, ¼ cup Maltogen (Nestle's: provides malt sugars in a readily assimilable form, made from malted barley and other cereals) and ¼ cup Glucodin (Boots Company: pure medicinal glucose). All these items were blended with water.

Barbara O'Brien fed her Australian lorikeets on a mixture of the following: 50 g each of Farex baby cereal, wheat germ cereal and malted milk powder; 40 g of glucose; 150 g of wholemeal bread sprinkled with a mixture of ground rolled oats; 60 g of "chick bits" and 1 level tablespoon of kelp powder. This was made into a soup with 1 kg of brown sugar added to 1½ pints of hot water, plus vitamins, then diluted in a ratio of 3 parts of water to one part of the mixture (O'Brien 1985).

Europe

At Vogelpark Walsrode the lories are fed on a mixture of 420 g of oat flakes (baby food), 220 g of pollen, 33 g of yeast powder and 33 g of glucose or grape sugar. This is mixed with 1 liter of warm water with 5 g of gelatin, 10 ml of honey and 20 g of carrots, ½ peeled apple, ¼ orange, ¼ banana and ½ peeled cooked potato.

In the private collection of Dr. R. Burkard in Switzerland, the lories were given a choice of two different nectar mixtures, protein rich and otherwise. Most of them would take both types. The base mixture was of a thin consistency and contained five types of cereal (wheat, corn, oats, rye and millet), baby cereal, two or three fruits (apple, pear, orange, banana, etc.), dextrose, vitamins and spinach, chard or some other type of green food. The protein mixture was based on the above, with the addition of yoghurt, a protein powder (hydrolyzed bone meal) or raw egg (Burkard 1983). [Because of the danger of salmonella infection, the addition of raw egg is not recommended.]

U.K.

A simple recipe used successfully for many years by Pat King was equal parts of honey, Complan and Horlicks. Dulcie Cooke used to use the following recipe for *Charmosyna* species. Cooked brown rice and bananas were liquidized (2 tablespoons of rice to two medium-sized peeled bananas); peeled carrots and sweet apples were also liquidized. One pint of nectar contained one tablespoon each of rice and banana mix and apple mix, one tablespoon of medium brown sugar, half a tablespoon of glucose, one teaspoon of Australian honey and one tablespoon of a fruit-based baby food. (This should not contain coconut, which is too indigestible for chicks, or any citrus fruit.) Also added were one teaspoon of condensed milk (or Vita Food: Boots plc, skimmed milk product which contains only 3½ percent lactose and 21 percent protein.) The carrot mix was given every five days and ½ teaspoonful of ground soya bean flour was given every five days. These items were first mixed with cold water, then with hot water to make up one pint (Cooke 1991).

Nectar containers

Stainless steel is the best material for nectar containers. It is easily cleaned and disinfected; initially expensive, it will last a lifetime. Earthenware dishes are not recommended because if they are not smooth or if they become scratched, they are difficult to clean; bacteria will flourish. Glass is readily cleaned but too easily broken. Circular stainless steel dishes which fit into a ring which hooks on the wire of cage or aviary (Coop Cups) are ideal for most lories. The *Lorius* species can remove the dish from the ring; for these birds I use the steel dishes made by Losenbeck of Iserlohn, Germany. They clip into a holder which is wired or pop-rivetted to cage or aviary.

Food and water containers must be placed at perch level, not on the floor. At a low level there is a greater danger of contamination from feces. In any case, many lories do not like to descend to the floor. A feeding hatch or swing door is very useful in preventing overtame or aggressive birds from escaping or nipping at feeding time.

Nestboxes (Nest Sites for Captive Birds)

Most lories in cage or aviary use their nestbox for roosting at night so it should be left in position throughout the year. They will usually accept the nestbox offered, provided that the entrance is of a suitable dimension. It should be only just large enough; that is, as the bird enters, it should brush the sides of the entrance without having to struggle to get in. Wood is almost invariably used in nestbox construction, because it absorbs humidity and is a natural substance. All parrots need to be able to gnaw inside the nest; in many cases this stimulates breeding as it equates with nest preparation in the wild.

Various designs

Many lory breeders use L-shaped nestboxes or boxes containing a baffle. Others use ordinary vertical boxes. I use L-shaped boxes. For the larger species the extension part juts out at the top. There are several advantages. A bird entering the nest cannot drop straight down onto the eggs or young; this helps to prevent eggs being broken. The male can sit in the top part and look out the entrance without disturbing the female. Fully feathered young will climb into the top part some days before they leave the nest. This gives them a dry look-out, for the feces are squirted into the part below. The nest litter can easily be changed without disturbing them at an age when any disturbance could cause premature fledging. When a nest containing eggs or young is checked, the parent can move into the top part if it does not wish to leave the nest.

In a box with a baffle, or shelf, the shelf is placed just under the nest entrance. Roland Cristo described the advantages: "Using this box, the birds come in on a shelf, go to the back of the box, then down under the shelf they drop into the egg chamber. When leaving this chamber, they cannot go directly up to the hole, but must go to the back of the box, and then up to the shelf to get out. The egg chamber is only big enough for one bird. That way, if the male comes in to squabble with the hen, he cannot get down on the same level as the eggs, and the hen can protect her eggs and young better. When sitting on the eggs, the hen cannot see the entrance hole. Instinctively, she knows if she can't see the hole, a predator looking in can't see her. This is very important with nondomesticated birds" (Cristo 1994).

Some breeders in California use a nestbox with an extended entrance tunnel set at an angle of 45° and a small nest chamber for some *Charmosyna* species. This design was pioneered by Carl McCulloch. In his design, an offcut of timber with bark attached covers the part of the box that the birds see, to make it appear more natural. This can be screwed inside the cage. A pair of wild-caught Stella's which had not nested in nearly three years, did so when offered this type of box. It also proved successful for a pair of Josephine's which had laid on several occasions but had never previously incubated their eggs (Schroeder 1994). After using a larger version, Schroeder suggested the size shown in Figure D on previous page.

In Australia, small lorikeets are often offered small wooden boxes hung at an angle. Rachel Lewis breeds varied, little and purple-crowned lorikeets in these. The measurements for the varieds' nestbox are 30 cm (12 in.) high and 12.5 cm (5 in.) square, hung at an angle of 30°. It is made of pine 2.5 cm (1 in.) thick, with a 5 cm (2 in.) PVC entrance spout (Lewis pers. comm. 1996).

In South Africa, where sisal logs (from agaves, also called century plants, *Aloe* species) are readily available, they are put to excellent use as nesting sites for many parrot species. This idea could be copied in California and other areas where century plants grow. The log is made from the base of the plant, which dies after flowering. Some breeders there use the whole log, and find that excavating it is a stimulus to breeding for their lories. Another breeder used the log only as the entrance to the nest, which was made of cement with a drainage pipe. This idea would not be practical in a cold climate. Given a choice of nests between a sisal log and an ordinary nestbox, many lories chose the sisal, and some alternated their nests between both types.

A most important practical point concerns the ladder inside the nest, which enables the bird to reach the bottom part. It must be able to do so easily or eggs or young could be trampled or, in a deep nest, the bird could even be permanently trapped. There are two ways in which a ladder can be fixed. One is by screwing pieces of wood inside; it is essential to check these regularly as they could be gnawed away. Nails should not be used as they could be a hazard if they were exposed. The other method is to staple a strip of welded mesh inside. This is the method I normally use, although I am always aware of the fact that a bird could be trapped in the mesh by its nails. It has never happened to one in my care

but it is a potential hazard. Some thought should also be given to the inspection door. This should **never** be above the sitting bird as it will cause panic. A hinged door which opens from the side, or a door which slides from left to right (or vice versa), is the most practical. The position of the door depends on inspection access. This is best achieved from a service passage at the back of the aviary. In a walk-in aviary, the necessity of entering in order to inspect the nest is one of the worst mistakes in aviary design. It causes stress to the birds and may result in injury to the keeper!

The perch at the entrance can be made from dowel or, preferably, from a harder wood, as dowel is easily destroyed.

Suggested sizes

Suggested sizes are shown in figures A, B and C. Size A is suitable for all *Chalcopsitta*, *Eos*, *Pseudeos* and *Lorius* species, also the larger *Trichoglossus*. Size B can be used for species the size of perfect and Stella's lorikeets and Size C for Josephine's and all smaller lorikeets. The diameter of the entrance hole should be 5 cm (2 in.) for species up to the size of a perfect and up to 7.5 cm (3 in.) for the larger species. A hole can be cut in the wire of the cage to coincide with the nest entrance and the box is hooked on to the mesh using L-shaped hooks which screw into the box. They must be of sufficient strength to support the weight of the box. Alternatively, in the case of cages which have a feeding area which protrudes beyond the general area of the cage, the nestbox can stand on the part which juts out. However, the nestbox should not be above food or water con-

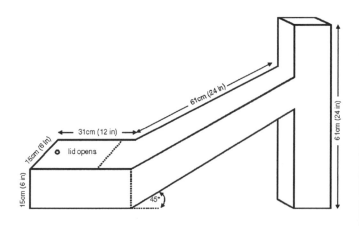

Figure D.

tainers which might be contaminated by feces draining out of the box.

Standardization of nestboxes is advisable. If one has to be changed, the birds will immediately accept the new one if it is of the same size and design. Also, every nestbox will fit the cage without having to make any modifications.

Keeping the interior dry

One of the most important considerations in a lory nestbox is keeping it dry. There are a number of ways to achieve this.

1. Drill holes in the wooden base.
2. For small lorikeets use spun plastic filter material, as used for fish pond filters, in the base.
3. Make an open-topped box to fit into the bottom of the nest and design the box so that this can easily be removed. It can be exchanged for a clean one, as needed. Roland Cristo uses such removable boxes.

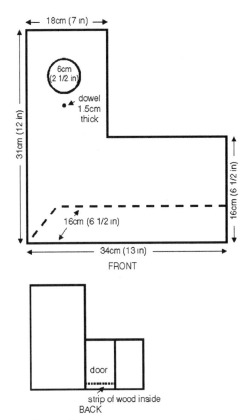

Figure B.

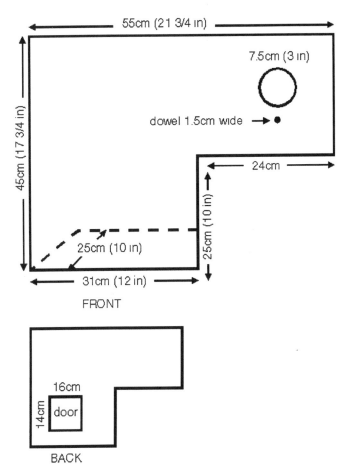

Figure A.

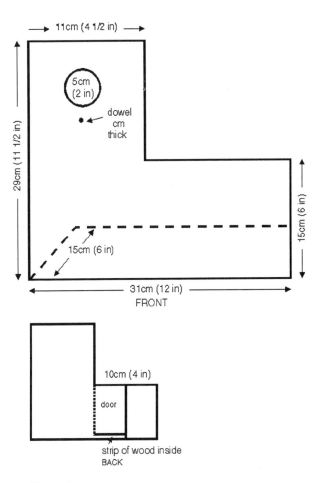

Figure C.

81

Dick Schroeder uses trays with the sides 10 cm (4 in.) high.

4. In a warm climate, the bottom of the nest can be made of welded mesh. Plant material is placed above this, and the wood shavings on top. Such boxes are in use at Loro Parque, Tenerife, designed by Roger Sweeney.

5. At San Diego Zoo, curator Ed Lewins designed a natural-looking nest, which fits into the decor of the planted aviaries on exhibit, and is also practical. He found tall, narrow logs and placed inspectable nests inside them. The interior is PVC pipe; the top is closed with a drain cover, the slits in which have been sealed. The bottom part of the nest is also a drain cover but this is not sealed, thus allowing the liquid feces to drain out. Wood shavings are placed on top of the drain cover. An entrance was made in the log to coincide with an entrance drilled in the PVC pipe. An inspection door and a welded mesh ladder were also added. An identical PVC pipe nest is made so that when the nest interior becomes soiled, changing it for a clean one can take place in seconds. The soiled nest is easily cleaned and disinfected. Another natural-looking nest at San Diego Zoo was that for blue-crowned lories. A nestbox was fitted into the back of a staghorn fern (*Platycerium bifurcatum* or *P. grande*), thus hidden from public view. The birds entered the nest through a hole in the staghorn.

Nesting material

Pine shavings or other shavings from untreated wood are generally used. Fine sawdust is not suitable. It could get into the mouth of the chicks, or into their eyes, causing an irritation.

Natural nest sites

In the wild, *Charmosyna* and perhaps some other lories nest in epiphytes (any parasitic plant which grows on another). San Diego Zoo is the only place where I have seen the attempt to emulate this. A pair of red-flanked lorikeets, on exhibit in a densely planted aviary, was given a staghorn fern (as were collared lories, which destroyed theirs; this species does not nest in such a site in the wild). The red-flanks excavated a hole in the middle of the clump,

from the front. In this nest they had hatched two clutches and reared young by the summer of 1996. The disadvantage of such a site is, of course, that it is impossible to inspect.

Natural logs can also be used as nestboxes, after modification. An inspection door should be made in the side and a solid roof added. A firm but porous base is essential. The entrance can be natural or made with a saw. Logs can be hung between 45 degrees and the horizontal, or vertically.

Location

A nestbox must be in a site where it can be inspected easily, preferably without moving it. The location must be planned before the boxes are made, so that the inspection door is in the correct position. In an outdoor aviary the nestbox must be placed under cover from the elements. In cool climates it is best positioned in the sheltered portion or indoor part of the aviary. Whereas most lories will enter a box regardless of which direction it is facing, some will not—and this could prevent them from breeding. A Swainson's lorikeet laid an egg in the nectar dish; she had never entered the nestbox. The owner therefore provided a second nest. This was treated with equal disdain. Having read that birds (in the Northern Hemisphere) prefer a box that faces north, he altered the box to face in that direction. To his amazement, the female entered within three minutes. She laid her second egg there (Low 1975).

Cleaning

Nestboxes should be thoroughly cleaned and disinfected after each clutch. If the occupants are not breeding, nesting material should be changed at intervals, if necessary. Wet nests of nonbreeding birds sometimes occur in the small *Charmosyna* species. This is probably because in the wild they nest in epiphytes or other plant material. As nearly all lorikeets roost in their nests, it is very important that these are kept in a dry and clean condition; therefore, whether or not the birds are breeding, nestboxes should be inspected at regular intervals. Lories roosting on damp material in a cold climate would be very susceptible to disease.

Nest Pairs

Some lory breeders believe that the two young in

the same nest are always a male and a female (like "pigeon pairs"). In theory, 50 percent of young produced over a substantial period will be males and 50 percent females but two in the same nest may be of the same sex. In nine nests of Tahiti blue lories hatched at San Diego Zoo where both young survived long enough to be sexed, five nests consisted of a male and a female, two nests contained a couple of males and two nests had two females. In six nests of blue-crowned lories at the same zoo, five consisted of a male and a female and in the other both young were females (San Diego Zoo species data sheets). In a number of instances of birds breeding in my care, two young in the nest have been of the same sex (usually females).

Nest Sites

Most parrots nest in cavities in trees—yet a surprising number of lory species use alternative sites. Very little data is available on the nest sites used by lories; nests of a number of species are either unknown or have never been described. Even in Australia few studies of lorikeets' nests have been made. However, nests of *Trichoglossus haematodus fortis* on Sumba have been described in some detail (see page 251). Most were in tree cavities, but three were in the rooty undersides of large arboreal epiphytes. Thane K. Pratt (coauthor of *The Birds of New Guinea*) of the Dept. of Biological Sciences of Rutgers University, U.S.A., was perhaps the first to point out that several species of small lorikeets do not always nest in tree cavities. He suggested that most *Charmosyna* species, also Goldie's and whiskered lorikeets, use the thick moss cushions draped over tree limbs in the forest canopy (see Whiskered Lorikeet, under Nesting). These "cushions," which can be several feet thick, hang down over the sides of branches and support small gardens of orchids and ferns. When observed from below, they appear to be dry on the underside. Thane Pratt observed striated, Wilhelmina's, fairy, Josephine's and Papuan lorikeets exploring them. On several occasions he saw fairy lorikeets climbing into one such clump, and he believed that they were nesting there. In New Guinea, on Mount Bosavi, Josephine's were frequently seen leaving an epiphytic clump from which chicks could be heard calling. According to the people of the Huon Peninsula, Papuan lorikeets chose another type of nest site—the massive accumulation of dead leaves in the crowns of palmlike pandanus trees (Pratt 1982).

An unusual nesting habit occurs among the olive-green lorikeets (*T. h. flavicans*) of the Admiralty Islands. On at least three islets, Poy-yai, Tuluman and Niakuni, they nest (and roost) on the ground. Poy-yai is less than 0.5 ha (1¼ acres) in extent and rises no more than 10 m (34 ft.) above high water level. This rocky islet is entirely covered by thick tree, palm and shrub vegetation. Mary LeCroy and W. Peckover visited Poy-yai on July 27 and saw many apparent excavations, from bare areas sheltered at the base of a tree or within a rock crevice, to "burrows" 60 cm (2 ft.) deep under a rock overhang or 1 m (3 ft. 3 in.) deep in a rock crevice. The burrows were horizontal or sloped slightly downward. It was obvious that some excavation had been necessary; dirt trampled in front of the entrance was probably deposited there by the excavating birds; actual digging was not observed. No sign of tree-hole nesting was apparent. There were few trees large enough to provide cavities. Two nests containing eggs were found, one with two eggs, the other with one. In both cases, the eggs were lying on bare ground in a natural rock cleft. There was no evidence of nesting material in any of the excavations. The lorikeets were wary and did not visit their nests while a human was within sight of them, although birds did go within 1.5 m (5 ft.) of the entrance on several occasions.

The clefts and crevices in the rocks were also used for roosting. By the time of departure from the islet, at 6:00 P.M., many lorikeets had flown in from nearby Manus island (about 0.5 km distant) and had begun to descend into the lower vegetation. Seven of these birds were trapped in burrows two nights later, and were released after being photographed (LeCroy, Peckover and Kisokau 1992). See the photograph on page 123.

Australia

In December, 1994, I watched musk lorikeets in the Black Ridge State Forest in the Cherrypool area of Victoria. The yellow box was flowering and several pairs of musks were going in and out of cavities in trees. Based on my knowledge of nest sites of this species at Swan Vale, New South Wales, I was uncertain whether these cavities were nests, as they looked so different. The entrances were much larger. I photographed them and sent a slide of one to John Courtney. His answer was interesting because it compelled him to consider the normal appearance of lorikeet nests. He commented:

> The nest aspect is totally unlike anything that I have seen before, with any lorikeet species. Lorikeet nests that I have seen go into the side of a tree trunk, or limb, so that the entrance either does not catch rainwater, or very little in those in which the lower edge of the nest entrance protrudes out a little. A small percentage of nests observed by me over the years are an exception to this rule, in that the entrance is a small neat hole directly on top of a limb. However, in these, the surface of the limb near the entrance slopes downwards away from the nest entrance, so that no rainwater can gather and run into the nest. (An inch of rain falling directly into a tiny hole and spreading out over the entire bottom of the nest, containing several inches of wood-dust, would do no more than supply perhaps much needed moisture.) Your slide shows a reasonably large catchment area scooped out of the trunk, also around the edge, which would direct water into the hole. I think the birds were drinking.

He made a few other interesting points about nests of little and musk lorikeets. One is that the interior is always clean and free from excreta. Nests contain a good quantity of very fine, reddish wood dust of a consistency that readily trickles through the slightly opened fingers, with only a tiny amount of moisture being detectable. The other point concerns the presence of a resident skink in every nest observed: "These little lizards are sometimes observed partly out of the entrance, sunning themselves. I saw, and photographed, one doing this then, only seconds later, the parent little lorikeets landed at the entrance to the nest, while the young were just inside. The lizard disappeared back inside, past the young. I do not know whether these little lizards perform any function useful to the lorikeets."

The third point is that after the death of a tree containing a nest, the nest is usually deserted within a couple of years. John Courtney suggested that there might be a change in the microenvironment within the nest chamber, perhaps a change in moisture content or humidity (Courtney pers. comm. 1995).

John Courtney is perhaps the only person in Australia to have carried out long-term research on the nesting habits of little and musk lorikeets. His study spans a period of 40 years in the same location. In a small area near his house in Swan Vale, Glen Innes, was the perfect combination of very suitable food trees (yellow box and white box) and three species of similar trees, Blakely's red gum, smokey gum and white gum, which are their preferred nest sites. They have specific requirements where the nest entrance is concerned. Many have been measured by him over the years. Examples are as follows. On November 6, 1964, John Courtney measured a musk's nest in a white gum (*Eucalyptus viminalis*). The entrance hole was in the side of the trunk and measured exactly 1½ in. (about 3.7 cm) wide and 1 5/8 in. (about 4 cm) from top to bottom. There appeared to be a cavity in the center of the tree about 9 in. (23 cm) in diameter. On October 25, 1964, he measured a little lorikeet's nest. The entrance was perfectly round, not oval, and measured exactly 1 1/8 in. (about 2.7 cm) to 1¼ in. (about 3.1 cm) in diameter, depending on the serrations in the bark around the hole. It, too, was in the trunk of a white gum. The nest measured 21½ in. (51 cm) from the entrance to the back of the nest cavity, and the cavity was judged to be as big as a closed fist. It was reached by a long, narrow tunnel (horizontal).

John Courtney commented:

> It looks as if the difference in diameter between the nest entrance of the musk and little lorikeets is about 3/8 in. (about 9 mm). This does not seem very much, but is probably of crucial importance in preventing musks from taking over little lorikeet nests. I have noticed that both species often have to squeeze and wriggle a bit to enter or leave their nests. The brooding musk lorikeet of the nest mentioned above, used to turn on its side and wriggle a lot to get out. The measurement of 1 5/8 in. (4 cm) from top to bottom (vertical) of the entrance was wide enough to let the bird's shoulders come through, and so it had to turn side-on. The bottom of this nest contained many fresh *Eucalyptus* leaves, but I am doubtful if musk lorikeets would carry leaves. It is more likely that another bird or a possum placed them in another hole further up the trunk, and they fell down (Courtney pers. comm. 1980).

As recounted under **Little lorikeet** (Nesting), during the 35 years up to 1989, John Courtney found 45 to 50 nests of this species, some of which had been active for 25 years or more. He is probably the first person to record such extended occupancy. Of course, that does not mean that the same pair occupied the nest throughout the period, but they probably did so for many years, until either one of the pair died or the nest site was no longer usable. In his diaries, John Courtney recorded the end of some of the nests which he had known for years:

March 28, 1988. Disaster with Stonebridge Gully nests. The limb in which the musks nested has broken off and fallen down, and where the base of the limb tore off, it appears to have opened up the nest cavity of the little lorikeets. The end of it, too. It was known to have been in use in September 1962—and was an old nest then.

September 20, 1991. I spent all morning fighting grass fires around my musk lorikeet nest trees at Old School Site, in a determined effort to save the trees from being burnt and destroyed. The fires were deliberate, to get rid of long grass. After the terrible fires had passed, a musk was observed at one of the nests which contained small young. The very thick, choking smoke, probably killed the young. (I climbed the nest two weeks later: it was empty.)

On this occasion the nests were saved—but only due to John Courtney's intervention.

March 3, 1995. I discovered to my sadness that a little lorikeet's nest had been bulldozed by Council road maintenance workers—quite unnecessarily.

In 1989, only four or five of these nests had survived. By September, 1996, only three nests were occupied, "with brooding littles popping their heads out to look at me, then going back in." With sadness he recorded: "They are the only little lorikeet nests that I can locate in an area of some 40 miles long by ten miles wide."

What a contrast to the November day, 22 years earlier, when he visited Elsmore, 12 miles west of Swan Vale! In his diary for November 8 he recorded:

I noticed a lot of yellow box trees in bloom around Elsmore. Coming back about midday I pulled up opposite the Common to watch large numbers of little lorikeets feeding on yellow box in a paddock. In the Common were numbers of smallish red gum trees. In a quick walk through I located what appeared to be four little

lorikeet nests. All were in two Blakely's red gum trees, one dead and one green, 5 yards apart. The small dead tree contained (in the trunk) about 18 ft. up, one eastern rosella nest. About 3 ft. further up, a pair of little lorikeets were feeding young in a tiny hole. A further 2 ft. up in a small spout a female redrump was brooding. In the green Blakely's red gum, five yards away, a pair of little lorikeets were feeding young in a long, dead horizontal spout. About 4 ft. or 5 ft. down, where the green wood started, a little lorikeet entered a hole in the green wood. In an adjacent long, dead hollow spout about 5 ft. away, and 1 ft. higher than the nest of the pair feeding young, a third pair of little lorikeets were entering and leaving a nest.

It is not only in New South Wales that lorikeet numbers are declining due to lack of nest sites. In the box-ironbark forests, a diverse array of vegetation communities in Victoria, north of the Great Dividing Range, the situation is no better. The selective clearing of all mature trees in the 1930s and 1950s means that now "the large mature trees with hollows which are so important for many species of bird and mammal are in extremely short supply." It is estimated that box-ironbark forests once covered approximately one million hectares in Victoria. Since European settlement, 75 percent has been totally cleared (Harley 1995). The implication for lorikeets and other hole-nesting species is frightening. Because Australia's lorikeets are still so numerous, it may be difficult for many to envisage that their populations will suffer a gigantic decline. But without suitable nest sites this is inevitable.

Nomenclature

A system for naming all organisms has been in use for more than 200 years. Each bird, animal or plant, etc., is arranged by class (birds are Aves), order (parrots are Psittaciformes) and family. Lories were placed in the family Loriidae, as in the first and second editions of *Parrots of the World*. By the time the third edition was published in 1989, systematists had revised their thinking and lories were considered to be members of the family Psittacidae and subfamily Loriinae. (Note that the names of families end in "idae" and the names of subfamilies in "inae".) Forshaw (1989) defined a family as a primary taxonomic category and noted that parrots are usually placed in a single family, thus reflecting their homogeneity (similarity). He defined a subfamily as a secondary category interpolated between family and genus to differentiate broad groupings of apparently related genera.

There is a further division, that of tribe (the names of which end in "ini"), to identify subgroupings of more closely allied genera. The Psittacidae is now divided into three subfamilies: lories (Loriinae), cockatoos (Cacatuinae) and all other parrots (Psittacinae). These families are further divided into 11 tribes.

The order Psittaciformes is well defined morphologically (by its form or appearance) but the primary evolutionary lineages within it are not so clear-cut. In all studies to date, two distinct assemblages have been recognized: the cockatoos and the lorikeets (Christidis et al. 1991). In every respect (appearance, behavior, chromosomes, etc.) the cockatoos are the most distinctive of the parrots and clearly seem to warrant their status as a subfamily. However, rather few studies have been carried out as yet, but work on chromosome analysis, DNA-DNA hybridization and other methods which indicate evolutionary relationships has increased in recent years. Knowledge is slowly being expanded and a decade hence we can expect a much clearer picture.

Meanwhile, one might ask the question, are lories and lorikeets sufficiently distinct from the Psittacinae to warrant their own subfamily? Indeed, what criteria should we use to try to judge this? One criterion might be: Has any member of the Loriinae hybridized with any parrot in a different subfamily? If we ask this question regarding the cockatoos, the answer is very clear: no. (We must disregard supposed records of many decades ago with no supporting evidence, involving cockatiels, which were then commonly regarded as parrakeets).

So what is the answer to this question concerning lorikeets? It might well have been no: no lory is known to have hybridized with a species outside the Loriinae. But in 1980, the production of a hybrid in Australia was reported which, if believed, might necessitate reconsidering present classification. This hybrid has been mentioned on several occasions in the avicultural literature (e.g., Smith 1985) and has also been seen by well-respected and highly experienced aviculturists who believed that it was genuine.

A male rainbow lorikeet was kept in the same aviary as a pair of Australian king parrakeets (*Alisterus scapularis*). The male king died. A week later the female apparently was laying, although how long after the death of the male the first egg was laid is not known. Two of the four eggs were fertile and one hatched. The chick was reared by the female king and the male rainbow lorikeet. In appearance and behavior it was said to have characteristics of both species. Some years ago I saw a photograph of this bird which seemed to substantiate this extraordinary happening.

If this was a genuine hybrid between a rainbow lorikeet and a king parrakeet, it would seem to indicate that members of the subfamily Loriinae are quite closely related to some members of the subfamily Psittacinae. This is not difficult to believe when we look at the swift parrakeet (*Lathamus discolor*) which, in appearance and habits, could easily be mistaken for a lorikeet, although not when a close examination is made of its behavior. Lendon (1973) considered that the swift parrakeet might be "an aberrant southernmost representative of the Opopsittidae [fig parrots]." He commented: "This species behaves very much as do lorikeets inasmuch as it is predominantly a nectar-feeding bird and it clambers about amongst foliage in very much the same way. However, it has not the typical, jerky movements of the lorikeets, mated pairs do not preen each other nor do they rest close together as do lorikeets, and it is said that the tongue is not as extensively brushed as are those of typical lorikeets."

The latter point is not significant as the brushes of the tongue of the swift parrakeet may be better developed than in, for example, iris and Musschenbroek's lorikeets. The extent to which the brushes are developed is determined by the species' food source.

Lories are accepted as an easily recognizable group within the parrots. However, the species on the fringe of that group, e.g., Musschenbroek's and iris lorikeets, have a heavy bill and poorly developed brushes on the tongue. These features led to Salvadori (1891) classifying them with the fig parrots. He placed both species, and these only, in the genus *Neopsittacus*. It is for this reason that they, and *N. pullicauda*, were not included in Mivart's *A Monograph of the Lories*, published in 1896. Based on the shared characteristics of the underwing stripe (present in some lory species), shaft-streaked facial feathers and pointed tails, the belief still exists (e.g., Courtney 1996) that in terms of evolutionary relationship, lories and fig parrots (genera *Opopsitta* and *Psittaculirostris*) are close. Lories are probably closer to fig parrots than to any other group of parrots. The chicks of *Vini* and *Charmosyna* do share the head shape of *Psittaculirostris* but they are quite unlike those of *Neopsittacus*, *Trichoglossus iris* and all other lories.

The belief has been promoted in several articles in avicultural journals published in the U.S.A. in 1995 and 1996 that eclectus parrots might be classified as lories. This theory is expounded due to perceived similarities in behavior and plumage. In fact, there are absolutely no similarities and it is difficult to see how the theory ever arose. Perhaps the confusion dates back a century, when eclectus parrots had the generic name of *Lorius* and present-day *Lorius* lories were called *Domicellus*. Before the end of the nineteenth century eclectus were given the common name of lory and the naturalist John Gould referred to lories as parrakeets. One final confusion concerns the genus *Loriculus*, the small hanging parrots. In some parts of their range they are sympatric with (occur in the same area as) lorikeets. In older ornithological literature the common name of lorikeet was sometimes used for hanging parrots. The mistake is still made on occasions (e.g., Lambert 1993b)—but rarely now.

Finally, mention should be made of the distinction between lory and lorikeet. A lory has a short tail which is not pointed; a lorikeet has a longer, tapering tail. The words equate approximately to parrot and parrakeet. In Australia, however, the term parrakeet is not used, although lorikeet is.

See also **Systematics**.

Paralysis (Cramps)

Paralysis, which usually affects the legs, is not uncommon in wild and captive lorikeets. There appear to be several different causes but immediate treatment should take place for maximum chance of survival. Typically, the bird is unable to perch and is lying face down with legs and wings outstretched and feet clenched.

Possible causes include:
    1. Dietary deficiencies. Wilson (1994) describes the condition in wild lorikeets in

southeast Queensland, mainly rainbow and, to a lesser degree, scaly-breasted lorikeets. The birds affected are usually young or subadult. They are often found on the ground, being unable to fly; they are in poor condition and thin, with only 30 to 60 percent of the normal muscle mass. The muscle mass of the leg is usually also reduced in size. Many are also suffering from Psittacine Beak and Feather Disease (PBFD), lice, fungal infections, enteric coccidiosis and *Capillaria* infestation. Some show partial wing paralysis. The legs are rigid and the toes clenched. The name Lorikeet Nutritional Myopathy (LNM) was proposed for this condition. The number of wild lorikeets needing attention at Currumbin Sanctuary reaches a peak during September to January; the main breeding season there is from August until January. From July, 1993, until June, 1994, 618 rainbow lorikeets and 172 scaly-breasted lorikeets were presented to Currumbin Sanctuary—either sick, injured or orphaned. About 60 percent were juvenile birds; incidence of leg paralysis was over 50 percent. Wilson commented: "This is obviously a disease of young, growing birds which are affected at a time of maximum nutritional requirement, when essential nutrients are needed for growth, muscle development, feathering, etc."

In a number of birds on which post-mortem examinations and histopathology were carried out, no pathological changes could be found in any areas of neurological tissue. Instead, it appears that the muscle fibers were affected; striated muscle fibers degenerated and became necrotic, losing their cross striations and exhibiting varying levels of inflammation and fibrosis; some became mineralized. In advanced cases there was widespread degeneration and pallor of muscle fibers. The scenario was similar to white muscle disease— selenium/vitamin E deficiency.

Treatment

Affected birds were housed in hospital

cages. Limited physiotherapy, in the form of limb massage and extending and flexing the legs, was carried out. Birds that showed signs of weight loss were supplemented with PolyAid (Vetafarm) using a crop needle. Those suffering from internal coccidiosis were given Toltrazuril 25 g/l (Baycox-Bayer) at 1 ml/liter for five days. Injectable Vitamin E (Hardcock's Vit. E—International Animal Health Products—400 mg Alpha Tocopherol/ml) at a dose of 0.1 ml intramuscularly repeated weekly, was given to some birds. They showed no greater improvement or improved prognosis compared to those receiving the oral vitamin E powder (White E-Vetsearch). This was added to the nectar at approximately 80 g/liter, providing 5 i.u. Vit E/ml of nectar. (The nectar was made from 9 parts of water (1 liter) and 1 part honey. To this was added 60 g of Complan (Boots Company). Wilson suggested dietary fat may increase the absorption of vitamin E from the gut; therefore the procedure was changed to orally dosing birds with vitamin E and canola oil mixture and withholding food for two hours after dosing.

Prognosis

Selenium/vitamin E injections were tried with no better results than injectable vitamin E alone. Some birds developed injection reactions and abscesses at the injection site when Selenium/vitamin E was used. Blood values of normal wild birds were compared with those of affected birds. Prognosis was directly related to serum CPK levels: the higher the level the worse the prognosis and the longer the recovery period. A CPK level of approximately 70,000 u/l appeared to be a cut-off point. Below this level birds usually recovered in about 2 or 3 weeks. Above this level few recovered; the legs remained rigid with locked joints and clenched toes.

Cause

The increasing human population in the area results in a dwindling natural food supply. However, thousands of people provide honey water or sugar water in their back yards. This food is nutritionally incomplete. The natural food, pollen, provides necessary oil solubles such as vitamin E. Without it, chicks will show severe growth and development problems.

2. Suspected virus. In the Sydney area, Dr. W. Hartley of Taronga Zoo Wildlife Pathology Register, had different findings (Wilson 1994). Lorikeets that had exhibited signs of leg paralysis were examined histologically. This showed a spongy appearance of the ventral horns containing the motor neurones of the spinal column, with demyelination and death of many neurones. Dr. Hartley suspected that this form of paralysis was viral and perhaps associated with a B-vitamin deficiency.

3. Reason unknown. In Melbourne, avian veterinarian Stacey Gelis reported similar symptoms in musk lorikeets presented to the clinic and peaking in numbers in 1992. Affected birds showed severe neurological signs: ataxia and head tilt. They died, despite antibiotic, anti-inflammatory, multivitamin and fluid therapy. Gross postmortem, histopathological and toxicological studies failed to reveal any cause of death (Gelis pers. comm. 1994).

I saw the symptoms described in the first paragraph in three dusky lories immediately after surgical sexing, apparently induced by the stress of handling. Two died soon after. The third survived; its toes gradually unclenched but thenceforth all the toes faced forwards. However, its mobility was not affected.

Perches

In cage or aviary, lories spend most of their day on perches—except when they are breeding. Unsuitable perching is equivalent to the human discomfort of badly fitting shoes; yet many parrot owners are negligent in this respect. Owners must ensure the following points are adhered to.

1. Perches must be clean. Keep an interchangeable set so that a perch can be removed immediately when it becomes dirty. In a small cage a perch will need to be cleaned every other day; in an aviary perhaps twice a month—or changed.

2. They must be of the correct thickness. A thin perch is usually more uncomfortable than a thick one; however, with natural branches which are easy to grip, size is less important. Suitable dowel sizes are as follows:
   - 1 in. diameter (about 2.5 cm): black-capped, yellow-backed, purple-naped.
   - ¾ in. (about 1.7 cm): *Chalcopsitta*, *Eos*, *Pseudeos*, smaller *Lorius*.
   - ½ in. (about 1.3 cm): *Trichoglossus*, except Meyer's and Mount Apo, also Stella's and Josephine's.
   - 1 cm: Meyer's, Mt. Apo, smaller *Charmosyna*, *Vini*.

3. Perches have to be of the correct hardness. Very hard wood, such as manzanita, which becomes shiny with wear, should never be used. It becomes difficult for the lories to grip it easily. Thin twigs of soft wood such as willow are unsuitable for permanent perches but ideal for providing enjoyment. Thicker branches of willow, poplar, eucalyptus, casuarina and fruit trees (such as apple), make the best perches. Dowel is a good (but expensive) alternative.

4. They must be carefully cleaned before use. Fresh-cut perches, to which wild birds have had access, could contain eggs of parasitic worms. Wash with a pressure washer or scrub before use.

5. They must be correctly sited; they should stretch the width of cage or aviary, and be placed near the ends to encourage flight, with the larger species. Different levels, including small branches placed on the floor of the cage, are appreciated. Some lories enjoy perching on a branch on the cage floor. Perches do not have to be horizontal; vertical ones are also enjoyed. Use imagination!

6. Perches must be available! Be sure to replace them immediately if they fall down. In a range of eight aviaries of one lory keeper, I noted that only three aviaries had perches in the outdoor section; the other birds had to cling to the wire mesh. This is no way to keep any perching bird—yet it often occurs as a result of laziness. Perches must be firmly fixed. Perch holders come in a multitude of ideas. I clip U-shaped pieces of welded mesh to the side of the cage using C clips and C clip pliers. Builder's merchants often stock suitable appliances made for other purposes. Replacing perches in suspended cages can be difficult. One breeder in California had a good method. He used two long, metal rods, curved at the end, which assisted him to remove or insert perches.

7. Special perches are made by several companies in the U.S.A. (such as Just Selling, Inc.) which are designed to trim beak and nails. They are made of a hard material with a rough feel, they are effective and birds will use them. They are useful for pet birds whose beak and nails to not receive enough wear and help to prevent repeated trimming.

8. Perches have to be positioned where the birds are least likely to contaminate food or water with their feces. Bear in mind that lories shoot their droppings a considerable distance. Where cages are closely spaced, their droppings could even reach the food or water containers in the adjoining cage.

Pets

As affectionate as a cockatoo, as colorful as a macaw and as playful as a puppy, the larger lories make enchanting pets—for those who can stand the mess. They are not for the houseproud. However, one can easily adapt the area around a lory's cage to make cleaning easy. On the floor by the cage of my companion black-capped lory I placed sheets of plastic carpet protector which one can buy in hardware stores. On the wall at the back of his cage is polythene sheeting. Both are easily wiped clean. In fact, some people might prefer this task to that of clearing up seed and more solid foods. The larger lories are intelligent, inquisitive and show a degree of affection which rivals any parrot and exceeds most. They can be noisy—as can all parrots—and the voice can be extremely harsh. Listen to an adult at full blast before deciding if a lory is the right pet for your household.

The dusky is one of the least expensive and most often available species; the red used to fall into the same category. The *Chalcopsitta* species are slightly to considerably more expensive; however, hand-reared birds show a degree of affection which is without parallel. They can make totally enchanting pets. Most parent-reared birds obtained when young will also become very tame and affectionate in due course if kept close to people, in a cage rather than an aviary; however, this may take several months. *Lorius*, especially yellow-backs and black-caps, are popular for their ability to mimic and for their beauty. All the large lories are intelligent and extremely observant. They are fascinating and beautiful.

The type of cage obtained needs some thought. Those with a plastic base which clips on to the upper part of the cage are very useful to catch the liquid droppings. Newspaper can be placed on the bottom. A false bottom of welded mesh will prevent the occupant from tearing up or playing in the newspaper. The size of welded mesh should be suitable for the species; it must not trap the feet or claws. However, most of the cages with a detachable plastic base are not really large enough for the larger species, unless they spend long periods outside the cage.

Most lories will appreciate something on which to play inside the cage. Rope swings and ladders give plenty of scope for entertainment and are entirely safe. There are countless excellent toys designed for parrots. A small wooden hammer on a chain, suspended from the roof by a dog clip, entertains one of my black-capped lories for hours. The cardboard centers from toilet rolls can also provide short-lived amusement. A single lory will also appreciate a "rubbing stick." Lories spend hours in mutual preening. When they have no companion to preen their head feathers, a branch from a suitable tree (willow, fruit, etc.) can give them a lot of pleasure. Cut a length and fit it firmly into the bars of the cage so that one end is free and facing upwards. Especially if the end is tapered a little, it forms a perfect point on which to rub the head. Many lories also like to sit on a small branch placed on the floor of the cage. (See also **Environmental Enrichment.**)

If a lory is tame, don't clip its wings. It needs to be able to fly. Give it a daily period of totally supervised freedom outside the cage in a safe environment with doors and windows securely closed. Take care, however, because many lories are extremely swift and strong flyers and, once outdoors, especially if startled, could quickly be so far from the house that they would be lost. Make a wooden frame covered with hardware cloth (welded mesh) to fit over an open window so that your bird can safely exercise out of its cage on warm days. Remember that lories are very inquisitive and soon get into trouble if they are left alone for a minute or two. It is not usually advisable to let the larger species out in a room where there are other parrots as they can be aggressive.

Lories must have access to a bath daily and will splash water around the cage area. Again, the plastic and polythene sheeting protects furnishings and wallpaper. Alternatively, a sheet of PVC plastic can be cut to fit on to the back and sides of the cage; the sections are held in position with wing nuts, which can easily be unscrewed for cleaning.

The small lories such as Goldie's make less mess and less noise. Unless hand-reared and exceptionally tame, it is recommended to keep them in pairs. They are not usually as affectionate and playful as the larger species. Those ranging in size from green-naped upward make entertaining pets. One occasionally reads items in parrot magazines where the emphasis is on pet birds, from contributors who have discovered the charm of a pet lory. One such item was from a lady who first encountered her green-naped lorikeet in a local pet shop, lying on her back, playing with toys. She went away and read all she could find about lorikeets. The advice was almost unanimous that they are aviary birds and not suitable for the house. Despite this, a week later she returned to the shop to buy the green-naped. Her husband made a "splashguard" for the front of the cage and she quickly "toilet-trained" her bird. Soon the Green-naped was saying "Pretty boy" and "Hello." She was let out of the cage daily to fly around with other birds and palled up with a budgerigar! The owner concluded her article by highly recommending lorikeets as pets as "they are very lively, very pretty and cheer you up. I wouldn't be without her!" (Penny 1996). Lively, pretty and amusing perfectly sum up the attributes of pet lories. One other point should be mentioned. The pet green-naped had produced 20 eggs in just over a year. My guess is that the owner was removing these eggs, causing the green-naped to lay again. If a pet lory lays, place some wood shavings in a corner of the cage as a depression for the eggs. Allow her to sit on them for as long as she wishes (probably three weeks or more). This will ensure that she does not become a continuous layer.

There has been a greatly increased demand for lories as pets in the U.S.A. in recent years. Indeed, Schroeder (1991) assessed that more were kept as pets than for breeding. However, a limited number of species are sold for this purpose (see **Beginners, Species for**).

Some lories and lorikeets are excellent mimics of sounds and words. A violet-necked lory (*Eos squamata*), hand-reared from the egg, was given to the rearer's 83-year-old mother. She taught him about 75 words—repeated in a very loud clear voice (Simmons 1994). However, it is the larger species, especially the *Lorius*, which are most renowned for their ability to mimic. The tale of one such bird, a black-capped lory, was recounted amusingly by a lady identified as "Olive St. A. S." (1938). She related how he was bought in a Chinese bird shop "and from the first he hated me":

> Alfred's vocabulary was very free, and fortunately for most of his hearers was in a Malay dialect not very widely spoken. But there was no doubt as to the import of most of the raucous remarks that used to be flung at my innocent and inoffensive head, so packed with meaning were they. Let anyone of the male sex approach him though. What a transformation! Gone was the flashing eye and snapping beak, and with what honeyed words did he spread his burnished quivering wings and bow his cheeky head in humble admiration.
>
> Those who could understand Alfred's speech assured us that he was unusually clever, and I have seen Malays sit for hours listening to his flow of conversation and, judging by the smiles of his audience and slightly anxious side-glances in my direction, one assumes that his type of humour must have been rather Rabelasian at times. He never learned to speak one word of English, and his little brain must have been so packed with other things that it could not absorb any more knowledge.

"Talking" ability is rare or nonexistent in the small lorikeets but has been recorded in *Eos*, *Trichoglossus* and *Lorius* species. A rainbow lorikeet (*T.*

h. moluccanus) had become an exceptionally talented mimic by the age of 18 months, quickly learning new phrases. One day his owners taped him for half an hour, during which time he talked non-stop and repeated a wide range of phrases. He also mimicked the sounds around him, such as water leaving the washing machine and the squealing of a guinea pig. He suddenly added "Whacko!" to his vocabulary. As this word was never used in the household, he must have learned it from a child on an adjoining property. At one time he showed a preference for resting on a piece of flannel at night, rather than on a perch. His owners made him a flannel hammock; he regularly used it at night and did not like to be disturbed. He would shout out: "What are you doing? Get out!" if he was woken up! (Oliver 1976).

Planted Aviaries, see **Aviaries and Cages**

Plant Stimulants

It was difficult to know how to label this entry. Lories find the juices of certain twigs, plants and fruits so stimulating that they bite the plant or fruit and then rub the beak through the plumage in a frenzied manner. It is reminiscent of how passerine birds behave with the formic acid from ants. The birds pick up the ants and apply them to their feathers, or allow the ants to invade their feathers. They act in a frenzied manner; this is followed by preening, bathing or even lying spread-eagled with the wings thrust in front of the body. The reason is uncertain. Lories behave in an excited manner when rubbing the plant juices on their feathers and even on their feet; however, only certain substances incite them to behave in this way.

An excellent example was described by Fred Bauer in respect to the madrone tree (*Arbutus menziesii*):

> When the green, flowering Madrone twigs were given to lories a truly remarkable display occurred. Remember that these birds had never seen a Madrone twig before and there are no relatives of this tree anywhere in the native range of lories, the only other member of the family being the Strawberry Tree (*Arbutus unedo*) of the Mediterranean. Within a few seconds after first contact, you will see a concentration as if a light bulb is switched on in the

bird's awareness. The tongue begins wild flicking, eyes dilate, and vigorous preening ensues. A frenzy develops, the bird rapidly masticating the tissue of the new bright green growth of the twig, ignoring any older bark and leaves, then transferring the liquid secretion first to its plumage and then to its feet until both glisten with it. The process begins to wind down with the bird often lying on its side momentarily as if exhausted. Like any passion, satiation dulls the response. After fifteen or twenty minutes, completely normal behavior returns. Or is it that the need has been met?...We treated our lories to this ritual for years with no apparent harm, much obvious pleasure, and the resulting beautiful glossy legs, feet, and feathers (Bauer 1995).

I can induce similar, but less frantic, behavior at will by giving freshly picked guavas to the *Charmosyna* lorikeets. Then follows five minutes or more of frenzied preening activity. They bite the guavas, then preen frantically, especially under the wings and by running the beak along the tail. I have also observed them touching their feet with the beak after biting the fruit. They become excited at the mere sight of a guava. Most other lories do not react in the same way: they eat them at once! I have seen this behavior less often in *Chalcopsitta* species. Some young *insignis* "anted" upon being given *Casuarina exquisetifolia*, immediately after biting at bark and needles. Most lories show some reaction when given *Casuarina* or other plants with juices they find exciting, but their reaction is not as pronounced as in the *Charmosyna* species.

Bosch (1993) commented of Goldie's lorikeets: "It is so characteristic of lories to rub and smear the juice and taste of different parts of plants into the whole plumage and over the legs, which I have not yet seen with these lories." Neither have I...

J. E. Wallace of Mount Surprise, Queensland, Australia, wrote to me about the behavior of his lorikeets—rainbows, red-collared, scaly-breasted and varieds—behavior which was not observed in any other parrot in his collection. "Growing in our garden are *Callistemons*, namely Captain Cook or weeping bottlebrush (*C. viminalis*) and other varieties. When we give our lorikeets cuttings from these plants, they just feed on the flowers and then very meticulously crush the seed pods and rub the crushed extract all over and through their feathers, as if the extract were an insecticide. It smells strongly of eucalypt."

Plumage

Look up any definition of the word lory and you will find reference to one fact, apart from the statement that it is a parrot. The *Concise Oxford Dictionary* states: "Kinds of bright-plumaged parrot-like bird." *A Dictionary of Birds* defines lories, under the heading of Parrot as "medium-sized to small birds with extremely colourful plumage." In *Parrots of the World* Forshaw (1989) states that lories "have tight, glossy plumage and most are brilliantly colored."

The feature of a lory which is most remarkable to most people is the brilliance of its plumage. Of all the world's birds, probably the lories and the pheasants are unique in having so many multicolored species of dazzling and varied hues. Think of almost any combination of colors (excluding black and white) and you will find them on at least one lory. Their plumage is truly remarkable. Forshaw mentions the "tight, glossy plumage." These qualities somehow accentuate the exquisite color combinations. The glossiness reflects the light in a way that the feathers of, for instance, an Amazon parrot (genus *Amazona*) do not. For example, the blue-black on the abdomen of a black-capped lory can look blue or black. This point should be borne in mind when trying to identify lories from photographs. One does not see precisely the same color every time; and much depends upon whether the bird is in sunlight. The rajah lory appears as a dull black bird in poor light—but view it in the sunshine and the black lights up with purple and vinous reflections.

As if the intensity and countless combinations of colors were not enough, some species have developed additional adornments. The enormously elongated central tail feathers of the Papuan (Stella's) lorikeet are longer than the body in mature specimens. They stream behind it in flight. Some of the closely related *Vini* lories have elongated feathers of the nape. The long, glossy deep blue feathers of the nape of Kuhl's lory, which are partially erectile, are like a thick mane. Perhaps total perfection of form and coloration is achieved in the collared lory. It has a cape of elongated feathers of almost luminous green—an adornment which is unique in a lory.

After observing collared lories in the wild, I was convinced they were the most beautiful birds I had ever seen. I was not the first. Sydney Porter, who had traveled the world indulging his interest in parrots and other birds, described collared lories as "the loveliest" of the whole parrot tribe:

> They are indeed like some gaudy blossom plucked from a flamboyant tropical tree. One cannot imagine these radiant creatures coming from any other place than from some fair tropical isle in the Southern Seas. What fitting jewels to grace the tropical greenery of those far away sunlit emerald isles, with a cap of the deepest purple like the intenseness of a tropical night, glittering green like the brilliant succulent grass, and a scarlet that makes even the glow of the hibiscus look dull (Porter 1935).

Another feature of the plumage which is rarely seen in any other parrot species is the streaked feathers of many lorikeets. In some *Trichoglossus* and *Vini* species the shafts of the head feathers are streaked, producing an intricately beautiful appearance. In others, especially in the genus *Chalcopsitta* and in the striated lorikeet, the narrow feathers, streaked with a contrasting color, are highly decorative. In the striated lorikeet, the yellow streaking on the underparts is almost luminous in its contrast with the otherwise green plumage.

But every genus of the Loriinae has its species of outstanding beauty or characteristics of plumage which enable most of its members to be recognized. Two *Chalcopsitta* species have spiky, elongated head feathers which frame the head when erected; and one of them has a color scheme which may be unique on a bird: brown and gold and violet. The *Eos* lories are glossy red and blue (and black) with no green in the plumage; the *Trichoglossus* have green wings, variable head coloration and upper breast barred with red, yellow, black or green. The *Lorius* are green above and red or mainly red below. The genus *Charmosyna* is large and can be divided into those with red underparts and those with green underparts. The red birds are stunningly beautiful and the females have yellow rumps. What a contrast! One species, Stella's, even has two color phases, red and melanistic. It is the only lory (I am not sure that the dusky qualifies as only one area of the plumage is altered), and one of only two of all the members of the parrot kingdom, to have more than one color phase (as distinct from a color mutation). The small green *Glossopsitta* species are less flamboyant but prettily marked, and with red on the head. Then there are the tiny *Vini* lories, two of

which are garbed entirely in blue and white with contrasting orange or red beak. Exquisite! And the two small *Neopsittacus* species are breath-taking when they hold open their wings (as they often do) to reveal the scarlet which matches their breast. And one of the tiniest of all, the only member of its genus, the whiskered lorikeet, is unique. It has 14 tail feathers—two more than any other parrot! And the underside of the tail is a glorious shade of rose pink.

Who could deny that lories are the most beautiful birds in the world?

Poisoning, Heavy Metal

Because lories have a tendency to lick surfaces, they are equally susceptible to heavy metal poisoning as are the larger, more destructive parrots. Lead and zinc are the metals with which they are most likely to come into contact. One lory breeder lost the majority of his 20 pairs of lories when they were placed in cages which had been treated with paint containing lead. Some were saved by giving Penicillamine tablets (55 mg per kg of bodyweight). These were crushed and added to a small amount of nectar. (This product is also available in suspension.)

Most galvanized welded mesh available these days is of good quality, but beware of cheap imported mesh. In Australia, imported mesh from China cost several parrots their lives due to lead poisoning. Faulty spot welding and galvanizing causes a soft slag to accumulate at the weld points, which is easily picked off. If even a small area of galvanizing on a roll is faulty, or if the mesh has been galvanized after the cage has been constructed, it could result in a problem due to a bird swallowing small, loose pieces of galvanizing. Cages constructed of new wire should therefore be treated as follows, before use. Any rough edges should be filed smooth. The cage should then be sprayed with acetic acid which has been diluted, 10 ml in 1 liter of water; the next step is to wash it with a pressure washer. Finally, it should be sprayed with sodium bicarbonate (100 g in every 2 liters of water).

My own knowledge of treating heavy metal poisoning unfortunately occurred after a pair of Josephine's lorikeets were placed in a newly galvanized cage which had been brushed with a wire brush, washed in vinegar, then pressure washed and left outside in frost and rain for two weeks. The birds were placed in the new cage at midday. Early next morning the male was drinking more water than usual, his droppings were white and he was fluffed up. I immediately moved both birds back to their old cage, put a heat lamp on them and consulted an avian vet. As there was the strong possibility that he was suffering from zinc/lead poisoning, the advice was to inject both birds with vitamin B_1 in the form of Duphafral Extravite and with Calcium EDTA. Only the Duphafral was available that day from a local veterinarian. By midmorning the male was coughing regularly and the female also looked sick. I suspected that a particle of galvanizing was lodged in the male's throat or trachea. He was also started on antibiotic injections. By midday I was alarmed to see that the upper part of the female's upper mandible had turned white. This is a sign of a liver and/or kidney problem so I was then sure that she had absorbed lead/zinc into her system. The following day the calcium EDTA arrived in the post and the full treatment commenced. As the male was also on antibiotics, I had to give him three injections. Josephine's are small birds (with a small area of breast muscle) and this was a cause of concern. The directions were to inject 0.1 ml of Duphafral daily until the symptoms disappeared and to inject 0.015 ml (about 2 units) of Calcium EDTA once daily for five days, then by mouth until the poison had left the body—possibly up to 14 days in some cases. That day the male's coughing became worse and, sadly, next morning he was dead. By that day the female was almost fully recovered and her beak color had returned to normal. I did not continue the treatment after that day as I believed that the zinc/lead had left her body (calcium binds metal) as her beak was orange again. She made a total recovery and laid her first eggs about six months later, with a new male.

Gelis (pers. comm. 1996) suggests that birds severely affected with heavy metal poisoning may need intensive supportive veterinary care in the form of extensive intravenous fluid therapy, heat, anticonvulsants, etc. Cases which are not responsive to initial treatment, or which relapse, require X-rays to be taken (if not taken initially) to ascertain the number, size and location of heavy metal particles. Sometimes these particles can be forced out of the body more quickly if the bird is force-fed bulking agents (e.g., peanut butter, psyllium). In some cases the particles may need to be removed surgically.

Pollen, see **Nectar**

Popularity, Avicultural

Which are the most commonly kept and bred species in Europe and the U.S.A.? To provide an indication, several different published sources were consulted. The results did not produce any surprises but are, nevertheless, interesting. In the tables below, the Breeding Register of the Parrot Society in the U.K. is contrasted with the ISIS records of zoos in the U.S.A. and with the returns for 1988 from members of the German association of bird breeders, AZ. In each case, the top 10 species are listed in numerical order, the most numerous species being number one. Only the 10 most numerous species are listed (most numerous meaning the numbers of people or zoos who keep that species, but the "young bred" column relates to the actual numbers bred).

Lory Popularity

	Parrot Society (U.K.) '92	Young bred
1	T. h. haematodus	T. h. haematodus
2	P. fuscata	T. goldiei
3	T. goldiei	P. fuscata
	C. p. goliathina	
4		L. garrulus
5	L. garrulus	Eos bornea
6	Eos bornea	C. p. goliathina
7	Eos reticulata	L. lory
8	L. lory	Eos reticulata
9	T. h. moluccanus	T. h. moluccanus
10	T. euteles	T. h. capistratus

	ISIS (zoos world) '92	Young Bred
1	Eos bornea	T. h. haematodus
2	T. h. haematodus	T. goldiei
3	L. garrulus	L. garrulus
4	E. reticulata	T. h. moluccanus
5	P. fuscata	C. scintillata
6	T. goldiei	C. placentis
	T. h. moluccanus	
7		T. or P. versicolor
8	L. lory	C. pulchella
9	Eos cyanogenia	Eos bornea
		G. concina
		T. h. rubritorquis
10	T. h. rubritorquis	

AZ (Germany) '89		Young bred
1	T. h. moluccanus	T. h. moluccanus
2	T. h. haematodus	T. goldiei
	L. garrulus	
	L. lory	
3		T. h. haematodus
4		L. lory
5	T. goldiei	L. garrulus
6	T. euteles	T. euteles
7	Eos bornea	C. duyvenbodei
	C. p. goliathina	
8	P. fuscata	C. p. goliathina
	T. ornatus	T. h. capistratus
	C. scintillata	
9		T. ornatus
10		Eos bornea

In 1994, the EEP Parrot TAG (European zoos) surveyed zoos regarding the numbers and species of lories in their collections. The zoos who responded recorded a total of 1,229 lories. The 12 most popular species are listed in order of abundance, with the numbers of birds held:

EEP Lory Popularity

1.	Lorius garrulus (all subspecies):	106
2.	Trichoglossus h. haematodus:	84
3.	Eos reticulata:	72
4.	Pseudeos fuscata:	70
5.	"Rainbow" (Trichoglossus haematodus):	62
	Charmosyna papou:	62
7.	Eos bornea:	56
8.	Chalcopsitta atra (all subspecies):	51 *
9.	Chalcopsitta scintillata:	47
10.	Trichoglossus h. moluccanus:	43
11.	Trichoglossus goldiei:	42

* This unexpected result is probably explained by the high numbers in one collection.

In Australia, numbers of exotic lories are small but are rapidly growing. In 1995, the first legal importation occurred. There are no figures of numbers held. Some idea of the captive abundance of native species can be gauged by registration returns. The figures produced by the National Parks and Wildlife Service (NPWS) in 1987 and 1992 are an example. It is generally suggested that one in three aviculturists in New South Wales is registered, therefore the figures do not show the complete picture (Stuckey 1994). Numbers of lorikeets registered in New South Wales in 1992 for species of which fewer than 500 individuals were reported, are as follows:

Australian Lory Popularity

Little lorikeet	378
Purple-crowned lorikeet	353
Varied lorikeet	276

As a comparison, there were 206 red-tailed black cockatoos and 509 Cloncurry parrakeets.

The status of Australian parrots in South Australian aviaries in 1985 was assessed by using computer records of the National Parks and Wildlife Service. In that state, people holding more than 9 protected birds must obtain a Class 2 permit issued under the National Parks and Wildlife Act, 1972, and must submit quarterly stock returns. The Avicultural Society of South Australia requested that the gross totals for each species for the three months ending December 31, 1985, should be made available to them. The following data was thus obtained (Shephard and Welford 1987) and is compared with the equivalent figures for 1991. In that year, however, all current permit holders were included, i.e., those holding fewer than 9 protected birds (Shephard, Pyle and Fairlie 1991).

Only one lorikeet was included in the list of common Australian parrots, determined by total stock greater than 500 in South Australian aviaries of people holding Class 2 permits. This was the rainbow lorikeet. It did not feature in the list of species for which more than 200 young were reared in the given period. Four lorikeets were shown in the list of rare Australian parrots, that is, species with a total stock of less than 200. They were purple-crowned lorikeet (174 birds), musk lorikeet (142 birds), little lorikeet (45) and varied lorikeet (24).

The table below shows the total stock of each lorikeet species at the beginning and end of the period, and the number bred. There was no information for the red-collared lorikeet for 1985.

Australian Lorikeet Stocks

Species	Total stock 1/10/85	31/12/85	% Increase	No bred	Total stock 4/91	% Increase 1985–91
Rainbow	1138	1148	1%	93	2292	–
Red-collared	figures included under Rainbow				363	–
Purple-crowned	149	179	17%	35	356	100
Musk	138	142	3%	9	431	204*
Scaly-breasted	129	132	2%	14	266	104
Little	39	45	15%	5	87	93
Varied	16	24	50%	0	74	208*

* Increase probably mainly due to increased importation from other states.

The conclusion made was that "the lorikeet family as a whole requires the urgent attention of South Australian aviculturists." It was noted that while there was a considerable number of rainbow lorikeets in South Australian aviaries and 93 were bred, 61 birds died. No varied lorikeets were bred during the given period because their main breeding season is April to August.

Predators

To which species do lories fall prey in the wild? Birds of prey are capable of catching lories and I witnessed a dramatic example of this in New Guinea. My diary extract read as follows:

> December 7, 1994. Below Tari Gap at about 2,300 m (7,500 ft.)...We came to an area where several species of lorikeets were feeding in tall trees. Our necks ached with the effort of peering 100 ft. (30 m) upwards. All morning Stella's (*C. p. goliathina*) and Musschenbroek's were flying around; we hardly went five minutes without seeing them. The weather was warm but overcast. As we went down a little lower we came across trees where Stella's, Musschenbroek's, Goldie's and a Josephine's were feeding. Emeralds flew over more rarely. Suddenly a harpy eagle (*Harpyopsis novaeguineae*) flew over and Goldie's (the most difficult to observe but recognizable by their call) scattered in all directions. I counted 26 and David, behind me, saw eight more. We climbed up a bank for a better view of lorikeets and ate our lunch up there. After a while we sat down at the roadside as our necks were aching, and our eyes, from looking upwards. Suddenly we jumped up as the harpy returned with a bird in its talons. Although it was high above us, we could see by the tail that it was a Stella's.

Renowned Australian ornithologist Len Robinson witnessed a dramatic kill. He, too, recorded the event in his diary (Robinson pers. comm. 1996):

> October 17, 1958. Katherine, Northern Territory. I cut across the end of the aerodrome and spent some time in the small open timber (eucalypt mainly) nearby. I was quite pleased to see several yellow figbirds and to see a black falcon (*Falco subniger*) chase a varied lorikeet. The lorikeet hurled itself into the long grass and before the falcon could turn and pick it up, a large hawk, a black kite (*Milvus migrans*) snatched the lorikeet from the grass and flew off.

Anthony B. Rose of New South Wales also witnessed a kill. He recorded (pers. comm. 1996):

> On August 7, 1995, alongside our house at Forster at 1500 hours, I heard pied currawongs yelling and a lorikeet squealing on a path near a flowering bottlebrush (*Callistemon* species). It was being used by lorikeets which were feeding on the nectar. Expecting to see a neighbor's killer cat I crept up under cover to within one meter and had a full view of a collared sparrowhawk (*Accipiter cirrhocephalus*) (female judging by its size) on a dying scaly-breasted lorikeet (*Trichoglossus chlorolepidotus*). It stared at me then flew with its prey to a front garden; a car on the road put it up and it flew in a half circle carrying the lorikeet to a reserve of tall trees, pursued by pied currawongs.

On another occasion, on August 3, 1988, Anthony Rose saw a whistling kite (*Milvus sphenurus*) being escorted by seven scaly-breasted lorikeets; they left and an Australian magpie took over. The kite did not appear to be carrying anything and he doubted that it would be swift enough to catch a lorikeet. Roy Dunn of Nimbin, New South Wales, saw a peregrine falcon carrying prey in the winter of 1996. It alighted on a power pole at the end of his garden. With binoculars he was able to identify the victim as a rainbow lorikeet (Dunn pers. comm. 1996).

Marchant and Higgins (1993) record the following raptorial predators of Australian lorikeets—and their prey species: Australian hobby (*Falco longipennis*), scaly-breasted, purple-crowned and little lorikeets; peregrine falcon (*Falco peregrinus*), rainbow, musk, purple-crowned and little; brown goshawk (*Accipiter fasciatus*), musk (Tasmania); collared sparrowhawk (*Accipiter cirrhocephalus*), little and purple-crowned lorikeets; red goshawk (*Erythrotriorchis radiatus*), varied and red-collared; black-breasted buzzard (*Hamirostra melanosternon*), red-collared lorikeet.

In Fiji, the nearly extinct peregrine falcon commonly catches collared lories. Ryan (1988) relates that Fergus Clunie analyzed the pellets of a peregrine near Joske's Thumb. The majority contained remains of this lory, also fruit bats. He commented that he had enthused over the lory's flying ability on a number of occasions and was "full of respect for the peregrine which makes this species a major part of its diet. It is presumably very tasty as there must

SOLOMON ISLANDS

E II R

Yellow-bibbed Lory
Lorius chlorocercus

$1.20

$1.20 stamp of the Solomon Islands depicts the yellow-bibbed lory. See page 14.

Photo: Ron Moat

ENCYCLOPEDIA OF THE LORIES
PHOTO ESSAY

The headdress of this Huli medicine man contained the long white tail feathers of King of Saxony's bird of paradise and part skins of Josephine's and Musschenbroek's lorikeets. See page 67.

A

B

C

A Lory aviaries at Loro Parque, Tenerife. More than 30 species are exhibited there. See page 16.

B The author's 5m-long outdoor flights, constructed from aluminum angle. See page 16.

C Sharon Casmier's lories in Washington are kept in a greenhouse-type building with large sliding doors. See page 17. *Photo: Sharon Casmier*

D Pebble floors are easily cleaned with a hose or pressure washer. See page 17.

D

A child's painting of a blue-crowned lory featured on a Tonga calendar. See page 39.

Yellow-bibbed lory anointing its plumage with the sap from casuarina. See page 92.

The larger lories, such as the yellow-streaked, use a foot to steady items of food. See page 52.

Head butting is a common component of display in *Trichoglossus* lorikeets, such as these Meyer's. The male is butting. See page 45.

A stainless steel dish attached to chains with a key ring and hung by a dog clip makes a perfect bath for the larger lories. See page 21.

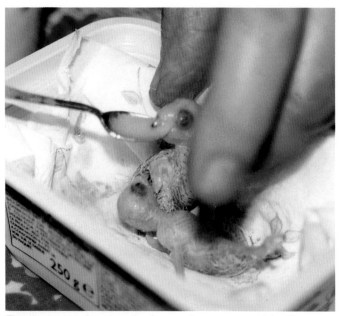

The tiny spoon from a pair of sugar tongs was used by Rosemary Wiseman to feed these fairy lorikeet chicks. See page 61.

Daily weighing on scales with increments of 1 g provides an early indication if a chick is not thriving. See page 62.

The author uses a spoon with the sides bent inward. See page 61.

In California 2 oz. paper or plastic cups are used by some breeders to feed chicks. See page 61.

A hybrid rainbow x scaly-breasted lorikeet (upper bird) typically has the head blue. See page 64.
Photo: Charles Hibbert

Wax moth larvae are a favorite food of captive Musschenbroek's lorikeets. See page 71.

Stainless steel and plastic containers suitable for nectar. Those (left) which fit into a ring which hooks on to welded mesh are widely used. See page 79.

Musschenbroek's lorikeet aged 54 days. See page 395.

Sheets of PVC have been screwed on to the sides of this cage to contain the droppings. See page 90.

The skin of a Josephine's lorikeet, mounted on a stick, made a bizarre head adornment. See page 67.

The inverted L-shaped nestbox is used by the author for the larger species. See page 81.

Nest boxes with a pull-out drawer and a long entrance tunnel are used by Willem Grobler in South Africa.

At San Diego Zoo a nestbox for blue-crowned lories (*Vini australis*) is hidden inside a staghorn fern. Front view and back view are shown here and above. See page 82.

In South Africa Jan-Hendrik van Rooyen gives his lories a choice between a sisal log (left) and another site. See page 80.

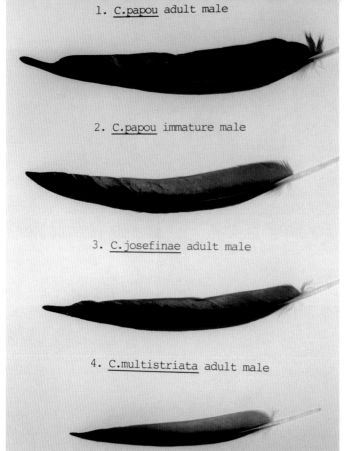

1. <u>C.papou</u> adult male

2. <u>C.papou</u> immature male

3. <u>C.josefinae</u> adult male

4. <u>C.multistriata</u> adult male

Adult Stella's and Josephine's lorikeets have attenuated (lengthened) tips to the primary feathers. Compare feather No. 1, an adult Stella's, with No. 2, from an immature bird of the same species. No. 3 shows that of an adult Josephine's and No. 4, for comparison, that of an adult striated lorikeet, a species which has only slightly lengthened primaries. See page 379. *Feathers supplied by Ruediger Neff*

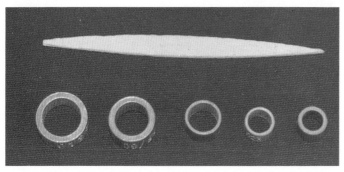

Rings suitable for lories and lorikeets, from left to right: 8.5 mm, 7.5 mm, 6.5 mm, 5 mm and 4.5 mm. See page 150.

The brushes or papillae are clearly visible on this red and blue lory's tongue. See page 154.

Shaft-streaked feathers such as those on the head of the green-naped lorikeet, are found in many species. See page 93.

The elongated nape feathers of the Rimatara (Kuhl's) lory are one of the most beautiful adornments of any lory. See page 92.

A trapper on Seram with a recently caught purple-naped lory tethered to a perch. See page 180.

Photo: Armin Brockner

Red lories caught in Seram being taken by boat to the market in Ambon. See page 222.

Photo: Armin Brockner

Walk through lory exhibit at San Diego Wild Animal Park. See page 162.

Photo: David Hancock

The pleasure of proximity. Mia Hancock at San Diego Wild Animal Park lory exhibit.

Photo: David Hancock

Some chicks are independent at a very early age. These rajah lories aged only 41 and 37 days were already weaned. See page 159.

The Species

Breeding pair of black-capped lories (*Lorius lory lory*) at Palmitos Park with a fostered *Lorius l. erythrothorax*.

Four-year-old *Lorius l. erythrothorax* mating. They were kept without nest boxes in a large colony of various lories on exhibit at Palmitos Park.

Lorius lory cyanuchen. Photo: Susanne Müller

Immature *Lorius lory erythrothorax*. Note the complete neck ring of a young bird.

Lorius lory cyanuchen photographed by Thomas Arndt on Biak.

Right, centre: Immature *Lorius lory erythrothorax* (and other *lory* subspecies) differ from adults in having the greater under wing coverts broadly tipped with black. Also, the lesser under wing coverts near the edge of the wing are blue not red.

Right: Newly hatched black-capped lory (*Lorius lory lory*) at Palmitos Park.

Figure A

Figure B

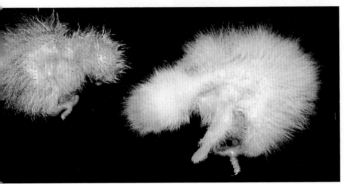

Figure C

Figure D

The appearance of chicks provides a clue to evolutionary relationship. *Lorius* and *Eos* are two closely related genera; young chicks look almost identical. Figure A: The six-day-old *Lorius lory erythrothorax* on the right could be distinguished from the seven-day-old *Eos bornea* only by the gray down on its forehead. Figure B: Eight days later the same two chicks retain their almost identical appearance. Figure C: Contrast the same *Lorius lory erythrothorax* at seven days with a nine-day-old Stella's lorikeet (*Charmosyna papou goliathina*). The two genera are not closely related and the chicks look different, especially in head shape. Figure D: Seven days later the difference between the two is even more pronounced.

Purple-bellied lory (*Lorius hypoinochrous devittatus*) owned by Anne Love in Port Moresby.

Lorius lory erythrothorax at 31 days; head, wing and tail feathers are just starting to erupt.

White-naped lory (*Lorius albidinuchus*). The tiny white nape patch gives this species its common name.

Photo: Bruce Beehler

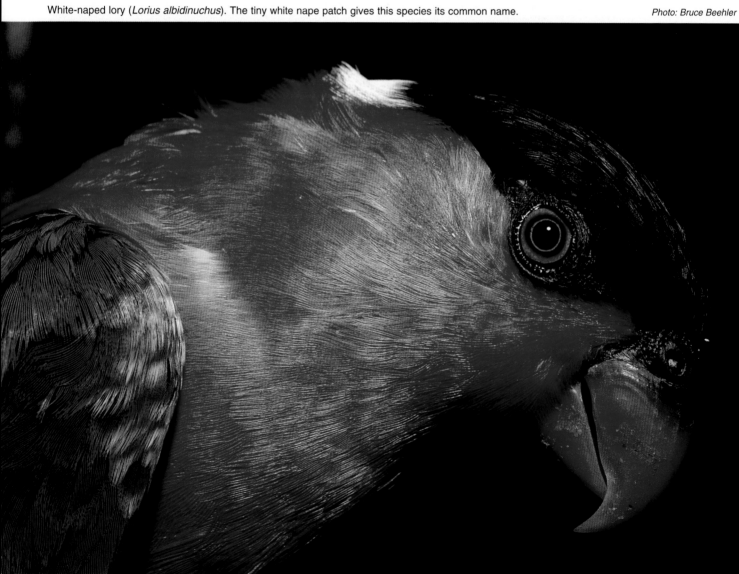

The yellow crescent on the upper breast and the silvery mauve color at the bend of the wing distinguish the yellow-bibbed lory (*Lorius chlorocercus*).

Immature *chlorocercus* have less yellow on the upper breast but the black patches on the side of the neck are conspicuous.

Lorius domicellus aged 29 days at Jurong BirdPark. *Photo: Imelda Santos*

Purple-naped lory (*Lorius domicellus*), a species which is declining due to trapping and deforestation.

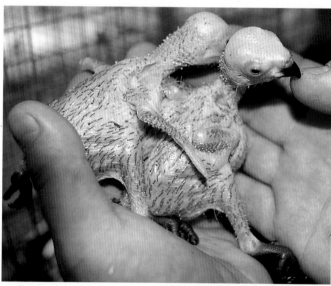

Chattering lory (*Lorius g. garrulus*) chicks.

A rare yellow mutation of *L. g. flavopalliatus*. *Photo: Patrick Tay*

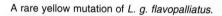

Le Perroquet Lori-Radhia.

A yellow mutation, probably *chlorocercus* or *domicellus*, depicted in a nineteenth-century book, described as Le Perroquet Lori Radhia.

Reproduced by kind permission of Memories of the East, Singapore.

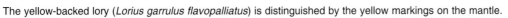

Chattering lory (*Lorius garrulus garrulus*) is a declining and endangered species.

The yellow-backed lory (*Lorius garrulus flavopalliatus*) is distinguished by the yellow markings on the mantle.

Black lory (*Chalcopsitta atra atra*) from Irian Jaya, Batanta and Salawati.

Three-month-old rajah lories (*Chalcopsitta atra insignis*) clearly showing the red underwing coverts. The white skin surrounding the eye is typical of immature birds.

Chalcopsitta atra spectabilis—the type specimen. It was almost certainly a hybrid between *insignis* and *scintillata*.   Photo: H. van Keulen

Two rajah lories, both females, aged two and a half years.

C. duyvenbodei chicks hatched at Palmitos Park, aged 15 and 16 days. The second down growing beneath the skin is clearly visible.

Chalcopsitta duyvenbodei at Jurong BirdPark.

The yellow-streaked lory (Chalcopsitta scintillata scintillata) has red underwing coverts.

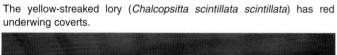

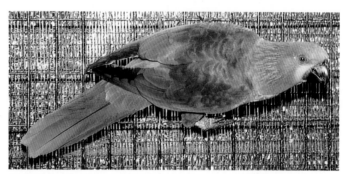

Wild-caught green mutation of Duyvenbode's lory (Chalcopsitta duyvenbodei)—an extremely rare phenomenon. It was owned by Patrick Tay in Singapore.   Photo: Armin Brockner

C. s. scintillata aged 30 days, hatched at Palmitos Park.

Chalcopsitta cardinalis chicks hatched at Loro Parque, aged 18 and 19 days.

The black-winged lory (*Eos cyanogenia*) comes from the Indonesian island of Biak.

Cardinal lory (*Chalcopsitta cardinalis*) was almost unknown in aviculture until the early 1990s.

Eos cyanogenia chicks aged ten and eleven days, from a prolific pair at San Diego Zoo.

The electric blue streaked feathers are unique to the blue-streaked lory (*Eos reticulata*).

Male *Eos reticulata* displaying to the female at Loro Parque, Tenerife.

Chicks of *Eos reticulata* hatched at Birdworld, Surrey, UK.

Photo: Roger Sweeney

Above: Immature *Eos histrio talautensis* bred by Jos Hubers in the Netherlands. *Photo: Jos Hubers*

Left: The red and blue lory (*Eos histrio talautensis*) is an endangered species with a population estimated at fewer than 2,000 birds in the mid-1990s.

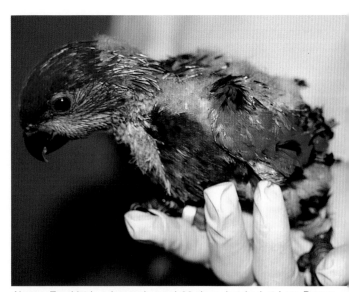

Above: *Eos histrio talautensis* aged 38 days, hatched at Loro Parque, Tenerife in 1996.

Left: Red and blue lories (*Eos histrio talautensis*) in a bird dealer's shop in Singapore in 1992.

Red lories (*Eos bornea bornea*) enjoying life in the Waterfall Aviary at Jurong BirdPark.

Eos bornea cyanonothus is recognized by its darker shade of red. This bird belongs to Gavin Zietsman in South Africa.

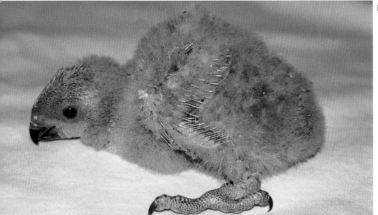

A

Development of an *Eos bornea* chick:

Figure A: 7 days old and weighing 27g;

Figure B: 25 days, and weighing 102g, the first feathers are erupting (on head and wings);

Figure C: 35 days, weighing 139g;

Figure D: 42 days, weighing 161g;

Figure E: 66 days and 161g.

(Photographs of this chick aged 7 and 15 days appear on page 107.)

B

C

D

E

The blue-eared lory (*Eos semilarvata*) is a little-known species.

The identifiction of this *Eos* lory is uncertain; it might be *Eos squamata squamata*. See page 227.

Photo: Thomas Brosset

Above: *Eos squamata riciniata* aged 41 days, hatched at Palmitos Park.

Right: Violet-necked lory (*Eos squamata riciniata*) pair at Loro Parque.

Below: Habitat of *Eos semilarvata*—the mountains of central Seram.

Photo: David Field

The Obi lory (*Eos squamata obiensis*) occurs only on the north Moluccan island of Obi.

Pseudeos fuscata aged 44 days and weighing 138g; it was fully feathered except under the wings.

Both color phases of the dusky lory (*Pseudeos fuscata*)—orange and yellow.

Trichoglossus ornatus aged 50 days, hatched at Palmitos Park.

The ornate lorikeet (*Trichoglossus ornatus*) is one of the first lorikeets to be recorded in captivity. It was kept in China during the Tang Dynasty (618–906 A.D.).

Trichoglossus h. haematodus at Lorikeet Landing, San Diego Wild Animal Park.

A

B

C

A Trichoglossus h. caeruleiceps is recognized by the extensive blue on forehead and sides of head and by the very narrow breast bars.

B Edwards' lorikeet (Trichoglossus h. capistratus) has the breast and underwing coverts yellow—and both may be marked with orange.

C T. h. djampeana is an extremely handsome bird with its unbarred breast and intense colors.

D From New Caledonia, T. h. deplanchii has more blue and less brown on the head than T. h. massena; it is also larger.

E The olive-green lorikeet (T. h. flavicans) is unique in that in the same population individuals are either olive green or midgreen. These two (on facing page) were photographed on Manus Island.

Photo: William Peckover

D

This young Forsten's lorikeet (*T. h. forsteni*) is typical of immature *haematodus* in general in having only slightly duller plumage than adults; the beak is brown.

T. h. massena is widely distributed throughout the Bismarck Archipelago, the Solomon Islands and Vanuatu.

T. h. forsteni aged about 14 days, hatched at Loro Parque.

Mitchell's lorikeet (*T. h. mitchellii*) (photographed in the collection of Gavin Zietsman), now occurs only on the island of Lombok, and is believed extinct on Bali.

Several mutations have been developed in *T. h. moluccanus*: left and right, normal plumage, second from left, olive mutation, center cinnamon and next to it, blue-breasted. Photographed in Australia.
Photo: Roland Seitre

Lutino *Trichoglossus h. moluccanus* in Australia. *Photo: Syd and Jack Smith*

The red-collared lorikeet (*T. h. rubritorquis*) from Australia is distinguished by the orange breast and nuchal band.

T. h. rosenbergii from Biak is one of the most distinctive and handsome of the subspecies of *haematodus*.

The perfect lorikeet (*T. euteles*) is the only non-Australian species to lay more than two eggs in a clutch.

Weber's lorikeet (*T. h. weberi*) is in danger of dying out in aviculture.

Trichoglossus chlorolepidotus chicks aged about one week, bred by Luch Luzzatto in South Africa.

Lutino and normal *Trichoglossus chlorolepidotus* hatched at Vogelpark Walsrode in Germany.

Photo: N. Neumann/Walsrode

Uniquely colored, the Ponapé lorikeet (*Trichoglossus rubiginosus*) comes from Pohnpei in the Caroline Islands. Photo: J. M. Kenning

The scaly-breasted lorikeet (*Trichoglossus chlorolepidotus*) is a study in green and yellow until the wings are opened to reveal orange-red underwing coverts.

Yellow and green lorikeet (*Trichoglossus f. flavoviridis*), seen here at Loro Parque, is rare in aviculture.

Trichoglossus f. flavoviridis aged about five weeks, bred by Rex Duke in South Africa.

The typical leaning-away component of *Trichoglossus* display is demonstrated by these Meyer's lorikeets (*T. flavoviridis meyeri*).

Trichoglossus flavoviridis meyeri aged 29 days, weighing 40g.

Family of Mount Apo lorikeets (*Trichoglossus johnstoniae*) at Loro Parque in 1994.

The iris lorikeet (*Trichoglossus iris*) from Timor is a threatened species which deserves more attention from aviculturists.

Trichoglossus iris aged 28 days and weighing 49g.

Pair of varied lorikeets (*Trichoglossus versicolor*), part of a successful breeding colony owned by David Judd in Victoria.

Varied lorikeets (*Trichoglossus versicolor*) aged six days. Note the sparse down indicating their tropical origin (northern Australia).

Photo: Rachel Lewis

Young *Trichoglossus versicolor* bred by David Judd.

Photo: David Judd

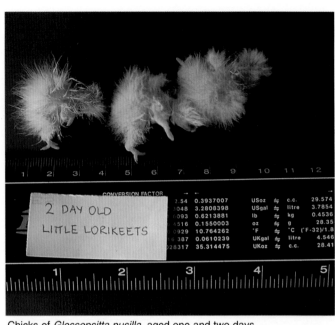

Chicks of *Glossopsitta pusilla*, aged one and two days.
Photo: Rachel Lewis

Goldie's lorikeet (*Trichoglossus goldiei*), a small species from New Guinea

Little lorikeets (*Glossopsitta pusilla*) at their nest at Swan Vale, New South Wales. *Photo: John Courtney*

Musk lorikeet (*Glossopsitta concinna*) at its nest at Swan Vale, New South Wales.

Photo: John Courtney

Musk lorikeet (*Glossopsitta concinna*) at 14 days weighing 32 g.

Photo: Alan Dear

Musk lorikeets (*Glossopsitta concinna*) aged 44 and 46 days, at Vogelpark Walsrode.

Photo: N. Neumann/Walsrode

The purple-crowned lorikeet (*Glossopsitta porphyrocephala*) is distinguished by its small size and yellow-orange forehead.

One of the most beautiful of all lories, the collared (*Vini solitarius*) from Fiji.
Photo: Ron Moat

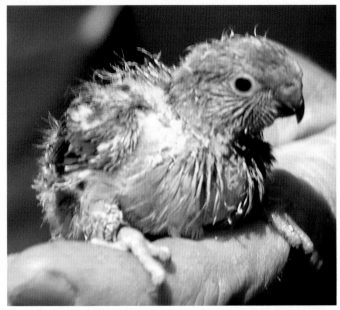

Purple-crowned lorikeet chick. *Photo: Rachel Lewis*

Vini solitarius chicks hatched at San Diego Zoo, aged four weeks. *Photo: Gwendolyn Campbell*

Bred at San Diego Zoo, the *Vini solitarius* on the left acquired melanistic plumage when it was about two years old.

Vini australis hatched at the Tongan Wildlife Centre.

Photo: Claudia Matavalea

The blue-crowned (*V. australis*) is the most widely distributed of the *Vini* species.

Rimatara or Kuhl's lory (*Vini kuhlii*) is exquisitely colored and has elongated nape feathers.

Left: The Tahiti blue lory (*Vini peruviana*) is now extinct on most of the 20 islands on which it was known to occur.

Right: Stephen's lory (*Vini stepheni*) occurs on Henderson Island in the southcentral Pacific. This is the easternmost limit of distribution of the Loriinae.

Photo: Rosie Trevelyan

Immature *Vini peruviana* are dull blue and blue gray. *Photo: R. H. Grantham*

Vini peruviana spends much time feeding and resting in coconut palms.

The ultramarine lory (*Vini ultramarina*) is the subject of a translocation program to try to save it from extinction.

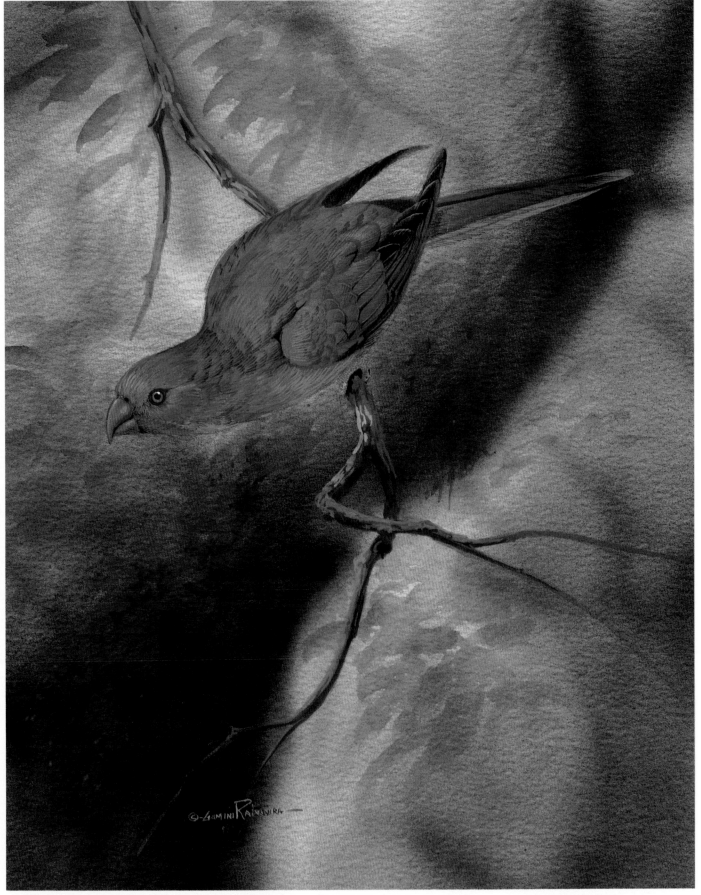

Red-chinned lorikeet (*Charmosyna rubrigularis*) is a little-known bird from New Britain and New Ireland.   *Painting: Gamini Ratnavira*

The palm lorikeet (*Charmosyna palmarum*) occurs in mountain forests on Vanuatu. *Illustration: Richard Parks; reproduced with his kind permission*

Meek's lorikeet (*Charmosyna meeki*) is found in the Solomon Islands. *Illustration: Richard Parks; reproduced with his kind permission*

The red-throated lorikeet (*Charmosyna amabilis*) is a rarely-sighted inhabitant of Fiji's forests.

Illustration: Chloe Talbot-Kelly; reproduced from Birds of Fiji, Tonga and Samoa *by kind permission of Dick Watling.*

It is not certain whether the New Caledonian lorikeet (*Charmosyna diadema*) still survives.

Illustration: Thomas Arndt; reproduced with his kind permission from Lexicon of Parrots

The Buru lorikeet (*Charmosyna toxopei*) is a little-known bird which has rarely been observed.

Illustration: Thomas Arndt; reproduced with his kind permission from Lexicon of Parrots

Striated chick aged 26 days, bred by Ruediger Neff. *Photo: Ruediger Neff*

The striated lorikeet (*Charmosyna multistriata*) is notable for the unusual bill coloration.
Photo: Ruediger Neff

Sometimes called pygmy lorikeet, *Charmosyna wilhelmina* weighs only 20 g.   *Photo: Gerhard Hofmann*

Male red-spotted lorikeets bred by Carol Day.

Photo: Carol Day

Female *Charmosyna rubronotata* aged eight weeks, two days before fledging.

Photo: Carol Day

Red-spotted lorikeet (*Charmosyna rubronotata*) female above, male below.

Photo: Ron and Val Moat

Red-flanked female (*Charmosyna placentis placentis*) with newly fledged youngster at Palmitos Park.

Charmosyna p. placentis aged 22 and 24 days, weighing 40g and 41g with the crop full. Their down has been plucked by the parents.

Duchess lorikeet (*Charmosyna margarethae*) from the Solomon Islands.
Illustration: Richard Parks; reproduced with his kind permission.

Charmosyna p. pulchella aged five and seven days; note the dark lines of second down growing under the skin in the eldest chick.

Charmosyna pulchella pulchella, female with newly fledged youngster, at Palmitos Park.

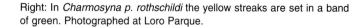

Above: Fairy lorikeet (*Charmosyna p. pulchella*) chicks at Palmitos Park. Male (right) and female can be distinguished by the color of the flank feathers, red and yellow respectively.

Right: In *Charmosyna p. rothschildi* the yellow streaks are set in a band of green. Photographed at Loro Parque.

One of the author's favorite pairs: Josephine's lorikeets on the day their young (a male and a female) left the nest. Note the light beak of the adult female.

Above: The same Josephine's chicks at 37 and 35 days, weighing 73g and 65g, both with full crops.

Left: Two Josephine's chicks and (right) one Stella's (*goliathina*) aged 27, 25 and 27 days, weighing 52 g, 42 g and 78 g.

Four-year-old melanistic male *goliathina*, hatched at Palmitos Park. It is just possible to observe the elongated primary.

Red-phase Stella's lorikeet (*Charmosyna papou goliathina*) owned by Leif Rasmussen, in Denmark.

Two red female *goliathina*, aged eight weeks. Note the short tail.

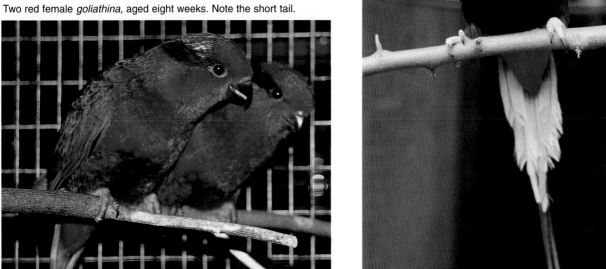

Oreopsittacus a. grandis photographed at Mount Scratchley.

Photo: William Peckover

O. a. arfaki female aged 47 days. A male of the same age has the white markings purer white.

Photo: Ruediger Neff

Breeding pair of whiskered lorikeets (*Oreopsittacus a. arfaki*) in the care of Fred Barnicoat in South Africa.

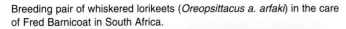

Note the dense down of this 23-day-old *arfaki* chick. Photo: Ruediger Neff

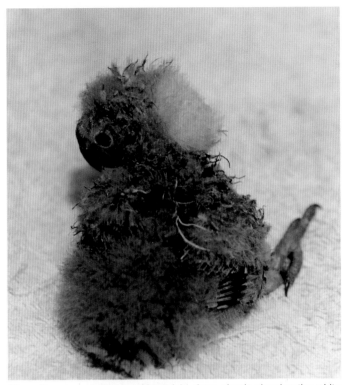

Neopsittacus musschenbroeki aged 31 days, clearly showing the white patch of down on the nape.

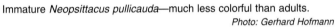

Immature *Neopsittacus pullicauda*—much less colorful than adults.
Photo: Gerhard Hofmann

Male Musschenbroek's lorikeet (*Neopsittacus musschenbroeki*) from the highlands of New Guinea.

Emerald lorikeet (*Neopsittacus pullicauda pullicauda*) from southeastern New Guinea. *Photo: Gerhard Hofmann*

Above: Forest at about 2,400 m (7,900 ft) at Ambua, in the Southern Highlands of New Guinea, habitat of Stella's (*Charmosyna p. goliathina*) and Musschenbroek's (*Neopsittacus musschenbroeki*) lorikeets.

Below: Habitat of the red lory (*Eos bornea rothschildi*) on Seram.   *Photo: Armin Brockner*

The protection of lories and lory habitat is the hope of the author. Only through understanding these birds and their need for safe and protected habitats will wise conservation practices come about.

be easier birds for peregrines to catch!" In the Tabubil region of Papua New Guinea, Phil Gregory has also witnessed predation by a peregrine falcon. He saw one catch a green-naped lorikeet in flight. The bird was carried off, screeching noisily. He commented that direct predation of lorikeets is seldom witnessed (Gregory pers. comm. 1996).

How common it is for a raptor to take a lorikeet is unknown. Many ornithologists who have spent hours in the field have never seen this occur. Josephine's and Papuan lorikeet females have the rump brilliant daffodil yellow, which is extremely conspicuous in flight. This suggests that predation by raptors in the highlands of New Guinea does not have a serious impact on the populations of these lorikeets.

Protein, Dietary

Little scientific research has been carried out on the dietary requirements of captive lories. The level of protein needed for maintenance and for breeding is unknown; in any case, it may vary with the species. American aviculturists Leann Collins and David Sefton suggested that the "proper ratio" is 25 percent. This is based on the assumption that eucalyptus pollen is "the lory's primary food source" and that the protein content of the pollen is 24–26 percent (Collins and Sefton 1994). This argument is flawed for several reasons. Eucalyptus pollen is the main food source only for some species and at a certain time of the year, i.e., when they are breeding. It virtually assumes that pollen forms 100 percent of the diet, because probably few of the other food items taken by lories would have a higher protein content. (When not breeding, they have less need to seek high protein foods.)

Using this assumption they added a 90 percent bodybuilder protein powder to the diet of their red lories "to achieve the proper ratio of 25%." They described the results as "dramatic. Within days, there was a change in beak color, from an opaque yellow to a healthy orange red...The increased protein also produced an immediate molt (the birds had not molted in a year and a half.) The improved feather color and condition were just as dramatic." The previous diet consisted of an "expensive lory nectar." (It was not stated whether this was formulated by avian nutritionists or by, for example, a breeder who had marketed his own formula.)

It is worth commenting on the facts that the red lories' beak color had been pale and that they had not molted for 1½ years. Both facts indicate that they were either suffering from dietary deficiencies or that there was a health problem among them. As they responded so quickly to a change of diet, a dietary deficiency is more likely. (I once saw a young yellow-backed lory whose plumage was mainly pink. However, the new feathers which were appearing were red, as the diet had been changed to a more nutritious one.) The new diet offered to the lories of Collins and Sefton consisted of a high protein, nonmedicated poultry food (gamebird starter), to which was added cooked sweet potatoes, mashed to resemble a thick soup. A grain mixture consisting of equal parts of brown rice, pearl barley, wheat and oats was cooked in a rice cooker, then feed corn was added after being cooked in a pressure cooker. The combined mixtures (the proportions of each were not given) apparently gave 24–26 percent crude protein. Blood profiles were monitored and "remained normal." (Over what time period was not stated.) Fruit was given "periodically" at the end of the day but consumption was controlled because "the babies' weights are lower if fruit is given while being fed by the parents."

This is true because fruits have a very low protein content (but, in my opinion, this is no reason to withhold fruit from pairs rearing young). On this diet their pairs have produced eight young, or as many as 12, per year. Chicks are fed by the parents for the first two or three weeks only. Collins and Sefton offered a warning: "As a caution, I must add that the increased protein is implicated as a contributing factor in the death of several lories."

I believe that this diet, if widely used, has the potential to be extremely detrimental to the future of lory breeding. There are several points which should be made:

1. The food described, made from poultry food and cooked cereals, has nothing in common with pollen except the percentage of protein of some types of pollen. This is almost irrelevant because the total protein is not necessarily a measure of what the bird utilizes. Proteins are composed of various amino acids which form a chain. Ten of these are termed essential because they cannot be produced in the body. If a certain protein has a lower proportion of an essential amino acid than is

ideal, this amino acid is termed limiting. The presence of a limiting amino acid restricts the use of the entire quantity of amino acids because its depletion will eventually hinder protein synthesis in the body. As dietary amino acids are consolidated into the specific proteins required by the body, protein synthesis ceases when the limiting amino acid is exhausted. The remaining amino acids will be used as an energy source rather than for tissue growth, immune function and metabolic compounds. Therefore, a diet containing 25 percent crude protein but severely deficient in amino acids would, in theory, be of no more value than a diet consisting of 12 percent of very high quality protein (Brue 1994). A "homemade" mixture is more likely to be deficient in essential amino acids than one formulated by an avian nutritionist.

2. A food containing 20 percent protein of good quality would undoubtedly assist in the production of strong, healthy chicks. But to maintain adults on such a diet year-round would almost certainly prove to be overstimulating and result in continual breeding. Indeed, Collins and Sefton state that their red lories produced up to 12 young per year. For one pair of lories to produce 12 young in a single year and feed them for two or three weeks would require six clutches of two, and would occupy them for the entire 52 weeks of the year. To allow lories to breed intensively, without a break, will shorten their lives and make them susceptible to death from infections which would not debilitate other birds. Also, their breeding span, in terms of years, would be shortened. What would happen to them when they stopped breeding? Would they be sold as breeding pairs to innocent purchasers? I believe that lories should not be permitted to rear more than three nests of young annually, or their stamina will suffer and the quality of young reared may deteriorate.

3. By suggesting that red lories should be offered a diet containing 25 percent protein, Collins and Sefton make no allowance for the fact that the maintenance diet, that of nonbreeding birds, should be lower in protein—probably half the level suggested. They warn that the increased protein is implicated in the deaths of several lories. Their diet, if widely followed, could result in the death of hundreds of lories. Excess protein in the food cannot be excreted and is stored in the kidneys in the form of urates. These build up, causing gout (resulting in lameness). As the urates build up further, they can be deposited around other organs, eventually resulting in the bird's death. Therefore, an excessive level of protein in the diet is highly dangerous and frequently fatal.

Psittacine Beak and Feather Disease (PBFD)

This disease is endemic in wild parrot populations in Australia, especially in cockatoos and rainbow lorikeets. In captive parrots it has been found worldwide, in parrots of almost every genera. The symptoms may vary slightly according to the species; for example, in cockatoos excessive beak growth often occurs. This does not happen in lorikeets, in which the main symptom is a varying degree of loss of plumage (from total loss to failure of tail and flight feathers). Alternatively, the feathers may be deformed and weak. Affected birds are highly susceptible to disease as the immune system is damaged. This disease has long been known in budgerigars under the name of French molt. It is caused by a circovirus. There is no cure. The virus can be transmitted in the egg, thus hand-rearing from the egg does not prevent its spread. It can also be transmitted from dust in rooms where PBFD positive birds are kept. In other words an infected bird places at risk all other birds in the vicinity. Euthanasia is therefore advised, except in the case of a single, cherished pet bird, where high standards of hygiene must be practiced.

Fortunately, it is rare in lorikeets outside of Australia. In that country, and elsewhere, testing can be carried out to confirm if the disease is present. It is a DNA test, performed on a drop of blood. In Australia, a haemagglutination (HA) and a haemagglutination inhibition (HI) test are used to diagnose PBFD on blood and feather samples. An inactivated vaccine is currently being developed there. In Queensland, veterinarian Dr. Adrian Gallagher sees many PBFD-infected rainbow and scaly-breasted lorikeets in his practice. The majority of these birds test positive during active disease from feather tissue but can test negative in chronic cases. The virus does not appear to remain in feather tissue in chronic cases but it can be recovered from other tissue. He believes that the virus affecting lorikeets may be a distinct form of PBFD or another related virus—but there is no evidence of this yet. Some of the affected juvenile birds spontaneously regrow the tail and flight feathers within two months and appear grossly normal, but these birds are thought to be carriers of the disease. The previous thinking was that they were immune and had eliminated the disease from the body. This has been disproved by testing. Such birds are a threat to the wild population when they are released into the wild by well-meaning people.

Juvenile birds affected in a different way develop progressive PBFD, as seen in other avian species. These birds can live for two or three years but eventually succumb to the virus or to a complicating disease (Gallagher pers. comm. 1996).

Quality of Life (Aviculture)

Few birds are so naturally endowed with a *joie de vivre* as lories. They are playful, acrobatic, inquisitive and intelligent. By keeping them in cages (enclosures of, say, 2 m [around 6 ft.] long and smaller), we are depriving them of the opportunity to fly and play. We are also depriving ourselves of the opportunity of seeing them at their best. I would obtain no pleasure in keeping birds in cages of this size, with the exception of Meyer's and Goldie's lorikeets who do well in them. Too often these days the maxim of bird keepers, on the subject of accommodation for their birds, seems to be to crowd as many cages as possible into the available space. Little thought is given to the comfort of the occupants. I keep only a small number of lories; the larger species have outdoor flights 5 m (15 ft.) long with 2 m (6 ft.) long indoor sections. It would be possible to house three times the number of pairs in the space allocated—but this would give me no satisfaction.

Lory keepers should also consider the importance of keeping their birds in close proximity to their own house. If you keep birds for pleasure it is pointless keeping them in a lock-up building elsewhere because the zoning regulations in your area prohibit keeping them on your property. Either forget parrot-keeping or move! It is not in the interest of the birds' welfare to house them where you cannot see or hear them from where you live. They need constant attention in order to protect them from disasters such as fire or other damage to the building, and to ensure you are on hand if one has an accident or becomes sick.

My birds are fed twice daily—or thrice when they have young. It would be possible to feed them only once daily, but they look forward to mealtimes, just as we do, to break up the routine of the day. Their feeding is time-consuming and expensive; items fed, apart from the basic nectar mixture, are varied as much as possible. Parrots enjoy their food as much as we do. That is one reason why I refuse to feed my nonnectar-feeding parrots on pellets as 100 percent of their diet. One person who appreciates this fact very well is zoo-owner extraordinaire John Aspinall whose two zoos in Kent (U.K.) are renowned for their breeding successes with gorillas and other rare animals. In *The Passion of John Aspinall*, the biographer described Aspinall's view of this subject:

> ...if you should find yourself imprisoned in a grand hotel in the south of France, for example, with all comforts, but given exactly the same food every day, a measured lump of bread and a bowl of soup, then you would suffer from the restrictions placed upon your enjoyment of food. Incarceration has already reduced your behavioural opportunities drastically, so that the food you eat becomes more important, not less, than it was in freedom. It is the one aspect of your life which may still be enlivened by variety.

Although Aspinall was referring to the importance of a varied diet for gorillas and other mammals, his philosophy could have been written for parrots, so apt is it. A good aviculturist is one who can put himself in the place of his or her birds and ask: "What else can I do to make my birds happy and healthy?" The knowledge and the practical components of bird keeping are so readily available today that there is absolutely no excuse for not keeping birds well. But there is something else that is essential: commitment. Without it, all the money and knowledge in the world are worth nothing.

Reproductive Span

The reproductive span of a bird covers the time it reaches sexual maturity until it is no longer capable of fertilizing or producing fertile eggs. Much information is available regarding the age at which lories become sexually mature, but little has been recorded about the age at which they are no longer capable of breeding. In part, this is because the age of many breeding birds is unknown; but it is also due in part to the fact that lory breeding has such a short history. In any case, this lack of data is compounded by the fact that far too few aviculturists have recorded details of reproductive span.

Sexual maturity

The age at sexual maturity varies according to the species. Certain *Trichoglossus* and *Charmosyna* species appear to mature earlier than the larger lories. For example, I saw a *C. placentis* which hatched on June 12, 1994, attempting to mate with a female on January 20, 1995; this bird was only six months old and not yet in full adult plumage—some yellow feathers remained in his ear coverts. Leif Rasmussen, a very experienced breeder of this species, found that it usually matures at nine months. Carol Day's red-spotted lorikeets start to breed at about 13 months. A parent-reared melanistic female Stella's lorikeet hatched at Palmitos Park on August 12, 1991, hatched her first chick on May 24, 1993. At Loro Parque, Tenerife, a Josephine's lorikeet hatched in January 1995, fledged her first chick in August 1996, thus she was 16 months old when she laid (Sweeney pers. comm. 1996). However, a female Josephine's in my care, hatched at Palmitos Park in March 1990, laid her first egg when she was 12 months old and hatched her first chick at 15½ months.

Other small species mature very early, especially the whiskered lorikeet. Females lay as early as 7–9 months old and may even breed with the red feathers of the immature plumage still evident on the forehead. Males of this age may be infertile and should be held back until the age of 12 months (Neff pers. comm. 1996). At 6–7 months old most whiskered have finished molting and both sexes then start to display—bobbing their heads and uttering a courtship song. Little lorikeets are sexually mature at about 12 months. At San Diego Zoo, two Tahiti blue lories laid first when they were two years old and, also at San Diego Zoo, two blue-crowned lories produced their first chick when the male was 18 months old and the female 22 months.

Ken Russell's perfect lorikeets, the first recorded breeding of this species in 1970, hatched chicks when they were 9 months old. G. van Dooren's perfect first laid when she was 15 months old. Some subspecies of *Trichoglossus haematodus* mature early. A female Weber's hatched in May, 1984, hatched her first chick in April, 1985. On the other hand, in Stan Sindel's experience, red-collared lorikeets (*T. h. rubritorquis*) are not sexually mature until the age of 18 to 24 months. My female Meyer's hatched in February, 1976, laid for the first time in February, 1978, and hatched a chick in the following month. Two pairs of Mt. Apo lorikeets produced infertile eggs in their first clutches and fertile eggs in their second clutches when they were 33 and 35 months old (Van Oosten pers. comm. 1996). The female of a young pair of iris lorikeets owned by Rosemary Wiseman laid when only 12 months old; my own females do not lay until 2 years. Rachel Lewis's 2 pairs of varied lorikeets were both just under 2 years when the first pair laid and reared young. The second pair, slightly younger, had not made a breeding attempt. A pair of purple-crowned lorikeets in the aviaries of Peter Odekerken started to breed when the female was seven months old. One of the three eggs was fertile and the parents made an excellent job of rearing the chick. A red lory (*Eos bornea*) hand-reared by the same lory enthusiast laid for the first time when she was 14 months old; the eggs were fertile (Odekerken verbal comm. 1996).

Less has been recorded about the larger lories (size of *Eos* and above). Some start to breed at 3 years; others, at 4 years; yet others mature as early as 2 years. Bobbie Meyer of California told me that a pair of Duyvenbode's, both hand-reared, produced their first chick when they were 2 1/2 years old. On the other hand, in South Africa Arthur Kemp has 3 parent-raised pairs. All were 4 years old, he told me, when the females laid for the first time. Louise Ethridge of North Carolina related to me how one captive-bred black lory hen, whose hatch date she knew, laid for the first time just before she was 2 years old. She was kept alone, awaiting a male, and by the time she was 2½ years old, had laid eight eggs. Another captive-bred black lory belonging to Mrs. Ethridge was said to be 2 years old when she laid fertile eggs; she reared young from the third or fourth clutch.

Age limits recorded for breeding

A question which many breeders will be asking

within a few years (when they are having difficulty in replacing some of their breeding pairs) is: At what age does a lory cease to breed? In Australia, Gordon Dosser bred from a female rainbow lorikeet until she was 25 years old, and from a female red-collared lorikeet until she was 24 years old; both birds were still alive a year later. Among my iris lorikeets, a female which died at 21 years ceased to lay after she was 18 years old (although she had not been fertile for some years) and a 12-year-old male is still fertilizing eggs. One would expect the life-span to be longest in *Lorius* and *Chalcopsitta* species (as they are the largest), and that males might breed until they are about 30 years old. Females, as they usually play a more demanding role in reproduction, have a shorter breeding span. Marc Valentine obtained the male of his breeding pair of *Lorius g. flavopalliatus* in 1976. In 1996, this bird was still breeding and "looked like a three year old" (Valentine verbal comm.). Arthur Kemp's Edwards' lorikeets were obtained in 1974. In 1996 they were still breeding (usually producing 2 females in each clutch), but the eggs were removed as they had started to pluck their young 5 or 6 years previously. Because the eggs were removed, the female laid 5 or 6 clutches annually. Also in South Africa, Luch Luzzatto had, in 1996, a male red lory which was obtained 24 or 25 years previously. It was still breeding. In one pair of his scaly-breasted lorikeets, the male was 16 or 17 years old and the female had been laying for 12 years. A male dusky lory was 17 years old and still producing young—4 in the current year. Another 17-year-old male dusky, hand-reared by Jay Kapac in 1977, was still breeding in 1996. In California, Dick Schroeder had a male Stella's, obtained as a wild-caught adult, that died after 16 years in his care. It was fertilizing eggs until a year before it died. It may have been quite old as it had started to lose its head feathers. These figures reflect the time at which lories not previously available started to breed in aviaries, rather than being noteworthy examples of longevity. The same birds could still be breeding in 2006.

Longevity

Much remains to be learned on this aspect, as few records have been kept and/or published, either for wild or captive birds. A wild rainbow lorikeet was found dead near the banding site, almost 11 years after it was ringed (Anon. 1978). Examining the records of one species in a collection where meticulous records are kept proved interesting. The taxon report sheets kindly provided by San Diego Zoo for the Tahiti blue lory were the source. I divided the birds into males and females, captive-bred and wild-caught. I expected most of those over 10 years old to be males; this was indeed the case. Because of the aggressive nature of males, I suspect that some females died of stress without injury (see page 338). Of 50 Tahiti blue lories which, by early 1995, had survived to adulthood, 10 exceeded 10 years. The figures below indicate age in years or years and months; w/c indicates wild-caught and c/b captive-bred.

Table 1.11   Tahiti Blue Lory Longevity

	Males	Females
w/c	14	16
w/c	11	
c/b	15.8	12.9 *
c/b	14.11 *	
c/b	14.11 *	
c/b	11.8	
c/b	10.6 **	
c/b	12.5	

* still alive
** hatched in the author's collection. All other captive-bred birds hatched at San Diego Zoo.

Ringing (Banding)

All chicks should be closed ringed for identification purposes. This is especially true of the species which are rarer in aviculture as it helps to prevent related birds being paired together, assuming that the ring carries the breeder's initials or some other significant legend. It also proves that the bird was captive-bred. I have ringed lory chicks over a period of 10 years and have never known a parent bird to object to a ring or even show any sign that it has noticed it. However, all the breeding birds were ringed although I doubt whether this was significant.

Chicks are ringed at about, or just before, the time their eyes open; their age at ringing will depend on species, diet and the capability of parents or hand-rearer; of course, disease could slow their growth. Generally, lories are ringed at about 16 to 18 days; the age has varied between 9 and 25 days, using the ring sizes shown below. For example, I have ringed black-capped lories with 8 mm or 8.5 mm rings at between 13 and 25 days. The chicks ringed at early ages were reared by parents who fed well; the later ages were with parents who did not feed so well or, in the case of 25 days, one which was reared by rajah lories. One chick which was missed was ringed at 24 days with a 9.5 mm ring. This is not desirable as, although it is not possible to remove the ring when the bird is full grown (except, of course, by cutting), it is a loose fit. This is a hazard as it could cause a bird's leg to become trapped on wire or branch. The correct size is a ring which leaves just a little space between leg and ring. It must not fit tightly. In some instances, a ring ½ mm smaller than the sizes listed might be used, and ringing would then take place about two days earlier. Note that single chicks often grow faster. I ringed a single Meyer's lorikeet (with correct sized ring) at only nine days.

The method is to slip the ring over the three outer toes and gently pull the fourth toe through using a wooden toothpick. Record all ring numbers at the time of ringing in a record book and/or on a computer. Incorrectly recorded or mislaid ring numbers can cause a lot of time to be wasted later on.

Table 1.12   Recommended Lory Ring Sizes

Chalcopsitta	7.5 mm
Eos	7 mm
Pseudeos	7 mm
small Trichoglossus	5.5 mm (up to euteles)
T. chlorolepidotus	6 mm
other Trichoglossus	6.5 mm
Lorius chlorocercus	7.5 mm
other Lorius	8 mm or 8.5 mm
Glossopsitta concinna	5.5 mm
other Glossopsitta	5 mm
small Charmosyna	4.5 mm (up to multistriata)
other Charmosyna	5.5 mm
Neopsittacus	5.5 mm and 4.5 mm
Vini	5.5 mm
Oreopsittacus	3.5 mm (three point five)

In Canada, where a limited number of lory species are available, the Avicultural Advancement Council of Canada recommends the following sizes: blue-streaked, violet-necked, dusky and green-naped, 7.04 mm (size R); red lory, 7.57 mm (size S); large lories, including chattering, 8.69 mm (size T).

Roosting Habits

Many of the observations of the roosting habits of lorikeets relate to *Trichoglossus haematodus*. This is because they roost in large, noisy and conspicuous groups, and because many roosting sites are in the vicinity of towns. I observed one such roost on November 29, 1994, at Port Augusta, 60 km (37 mi.) north of Cairns, in Queensland. This is a resort town with a marina, in front of which there is a complex of shops and restaurants. Dusk was near as I reached the car park with a couple of friends. Not very high overhead dozens of rainbow lorikeets (*T. h. moluccanus*) were flying in twos and threes, landing in the crowns of palm trees, then taking off again. We found a restaurant overlooking the marina and unexpectedly seated ourselves for the best entertainment I ever watched over a meal. The road along one side of the marina was lined with 20 palm trees, into which rainbow lorikeets were streaming in dozens. When we sat down I could count about 30 birds in each of the nearest trees and, as we ate, at least 200 more came in, bringing the total to a minimum of 800 lorikeets. Their incessant chattering was kept up until one hour or more after sunset, as birds constantly jostled for position, occasionally taking off to circle around, returning almost at once to the trees. I enjoyed the chicken satay and rice, but it was the lorikeets which made the meal so memorable!

Peter Odekerken (1993a) recalls a roosting site of rainbow lorikeets in Townsville, Australia's largest tropical city, 300 km (186 mi.) south of Cairns. It was next to the Travelodge Motel. The lorikeets were so noisy during the night that it was necessary to close the doors and windows in order to hear the television!

In New Guinea some roosts of green-naped lorikeets (*T. h. haematodus*) can number up to a few thousand birds. Coates (1985) described how "birds converge to roost in parties and flocks from all directions but do not fly into the actual roosting tree until close to nightfall. All members call excitedly and an incredible sound emanates from a large roost. It becomes quiet when all birds have settled down to sleep." In the Port Moresby suburb of Gordon Estate, this species roosts only in tall eucalypts, according to Coates. If disturbed during the night, they settle in other eucalypts in the vicinity, sometimes quite low ones which are unsuitable, rather than resort to other types of trees. When roosting during the day, they commonly use the large, heavily foliaged rain trees. There must be a good reason why they do not use these trees at night...

In 1984, the roosting activities of green-naped lorikeets at Wau, Papua New Guinea, were studied. About 450 birds used the roost there. Evening flights to the roost continued for over two hours until 6:30 P.M., and morning flights lasted half and hour until

6:00 A.M. The actual roosting tree did not remain the same; from time to time a different tree was used. This was interpreted as antipredator behavior, as was the fact that the roosting tree had vertical branches with no contact with neighboring trees (Utschick and Brandl 1989).

Ornithological literature contains a number of references to communal roosting sites, but I am aware of none regarding the roosting habits of a single pair. Once again I am indebted to John Courtney for his meticulously recorded observations. In June 1992 (June 1 is officially the first day of winter in Australia) he kept watch to discover how much time little lorikeets spend at their roosting sites in mid-winter. He watched them at their nests, where they roost for most of the year. His field diary reads as follows:

June 6, 1992, 4:30 P.M. I arrive at little lorikeet nest sites, Wet Gully, Swan Vale. At 4:49 P.M. the "forked tree" nest site pair arrived and landed four to five metres above their nest. At 4:50 P.M. they descended in hovering flight and entered the nest for the night. At 4:55 P.M. a second pair arrived at and entered their nest in a giant Blakely's red gum 40 m to the west. The sun set at about 5:15 P.M.

June 7, 4:00 A.M. I arose and set off for the little lorikeet nest sites, to note when they emerge. At 5:15 A.M. it was still pitch dark (no moon). At 6:10 A.M. it was light enough to just discern at a distance of 20 m the nest entrance of the forked tree nest. At 6:27 A.M. it was possible to see quite well; small passerines were moving about, and a little lorikeet's head was seen at the nest entrance. At 6:28 A.M. it launched itself straight out of the nest, flying swiftly to the northeast, and was followed 2 m behind by the other member of the pair. They were lost to view in the distance. My attention then turned to the nest in the giant Blakely's red gum, which was more shaded, with less light falling on the entrance. At 6:33 A.M. a little lorikeet's head was seen but, on seeing me, the bird disappeared inside. At 6:37 A.M. a little lorikeet was seen to emerge from a third nest nearby, in the tall leaning Blakely's red gum. It hung there, head downwards. (Yesterday this third pair had entered earlier than the second pair, while my attention was on the first pair.) At 6:38 A.M. both pairs exploded off together.

The first pair had spent 13 hours 38 minutes in their nest during the night. The second pair had spent 13 hours 43 minutes in their nest, possibly delayed slightly by my presence. The third pair had spent longer than 13 hours 43 minutes, because they had already entered their nest before I took up position the previous evening. At 6:56 A.M. the tip of the leading edge of the sun became visible on the mountainous horizon of the northeast. There was a light covering of frost on the ground. At 7:25 A.M., when I reached home, about 1 km north, the temperature read -2.1°C (28.2°F).

June 21, 1992 (the shortest day of the year). The forked tree nest pair entered at 4:35 P.M. and 4:36 P.M. The tall leaning Blakely's red gum pair entered at 4:45 P.M. and 4:46 P.M. At 4:50 P.M. a pair of little lorikeets arrived and flew noisily around, disturbing the pair that had already entered, causing them to emerge again. The four birds flew around together, then the pair that had come out went back into their nest at 4:51 P.M. The pair that had caused the disturbance flew to their nest in the giant Blakely's red gum at 4:53 P.M. At 5:15 P.M. there was still enough light for lorikeets to fly around, but at 5:30 P.M. it was deep dusk. In conclusion, in winter little lorikeets spend longer in the nest than they do foraging. They did, however, make absolute maximum use of the daylight hours available.

Sexual Dimorphism

The majority of lories are monomorphic, i.e., appearance of male and female are similar. But there are subtle differences which the practiced eye can detect (or imagine it detects!). Rutgers (1965) wrote that the female ornate "may be slightly smaller and with a more ladylike expression"! Perhaps what he meant to infer was that she looks more feminine—and this is true of most female lories. As a general rule they appear slightly more slender and weigh a little less than the male of the species. The width of the upper mandible is narrower, the frontal rise over the forehead is less pronounced and the overall size is slightly smaller. This is usually quite noticeable in young just before they leave the nest—if there are two for comparison.

Another point to observe is the stance on the perch: males usually stand taller, whereas females have more of a tendency to squat on the perch. In a few species, such as the varied, the brighter plumage of the male is an indication of sex and in another, the little, very mature males are usually, but not always, more colorful than the females. In a number of species, such as the yellow-streaked, it would be easy to mistake the brighter plumage of a certain subspecies (in this case *rubrifrons*) for sexual dimorphism, if presented with one bird of each subspecies.

The table below shows how many species in each genus are sexually dimorphic.

Lory Sexual Dimorphism

Genus	Number of Species	Number sexually Dimorphic (pronounced)	Unknown
Chalcopsitta	4	none	
Eos	6	none	
Pseudeos	1	none	
Trichoglossus	10	none	
Glossopsitta	3	none	
Neopsittacus	2	none	
Lorius	6	none	
Vini	6	none	
Charmosyna	14	8	2
Oreopsittacus	1	1	
Total	53	9	2

It is interesting to note that sexual dimorphism is a pronounced feature of only two genera. In the genus *Charmosyna*, the largest of the lorikeets, more than half the species are sexually dimorphic but in some the females are more colorful than the males.

There is another interesting fact of dimorphism in nest feather of the species in which the female is more colorful (i.e., she has the back and rump yellow or greenish). Most individuals can be sexed in the nest by the rump color; however, there are exceptions and rarely the sex is not certain until the first molt. This has been recorded in *josefinae*, *papou* and *pulchella*.

Sexual Maturity, see **Reproductive Span**

Small Island Endemics, Captive Breeding of

Difficulties are often encountered in captive breeding certain lory species from small islands. For example, Mitchell's lorikeet is a subspecies of one of the most free-breeding of all lories, yet attempts to breed it usually yield poor results. (See **Mitchell's Lorikeet,** General.) This is not always the case. Jan van Oosten in the U.S.A. had a very prolific pair of wild-caught Mitchell's. Some years later when trying to obtain unrelated stock, he re-

peatedly found that available birds all carried his ring. Obviously, inbreeding was occurring—and this compounds another problem often encountered when trying to maintain species in aviculture, that of a small founder population, often with insufficient genetic diversity to sustain a captive-breeding program. This underscores the necessity of incorporating as many of the original wild-caught birds, which are less likely to be closely related, into the breeding population. It emphasizes the importance of ringing all lories in order to distinguish the available bloodlines, thereby reducing the chance of weakening the gene pool through inadvertent inbreeding. For this reason all wild-caught individuals of species imported in limited quantities should be sold by the importer only to breeders, and not directly into the pet trade.

During several years of breeding Tahiti blue lories, I encountered various problems—extreme aggressiveness of males, enormous susceptibility to stress—which I have not experienced in any other lory. (See **Tahiti Blue Lory**, General.) My theory is that species (or subspecies) endemic to very small islands had small founder populations when first colonizing the islands, and are already highly inbred. This lack of diversity in their genetic make-up may account for various problems encountered by aviculturists. This is a warning! One must not assume that just because the green-naped lorikeet (*Trichoglossus h. haematodus*), for example, is an extremely prolific breeder in aviaries, the same will apply to all of its subspecies, such as Mitchell's lorikeet (*T. h. mitchellii*) from the island of Lombok, and Bali (where it is probably extinct). It would be easy to reach the conclusion that endangered lory species could easily be saved by captive breeding because, generally speaking, lories are easy to breed. Some might be, but—as has already been found—others present problems seldom met with in species or subspecies with a large geographical range.

Societies and Groups

In Europe, Australia and North America there are several societies for lory breeders, some of which publish their own newsletters. The societies known are listed below. However, as officials seldom spend more than two or three years in office, names and addresses are not given. Contacting the editor of the leading avicultural magazine of the appropriate country is a good way to obtain that information.

International Lory Societies

Australia	The Grass Parrot and Lorikeet Society of Australia
Belgium	Belgische Vereniging van Loris en Loricule
Canada	Canadian Lory Group
Denmark	Dansk Lori Klub
France	Lori Club de France
U.K.	The Lory Group (of the Parrot Society)
U.S.A.	American Lory Society, International Loriinae Society

Systematics

Before I try to offer some explanation regarding the classification of the lorikeets, the words of J. D. Macdonald (*Birds of Australia*, published in 1973) should be borne in mind:

> The notion that each kind of bird was specially created, and therefore a true entity, is not now widely held although perhaps tacitly accepted by those who have had no reason to give the matter thought. It is important to note that scientific names are affected by changing views on relationships. Contrary to what Linnaeus may have intended, they are rather loose pegs, to the exasperation of naturalists who want to hang their information on something stable.

Taken as a group, this is certainly true of the lories. Some species have been assigned to five or six genera in 150 years and, more importantly, to two or three different genera since 1937, when Peters' *Check List of Birds of the World* was published. I will try to explain the difficulty in assigning certain small lorikeets to any genus. Lorikeets evolved comparatively recently and, unlike the older groups of parrots, the relationships of the various species is not yet clear. The problem may be resolved before long with the help of DNA analysis. Research of this nature has not yet been carried out on all lory species.

The fact is, especially with small lories, that although they are sufficiently distinct to separate at species level, the distinctions between groups of species seem too trifling to place them in different genera if the species apparently most closely related are compared. To try to clarify this, imagine that a skin of each species of small lorikeet, excluding the *Vini* and *Charmosyna* (which form fairly obvious groups) were laid on a table in front of someone who knew all these species in life. (No such person exists.)

This person was then asked to arrange them in order of relatedness, then to place in groups those believed to belong in the same genus. Suppose that he or she made a straight line of skins, stating that each one was closely related to the next. An observer might then ask: "Does this mean that they all belong in the same genus?" The answer would be "No." If the first and last skin were placed side by side, along with considerations of the behavior of the living bird, it would seem obvious the two species would not belong in the same genus, because they are very different in appearance. The problem, then, is where to divide the genera.

Smith (in prep.) states: "Systematics arranges species in such a manner as to show relationship. The traditional way of doing this was to sift through museum skins and, whenever possible, supplement this knowledge by comparing anatomy and bones. Use is now made of behavior, differences in the proteins of egg-white or muscle, changes in DNA and, particularly when trying to determine early origins, distribution."

The problems involved in trying to decide in which genera certain species usually classified as *Trichoglossus* or *Glossopsitta* should be placed are described in the species accounts. Yet protein electrophoresis (Christidis et al. 1991) shows that the members of these genera (and of *Psitteuteles*, for those who recognize this genus) are so close as to be virtually indiscernible. Indeed, the same study showed that the genetic distances between these genera and between *Lorius*, *Neopsittacus* and *Oreopsittacus* are so slight that all could actually be members of one genus! It is hard to imagine placing the black-capped lory, for example, in the same genus as the whiskered lorikeet, but judged solely on protein electrophoresis, this is indicated! Dr. R. Schodde, in a letter to John Courtney dated July 18, 1990, stated that "taxonomists have had so much trouble in breaking the lorikeets up into genera on morphological traits and [this] raises the possibility of many parallelisms in behavior that may not be as indicative of phylogeny as appears at first sight."

Several characteristics might be used to argue the case for "lumping" various species of small lorikeets. One of these is the juvenile food-begging call (see **Call, Juvenile Food-Begging**). The way in which lories and lorikeets are currently classified results in anomalies. A study of the taxonomy of the whole group is needed, plus some radical thinking

The tongue of lories and lorikeets has numerous small projections called papillae.   *Drawing: Rachel Lewis*

struments in nature. Much of the time it appears like that of many other parrots. However, when the bird is feeding or testing an object, the elongated papillae on the tip are evident. In tame birds this is easily demonstrated by allowing them to lick honey or some other sweet substance from one's finger. Alternatively, hold a piece of fruit or a favored food item just outside the cage, near enough for them to stretch out the tongue to reach it.

The papillae are "small surface projections consisting of essentially two parts: a stratified squamous epithelium including a basal generative region, and a cornified spine; a dermal vascularized papillae forms the core" (Pagel and Greven 1990). The "brushes," as the papillae are often called, are developed to a varying degree according to the species of lory. In those in which pollen and nectar play a large part in the diet, such as the *Charmosyna* species, they are developed to the greatest degree. In iris and Musschenbroek's the brushes are so poorly developed that they are very difficult to observe. In captivity, these two species show more interest in seed than in nectar and could even survive without nectar. In the *Charmosyna* and *Chalcopsitta* species the tongue appears to be longer and is often stretched to its full length outside the beak and even "waved" around the beak.

The papillae develop at an early age. For example, they were evident in a 22-day-old Stella's lorikeet when it was preening itself (report Jurong BirdPark, I. Santos 1994). Pagel and Greven (1990) examined macroscopically the tongue tips of chicks of chattering and dusky lories and Swainson's (rainbow) lorikeet. After hatching, the flesh-colored tongue is covered with a smooth horny layer. Papillae can be seen clearly through this covering between the seventh and ninth day. Piece by piece molting of the keratinized layer, which sets free the papillae, begins at 22 days for the rainbow lorikeet (*T. h. moluccanus*), at 33 days for the dusky lory (*Pseudeos fuscata*) and at 35 days for the chattering lory (*Lorius g. garrulus*).

The *Vini* species also have very well developed papillae. Porter (1935) made an excellent description of the tongue of the collared lory:

> ...it is very long and fleshy; at the tip is an arrangement like a miniature sea-anemone, composed of many fleshy tentacles which, when the bird is not feeding, are folded in a

to drastically reduce the number of genera currently in use. For example, *Psitteuteles* and *Glossopsitta* might both be included in *Trichoglossus*.

Parrot evolution is the subject of a study at the University of California in Davis. This study confirms what we already know: lories are a coherent group, evolved from a common ancestor. Biologists call such groups "monophyletic" (monophyletic means from one origin and polyphyletic means from multiple origins). Studies such as theirs, using DNA, promise to be the most decisive in determining relationships at the family level among the parrots (Toft, Langley and Brown 1994). See also **Nomenclature**.

Tongue

A lory's tongue is one of the most fascinating in-circle just as a sea-anemone in repose. When feeding or licking one's fingers, these are unfolded and have a peculiar muscular action, and are used for gathering up the honey and pollen of the flowers upon which the bird feeds. Each tentacle can be controlled separately. After death they contract and cannot be seen. I have seen the tongues of many of the Lories, but I don't think the brush tip reaches such a development as in this one. With the birds being tame it is very easy to examine this strange organ.

The tongue color varies according to the species. Generally speaking, in those with heavy deposits of pigment, such as black and Duyvenbode's lories, the tongue is black (as is the bare skin surrounding the eye and beak, as well as the feet). In those with little pigment (such as the red *Charmosynas*) it is pink, as are the feet. In the iris lorikeet, which could be described as having a medium amount of pigment, it is gray. However, in blackcapped lories the tongue is pink but the feet are dark gray. In all species the papillae are contained in a little horny sheath. In most species this sheath is black, but in the red *Charmosynas* it is pale gray.

An interesting aspect of the way in which a lory uses its tongue is how it vibrates it—too rapidly for the human eye to observe. This is most easily comprehended when a tame bird vibrates its tongue against one's finger. I imagine that this action would allow it to very rapidly collect the tiny grains of pollen from flowers. To my knowledge there is only one other parrot capable of vibrating the tongue in this way—Pesquet's parrot (*Psittrichas fulgidus*). This large red and black parrot from New Guinea is, in my opinion, the only parrot which is not a lory which is lorylike in several aspects of its behavior, such as the way it moves.

Lories are not the only parrots whose tongues are adapted to utilize a specialized food source. In the swift parrakeet (*Lathamus discolor*) and in the kaka (*Nestor meridionalis*), the tongue is modified for feeding on nectar, whereas the lory's tongue probably makes it a more efficient gatherer of pollen. Homberger (1981) states that the brush-tipped tongue of the swift parrakeet "has been used by several authors as an argument for the close relationship of *Lathamus* with the Loriinae" but:

> A careful study of the fine structure and arrangement of the papillae on the tongue, however, revealed major differences between the

Loriinae and *Lathamus* in these structures which have no influence on the adaptiveness of the papillae for pollen feeding, thus indicating that the brush-tipped tongue in the Loriinae and *Lathamus* is a convergent, non-homologous character.

Trade

In the early years of the twentieth century lorikeets from Australia, especially the rainbow and the red-collared, were quite often available. In 1960 Australia ceased to export its fauna. Large scale trade in lories and lorikeets from Indonesia did not commence until the early 1970s. Until then, most Indonesian and Papuan lories were unknown or rare in aviculture. This trade continued for 20 years; the numbers caught and the careless method of handling birds was unacceptable, resulting in a very high mortality rate. For example, Milton and Marhadi (1987) stated that during periods of inclement weather "mortality levels approaching 40% may occur in species like *Eos squamata* during interisland transport if birds were not protected."

Another sad aspect of this trade was that during the 1970s some species were captured in huge numbers. Dealers in Europe and the U.S.A. and probably in other countries such as Japan, received such large numbers that prices were very low. There were two negative aspects to this. Some dealers did not care for them properly, and many birds died. Of those which survived, many ended up in pet stores and were bought by people who had no idea how to look after them. They were colorful and they were cheap; they were also short-lived. It was a very wasteful trade.

It did allow genuine aviculturists to acquire them for the first time, to breed them and to establish them. Probably 90 percent or more of lories offered for sale in Europe and the U.S.A. and 100 percent of those in Australia today are captive-bred. Lories are not usually difficult birds to breed and there should be no need to catch more wild birds. Unfortunately, however, the export trade continues, fueled not by demand for wild-caught birds but by income on the part of catchers and exporters on their islands of origin. The red and blue lory is one example. It is endangered by such trade and was placed on Appendix 1 of CITES (See **CITES**) for this reason—but it seems unlikely that this formality alone will result in cessation of trapping. For some rare species, such as the purple-naped lory, there is a large local demand, and few birds are exported.

In the 1980s, concerned that trade in some island species, such as the purple-naped and blue-streaked lories, would result in their extinction, regulations were passed to prevent such species from entering western Europe. (See table for full list of species whose import is prohibited.) Soon after, the commercial importation of wild-caught birds was made illegal in the U.S.A. The closure of these major markets did not cause trapping to cease; it just diverted trade to countries such as Japan and some of those in eastern Europe who were totally unconcerned about the effects of trade. In some of these countries there was no tradition of aviculture and very little knowledge. Here again mortality would be high. Most birds would be bought to be caged singly as pets, without ever having the opportunity to breed. The same applies to the domestic trade. In parts of Indonesia where lories occur, this involves a large number of birds of certain species. Not until 1995 did Indonesia invoke legislation to regulate this trade. As yet, it is too soon to know how effective this is, bearing in mind the scale of the illegal trade. However, once Indonesia places an export ban on a species, it means that wild-caught birds cannot be legally imported into countries which are signatories to CITES. Illegal export from Indonesia is difficult to control; it is impossible to monitor all of the countless islands and points from which birds could be exported. As long as money is to be gained from trapping, this activity will continue—and unless every country worldwide becomes a signatory to CITES, the trade in wild-caught birds will not cease.

Many researchers believe that there are good conservation arguments in favor of sustainable harvesting, in which a certain percentage of the wild population of some species could be removed annually without having a negative impact on the species as a whole. While this may be possible, in my opinion, it totally misses the point: trade in nearly all wild-caught parrots should be illegal for the suffering which it causes to individual birds and for the poor quality of life which too many of the survivors will have to endure. Birds which are reared in captivity are generally tame and adapt well to life in a cage. No one will convince me that it is not a cruel act to remove an adult parrot from the wild. This is especially the case for birds which may be rearing

young at the time of capture. However, I am well aware of the argument that sustainable trade might in fact promote conservation. Michael Poulsen asks:

> Is it fair to stop local people on Halmahera selling parrots without compensation? Perhaps they should be allowed to catch a sustainable number of parrots and be given a fair price for them. Monitoring of wild populations and control of the trade will be needed to make sure that the use is sustainable, but the parrots are sufficiently valuable to pay for this. People on Halmahera, and the rest of Indonesia too, would certainly be more supportive of habitat conservation if they could see economic benefits. We have to ask whether a total ban on catching and trade would help the parrots or is it only helping the people who would make huge profits on illegally traded parrots before shipping them off to some far away country? (Poulsen 1996).

These "far-away countries" are those where aviculture is not yet established—where few lories are bred in captivity. By the mid-1990s it was no longer economically viable for dealers in Europe to import lories because they could buy all they needed from breeders, even though they could offer only a low price. In late 1996, a major bird importer/exporter in the U.K. told me that it was cheaper to buy from breeders than to import wild lorikeets. He was exporting 90 percent of the young he bought, mainly to Portugal and other European countries.

At that time the blue-streaked was not considered rare in aviculture. In 1993 a survey on Yamdena, in the Tanimbar Islands, showed that the blue-streaked lory was common—and not endangered by trade as had previously been thought. In the previous year this species was placed on Appendix 1 of CITES and the catch quota had been reduced to nil. Following the survey, it was suggested that catch quotas might be re-introduced and that consideration should be given to a bird catcher registration scheme to obtain a more reliable way to estimate numbers being trapped. (At the time of writing, there was still a nil export quota for the blue-streaked lory.) The only purpose of trapping these birds would be to provide income for trappers and exporters. There are enough captive-bred birds in countries where aviculture is practiced; wild-caught birds are not needed. Indeed, by the mid-1990s, many breeders had problems in trying to sell blue-streaked and other captive-bred lories.

See **Trade** headings under entries for following species: **Cardinal Lory**; **Blue-streaked Lory**; **Red and Blue Lory** (*E. h. talautensis*); **Red Lory**; **Violet-necked Lory**; **Purple-naped Lory**; **Chattering Lory**. See also **CITES, Conservation.**

Import/export bans and quotas

At the completion of this book in late 1996, trade in many species of lories and lorikeets was controlled under CITES legislation. In Europe this is implemented by the European Community Council regulation 3626/82. In particular, this regulation places many CITES Appendix II species on Annex C2. For these species Article 10.1.b states that an import permit will be issued only if (1) capture or collection is not harmful to the wild population; (2) specimens are legally obtained from country of origin; (3) housing is suitable for the species; (4) it does not impede any other requirements relating to the conservation of the species. Where the E.C. considers that trade is not compatible with these requirements, it may impose Europe-wide species-specific or country-specific bans on import. Thus the European Union (E.U.) prohibited the importation of some species, such as the blue-streaked lory, before Indonesia banned its export. There are export quotas for some species from Indonesia, and these quotas are revised annually. Species covered by controls are listed below, except those native to Australia; there is an export ban for all Australian birds except in some circumstances to bona fide zoos.

Lory Import Status in Europe

Species	Situation at the end of 1996
Lorius lory	E.U. import ban since 1987
Lorius albidinuchus	E.U. import ban since 1988
Lorius chlorocercus	E.U. 1996 annual quota from Solomon Islands: 300
Lorius domicellus	E.U. import ban since 1987
Lorius garrulus	E.U. import ban since 1987
Chalcopsitta atra	Indonesian export ban since 3/95
Chalcopsitta duyvenbodei	E.U. 1996 annual quota from Indonesia: 200
Chalcopsitta scintillata	E.U. 1996 annual quota from Indonesia: 400
Chalcopsitta cardinalis	E.U. 1996 annual quota from Solomon Islands: 500
Eos cyanogenia	E.U. import restriction since 12/89; export moratorium from Indonesia since 1994
Eos reticulata	Indonesian export ban 3/95
Eos histrio	CITES Appendix I - no commercial trade allowed
Eos bornea	Indonesian export ban 3/95
Eos semilarvata	Indonesian export ban 3/95
Eos squamata	Indonesian export ban 3/95
Trichoglossus ornatus	E.U. import ban since 1987
Trichoglossus haematodus *	E.U. 1996 annual quota from Indonesia: 2,000
T. h. massena	E.U. 1996 annual quota from Solomon Islands: 500
Trichoglossus euteles	E.U. import restriction since 1989; Indonesian export ban since 3/95
Trichoglossus rubiginosus	E.U. import ban since 1986
Trichoglossus flavoviridis	Indonesian export ban since 3/95
Trichoglossus johnstoniae	E.U. import ban since 1991
Oreopsittacus arfaki *	Indonesian export ban since 3/95
Charmosyna palmarum	E.U. import ban from Solomon Islands since 1990
Charmosyna meeki	E.U. import ban from Solomon Islands since 1990
Charmosyna toxopei	Indonesian export ban since 3/95
Charmosyna multistriata *	Indonesian export ban since 3/95
Charmosyna wilhelminae *	Indonesian export ban since 3/95
Charmosyna rubronotata *	Indonesian export ban since 3/95
Charmosyna placentis *	E.U. 1996 annual quota from Indonesia: 400
Charmosyna diadema	E.U. import ban since 1990
Charmosyna amabilis	E.U. import ban from Fiji since 1988
Charmosyna margarethae	E.U. 1996 annual quota from Solomon Islands: 100
Charmosyna pulchella *	E.U. 1996 annual quota from Indonesia: 750
Charmosyna josefinae *	Indonesian export ban since 3/95
Charmosyna papou *	E.U. 1996 annual quota from Indonesia: 400
Vini	(all species) E.U. import ban from all states since 1986
Neopsittacus pullicauda *	E.U. import ban from Indonesia since 3/95
N. musschenbroeki *	E.U. import ban from Indonesia since 3/95

* These species also occur in Papua New Guinea.
In the U.S., CITES species can only be imported through a USF &W approved Cooperative Breeding Program.

Vocalization

Is vocalization inherited or learned in lories? Marc Valentine related to me the interesting history of a young parent-reared yellow-backed lory which had been removed from its parents and kept away from them, but near Amazon parrots. It vocalized like an Amazon. He kept it with its own species but it was not until it was three years old that it learned the full vocalization of the yellow-backed lory. If it had merely been mimicking the Amazon for amusement, as some lories do mimic other species, it would surely have learned its own language much sooner. Incidentally, this bird did breed successfully. Another case concerned a dusky lory which I hand-reared from the age of 22 days. It was reared with a red-bellied parrot (*Poicephalus rufiventris*), then kept with other lory species until it was 11 months old. It was with me all that time until it went to a new home and never saw another dusky lory.

At 11 months the only form of vocalization which it knew was a food-begging call.

These cases suggest that some lories have to learn the language of their own kind. This is not the situation with many other parrot species which, even if hand-reared from the egg without any association with their own species, vocalize normally from an early age. However, Bosch (1995), basing his observations on the fairy lorikeet, stated that the marked differences in vocalization of the two subspecies "are not...learned dialects but are inherited so that even hand reared birds are totally normal. Two *pulchella* reared by adult *rothschildi* tried to imitate their foster parents which could of course only reach a limited volume."

The vocalizations of Stella's lorikeet seem to be inherited, rather than learned, knowledge. One of the most distinctive and unusual sounds made by a lory is the "nasal note which increases in volume: *nreeennnNGG*, generally on one pitch" (Beehler, Pratt and Zimmerman 1986) which is unique to this species and to Josephine's lorikeet. A male melanistic Stella's which I hand-reared from the age of four weeks did not have visual or vocal contact with other members of its species until it was just over six months old. However, at 14 weeks of age it started trying to make this adult vocalization, which takes some months to master well (at eight months young birds cannot make the sound as effectively as adults). It learned to make the sound at about the same rate as parent-reared young.

Whereas all Amazons, for example, or all *Aratinga* conures, have calls which are recognizable as belonging to that group, even if the species can only be identified by an expert, vocalizations of lory species vary greatly. The little Meyer's lorikeet has a pleasant warble, not unlike that of a budgerigar, and the tiny whiskered has an almost finchlike twittering. In contrast all the *Chalcopsitta* species have calls so loud and so harsh that they literally hurt the ear of someone a few inches away. In other words, their vocal power is related to their size. Most lories have harsh voices which are not very melodious, yet the variety of vocalizations within the group is surprisingly wide. Some species, especially *Lorius*, have a wide range of calls, from almost melodious to loud and piercing.

One element of the sounds produced by lories and lorikeets is unique to this group of parrots—hissing. It cannot be described as a vocalization since it is not produced by the vocal chords but by blowing air through the open bill, or possibly in some cases, through the nasal openings. It is a prominent feature of the display of many species in probably every genus.

In the wild, the vocalizations of some lorikeets with a wide geographical range are known to differ according to the area or island they inhabit; in other words, there are lory dialects. For example, the Bellingshausen population of the Tahiti blue lory often ends its call with "a sort of goggle never heard from the Rangiroa birds" (Seitre and Seitre 1991).

See also **Calls, Juvenile Food–Begging**.

Watching Lories in the Wild

Watching lories in their natural habitat can be a difficult challenge or unbelievably easy, depending on the locality and species. In many Australian towns one can stroll down the main street and see small flocks of lorikeets feeding on flowering shrubs and trees. Rainbows, musks and scaly-breasts are easily observed; the tiny purple-crowns which usually keep high up in the trees are more difficult to see in the dense foliage. But their *zit zit* calls attract attention to their presence. On Buru, the tiny green Buru or blue-fronted lorikeet (*Charmosyna toxopei*) is so difficult to observe in what little forest still survives that nothing is known about its status (rare or endangered)—assuming that it is not extinct.

On Pacific islands also, where coconut palms flourish along the coast, lory watching can be easy. In Fiji, collared lories feed on the coastal palms, on the coconut inflorescences, as do the striking red cardinal lories in the Solomon Islands. They can be observed at some height but not until they descend to feed can they be closely studied.

In warm climates, early morning and an hour or so before dusk are the periods of greatest activity; few lories will be seen at other times. But in more temperate climates, such as the mountains of New Guinea, lories are active all day long, flying and feeding. There they are more difficult to observe because of the great height of the trees; binoculars or spotting telescope are essential.

After watching such exquisite and colorful birds going about their lives, seemingly untouched by man, one wishes that this really was the case.

Lories cannot fail to give joy to anyone who cares about nature's countless treasures. If only man valued these things more highly than the unnecessary items of this consumer age, mankind would once again be in tune with nature—instead of poised on the abyss of environmental catastrophe.

Water, see **Bathing, Drinking**

Weaning

Lories and lorikeets are the easiest of all parrot chicks to wean when hand-reared and the quickest to feed themselves when parent-reared. Weaning is the stage during which they progress from being dependent to becoming independent. Some hand-reared lories will feed themselves and become totally independent at a very early age, earlier than any other parrot of comparable development (bearing in mind that lory chicks leave the nest at an average age of nine or ten weeks, whereas many of the long-tailed Australian parrots, for example, spend only five weeks in the nest). The age at which hand-reared lories are weaned will depend on the method used and on the individual bird. Several lories of slightly differing ages may be reared together. If the oldest are offered warm food in a shallow container when seven weeks old, the youngest usually copy the older chicks in feeding themselves when they are as young as six weeks, or even five weeks, old. However, there is a problem associated with early weaning in groups. The chick's head feathers become matted with nectar and must be cleaned daily. Some chicks object strongly to this. I usually prefer to spoon-feed for a longer period than to face the tedious task of cleaning the feathers. A single chick weaned at an early age is not a problem as there are no other young to spill food on its head.

Lories are the only parrots I know in which it is possible for "instant" weaning to occur. From the moment they are offered food in a shallow container, some take it—and never again feed from the spoon. This is fairly rare; weaning is generally a gradual procedure. In a single bird, especially in *Chalcopsitta* species, some of which are less inclined to feed themselves at an early age than most other lories, weaning can be slightly more protracted.

Initially, to encourage a chick to feed itself, it can be fed from a small container, with the spoon just above the surface, then below the surface, so that the lory is actually drinking from the container, not from the spoon. I use a plastic lid from a coffee jar, as this is small and shallow, and offer warm food in this four or five times daily. Chicks cannot be expected to take food which has cooled until about the time that they would have left the nest. I usually provide the warm rearing food in one dish and warm nectar in another. Eventually they change their preference from the rearing food to the nectar. Fruit, corn and millet spray (depending on what is appropriate for the species) are also available. Young lories are very playful and may use these as toys at first, even lying on their back and juggling with them.

When young lories start to feed themselves they sometimes fill their crops to a much greater degree than I do when spoon-feeding them. For example, two young Josephine's, aged five weeks, each weighed 53 g and I was giving them 4 or 5 g of food at each feed. On the day they started to feed themselves, they filled their crops with 13 g of food. Most lories are gentle and affectionate during the hand-rearing period and greatly enjoy being handled. However, this was certainly not true of the Musschenbroek's I have hand-reared. They disliked being handled and bit hard from an early age! I therefore encouraged them to feed themselves. One was independent as early as 38 days (much to my relief!).

Selected totally at random, the weaning history of 2 single chicks was as follows: (1) Duyvenbode's lory: still being fed five times daily at 38 days; four times at 42 days; offered warm food in a small container at 57 days—filled crop with 36 g of food! Moved to cage from brooder at 68 days—independent. (2) Dusky lory: still being fed five times daily at 36 days; first offered warm food in container at 46 days—filled crop; heat turned off in brooder at 47 days; independent at 55 days. Moved from brooder to cage at 60 days.

In contrast, a chick reared with its sibling or other lories, will learn to feed itself at an earlier age. Even *Chalcopsitta* species will wean early if there are more than one. The photograph on page 104 shows two *insignis* at 37 and 41 days. They were independent then—and their feathers were only just starting to erupt.

As with all parrot chicks, the golden rule is never to force a chick into weaning but to encourage it to feed itself. A health problem, such as candidiasis, should be suspected in chicks which are hard to wean, because normally weaning is so easy.

Worms, Parasitic

Parasitic worm infestations are not a common problem in captive lories, especially now that so many are kept in suspended cages. Lories therefore have reduced access to worm eggs as these are squirted with the faeces to the ground below the cage. It should be noted, however, that even birds kept in suspended cages have occasionally been found to be infected with parasitic intestinal worms. In Australia, many lories are kept in aviaries with earth floors, and infestations with roundworms (Ascaridia species) and hairworms (Capillaria species) sometimes occur. These are the most common internal parasites of aviary birds and live in the bowel. Their presence can be confirmed by microscopic examination of a fecal sample. Consult a veterinarian if in doubt. A heavy infestation will cause debilitation and, if untreated, death. Several worming preparations (anthelmintics) have been used successfully to treat this problem. Fenbendazole (e.g., Panacur 25 [25 mg/ml fenbendazole] Hoechst Animal Health) has been used successfully at a dose of 50–100 mg per kg bodyweight for three days consecutively. This equates to 0.2–0.4 ml of Panacur 25 per 100 g of bodyweight. This should be administered directly into the crop. It might be given in a small quantity of nectar but unless all the nectar is taken this method will not be effective. This should be followed with a second dose two weeks later to kill the worms which have hatched in the meantime. Being a suspension, this preparation is unsuitable for use in the drinking water as it tends to settle to the bottom of the drinking container. It should not be used during periods of new feather growth (i.e., molting, growing chicks) as it has been found to cause abnormal feather development in some cases. It is not advisable to use the more concentrated preparation Panacur 100 (100 mg/ml fenbendazole) as fatalities have been recorded with this preparation, even if diluted. Alternatively levamisole (e.g., Nilverm Pig and Poultry Wormer-16 g/L levamisole hydrochloride 14 g/L levamisole, Mallinckrodt Veterinary: Coopers Brand) can be given directly into the crop (it is bitter) at a dose of 15 mg per kg bodyweight, repeated in two weeks. This equates to 0.1 ml of Nilverm (14

g/L levamisole) per 100 g bodyweight. Levamisole has a lower safety margin than other anthelmintics and doses should be calculated with caution.

Ivermectin has also been used successfully. The recommended dose is 0.2 mg per kg of bodyweight, given orally. In some cases, doses as high as 1 mg per kg have been required to eliminate roundworms satisfactorily. This chemical has the advantage of concurrently treating external parasites such as scaly mite (Knemidocoptes species). It should not be readministered within six weeks of initial dosing.

Tapeworms (cestodes) are seen in some wild-caught lories—even those which have been in captivity for as long as two years. Tapeworms require an intermediate host to carry their larval forms. The treatment of choice is praziquantel, 10–20 mg per kg bodyweight, repeated in two weeks. This is most readily available as Droncit Tapewormer for Cats (23 mg praziquantel-Bayer) or Dogs (50 mg praziquantel-Bayer). This equates to 0.5 to one cat tablet (or 0.25 to 0.5 dog tablet) per kg of bird. The tablet should be finely crushed and added to the nectar. Less nectar than is normal should be offered, to ensure that none of the crushed tablet remains in the container. Worms, if present, will be passed 24 to 36 hours later. Praziquantel has a fairly wide margin of safety compared with other anthelmintics.

It should be noted that most preparations described above are not registered for use in cage birds, thus responsibility for their use lies solely with the aviculturist. Although they have been used successfully by aviculturists and veterinarians, it is important that this point is made clear. It is preferable to use products specifically designed for use on cage birds where these are available. Where they are not available, the above-mentioned products (and their equivalents in each country) can be used, usually with satisfactory results. It is advisable to consult a veterinarian with avian experience, who can advise the appropriate product to use.

The aviculturist should also be aware that treatment of a heavily worm-infested bird might still result in its death because mass deaths of worms may result in intestinal obstruction. The concurrent use of bulking agents such as peanut butter or psyllium (e.g. Metamucil) may help to alleviate this. Severely debilitated birds may require veterinary supportive care in the way of fluid and antimicrobial therapy.

(The foregoing was contributed by Dr Stacey Gelis, BVSc(Hons), MACVSc (Avian Health).)

Zoos, Lory Exhibits in

Comparatively few zoos have realized that as a colorful and entertaining spectacle, a lory exhibit is unsurpassed. Also, it does not cost a lot to maintain. Probably the first zoo to show a lory aviary of mixed species as an attraction was San Diego, California. In the early 1970s, under the direction of the late K. C. Lint, many lories were hand-reared there. These tame young birds were placed together in a large aviary. Their playful and friendly antics and incredible variety of colors held visitors (including myself) spellbound.

Undoubtedly the most famous lory attraction in any zoo is Currumbin Sanctuary at Palm Beach, Queensland, Australia. The zoo actually evolved from the former owner's habit of feeding lorikeets in his garden. Hundreds came daily and he realized what a wonderful public attraction this would make. Today the wild rainbow and scaly-breasted lorikeets are still fed twice daily by hundreds of eager visitors. It was some years before zoos outside Australia started to copy this idea, using tame, hand-reared birds. One of the first was the Wildlife World Zoo in Arizona. In 1988, Marine World Africa U.S.A. in Vallejo, California, opened its lorikeet aviary. Many visitors consider it to be the most exciting exhibit in the park. The aviary was constructed on a slope using telephone poles and cables to support 7/8 in. x 7/8 in. nylon netting. It is divided lengthwise into a public feeding area and a retreat for the birds. The latter is a completely enclosed building measuring 3 m (10 ft.) x 4.2 m (14 ft.) x 2 m (6 ft. 6 in.) high. In 1995 it contained 42 lories of 12 species. The aviary is open to the public for two or three half-hour feedings per day. A double-door system permits only a certain number of people to enter the aviary at one time. They are given a brief talk on the lorikeets and pieces of apple or grapes to feed the birds. It was found that half an hour was the optimum period for each feed as after that the birds have eaten well and lose interest in the public. Also, their rapid digestive system means that the attendants have to pass out wet wipes!

Four or 5 liters of fruit are given to the public for each feed. The birds are fed Avico Lory Life nectar early in the morning; this is removed after 1 to 1½ hours. At 11:00 A.M. the first public feeding occurs. After the last public feed the birds receive an additional 6 liters of fruit (apple, grapes, papaya, banana and melon), 1 or 2 liters of liquid nectar and 1 or 2 cups of dry powder. Once a week they receive a mixture of cooked yams, carrots, corn and sometimes broccoli mixed with fruit or blended with sugar. All the lorikeets were hand-raised and most are under a year old when acquired. New birds are introduced as a group after having been quarantined together for 30 to 45 days. Any birds which prove unsuitable due to aggression are removed from the aviary. If pairs are formed and try to nest under tufts of grass or holes between rocks, they too are removed from the aviary and either set up for breeding or exchanged with a local breeder for young birds.

The assessment of the suitability of the various species for this type of exhibit is interesting. Lori Hill commented:

> The Perfects rank #1 in customer satisfaction and are aptly named. They are sweet little birds that never bite, they have outgoing personalities and come down readily to the public. They are also a favorite with children as their size is less intimidating. We have tried other smaller species such as Webers, Meyers, and Goldies but they just weren't able to compete with the larger birds and didn't do as well.
>
> The Green-nape Rainbows cause the least amount of trouble. Although these are the most numerous birds in the aviary they are...seldom involved in any "gang fights." They generally do well with the public.
>
> The Duskys, on the other hand, will sometimes band together and gang up on a particular individual. Usually a larger bird such as a Black lory will be the instigator and the Duskys will join in. For that reason we limit the numbers of Duskys...The birds from the genus *Eos* (Reds, Black-wings, Blue-streaks) have had a tendency to be a little more nippy than others but it's never been a big problem. We caution the public not to pet any of the birds, but these are the birds that are most likely to bite if touched.

Initially disease was a problem. Before the exhibit was opened to the public, the flock was infected with *Clostridium*, an anaerobic bacteria. A number of birds died in a very short period before an effective antibiotic was found. It was chloromycetin palmitate. The source of the infection was not

161

discovered. The outbreak occurred after very hot weather. There was a recurrence during the following year—also after a hot spell—but antibiotic therapy was commenced immediately and no birds died. On two occasions since, birds have been treated prophylactically following heat waves; no treatments had been necessary in the preceding two years. *Yersinia* is another potentially fatal bacteria, usually seen during the colder months; it is probably rodent-borne. Baytril given in the nectar was effective. A few cases of candidiasis had occurred. The problem was eliminated by regularly giving additional sources of vitamin A.

Lori Hill concluded:

> It is rewarding to see children, and adults that have never held a bird in their life, covered with lorikeets happily eating an apple, or playing with their buttons, zippers, sunglasses and cameras. It gives visitors an opportunity to form an attachment to these wonderful animals and perhaps a desire to learn more. Without such experiences where would our future aviculturists, biologists and environmentalists come from; and what would be the future of the world's birds? We hope that with more exhibits such as these across the U.S., we can help to make a positive impact on the future of aviculture (Hill 1995).

Another California zoo with a lorikeet aviary is the world-renowned San Diego Wild Animal Park. Their "Lorikeet Landing" opened in 1994. It differs from the Marine World Africa exhibit in several aspects. Only green-naped lorikeets are kept in the aviary and almost all of them were wild-caught birds. They had a month's training before the exhibit opened. It took some while for them to become really tame—but when I visited Lorikeet Landing in August 1996, I saw a scene like Currumbin reenacted with green-naped lorikeets. The birds were totally fearless and climbing over everyone in impressive numbers. At the door visitors buy a small paper cup of Nekton-Lori—then they are warned what to expect! In a second, as many as ten lorikeets have landed on the admirer's arms, head and shoulders. It is a large aviary with plenty of space for the public, and even a pond with waterfowl. There were about 40 people inside the aviary and all were entranced by the birds. The lorikeets which were not feeding were sitting around in groups in trees and lined up on low fencing. I was impressed by a couple of facts: only one species was used which to my eye somehow made the scene appear more natural, especially as this is a species often found in large flocks in the wild; and all the green-napes were in perfect condition. I asked Michael Mace, curator of birds, whether there had been any problems with aggression. The answer was: "No." Had any pairs tried to nest? Yes they had—in a very unusual site! Two pairs had usurped the ducks from their barrels and reared young inside!

Woburn Safari Park, Bedfordshire, U.K., is one of the latest zoos to feature lorikeets. In June 1995 "Rainbow Landing" was opened. It houses 50 birds, all bred in the U.K.—green-naped and Swainson's (rainbow) lorikeets. They fly in a 223 m^2 (2,400 sq. ft.) aviary, amid plants and running water. It was designed by the American company responsible for the lorikeet aviary at San Diego Wild Animal Park. Personnel from that company showed the staff how to train the birds to take nectar held by visitors in small cups. Woburn Safari Park was started by Lord Tavistock, grandson of the Lord Tavistock whose breeding successes with *Vini* lories are recounted elsewhere in these pages.

Generally speaking, few lories are exhibited in zoos. Lories are colorful, active and amusing, which are the qualities which appeal most to visitors. Yet it would seem that those responsible for acquiring zoo stock are not sufficiently familiar with them. For those zoos whose main justification is the rearing of threatened species, there are a number which can be obtained which fall into this category. (Examples are black-winged, iris and Mount Apo and the red and blue is likely to be available before long.) Only in zoos or bird parks which specialize in parrots can one hope to see more than two or three pairs of lories, except perhaps in Australia. There the expense of acquiring exotic parrots limits the number that can be exhibited—but the colorful native lorikeets are much appreciated by zoo-goers. The zoo which exhibits most lory species worldwide is Loro Parque in Tenerife, Canary Islands. It has every species and subspecies currently available in aviculture.

A look at the ISIS database at December, 1993, gives an indication of numbers and species or lories held in zoos. It listed stocks of every kind of zoo animal held in 465 collections in 53 countries on five continents. However, many zoos are not members of ISIS; Palmitos Park, Gran Canaria, is among these, and on that date held 120 lories in the collection. At that time ISIS listed 327 males, 283 females and 419 lories of unknown sex—a total of 1,029 birds. In comparison, there were more than twice as many cockatoos: 753 males, 614 females and 424 unsexed birds.

Of the lories in the ISIS database, 62 percent were captive-bred, 17 percent were wild-caught and the rest were of unknown origin. In the previous 12 months, 174 chicks had hatched from these birds, of which 32 had died during the first 30 days. The percentage of chicks to adults kept was therefore 16.9 percent, or less than one youngster for every five birds. (This low rate may reflect the fact that some of these birds were kept in colonies and the non-breeding young from previous years were still in the aviary.) As a comparison, 199 chicks were hatched by a total of 1,791 cockatoos, giving the percentage of chicks hatched to adults as 11 percent. This indicates that lories are easier to breed than cockatoos.

Thirty-one lory species were kept by the ISIS member zoos, of which 23 had hatched chicks in the previous 12 months. Five species had produced more than 10 young: rainbow (Swainson's) and red-flanked, 11 each; yellow-backed, 12; Goldie's, 25; and green-naped, 30. Five to 9 young had been produced by the following species: yellow-streaked, fairy, black-winged, musk, dusky, red-collared, varied and blue-crowned (*Vini australis*), although not all of these survived to fledge. It is therefore clear that few lory species are sufficiently represented, and perhaps none bred on a consistent enough basis, to maintain a self-sustaining captive breeding population in zoos alone. Most lory species would therefore not survive in captivity without dedicated private breeders.

In 1994, the EEP Parrot TAG (Taxon Advisory Group of European Zoos) surveyed 77 member zoos regarding the lories maintained in their collections. Returns showed a total of 1,229 birds of 30 species. However, only 5 species were represented by more than 50 individuals—the blue-streaked (72), dusky (70), yellow-backed and chattering lories (104), red lory (56) and green-naped lorikeet. The majority of species were represented by 20 or fewer birds, and in 73 instances only birds of the same sex were kept (Wilkinson 1995). This further demonstrates that most zoos are not committed to breeding these birds, and suggests that the cooperation of zoological institutions and private aviculturists might be vital to continued existence of certain species in captivity. The EEP Parrot TAG maintains two European Studbooks (ESB): one for the red and blue lory and the other for the Mount Apo lorikeet; studbook holder is Roger Sweeney, at Loro Parque, Tenerife.

World range of lories and lorikeets

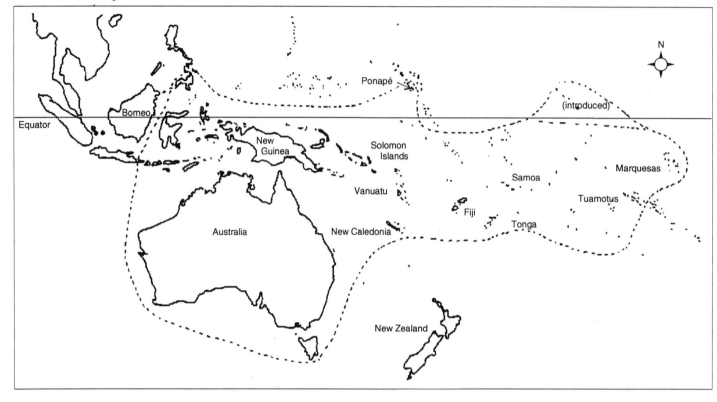

Part 2

Lory Species Accounts

In this species summary, the following abbreviations are used for countries and areas of origin:
A = Australia   **IND** = Indonesia   **NG** = New Guinea
NGI = New Guinea and/or islands in the vicinity
PAC = Pacific Islands   **PH** = Philippines   **S** = Solomon Islands

The principal habitat is indicated:
L = lowland   **M** = midmontane   **H** = highlands.
Weights refer to nominate race, where subspecies occur.
Estimated weights are followed by "**e**."

Lory Species

Scientific name	Common name	Weight/Origin/Habitat		
Genus: *Lorius*				
L. lory	Black-capped lory	230 g	NGI	L, M
L. hypoinochrous	Purple-bellied lory	220 g	NGI	L, M
L. albidinuchus	White-naped lory	160 g	NGI	H
L. chlorocercus	Yellow-bibbed lory	170 g	S	L, H
L. domicellus	Purple-naped lory	220 g	IND	L, M
L. garrulus	Chattering lory	220 g	IND	L
Genus: *Chalcopsitta*				
C. atra	Black lory	230 g	NG	L
C. duyvenbodei	Duyvenbode's lory	215 g	NG	L
C. scintillata	Yellow-streaked lory	210 g	NG	L
C. cardinalis	Cardinal lory	220 g	NGI	L
Genus: *Eos*				
E. cyanogenia	Black-winged lory	170 g	NGI	L
E. reticulata	Blue-streaked lory	160 g	IND	L
E. histrio	Red and blue lory	170 g	IND	L
E. bornea	Red lory	170 g	IND	L
E. semilarvata	Blue-eared lory	170 g e	IND	H
E. squamata	Violet-necked lory	110 g	IND	L
Genus: *Pseudeos*				
P. fuscata	Dusky lory	155 g	NGI	L, H
Genus: *Trichoglossus*				
T. ornatus	Ornate lorikeet	110 g	ND	L
T. haematodus	Green-naped lorikeet	130 g	ND NG PAC	L
T. euteles	Perfect lorikeet	80 g	IND	H
T. rubiginosus	Ponapé lorikeet	80 g	PAC	L
T. chlorolepidotus	Scaly-breasted lorikeet	85 g	A	L
T. flavoviridis	Yellow and green lorikeet	90 g	IND	H
T. johnstoniae	Mount Apo lorikeet	50 g	PH	H
T. iris	Iris lorikeet	65 g	IND	L
T. versicolor	Varied lorikeet	60 g	A	L
T. goldiei	Goldie's lorikeet	60 g	NG	H

Scientific name	Common Name	Weight/Origin/Habitat		
Genus: *Oreopsittacus*				
O. arfaki	Whiskered lorikeet	20 g	NG	H
Genus: *Glossopsitta*				
G. pusilla	Little lorikeet	45 g	A	L
G. concinna	Musk lorikeet	70 g	A	L
G. porphyrocephala	Purple-crowned lorikeet	45 g	A	L
Genus: *Vini*				
V. solitarius	Collared or solitary lory	75 g	PAC	L
V. australis	Blue-crowned lory	45 g	PAC	L
V. kuhlii	Rimatara or Kuhl's lory	55 g e	PAC	L
V. stepheni	Stephen's lory	50 g	PAC	L
V. peruviana	Tahiti blue lory	32 g	PAC	L
V. ultramarina	Ultramarine lory	40 g e	PAC	L
Genus: *Charmosyna*				
C. palmarum	Palm lorikeet	35 g e	PAC	H
C. rubrigularis	Red-chinned lorikeet	34 g	NGI	H
C. meeki	Meek's lorikeet	28 g	S	H
C. toxopei	Buru lorikeet	28 g e	IND	L
C. diadema	New Caledonian lorikeet	45 g e	PAC	H
C. amabilis	Red-throated lorikeet	34 g e	PAC	H
C. multistriata	Striated lorikeet	40 g	NG	M
C. wilhelminae	Wilhelmina's lorikeet	20 g	NG	L M H
C. rubronotata	Red-spotted lorikeet	30 g	NGI	L
C. placentis	Red-flanked lorikeet	36 g	NGI	L, H
C. margarethae	Duchess lorikeet	50 g	S	L
C. pulchella	Fairy lorikeet	38 g	NG	H
C. josefinae	Josephine's lorikeet	70 g	NG	H
C. papou	Papuan lorikeet	95 g	NG	H
Genus: *Neopsittacus*				
N. musschenbroekii	Musschenbroek's lorikeet	50 g	NG	H, M
N. pullicauda	Emerald lorikeet	35 g	NG	H

Genus: *Lorius*

The members of this genus are unmistakable: large, stocky lories with longish, broad tails. All species have the wings green, and red predominates in the body plumage. They are handsome birds. In immature plumage, in most (or all?) species, the greater underwing coverts (the outermost feathers only) are broadly tipped with black. This is confirmed in *domicellus*, *lory* and *chlorocercus*. According to their plumage, *Lorius* can be divided into two groups: (1) the black-capped species: purple-naped, black-capped, purple-bellied, white-naped and yellow-bibbed; (2) the single species without the black cap, the chattering (yellow-backed) lory.

These lories originate from New Guinea and surrounding islands, Indonesia and the Solomon Islands. In aviculture, the chattering and the black-capped are well known and cherished. One species, the purple-naped, is rare. The yellow-bibbed was almost unknown until the 1980s, but its numbers are increasing. The purple-bellied is almost unknown outside New Guinea and the white-naped has never been in aviculture. The *Lorius* species are exceptionally attractive aviary birds but their loud voices may deter some people from keeping these otherwise desirable lories. They make excellent pets, and some become good mimics. If kept singly they need plenty of attention to compensate for the lack of a partner of their own kind. They are highly intelligent and playful and will make use of every kind of swing or toy provided. Unlike some lories, they need at least 20 to 30 percent solid foods (fruits and vegetables and even a few sunflower seeds) in their diet, in addition to nectar.

Black-capped Lory *Lorius lory* (Linnaeus 1758)

Native names: Fore, *Korió*; Gimi, *Korió*; Daribi, *Sómu*

Description: Forehead, crown and nape are black; the rest of the head is red and a red band separates the nape from the mantle; in most birds the middle of the band is a bleached or washed-out shade of pink. The upper part of the upper breast is also red. A broad collar of dark blue joins the same color on the mantle and on the lower breast, extending to the underparts, under tail coverts and thighs. The shade of blue is brighter, more violet, on the lower abdomen, ventral area, under tail coverts and thighs. Rump and upper tail coverts are crimson, the bases of the feathers being gray and green. The wings are green with a bronze patch on the primary coverts. Underwing coverts are red and there is a broad band of yellow across the underside of the flight feathers. The tail is red above at the base, dark blue at the tip; the underside is gray-tinged yellow. The beak is orange; the iris is orange with a narrow inner ring of pale yellow, a narrow ring of black and a wide ring of orange. The legs are black.

Length: 29–30 cm, nominate race; *jobiensis* and *cyanuchen* are slightly larger and the other subspecies are smaller.

Weight: 230–250 g, nominate race; about 180–240 g in *erythrothorax*.

Immatures: These birds of the nominate race differ from adults in having the dark blue of the upper breast extending higher up, as far as the throat in some individuals. Up until the age of 18 months or so, the area between throat and upper breast gradually loses the dark blue feathers. The bill is brown on fledging, gradually becoming dark yellow with dark gray at the base of the upper mandible. The iris is dark brown. An immature *erythrothorax* looks rather different from the adult, as it has a complete collar of dark blue, while adults of this subspecies have a half collar which does not meet below the throat. By about ten months of age the dark blue feathers of the throat are lost, so the collar is then incomplete, and remains so. In some young the collar is narrow; in others it may be 2 cm wide. The beautiful iridescent violet-blue of the post-torqual band is notable. The complete collar of dark blue is present in *cyanuchen*. In *jobiensis*, immature birds have duller plumage with green feathers on the flanks. The blue band which in adults extends from the lower breast to under the wings, is only partly formed (Tiskens 1993). In *erythrothorax*, and almost certainly in all subspecies, the lesser underwing coverts are blue toward the carpal edge of the wing. The greater underwing coverts are very broadly tipped with black (see illustration).

Sexual dimorphism: Plumage is the same in male and female but beak shape usually differs slightly. The upper mandible is normally broader, and more protruding, in the male; the female's beak therefore appears more rounded and less elongated.

Subspecies: Seven or more subspecies are generally recognized, but only six appear to be valid. Their identification has caused aviculturists confusion. In the U.K. Trevor Buckell studied skins in several museums to assist aviculturists in pairing together the correct races. He came to the conclusion that *salvadorii* and *viridicrissalis* were synonymous with *jobiensis*, but later had reason to believe that *salvadorii* was a valid form. *L. l. viridicrissalis* supposedly differs from *salvadorii* by the more blackish blue on the hindneck. Forshaw (1989) describes the underwing coverts as mainly black with some dark blue in males, and as dark blue in females.

The subspecies can be divided into two groups for ease of identification: (1) those with red underwing coverts: *lory*, *somu* and *erythrothorax* (the latter sometimes known as the red-breasted lory); (2) those with black or dark blue underwing coverts: *cyanuchen*, *jobiensis* (known as the Jobi lory) and *salvadorii* (Salvadori's black-capped lory).

Within these groups, the subspecies can be recognized by noting the following points.

1. In *lory*, the blue of the abdomen and upper breast extends over the back, forming a continuous area; *somu* is easily distinguished as it lacks any blue on the nape or breast, these areas being entirely red; *erythrothorax* has the abdomen blue, the breast red, and the blue of the collar extending slightly and irregularly onto the upper breast. The demarcation between the red of the breast and the blue of the abdomen is an almost straight line.

2. Although *jobiensis* agrees with *erythrothorax* in having the breast and interscapular area red and the blue of the collar extending slightly on to the upper breast, it differs from group one subspecies in having blue underwing coverts and no yellow under the wings; *cyanuchen* differs from *jobiensis* in lacking red on the nape. Thus there is only one red band, that on the interscapular region, thus this is the only subspecies with no red on the back of the head, as its name suggests (*cyanuchen* means blue nape). According to Buckell, *salvadorii* differs from *jobiensis* in having the abdomen black, with the odd green feather in some specimens (dark blue abdomen in *jobiensis*), in being smaller, and also in the shade of red, which is not the rose red of *jobiensis*.

Natural history

Range:

1. The nominate race is from Vogelkop, Irian Jaya (extreme western New Guinea) and the nearby western Papuan Islands of Waigeu, Batanta, Salawati and Misool (see **Gazetteer**). *L. l. erythrothorax* occurs in Irian Jaya, from the southern parts of Geelvink Bay and from the Onin Peninsula eastward to southeastern Papua New Guinea and the Huon Peninsula. Its range meets that of *somu* in the Fly River area. The southern New Guinea subspecies, *L. l. somu*, is known from the Karimui Basin and south of that area, from the mouth of the Purari River, from the Fly River and from the Lake Kutubu area (Forshaw 1989). Coates (1985) gives its range as the south side of the main ranges from the Ok Tedi area, Western Province, to the Karimui area of the Chimbu Province.

2. *L. l. jobiensis* is restricted to the island of Japen (formerly called Jobi) and the Mios Num Islands in Geelvink Bay, Irian Jaya.

L. l. cyanuchen is restricted to Biak and Supiori islands (see **Gazetteer**), Geelvink Bay, Irian Jaya. *L. l. salvadorii* comes from northern New Guinea, from Astrolabe Bay through the lowlands westward to the Aitape area and, if *viridicrissalis* is synonymized with *salvadorii*, even further westward, with a disjunct distribution extending as far west as Mamberamo River.

Habits: This is principally a bird of lowland forests and, in New Guinea, is uncommon above 1,000 m (3,280 ft.). It is seen in pairs or family groups; larger numbers congregate to feed but do not stay together. In virgin forest, where it occurs along with small flocks of green-naped lorikeets and dusky lories, the latter two species are found consistently several hundred feet higher (Diamond 1972). Beehler, Pratt and Zimmerman (1986) state that it occurs in small parties, foraging in the canopy of forest and forest edge. It usually flies at canopy level. Coates (1985) gives its habitat as primary forest, forest edges, tall secondary growth and partly cleared areas, being absent from monsoon forest, gallery forest and coconut plantations.

In 1991, Thomas Arndt searched for *cyanuchen* on Biak and reached the conclusion that its population was small. He saw several black-caps of other subspecies offered for sale in the market in Biak town, and one *cyanuchen*, but a dealer told him that the latter was seldom caught (Arndt 1992). However, John Tabak (pers. comm. 1995) bought 2 *cyanuchen* in Biak markets at different times and stated: "They are readily available in the local markets, being the most esteemed lory amongst Biak people on account of their character and reputation for speaking ability. They are common household pets. They and the sulphur-crested cockatoos are the main species of parrots traded in Biak." He commented that *cyanuchen* is still fairly common in the wild in northern Biak and on Supiori Island, which remain well forested, but that he observed it less often than black-winged lories and Rosenberg's lorikeets in January and February of 1995. However, it is not possible to obtain a permit to export *cyanuchen* from Biak without the personal consent of the national minister of forests.

Newman (pers. comm. 1994) observed *erythrothorax* in the Varirata National Park, 48 km (30 miles) from Port Moresby. He saw groups of two to four birds, plus a single specimen perched at the forest edge. Most sightings were in the area of the park headquarters and car park. I had a fleeting glimpse of three birds not far from the park entrance (coming from Port Moresby) in November, 1994. It was midday, so few birds were in evidence. Opened in October, 1973, Varirata is the first national park of PNG. (It was dedicated to present and future generations, "for wise use, education and understanding of the ways of nature.")

The black-cap is usually seen in flight, when it is conspicuous and noisy. Coates (1985) describes its vocalizations as follows: "The flight call is an oft-repeated variable loud musical whistled note, sometimes disyllabic or trisyallabic and often given as pairs of upslurred notes, sometimes interspersed with other notes. A variety of loud whistles and piercing squeaks are given when perched."

Its repertoire can contain pleasant and varied phrases, at least among captive birds; but there is considerable variation. Accounts by other ornithologists differ. Beehler, Pratt and Zimmerman (1986) describe "a short series of melodious whistles or squeals, more like the call of a wader (or Golden Myna) than of a parrot: *wheedle, wheedle*. Song is a long series of phrases, each of a few notes repeated over and over before beginning a new phrase. Sometimes gives a monotonous series of identical notes suggesting a goshawk."

Diamond (1972) has a different, perhaps slightly fanciful, but equally accurate account: "The commonest vocalization consists of two identical squeals in immediate succession, with a quality as of sleigh-bells when heard in the distance; this double squeal cannot be confused with the voice of other New Guinea parrots. In addition, a variety of high piercing squeaks and loud whistles may be emitted, particularly when the birds are perched."

Diamond suggested that, judging by the trees frequented, it feeds principally on flowers, secondarily on fruits. Others have recorded it feeding on the climbing *Freycinetia*. Coates states that it feeds in the canopy, mainly on pollen, nectar, flowers and fruits, also insects. Pollen and small insects have been found in the stomachs of birds collected in the Weyland Mountains, Irian Jaya.

Nesting: Little has been recorded. In the Markham Valley (west of the Huon Peninsula) a pair was excavating a nest hollow in a dead tree at the edge of

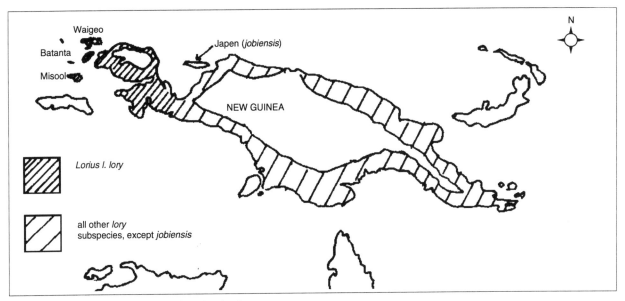

Distribution of black-capped lory (*Lorius lory*).

a garden, in October. One egg in the British Museum (Natural History) measures 27.0 x 22.0 mm (Harrison and Holyoak 1970). This is quite small—but there is considerable variation in eggs laid by some females. For example, three measured from one female *erythrothorax* at Palmitos Park were 29.0 x 24.6 mm, 30.0 x 24.1 mm and 27.8 x 24.0 mm. A second female laid one which measured 30.5 x 26.8 mm. One egg laid by a *salvadorii* measured 31.4 x 25.5 mm (van Dooren pers. comm. 1995).

Status/Conservation: Common to fairly common. The exception is *cyanuchen* from Biak, whose small population is believed to be declining. Due to its restricted habitat, this is the most vulnerable subspecies and conservation measures should be instigated. No such measures are known to be necessary for black-capped lories from other parts of the range. Some trapping for trade occurs, but large importations of this species have not occurred, presumably because it is not a flock species and tends to be wary. As from 1987, importation from Indonesia into European Community countries was prohibited.

Aviculture

Status: The subspecies *lory* and *erythrothorax* are well known; *cyanuchen* is almost unknown, and the other subspecies have occasionally been available.

Clutch size: Two

Incubation period: In the Canary Islands about 25 days, but as long as 27 days has been recorded for the first egg, also 26 days. For the second egg, 24 days has been recorded on several occasions. Paul Tiskens (pers. comm. 1994) found that in Germany, eggs of *erythrothorax* hatched after 26 to 28 days and *jobiensis* after 25 to 28 days (the difference presumably being due to when the female started to incubate). One egg of a *salvadori* hatched after 25 days (van Dooren pers. comm. 1995).

Newly hatched chick: Weight, 9 g; covered in long, white down, longest on the back and flanks; the down on the forehead and crown is gray; feet are pink with dark gray "heels" and the toes have black nails. The beak is dark brown.

Chick development: 3 to 6 days, feet turning gray; day 8, down feather tracts visible under skin of wings and body; day 9, gray head down only just discernible; days 10 to 13, ears opening; day 11, eyes slitting; day 13, eyes half open; day 16, gray second down erupting on wings and breast; day 19, second down erupting all over; day 23, first contour feathers erupting on head, breast and scapulars, red "blush" on cheeks (feathers developing under skin); day 28, head and underparts half feathered; day 50, fully feathered, beak dark brown. Chicks are ringed with 8.5 mm rings between 14 and 17 days; Tiskens (pers. comm. 1994) ringed chicks with 8 mm rings at 16 to 18 days.

Chick growth

Records the weights of two parent-reared chicks from different pairs at Palmitos Park. No. 1 hatched on 28/1/95 and No. 2, on 12/3/95. Both were the only chicks in the nest. The rearing food consisted of a mixture of Nekton-Lori, Milupa baby cereal and honey, with the addition of half a fruit twice daily, usually apple, pear or banana. These weights are contrasted with approximate weights of two parent-reared *jobiensis* in the collection of Andrea and Paul Tiskens in Germany (chicks No. 3 and No. 4) and one parent-reared *salvadorii* belonging to G. van Dooren in the Netherlands (chick No. 5).

Black-capped lory weight table

Age in days	No. 1	No. 2	No. 3	No. 4	No. 5
day hatched	9 e	–			10 fc
1	12 e	11 e			12 fc
2	14 e	14 ne			15 fc
3	17 fic	17 ne			17 fc
4	20 fic	23 fc			19 fic
5	24 nf	25 fic			26 fc
6	27 nf	26 e			28 fc
7	34 nf	37 fc			37 fc
8	40 nf	–			37 fic
9	57 vfc	42 e			45 fic
10	57 fic	57 nf		ringed	55 fc
11	58 nf	55 e			65 fc
12	55 ne	73 vfc			71 fc
13	62 fic	–			81 fc
14	76 ne	82 fc	60		91 fc
15	80 ne	93 fic			100 fc
16	76 fic	–			104 fc
17	82 e	98 fic			108 fic
18	90 ne	–			129 fc
19	132 fc	–			
20	125 ne	–			
21	–	–			
22	149 nf	–		130	
23	146 fic	–			
24	152 fc	150	150		
25	147 fic	162 fc			
26	167 nf	159 fic			
27	173 fc	164 ne		140	
28	170 fic	157 e			
29	169 e	182 fc	160		
30	180 fc	165 e			
31	173 ne	177 e			
32	170 e	182 ne			
33	203 fic	170 e			
34	191 e	173 ne		170	
35	209 fc	179 e			
36	193 e	–	190		
37	196 e	193 fic			
38	190 e	–			
39	215 fic	184 e			
40	197 e	–			
41	202 e	195 fic		200	
42	215 e	–			
43	221 fic	–	210		
44	223 fic	187 e			
45	216 ne	–			
46	220	fic	–		
47	214 e	–			
48	216 e	191 e			
49	224 e	–			
50	245 fc	–			
52	217 e	–			
53	–	186 e			
57	–	–		200	
59	–	–	210		

fc = full crop, fic = food in crop, nf = nearly full, vfc = very full crop, e = empty, ne = nearly empty

Age at independence of hand-reared young: Great variation recorded by author, from only 44 days in a single reared alone to 68 days in two reared together; most are feeding on their own before the age of seven weeks but may still require some food from the spoon.

Young in nest: In the Canary Islands, average 69 days; but between 61 and 77 days recorded. The shorter periods generally occurred during the warmest months. In the U.K., 65 to 70 days, but only 57 days for one particular pair recorded by Trevor Buckell.

General: The black-capped is popular among those who are able to keep the larger and noisier lories. It has color and character and is one of my own favorites. Its complex behavior and vocalizations make it a most interesting and amusing bird in the aviary. Tame birds are excellent pets and often become accomplished mimics and whistlers. Perhaps because of its scientific name (it was the first lory to be described) it somehow stands almost as a symbol for the whole family of lories. To me, this is the quintessential lory, possessing all the attributes which give this group so much appeal. Tiskens and Tiskens (1993) agree: "The Black-cap is one of the most beautiful and interesting of the lories, and ideally should not be missing from any collection of lories." However, they warn: "If you still have 'beloved' neighbors, then one should not wish for this lory since its call can be loud and shrill."

Despite this, black-caps are quite often kept as pets. Like most *Lorius* species, they can become very talented mimics of the human voice. They have such endearing mannerisms as to be irresistible as a pet to the true lory lover. However, he or she must be prepared for the frequent cleaning in the vicinity of the cage in return for the joy of having a black-cap share his or her life.

From the late 1970s, importations of black-caps

gradually became fewer; not enough were bred to maintain numbers and by the mid-1980s it was considered to be one of the rarer species. Prices rose. This, of course, added to its desirability, and more interest was shown in breeding it. Its numbers have since grown steadily.

In the breeding center at Palmitos Park the lories are housed in a large, high, airy, enclosed building, each pair confined to a suspended cage 2 m (6 ft. 6 in.) long. They were set up for breeding at the end of 1989. Between January, 1990, and March, 1994, 84 eggs were laid by several females of the subspecies *erythrothorax* and by one nominate *lory*. February and March are the most likely egg-laying months on Gran Canaria. During the period mentioned above, the number of eggs laid for each month was as follows: January–4; February–15; March–14; April–7; May–7; June–9; July–4; August–9; September–6; October–0 (molting); November–5; December–4.

Forty-one of these eggs produced chicks of ringing age or older. Fertility was high, and most chick losses occurred in the few days after hatching. Oblong or L-shaped nestboxes are attached to the outside of the cage, for ease of inspection. Females tend to be secretive about having eggs and will quietly leave the nestbox when someone is in the vicinity. Unless the nest is checked daily when laying is suspected, the laying dates may be unknown. I inspect the nest about once a week during the incubation period, and daily after the chicks hatch. With the exception of one female who was more aggressive, females would leave quietly when I opened the door, and return quickly. They ceased to stay in the box during the day when the young were about three weeks old, but usually came to watch when I was weighing chicks. They accepted the routine without any fuss, unlike many lories with chicks. The blackcaps are normally excellent parents. One female killed her first youngster on the day it hatched, but thereafter reared her young successfully. Occasionally two chicks hatch but, with the pairs in my care, more clutches produced only a single chick.

Many young were left with their parents for several months after fledging. On one occasion a single youngster remained in the cage for the entire rearing period of the young one from the following clutch; the two were eventually removed from the cage at the same time. I remember the day on which

the chick from the second nest hatched. Oddly, the whole family was in the nestbox; the young one must have been curious about the new arrival! In the warm climate of the Canaries, fairly spacious nestboxes can be used. This experiment might not have been successful in a small nestbox.

Their diet consisted of a mixture of Nekton-Lori and Milupa baby cereal with some honey added. Fruit was offered once or twice daily. Pear and banana at the correct stage of ripeness (not hard, but not overripe) and guavas are favored; apple, cactus fruits, loquats and other fruits in season are readily accepted. I occasionally gave them a few grains of dry sunflower seed. When offered branches of casuarina with small fruits (cones), the seeds were removed first, and then the entire branch was destroyed.

Courtship behavior was difficult to observe in the breeding pairs, except for the preliminary phase. This consists of rapid bobbing up and down with the feet anchored to the perch, and swaying from side to side. I have also observed males briefly holding open the wings to show the striking yellow area on the primaries. The behavior of my hand-reared and somewhat imprinted male, when he attempts to copulate with my hand, is described here, but I am not certain if this is entirely typical. After the preliminary phase described, he makes a rhythmic squeaky sound, sometimes blowing (roughly equivalent to hissing in *Trichoglossus*). This is followed by a drawn-out "*Fooh!*"

Experiences of breeders with other subspecies are of interest. In Germany, Andrea and Paul Tiskens had a female *jobiensis* who was paired with a male of the nominate race. The resulting young had the breast patch clearly defined as in *erythrothorax*. Finally, in 1991, they were able to obtain a male *jobiensis*. The other male was removed from earshot—but it was several months before the female would tolerate the new male or allow him near the nestbox. Mating was first observed in January, 1992. The female laid on March 20 and 24, but both eggs were infertile. She laid two fertile eggs in June. The first hatched on July 1 or 2 and the second, on July 4. The young left the nest for a short time on September 1, and by September 12 were spending most of the day out of the nest (Tiskens and Tiskens 1993).

A *jobiensis* owned by Trevor Buckell confirmed the value of surgical sexing. An enormous bird, hatched in 1982, it was assumed to be a male. It had been observed trying to mate with the female in a rather ineffective manner—but the eggs were always infertile. Then one day in 1993 the bird was surgically sexed—only to prove that it was a female! At the time, there were no male *jobiensis* in the U.K., so she was paired with an *erythrothorax*. Neither of the resulting young had any red on the underwing coverts (Buckell pers. comm. 1993).

The subspecies *cyanuchen* is almost unknown outside Indonesia. However, in Canada, John Tabak is the owner of a single bird. In 1994, while traveling in Indonesia, he obtained a young partner for this lory. When he reached Jakarta, he was denied CITES papers to export it as this subspecies is protected by the Indonesian government. He ex-

changed it for a black lory, for which a CITES permit was issued (Tabak pers. comm. 1995).

In 1937 the zoologist S. Dillon Ripley visited Biak. He was most impressed by a male *cyanuchen* whose "conversational powers were a constant source of wonder":

> He would rattle along for minutes on end in a mixture of the Biak language and Malay. The tone of his voice was low and so quaintly pitched that one could never fail to drop all work and stop to listen to his endearing chatter. He was very neat and tidy, constantly bathing and preening his vivid blue and red feathers in the half coco-nut shell tied to his perch. In this he delighted to bathe, becoming so drenched that he looked more like a bunch of badly-chewed rags than a trim little lory (Riplay [*sic*] 1938).

Purple-bellied Lory   *Lorius hypoinochrous* (G. R. Gray 1859)

Synonyms: Eastern black-capped lory, Louisiade lory

Native name: *Mallip* or *Malip*

Description: Forehead, crown and nape are black; wings are green and the thighs, lower abdomen and undertail coverts are dark bluish purple; much of the rest of the plumage is red. The underwing coverts are red (with the outermost feathers being partly black), and there is an extensive yellow band across the underside of the flight feathers. This description could apply almost equally to *Lorius lory*, to which this species is closely allied. It differs in several respects. A broad but faint collar extending from the lower cheeks to the upper breast is washed with dull purple, and much of this area is faintly shaft-streaked. Some individuals have a blackish band on the mantle. The tail is red above, hidden by the tail coverts, the exposed area being dark blue-green; the underside is dull olive yellow. A conspicuous feature of this lory is the startling white cere. The iris is orange-brown; the beak, orange; the feet, and the skin surrounding the eye, are dark gray.

Length: 28 cm (11 in.)

Weight: 200–240 g (Forshaw 1989); one female, 187 g (Beehler 1978)

Immatures: The birds have the iris brown and the bill brownish.

Subspecies: According to Forshaw (1989), *L. h. rosselianus* differs from the nominate race in having the red of the breast the same shade as the upper abdomen, whereas in *hypoinochrous* the breast is paler. Buckell examined skins of all three subspecies and stated that the red is a darker shade, similar to that of *Eos bornea cyanothus*. *L. h. devittatus* supposedly differs from the nominate race in lacking the black margins to the greater underwing coverts. However, what is not clear is whether the black margins are present in immature birds (as in *Lorius lory*). In the accompanying photograph of *devittatus* some of the greater underwing coverts are in fact tipped with black.

Natural history

Range: The nominate race occurs on Misima and Tagula (Louisiade Archipelago), in the eastern

172

Papuan Islands; *rosselianus* is confined to Rossel Island (Louisiade Archipelago) (see Gazetteer), and *devittatus* is found in southeastern New Guinea, west to the southern side of the Huon Gulf in the north and Cape Rodney in the south. Reports of this species in the Port Moresby area were in error, according to Coates (1985). It inhabits nearby offshore islands and eastern satellite islands. These include Goodenough, Fergusson and Normanby (these three comprise the D'Entrecasteaux Archipelago), as well as Bentley. It also occurs in the Trobriand and Woodlark islands. In the Bismarck Archipelago (see Gazetteer) it is found on New Britain, New Ireland, Long, Umboi, Sakar, Witu, Lolobau, Watom and New Hanover. In addition, the islands of Lihir and Tabar are often listed in the range of this species. In 1995, Peter Odekerken spent three weeks on Lihir; he did not see it there and neither did Ian Burroughs who was studying megapodes there. Furthermore, the local people had no knowledge of it (Odekerken 1995b).

Habits: This species frequents flowering trees, especially coconut palms, in pairs or small groups. Coates describes it as "noisy, conspicuous and active." It is often seen flying through the treetops at no great height. Distinctive in flight, its wing beats are rapid and shallow, and the wings appear relatively short and rounded. He noted: "The loud drawn-out coarse nasal wailing or whinnying call notes of this species are very distinctive." Beehler, Pratt and Zimmerman (1986) describe the call as "a single, unmusical, harsh rising then falling, note (reminiscent of a steam locomotive whistle or donkey's bray), often repeated by birds in a flock." They found it very different to the "musical whistles and squeals" of *Lorius lory*. The purple-bellied lory occurs in primary forest, forest edges, tall secondary growth, partly cleared areas and coconut plantations. On New Britain it was "commonly seen in the forest canopy and flying over the camps...usually...in pairs" (Orenstein 1976). It is a lowland and foothill species on mainland New Guinea, but elsewhere may be found higher—up to 750 m (2,460 ft.) on New Ireland, for example, and even as high as 1,600 m (5,250 ft.) on Goodenough Island. In the Ok Tedi area of the Western Province it is common, and usually observed in pairs. Once it was reported at 1,535 m (5,035 ft.), and is sometimes seen flying over the town of Tabubil (Gregory 1995).

Goodenough is only 28 km (17 miles) from the tip of Cape Vogel in Papua New Guinea. It is wet and mountainous. In 1976, Bruce Beehler visited Goodenough in search of a mysterious black bird. He did not find it—but while camping at 540 m (1,800 ft.) he made an interesting discovery: "Here I observed a remarkable phenomenon exhibited by two species of parrots, the Eastern Black-capped Lory and the Eclectus Parrot. Numbers of these birds roosted at night in the montane forest above our camp and spent the day foraging in the lowlands two thousand feet below. In the early morning I watched the birds sail down the mountain, wings folded, at high speeds, their bodies making remarkable rushing sounds when they passed over the camp. Near dusk, the birds made their way up, slowly, gaining altitude by flying in a lazy circling pattern, flapping higher and higher, like soaring hawks gaining altitude under a thermal. Why these birds chose to roost in the mountains and forage all day in the lowlands remains a mystery" (Beehler 1991).

Bruce Beehler also observed this lory on New Ireland where he found it to be common from the coastal lowlands up to about 750 m (2,460 ft.). He described its call as "a loud grating whistle that resembles 'whoaoa;' *albidinuchus* gives a weak rising whistle that is quite different...I observed the two species feeding and resting in the same tree without interspecific aggression. I normally saw *albidinuchus* in pairs but *hypoinochrous* was solitary" (Beehler 1978).

Coconut palms and the blossoms of *Plerandra* are favorite food sources of this species. On Normanby Island purple-bellied lories were observed in *Plerandra* shrubs inserting the tongue into blossoms or into the tip of unripe green fruits, where there was an indentation, but they were not seen to bite blossoms or fruits. Several lories were seen in the company of cockatoos and eclectus parrots in the crown of a tall emergent tree.

Nesting: Almost nothing has been recorded. There is one observation from Witu Island in the Bismarck Archipelago, at about midyear, of a pair nesting high in a huge tree. Kenning (1994) was told of a nest in a forest in a tree stump at a height of 6 m (20 ft.). On Misima Island, young birds removed from the nest were being hand-fed in July (Odekerken 1995b).

Status: Common and locally abundant. In 1992 and 1993, J. van Oosten found it to be common in eastern New Britain, especially in the plantations, where it was found with Massena's and red-flanked lorikeets. Observers in most localities have reported *devittatus* to be common or extremely abundant.

Aviculture

Status: Almost unknown outside New Guinea and islands.

Breeding data: Not recorded, but would almost certainly be identical to that of *Lorius lory*.

General: In England a pair owned by Mrs. Burgess, reported in 1922, were believed to be "new to aviculture." In 1933 Herbert Whitley of Devon also kept a pair. There appears to be no other report of this species until that from Chester Zoo, where it was bred in 1973 (under the name of Louisiade lory). My first and only sight of this lory was in November, 1994, when visiting Anne Love, a lady from New Zealand who was then residing in New Guinea. Her collection of lories included four *devittatus*. In beautiful feather, they were a delight to observe. I noted that their voices were harsher than that of *Lorius lory* and that their conversational notes were quite different. The couples were kept in separate cages and, in turn, each pair was allowed out of its cage to fly around the room. They kept mainly to the perches placed high above the cages and did not molest the other birds. In August, 1995, these four lories were reluctantly sent to the Bronx Zoo. This was due to their owner's impending departure to Australia (where the importation of this species is not permitted).

According to Kenning (1994), purple-bellied lories are "beloved house pets in Rabaul [New Britain] and regularly seen."

Stresemann's Lory, New Britain Lory *Lorius amabilis*

This "species" was named from a single specimen, a female, obtained at Nakanai in New Britain. It differed from *hypoinochrous* in lacking the black cap. Other differences were small: the center of the abdomen was dull purple-blue, tinged with green; and undertail coverts, dull purple-blue, marked with green. Length: 26 cm (10 in.). No other specimen is known. Either it was an aberrant specimen of *hypoinochrous* or a hybrid between *hypoinochrous* and an escaped lory of some other *Lorius* species.

White-naped Lory *Lorius albidinuchus* (Rothschild and Hartert 1924)

Description: The cap is black, the wings are green and most of the rest of the plumage is red. The small area of white on the nape is a unique feature. It is otherwise distinguished from *chlorocercus* by the lack of a yellow collar, although there are yellow markings on the sides of the upper breast. It resembles *chlorocercus* in having unusual silvery blue feathers at the bend of the wing. Underwing coverts are red and there is a broad band of yellow across the underside of the flight feathers. The tail is red, broadly tipped with green, the underside being tipped with bronzy yellow. The beak is orange; the iris is dull orange, followed by a ring of brown and a narrow inner ring of white. The cere and the skin surrounding the eyes are black.

Length: 26 cm (10½ in.)

Weight: Beehler (1978) recorded the weights of four females as 152 g, 158 g, 160 g, and 164 g. Males might be slightly heavier.

Immatures: The birds have yet to be described.

As in other members of the genus, the beak and iris would be brown.

Natural history

Range: New Ireland, Bismarck Archipelago (see Gazetteer).

Habits: It inhabits montane forest between 500 m and 2,000 m (1,640 ft. and 6,560 ft.). Usually observed in pairs, it is sometimes seen in the same tree as *hypoinochrous* with which it is sympatric between 500 m and 750 m (1,640 ft. and 2,460 ft.).

Beehler (1978) notes that *hypoinochrous* is heavier, with a larger bill; thus the two species are likely to exploit different food sources. The white-naped lory "was common throughout its altitudinal range. I recorded its call as '*schweet*' or '*schweet-schweet.*' It feeds upon fruit and flowers of the wild oil-palm that is found in abundance to the summit of the range."

Nesting: Nothing has been recorded.

Aviculture: It is unknown in captivity outside its native island.

Yellow-bibbed Lory *Lorius chlorocercus* (Gould 1856)

Synonym: Green-tailed lory

Description: Forehead, crown and the back of the head are black, the feathers of the occiput having the base green; the rest of the head is red. A broad crescent of yellow decorates the upper breast, extending upward almost to the throat (the yellow there being hidden by red feathers); the yellow on the breast is touched at each end by a patch of black which extends upward to the area behind the cheeks. The wings are green; the bend of the wing, including the underside, appears to be a most unusual shade of mauve-tinged silver; on closer examination the feathers are mauve at the base, silver elsewhere. There is a large area of pinkish red on the primaries, and the underwing coverts are blue. The feathers of the thighs are green and blue. The back and upper tail coverts are slightly darker red than the body. The basal half of the tail is red, the rest being green above and green-tinged yellow below. The rather small bill is orange, marked with gray at the base of the upper mandible. The iris is orange with an inner ring of white. The legs are very dark gray.

Length: 28 cm (11 in.)

Weight: About 170 g; 125–160 g (Engels and Philippen 1992); one adult captive male weighed 181 g.

Immatures: The birds have less yellow on the upper breast. All of the dozen young reared by the author had a very conspicuous black patch on the

side of the neck, in contrast to Forshaw's statement: "no black markings on sides of neck." The plumage is duller overall than that of adults, and further differs in that the blue of the outermost greater underwing coverts is broadly edged with black. The skin surrounding the eye is whitish; beak and iris are brown.

Natural history

Range: Eastern Solomon Islands (see Gazetteer): Guadalcanal, Malaita, Ugi, San Cristobal and Rennell.

Habits: Little has been recorded about this common species, presumably because it is more difficult to observe than the other lories of the Solomons. A canopy dweller, it occurs in forest and secondary growth at all altitudes. On Guadalcanal, Cain and Galbraith (1956) found it to be more plentiful in the hills than in the lowlands; it was most common in lower mist forest. On other islands, it was seen in and around coconut plantations. Very active in the uppermost branches when investigating flowers and epiphytes, it usually occurs in pairs or in groups of up to ten birds. In 1990, Jonathan Newman found it to be common in primary forest on Guadalcanal, where it replaced the cardinal lory. It occasionally visited tall trees surrounding gardens. On Makira it was abundant in all habitats. It was not seen on Kolombangara (Newman pers. comm. 1994). When

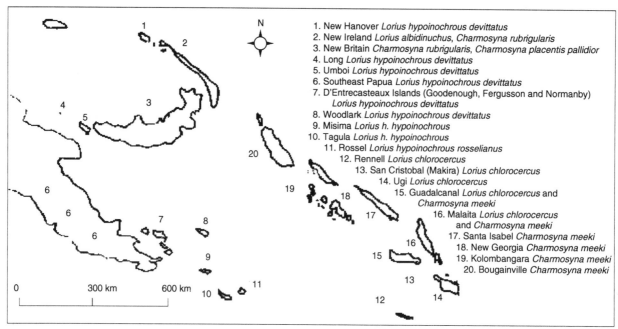

Some species of southeast Papua to the Solomons.

1. New Hanover *Lorius hypoinochrous devittatus*
2. New Ireland *Lorius albidinuchus, Charmosyna rubrigularis*
3. New Britain *Charmosyna rubrigularis, Charmosyna placentis pallidior*
4. Long *Lorius hypoinochrous devittatus*
5. Umboi *Lorius hypoinochrous devittatus*
6. Southeast Papua *Lorius hypoinochrous devittatus*
7. D'Entrecasteaux Islands (Goodenough, Fergusson and Normanby) *Lorius hypoinochrous devittatus*
8. Woodlark *Lorius hypoinochrous devittatus*
9. Misima *Lorius h. hypoinochrous*
10. Tagula *Lorius h. hypoinochrous*
11. Rossel *Lorius hypoinochrous rosselianus*
12. Rennell *Lorius chlorocercus*
13. San Cristobal (Makira) *Lorius chlorocercus*
14. Ugi *Lorius chlorocercus*
15. Guadalcanal *Lorius chlorocercus* and *Charmosyna meeki*
16. Malaita *Lorius chlorocercus* and *Charmosyna meeki*
17. Santa Isabel *Charmosyna meeki*
18. New Georgia *Charmosyna meeki*
19. Kolombangara *Charmosyna meeki*
20. Bougainville *Charmosyna meeki*

disturbed or in flight, yellow-bibbed lories gave "trumpeting calls."

When Jan van Oosten visited Guadalcanal in April 1992, his main interest was in seeing lories in the wild. Only one pair of yellow-bibs was observed. While following a trail down Mount Austin, he observed the pair searching the red flowers of a vine growing in a tree. He was 15 m (50 ft.) away and decided to walk over and pick the flowers. "The pair stopped what they were doing to peer down at me. Up to this point they were only making a soft chattering noise but as soon as I was close to the tree their chatter became louder. I picked a handful of flowers and retired back to where I had spotted them. They continued climbing up the vines and examining the flowers, stopping only long enough to drain the flower. After they had climbed to within ten feet of the top of what I estimated to be a sixty foot tree they flew off up the mountain. It was indeed a beautiful sight to see the sunlight on them as they flew away" (van Oosten 1993).

This lory feeds on pollen, nectar, seeds, fruits and insects. Cain and Galbraith found vegetable matter with a high proportion of seeds in the crop contents they examined. Newman (pers. comm. 1994) observed it on flowering trees (but never on coconut palms) and believed that it fed mainly on pollen. On Makira, he regularly saw yellow-bibbed lories on one species of tree, feeding on large (5 cm diameter) green fruits, on which no other species was seen feeding.

Nesting: Nothing has been recorded. Egg sizes have been recorded as 29.7 x 23.9 mm and 30.3 x 23.9 mm (Harrison and Holyoak 1970) and, from captive birds, several 32.0 x 22.0 mm (Engels and Philippen 1992) and one 30.8 x 22.7 mm (van Dooren pers. comm. 1995).

Status/Conservation: It is invariably described as common. No conservation measures appear necessary at present.

Aviculture

Status: Uncommon but increasing.

Clutch: Two

Incubation period: About 25 days; however, the second egg in the nest of a pair owned by G. van Dooren hatched 29 days after it was laid (van Dooren pers. comm. 1995).

Young in nest: About ten weeks.

Newly hatched chick: Weight, 7 g; dense white down; beak brown.

Chick development: Day 12, eyes slitting; day 18, midgray second down erupting on wings and breast; day 21, second down all over body; day 25, many of contour feathers erupting; 40 days, most of contour feathers free of quills except on crown, nape, back and some wing coverts; eight weeks, fully feathered, beak dark brown. One chick reared by van Dooren's pair was ringed at 15 days with a 7 mm ring.

Chick growth

Shows weights of one chick hand-reared from the egg by Philip Eglinton of Paraparaumu, New Zealand in 1993. It was reared on Lake's hand-rearing food.

Yellow-bibbed lory weight table

Age in days	Weight in grams
7	13
15	45
21	71
29	104
37	123
45	138
53	142

Another breeder in New Zealand has recorded peak weights of 170–180 g, and weaning weights as low as 130 g (Eglinton pers. comm. 1995).

Bosch (1988) recorded the weights of two parent-reared young as 135 g and 140 g at six weeks, and as 130 g and 150 g when they started to fly.

Aviculture: Only in the 1990s did this species start to become more generally available. In or about 1991 the Solomons permitted, for the first time, the export of certain bird species, on a quota system. The yellow-bibbed lory was among the few species included. However, the quota system was abused to such a degree (larger numbers than permitted being exported) that from October, 1996, the Solomon Islands were no longer allowed to export birds on a commercial basis. In the previous couple of years, the relatively large numbers of cardinal lories imported into Europe had resulted in the price falling substantially. In 1996, consignments of lories from the Solomons were still entering South Africa. In August of that year, newly imported yellow-bibs were being offered for sale at R3,000 per pair (equivalent to about $750 or £500); cardinals were advertised at R2,000 per pair and Massena's, at only R800 per pair. In the same issue of *Avizandum* in which this advertisement appeared, a well known bird park was offering yellow-bibbed lories (either long-established or captive-bred) at R3,900 per

pair. When prices of captive-bred birds are undermined by newly imported ones, the cessation of importation is long overdue! The U.S.A., however, received only a small number of wild-caught birds, but under a special cooperative breeding program, more may arrive. At the time of writing, approval was pending from the U.S. Fish & Wildlife Service for the International Loriinae Society and the Avian Ark Foundation to bring in 30 birds. These would go to breeders under terms of agreement arranged with the government of the Solomon Islands.

The small number of yellow-bibbed lories in Europe in the 1980s had originated either from those privately imported into Switzerland in the late 1970s or from Loro Parque, Tenerife. The first recorded breeding appears to be that which occurred in the collection of Dr. R. Burkard in 1980, when a single young fledged. Dr. Burkard had two birds believed to be males and one female. After three years the pair had made no attempt to breed, so the males were exchanged. The new male was interested in the female but she avoided him. The synchronous duetting which is apparently normal for this species was no longer heard. After some months the original pair was reunited. The birds then called, duetted, danced and hopped. This pair went on to produce many young (Burkard 1983).

Joe Kenning from Australia had obtained parrots in the Solomons for a few collections. He described the yellow-bibbed as the most popular pet parrot there, as it was hardy and attractive. It could survive on the usual basic diet of sweet potatoes, whereas cardinal lories fed in the same way would die within a year. It was probably popular also because birds of this species become good mimics and can learn to whistle tunes. The yellow-bibs which Kenning collected came from Malaita Island, to the north-east of Guadalcanal, as did all the parrots offered for sale on Guadalcanal (Kenning 1993-4).

Prior to the 1980s, the yellow-bibbed lory had been established in aviculture in New Zealand, more or less by chance. In or about 1971, 14 birds were offered for sale in a pet shop. Perhaps someone who had owned or worked on a boat had taken them into New Zealand, probably unaware that they could not be imported legally, and offered them to the first pet shop he had come across. Why else would a species so rare in aviculture be found in a pet shop? Fortunately, one pair was obtained by Fred Rix from South Auckland, who has bred a substantial number of young from them. In 1989, he told me that there were seven breeding pairs in New Zealand. In 1995, he bred a pied individual; half the underparts were yellow and the wings were heavily marked with this color. Pied birds are often the result of metabolic or dietary problems, but this was probably not the case as the bird's nest mate was also unusually marked, with the black neck markings extending right across the breast. In that season Mr. Rix bred 17 young from six pairs (Rix pers. comm. 1995). In 1990, the Government of New Zealand permitted the export of nonnative bird species for the first time in many years; thus a few yellow-bibs bred in New Zealand were imported into the U.K. At the time of writing there are increasing numbers in Europe resulting from direct importations from the Solomon Islands. In the U.S.A., yellow-bibs are rare.

Little has been written in the avicultural literature regarding this beautiful lory. Engels and Philippen (1992) recorded eggs laid, but none hatched. The first three clutches were infertile, laid when the pair was at least four years old and had been in the aviaries for three years. The fourth clutch was fertile, after various additions to the diet, but the eggs were abandoned shortly before they were due to hatch.

This species was in my care for the two years during which I was curator at Loro Parque, Tenerife. Breeding results were not good. Only a single pair produced chicks; the eggs from the other two or three pairs were infertile or, on occasions, destroyed. As I recorded in *Parrots, their care and breeding*:

> With one exception, the males were aggressive and delighted in biting at every opportunity. They were not popular with the keepers! All the *chlorocercus* were cheerful and cheeky, whistling a greeting and always trying to attract attention. The only non-aggressive male was that of a pair kept off-exhibit. He was a delightful bird, always friendly. The female was surely the most remarkable Yellow-bibbed ever to be in captivity. Together they produced 12 young during my 2 years there—with intensive management, for the female was totally blind. She would whistle to those she knew, usually from her nestbox as she spent so much of her life breeding. She moved around by touch, espe-

cially when on the aviary floor, where she would use her beak to feel her way (Low 1992a).

In 1987, five young were hand-reared from the egg. In 1988, the female laid five clutches of two eggs, the first egg of each clutch being laid on or by February 28, March 30, May 17, June 12 and July 27, respectively. In the first clutch, two chicks hatched under Forsten's lorikeets. In the second clutch, one egg was infertile and the other hatched under the parents on April 23. The chick was at once placed in the nest of yellow-streaked lories with a newly hatched chick. Both eggs of the third clutch had hatched under violet-necked lories by June 18. The eggs of the fourth clutch were moved to a pair of yellow-streaked lories on June 29 and hatched on July 7 and by July 11. The single chick of the fifth clutch hatched under green-naped lorikeets on August 22. All but one of these chicks were removed from foster parents after two or three weeks and hand-reared. The other one fledged with the yellow-streaked.

I found the hand-reared young fairly slow to become independent, although they would sample warm rearing food at an early age. The weaning period was gradual and the young were not independent until about 10 weeks. Even after they left the hand-rearing room, I continued to offer them warm rearing food until they were four months old, as they much preferred it to the thinner nectar. During the rearing period the usual fruits were eaten.

I have found this species more willing to sample a wide range of foods than almost any other lory. One male is remarkable for eating virtually anything which is edible! Yellow-bibs seem to need some items of more solid food, and relish a few sunflower seeds daily, or almost any extras. They eat such hard items as raw carrot but equally enjoy soft fruits and blossoms. Nectar should always be available; they usually take a sip between every few bites of solid food.

Philip Eglinton in New Zealand offers his yellow-bibs nectar and half an apple daily, plus other items such as orange, kiwi fruit (seeds only are eaten), seeding grasses, dock—which is preferred to green foods such as silver beet and puha (milk thistle)—Madeira cake and fresh eucalyptus branches. Even during incubation, when fresh branches are placed in the aviary, the male will call to the female who will immediately leave the nest. She starts to chew the leaves, then rapidly preens her feathers. The nestbox for this pair is made of untreated wood 2.5 cm (1 in.) thick; it is 50 cm (20 in.) high and the base measures 20 cm (8 in.) x 15 cm (6 in.) (Eglinton pers. comm. 1995).

It is likely captive-bred birds will prove easier to breed than their wild-caught parents. This is a very attractive lory and the demand for captive-bred young should be high as long as they do not have to compete with a flood of wild-caught birds. From the avicultural viewpoint, stopping the trade in the latter would be beneficial; aviary-bred birds would then maintain a price which makes them worth breeding.

Purple-naped Lory *Lorius domicellus* (Linnaeus 1758)

Synonym: Purple-capped lory

Description: Forehead, crown and the back of the head are black; the streaked, lengthened feathers of the nape are violet. The rest of the head, and the entire underparts, are red, except for the variable yellow band on the upper breast. This is less pronounced than in the yellow-bibbed lory in most individuals, but even more pronounced in others; yet in some it is only slightly indicated. The mantle is darker red than the body. The thighs are violet-blue. The wings are green with a bronze caste to the median wing coverts. The feathers on the bend of the wing are mingled with white, reminiscent of *chlorocercus*. Underwing coverts are blue, and there is a broad yellow band across the underside of the flight feathers. The tail is red at the base, dark brownish red at the tip. The underside is olive-gold-red, the feathers being margined with greenish olive. Spence (1955) described the tail as "deep ox-blood" on the tip and "fluorescent orange, shot red, below." The iris is dark grayish brown; in Tom Spence's pair, the female's iris had a faint light gray inner ring, while the male's was a "more brilliant light ring." The beak is orange, marked with brown or gray at the base of the upper mandible. The cere and the skin surrounding the eye are dark gray.

Length: 28 cm (11 in.)

Weight: 210–230 g

Immatures: The birds are variable. Kuah (1993) described two young; they had "an extensive spread of purple on the nape" and one lacked any yellow on the breast. Spence (1955) made a careful description of the young one from his pair, the first to be reared in the U.K. Its plumage was duller. "The purple of the nape is deeper in this individual and more extensive than in its parents, while the yellow chest patch is widespread and diffuse. The feathers of the mantle differ in having dull green bases to the red tips, while the tail has a faint bluish tip with the two central tail feathers drawn out like a pintailed snipe. The greatest difference is in the underwing-coverts, the posterior row of which is black-tipped and not wholly blue as in the adult." At about four months old, the young one's bill color started to change from black to orange.

Mutation: A breeder in Singapore acquired four birds with patches of yellow on wings, flights and thighs and red spots on the black crown. Acquired piedness in any green bird is not unusual due to various reasons (such as a dietary deficiency); however, these birds were all caught in the same area of Seram, according to the trapper (Kuah pers. comm. 1992). Photographs show that only one had a marked amount of yellow on it. The red spots on the crown are worth noting, as I have seen at least two other *domicellus* with this feature. One, in the collection of Francoise and Antonie Meiring in South Africa, also showed a lack of pigment: the cere was white, the skin surrounding the eye, pinkish; and the feet and legs, pinkish mottled with light gray.

Natural history

Range: Seram (see Gazetteer), Indonesia; formerly occurred on Amboina (Ambon). Its distribution is often said to include Buru; but, if it occurred there, it was probably due to accidental introduction through escaped pet birds.

Habits: The purple-naped lory inhabits primary and secondary forest, between 400 m (1,312 ft.) and 1,000 m (3,280 ft.). Taylor (1991a) carried out two surveys. During the second, only seven observations were made: "an alarmingly low number. Although it is a very brightly colored species it is well camouflaged when sitting quietly in trees. It is easily overlooked when not calling and may not be as rare as the number observed suggests."

Taylor (1991b) further suggested that lack of observations may be due to the shyness of the species. Local trappers use a decoy bird in a cage because purple-naped lories are so difficult to find. They make a trap using nylon threads woven around a branch. When searching for this species, Taylor took a caged bird with him. When it called it was sometimes answered by wild birds. Without this bird, he suspected that he would not have seen any. During both studies, only single birds were seen, with the exception of one pair.

Nesting: No information. Schönwetter (1964) described the average size of eight eggs as 32.0 x 25.5 mm, and the range as 30.5–33.6 mm x 25.0–26.6 mm.

Trade: Ornithologists who took part in the surveys on Seram in the late 1980s, and in 1991, concentrated their attention on the Moluccan cockatoo—a sort of flagship species for Indonesia, much as the hyacinthine macaw is for Brazil. The cockatoo has also suffered a terrible reduction in numbers due to trapping, and there are large captive populations worldwide. In contrast, the captive population of the purple-naped lory is small. I suspect that mortality is higher among recently trapped lories than in the Moluccan cockatoo which, despite its sensitive temperament, is a strong bird. It is sad to think of the "wastage" which has occurred in wild-caught purple-naped lories. As well as losses due to incorrect care and feeding before export, this lory is so popular with the local people that perhaps most of those trapped would be sold on Seram and Ambon. Brockner (1993b) recorded of Ambon: "We saw many just by walking through the streets and often several were kept by one person."

As the purple-naped lory has been rare in aviculture since the late 1970s, one assumes that the domestic pet trade "consumed" a large proportion of the trapped birds; kept as pets, with no opportunity to breed, these birds contribute nothing toward the establishment of the species in aviculture. In southeast Ambon, near the fishing village of Atiahu,

Brockner met a bird dealer. He obtained parrots, including red lories and green-naped lorikeets, from the local people and took them to the wholesalers in Ambon. In 1990 this dealer sold over 200 purple-naped lories and over 1,000 Moluccan cockatoos; in 1989 this cockatoo was placed on Appendix 1 of CITES —so the trade in them was totally illegal. At the time of Brockner's visit, the dealer had nine purple-naped lories on hand—"adults of breeding age which were then lost to the wild population." By this time, these lories were becoming rare; they commanded top prices and were quickly sold to the wholesalers.

Status/Conservation: Apparently already extinct on Ambon. Brockner (1993b) wrote of his 1991 visit: "Today only secondary vegetation is to be found on the densely populated island, apart from a tiny area of primary forest in the western part. There is, however, not enough to provide suitable habitat for a parrot." This lory is now considered to be an endangered species (Lambert, Wirth et al. undated). On Seram it is declining due to trade and habitat destruction. Brockner noted: "The rainforest around the towns of Amahai and Masohi in southern Ceram has been largely cleared. Our guide told us that Japanese companies had been systematically logging the area for years." Although the Manusela National Park covers 1,800 km² (about 10 percent of Seram), bird trappers penetrate deep within its boundaries to catch parrots. Brockner commented that it is virtually impossible for the park management to protect the birds. Financial help is needed to enable the parrots to survive on Ceram. The purple-capped lory is protected by law—but it appears that no attempt is made to enforce it. Lambert, Wirth et al. suggest that "increased protection of Manusela, including increased patrolling and the prevention of trapping in the park, are also necessary. An awareness campaign among the villagers in the region should be initiated in an attempt to reduce trapping levels."

If the purple-capped lory is to survive in its natural habitat, I believe that urgent action must be taken before its numbers, like those of the Moluccan cockatoo, are decimated by trapping. It seems that the following procedures are needed:

1. A local education program, to make people aware that trapping is illegal and will result

in extinction of the species if continued on the present scale.

2. Fines for illegal trapping.
3. More field work, to obtain information on its natural history, especially feeding biology, and to determine appropriate actions to protect it.
4. Enforced protection of habitat.
5. Participation of **all** who keep this species in the studbook kept by Armin Brockner, Altmannstrasse 25, Meckenbeuren, 88074 Germany. (Participation does not affect ownership.)

Aviculture

Status: Rare outside Indonesia and Singapore.

Clutch size: Two

Incubation period: Usually 24 days for first egg; 23 to 26 days recorded.

Newly hatched chick: Weight, 9 g. Covered in "long almost white down on back and humeral region [shoulders]" (Spence 1955).

Chick development: A chick hatched at Jurong Birdpark, Singapore, developed as follows (Santos pers. comm. 1994). "Day 11, eyes closed; entire body covered in long white fluffy down, below which was shorter, darker gray secondary down. Day 14, eyes slightly opened. Day 17, crown of head purplish when pins have just emerged from the skin; yellowish wispy down feathers apparent on the nape, parts of the wing coverts and the breast; red pins emerging from the skin on the breast; beak dark, almost black, but lower commissures [where mandibles meet] pinkish. Day 21, red pin feathers erupting on crop area. Day 25, red pin feathers emerging from the tail. Day 29, red patch around the eye very evident; feathers erupting from the pins on the wing and tail; red feathers irregular from the breast to abdomen. Day 37, red upper tail covert feathers prominent; all red feathers on the breast interspersed with green ones; feathers erupting from the forehead but still in pin-stage at crown; secondary feathers and coverts erupting from the pins; green tail feathers also erupting; cheek and ear coverts still in red pins; entire beak now black. Day 56, nape feathers purple-green; almost fully feathered except for some downy feathers scattered all over the back."

Chick growth

Records the weight of the chick described above (No. 1), hatched at Jurong Birdpark in 1994, plus three others hand-reared there. The diet consisted of blended peanut butter and well soaked Science Diet Canine Maintenance Pellets (23 percent protein), plus papaya puree. Ten percent of the chick's body weight was given per feed, five times daily. At day 47, feeds were reduced to three daily, 24 ml per feed; at day 69, 20 ml morning and evening and 10 ml in between; at day 87, 10 to 20 ml twice daily; at day 89, 20 ml in evening only; and at day 95, 12 ml in evening only (Santos pers. comm. 1994).

Purple-naped lory weight table

Age in days	Weight in grams			
	Chick No. 1	No. 2	No. 3	No. 4
11	61*			
14	64			
17	79	84*		
20	98	84		
23	118	110		
26	132	128		60
29	148	146	136	86
32	162	166	163	105
35	177	180	183	132
38	186	185	195	146
41	199	196	213	163
44	202	200	212	171
47	210	200	216	185
50	206	198	219	204
53	201	198	216	208
56	185	194	213	214
59	177	189	209	213
62	186	184	206	206
65	177	178**	200	201
68	–	171	194	197
72	–	171	189	192
75	–	166	186	187
78	–	173	183	186
82	–	171	187	182
85	–	171	193	180
88	–	174	189	179
92	–	169	187	177
95	–	174	–	175
98	–	169***	–	178
102	–	–	–	173
105	–	–	–	172
111	–	–	–	164

Note that chick No. 4 was retarded due to poor feeding by the parents, thus its development was slower.
 * food in crop when chick removed from nest ** first flight *** completely weaned

The weights of one parent-reared chick were given by Brockner (1992b) as follows: 20 days, 132 g; 35 days, 170 g; and 65 days, 230 g.
   Young in nest: 65 days, two in one nest (Brock-

ner 1992a); 95 days (Spence 1955). The normal period is probably about 70 days.

General: In 1989, a breeding conservation program for the purple-naped lory was set up in Germany by the Zoologischen Gesellschaft fur Arten- und Populationsschutz. After some inquiries, it learned of the following in captivity in 1990: 12 pairs in the Netherlands, ten pairs in Switzerland, two pairs at Loro Parque, Tenerife, ten or so birds in the U.S.A. and seven with Armin Brockner, the studbook keeper. By 1994, Armin Brockner believed that there might be as many as 50 pairs in Germany. The demand was growing, despite the high price. In addition, there were many birds, more than those listed above, in a few private collections in Singapore. I have seen 40 pairs in just two private collections and, at the same time, Jurong Birdpark had 20 or so birds. In Singapore, Patrick Tay was the first to breed the purple-naped lory, in 1990. As well as nectar and fruit, his birds received a mixture of ground-up cooked beans and rice, 2 tablespoons per pair daily. Kuah (1993) of Singapore mentions "almost 60 pairs" in one collection there without naming the owner. In December, 1992, Lawrence Kuah wrote of his collection: "It is really quite difficult to maintain a large group of lories commercially; with approximately 120 *L. domicellus* and all those of the other species, the total number is close to a thousand individual lories" (Kuah pers. comm. 1992). The remarkable thing is that he was still at school at the time (i.e., about 18 years old). However, one might ask whether it was right that a person should have had more individuals of an endangered species than everyone else worldwide (excluding pet keepers), in view of the fact he had so many he could not set them all up for breeding.

There seems to be no logical reason why the purple-naped lory should be more difficult to breed than any other *Lorius* species. By 1990, perhaps only one aviculturist had had consistent success—lory specialist John Vanderhoof in California. The relatively small number of birds which had reached Europe or the U.S.A. resulted in few breeding successes occurring before the mid-1980s.

Little has been recorded in the avicultural literature. Spence (1955) wrote:

My pair were imported for me under licence by Mr. Frost and along with 3 other individuals appear to be the only specimens brought to

Britain for many years. The birds were not united for many months and during this period the female became exceedingly tame and quite fantastically attached to me, caressing me, grooming my eye-lashes with her papillated tongue and regurgitating food into my mouth or ears quite promiscuously. After some premonitory pumping she would seize the commissure and, pulling my mouth round toward the shoulder she sat on, regurgitated macerated banana and condensed milk into me. Indeed, I even developed a taste for this strange pabulum for I felt her too sensitive a creature to offend by spitting out her gift! The male became very, very tame and when he moulted into adult plumage and the pair were brought together the 2 would sit, one on each shoulder, fondling my ears and murmuring love-words to me or as often deafening me with their incontinent shrieking.

This state of psittacine felicity was not to last long, however, for the 2 had more or less the liberty of the farm office and one day, after being called away, I returned to find them making confetti of a small fortune in National Insurance stamps. When I seized the male somewhat roughly to return him to his cage, he bit me savagely and has sought to do so ever since...

The female laid several clutches in a pie dish half filled with peat litter before the pair was united. The male ate the first egg he saw; the second was removed and replaced with a china pigeon egg. This ended his egg-eating. In June, the pair was placed in a garden aviary and they roosted in the nestbox from the very first night. The first egg was laid on June 23 and the second, on June 25. Both hatched. One chick was removed for hand-rearing after its toes were mutilated—but it died after being given banana. The other was reared and the pair nested again "indoors, with their first young one still beside them, unmolested." One chick hatched and was four weeks old when the breeding was recorded.

In Germany which, like Britain, has a long tradition of bird breeding, the first successful breeding did not occur until 1976 when young were reared at Vogelpark Walsrode. Then a few breeders there recorded isolated successes. Perhaps not until after Armin Brockner and his father acquired their first two purple-naped lories, in 1980, was the importance of breeding this species realized. In 1981 endoscopy showed both birds to be males. That year one was exchanged for a 1980-bred female. The female laid for the first time in 1984, but the eggs were infertile. Her next clutch, in March, 1985, was fertile. Twenty-six days after the first egg was laid there was a chick in the nest. Next day chick and egg had disappeared without trace. The female did not lay again until April 1987. Two young hatched in May and were reared. The pair repeated the success in 1988, in 1989, and in later years.

The main rearing food of this pair was lettuce and CeDe rearing food mixed with grated carrot and hard-boiled egg. Their feces were therefore firmer than is usual with lories, thus it was not necessary to clean the nestbox. Also, the parents gnawed strips of wood off the box so that the nest surface remained fairly dry. No display of affection or mutual preening was ever witnessed between male and female during the eight years they had been together.

Their display was described as follows: "Both sat stretched upright, diagonal to each other on the perch, making castanet-like noises with their beaks. Dipping up and down on the perch and rattling their beaks they would then fondle each other's nape feathers, quite frequently biting too. This behavior was also common when we fed them, cleaned the aviary or just stood watching" (Brockner 1990, 1992a).

Lawrence Kuah (1993) described some of the problems he had encountered in trying to breed this species. "They are extremely aggressive and only one pair can be kept in an aviary or cage because they will kill and harass other specimens of the same species kept with them. They are also dangerous to other birds and one of my birds managed to kill a fully spurred adult male Peacock Pheasant in the same aviary...In some rare cases, pairs can turn on and kill each other...Young chicks, up to 2 weeks old, were torn apart and dismembered by their parents after they had been inspected and photographed. Older chicks were sometimes plucked very severely and had the tips of their wings and toes bitten off."

Another problem had been feeding the liquid food in an overconcentrated form, "as a concentrated mixture tends to kill the birds after some time." And overcrowding was warned against, "as they catch enteritis and other diseases very quickly."

However, in one successful nest, the first chick hatched after 23 to 24 days and the second two days later. The nest was not inspected and the young ones were seen peering out at about 80 days. The male then became extremely aggressive. The two young were removed when they were about 86 days old. They fed themselves and soon became friendly and trusting (Kuah 1993).

What does the future hold for the purple-naped lory in aviculture? In 1990 Armin Brockner wrote that "it is practically certain that it will not survive without coordinated breeding efforts." Kuah (1993) concurred, noting that "aviculturists must make certain that proper records are kept and a stud-book maintained." In short, the fortunate few who now hold this species do so in trust for future generations of aviculturists. If they do not handle them in a responsible way, the purple-naped lory will be lost to aviculture.

Chattering Lory
Yellow-backed Lory

Lorius garrulus (Linnaeus 1758)

Description: Mainly scarlet; the mantle is a darker shade of red and is sometimes marked with a few yellow feathers. Wings and thighs are green. Bend of wing and underwing coverts are yellow and there is a broad pinkish red band on the underside of the primaries. The tail is scarlet on the upper side, with dark greenish upper tail coverts; the underside is pale red tinged with golden-yellow; the under tail coverts are scarlet. Beak and iris are orange, and the cere and skin surrounding the eye are gray.

Length: 30 cm (12 in.); birds from Obi are slightly smaller, as first pointed out by T. Buckell (van Dooren pers. comm. 1995). (This was noticed in *flavopalliatus* imported with *E. s. obiensis* and was confirmed by the examination of skins.)

Weight: About 220 g; 180–250 g for the nominate race and 190–220 g for *flavopalliata* (Wenner undated).

Immatures: They have the beak dark brown and the iris grayish. Bertagnolio (1967) described a *flavopalliatus* as having a dark bill and the cere and skin surrounding the eye light brown. There was "a trace of dirty yellow bordering the feathers of the hind crown." Wenner states that the eyes "are dark brown and become orange-brown about 2–3 months after they have left the nest...On top of the cere there are some very fine black feathers that look like hair. The outermost point of the periophthalmic skin ends in a line of black feathers toward the ear region. All of the red feathers are slightly more subdued in shade than the adult."

Mutation: A true yellow is known. In 1991, Patrick Tay of Singapore obtained two wild-caught males of this mutation. They were without doubt the most attractive mutation parrots I have seen. The pure bright yellow of wings, thighs, a patch on the mantle and some yellow on the tail made a striking contrast to the red plumage elsewhere. My photograph of one of these birds was published in *Cage and Aviary Birds* (February 22, 1995). In response, Hesford (1995) commented:

> The evidence suggested that these birds are striking examples of a true Yellow form. It was welcome confirmation that such forms can occur in Parrot species. Almost certainly the factor involved is a straightforward (autosomal) recessive clear gene, which prevents melanin deposition in the feathers, but (unlike the Ino) allows normal melanin pigmentation to develop in body tissue...Most so-called Yellow parrots are dilutes in which melanin deposition in the feathers is reduced. Others may be Cinnamons or Fallows.

The yellow yellow-backed lories were the subject of a strange coincidence. The day after I saw them I left Singapore. While at the airport I came across packets of greetings cards depicting parrots. The original lithographs had appeared in an antiquarian book. One of them showed a yellow mutation of a *Lorius*, probably *chlorocercus* or *domicellus* as there was a crescent shaped area of yellow on the breast. More difficult to explain was the yellow on the crown. This lory was described as "Le Perroquet Lori Radhia." It was depicted with yellow

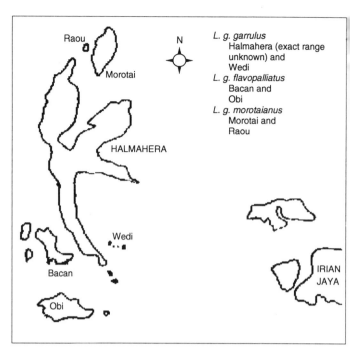

Islands inhabited by *Lorius garrulus*.

iris and pale beak—the latter suggesting that it was drawn from a skin.

Subspecies: The nominate race is described above. It does not usually have any yellow on the mantle. *L. g. flavopalliatus* is distinguished by the usually large patch of yellow on the mantle and by the brighter green wings. It is known as the yellow-backed lory. The subspecies *morotaianus* is said to have a duller and less extensive yellow patch on the mantle and darker green wings than *flavopalliatus* (Forshaw 1989). J. Hubers, T. Buckell and G. van Dooren examined skins of this subspecies in the museum at Leiden, Netherlands; they described the yellow patch as intermediate in size between the other two subspecies, but could see no difference in the shade of green (Hubers pers. comm. 1995).

Natural history

Range: The northern Moluccan Islands, Indonesia. The nominate race is from Halmahera (see Gazetteer) and the Wedi (also called Widi or Weda) Islands; *flavopalliatus* is from Bacan (Batjan) and Obi and *morotaianus* is said to occur on Morotai and on Raou (Rau) Island.

Habits: According to Lambert (1993a), this lory is primarily a forest inhabitant and a canopy-

184

dweller, contrary to some previous reports which stated that it is found in coastal areas. It occasionally descends to the lower canopy to feed. Lambert stated that it was not observed feeding on coconut palms along the coast, so either this behavior is seasonal or coastal populations have declined. Poulsen (in prep.) reported on the joint PHPA/BirdLife surveys on Halmahera. The first was carried out from June 29 through August 14, 1994, in the proposed protected Ake Tajawe area; the second, from February 16 through April 31, 1995, in the proposed protected Lalobata area. Information obtained at eight study sites suggested that the chattering lory is very common in undisturbed primary forest on rich soil. It was observed from 20 m to 700 m (66 ft. to 2,300 ft.), but has been recorded from as high as 1,040 m (3,410 ft.). In common with the umbrella cockatoo, there were only a few records from the forest on nutrition-poor ultrabasic soil near Buli and Fongli. Highest densities were recorded in primary and logged lowland forest on volcanic rock at Air Toniku, Air Oba and Miaf. It was also common in primary forest on sedimentary soils but was nowhere common in secondary forest. Uncommon in coastal lowlands where coconut palms occurred, it was nearly absent from forest close to habitation and from easily accessible localities such as agricultural land and forest trails. Its numbers increased at greater distances from roads and villages. Chattering lories were usually observed in pairs or small groups; the largest flock size seen was of 10 birds.

On Obi it is said that yellow-backed lories visit coastal coconut plantations during prolonged dry seasons. On Halmahera the chattering lory was not seen within closed forest and was rarely encountered in coconut plantations on the west coast of the northern part of the island. It was recorded at the edges of primary lowland and hill forest, and in mature secondary woodland up to about 400 m (1,310 ft.) (Bishop, in Forshaw 1989). Thomas Arndt spent ten days on Halmahera in December, 1992. He reported that "*Lorius garrulus* has apparently become rarer due to extensive trapping in the northern arm of Halmahera. I could only find pairs in the mangroves near Tobelo" (Arndt pers. comm. 1993).

On Bacan, *flavopalliatus* was found in tracts of forest by Milton and Marhadi (1987) during July to September, 1985. They did not find it near human habitation, a situation they suggested was related to trapping pressure. They described it as uncommon and believed that it would soon become rare. They stated that its low abundance on Bacan strongly suggests that temporary severe restrictions or complete bans on trade should be invoked to allow populations to recover.

Dr. Alan Lendon (coauthor of *Australian Parrots in Field and Aviary*), was attached to a hospital on Morotai from May to November, 1945. The yellow-backed lory was the most common parrot species he observed there. He also saw it on Raou. On both islands it was "almost always seen in pairs and was very pugnacious when two pairs met" (Lendon 1946).

Nesting: Breeding activity is believed to commence in September or October on Obi. On Bacan, pairs were observed investigating potential nest sites in logged forest at about 140–370 m (460–1,215 ft.) altitude. They favored holes on main trunks at about 20–25 m (65–82 ft.) in trees about 30–35 m (98–115 ft.) tall. One nest site was in the top of a broken palm trunk. On Obi, a young bird with two adults was observed at 550 m (1,800 ft.) in primary forest in mid-February (Lambert 1993a). On Morotai, a pair was seen investigating a hole in a dead tree in June, and fledged young were being fed in October and November (Lendon 1946). One egg was described by Schonwetter (in Forshaw 1989) as 25.8 x 21.8 mm; a misprint is suspected—31.8 x 25.8 mm would be more feasible.

Status/Conservation: Declining throughout its range, due to trapping and deforestation. Lambert et al. (undated) list all subspecies as endangered and give the threats to their existence as trade and hybridization, without any explanation regarding the latter. In 1995, field studies commenced in Halmahera. Project Halmahera was implemented by BirdLife International in collaboration with PHPA; much of the funding came from Fundación Loro Parque. Comprehensive species and habitat surveys were carried out in two regions, Ake Tajawi and Lalobata. The lory had suffered the effects of trapping more than the umbrella cockatoo. It was proposed to the government in 1995 that 350,000 ha of prime forest should form a reserve with national park status. This would be an important step in the conservation of the chattering lory, which is unlikely to be as numerous as Lambert's estimates (below) made three years previously.

Lambert (1993a) estimated the populations of each subspecies using pooled site-density estimates and estimates from transect data, assuming that this species occupies 10–25 percent of nonforest habitat. He conducted surveys on Bacan, Kasiruta, Halmahera and Obi, between October 24, 1991, and February 24, 1992.

Population estimates

	Minimum	Maximum
Bacan, Kasiruta, Mandiole	4,546	32,267
Obi Island group	6,746	16,144
Halmahera Island group*	31,220	220,009
Morotai*	3,848	27,120
Total:	46,360	295,540

*Estimates more tentative than those for Bacan and Obi.

On Bacan, there are still large areas of forest despite the fact that logging commenced in 1971. A positive step that will help to conserve the yellow-backed lory there was the creation of a reserve in the Gunung Sibela area in the 1980s; this area is therefore protected from logging. Milton and Marhadi suggested that with its almost undisturbed tracts of forest, this may be an area where the lory is still locally common, and will thus serve as a refuge for the species.

Trade (see below) is of major concern. Lambert suggested that more realistic export quota levels must be set, and attempts should be made to upgrade the limited knowledge of population dynamics and ecology by initiating field research. Poulsen (in prep.) stated that there are not yet any protected areas within its range and the threat of trapping remains high. He wrote: "We commonly found signs of trapping and met people carrying birds out of forest in all visited areas. We found it as a common pet in all visited villages and rarely used public transportation without meeting people transporting the species."

Lambert suggested the long-term possibility of changing the tradition of catching adults to one of either taking nestlings or removing and incubating eggs. The wisdom of this must be questioned in view of the difficulty of taking eggs from nest holes at a height of 20 m (65 ft.) or more without damaging the tree. Also, in inexperienced hands, mortality among chicks reared from the egg would be high. In any case, it is totally unnecessary (and no longer permitted in E.U. countries) to import this species into countries where aviculture is practiced. There are enough captive-bred young to meet demands. Banning the export trade could have the effect of increasing the domestic market. Instead of local people "consuming" large numbers of wild-caught birds as pets (which probably have a short life span), they must be shown how to breed from the birds which are available to them. A conservation education program is necessary to make them aware that they cannot continue to catch these lories; this is not a never-ending resource, but one which is being depleted rapidly. Will the fate of this species be similar to that of the Moluccan cockatoo, placed on Appendix 1 when it has nearly been trapped out of existence in the wild yet is still common in our aviaries? The time to act is now—while substantial numbers still remain in their natural habitat.

Trade: Accurate trade figures are impossible to obtain and bear little relation to official figures. Lambert (1993a) obtained information from parrot trappers on Bacan. There are two types: the professionals, whose major source of income comes from trapping, and those who occasionally catch parrots. Interviews suggested that more than 100 people regularly trap parrots, although only about 30 are professionals. Nearly all of these are villagers from Desa Gandasuli. Lories and other "target" species are attracted with decoy birds. They are usually caught with gum which adheres to their flight feathers. Young yellow-backed lories are occasionally taken from nests, usually by felling the nest tree. The North Moluccan suppliers send their birds to dealers on Ternate and Ambon, who sell them to exporters in Jakarta, Surubaya and Denpasar. Lambert noted that some suppliers trade illegally, using permits more than once, or letters of recommendation instead of real permits. They may disregard numbers or species on the permits. Most trappers in the North Moluccas specialize in yellow-backed lories and umbrella cockatoos. A few red-flanked lorikeets are caught but this is not a "target" species. Anecdotal evidence obtained from numerous interviews suggested that 15–20 percent of chattering and yellow-backed lories, and probably more than 25 percent of violet-necked lories (*Eos squamata*), die prior to shipment to other destinations within Indonesia. In comparison, 7–10 percent of umbrella cockatoos died at this stage.

Between 1985 and 1991, the numbers of chattering and yellow-backed lories exported increased, those of violet-necked remained fairly constant, and the number of umbrella cockatoos exported annually declined. Quotas are set for the number of each species which may be captured (including those for domestic purchase). However, trade in parrots throughout Indonesia is poorly monitored. The quota system is inadequate, and records probably represent only one-quarter to one-half of the actual trade. The quota for capture of *L. garrulus* in 1990 and 1991 was 5,900 birds each year. Numbers recorded as exported were 4,727 in 1990 and 3,526 for the first six months of 1991. The catch quota for Bacan in 1991 was a mere 250 *garrulus*, yet during the survey period (October 24, 1991 through February 24, 1992) 2,088 were seen in holding cages. Observations thus suggest that about 50 percent of the parrots exported were *Lorius garrulus* and 25 percent were umbrella cockatoos. At other times of the year the situation may be different.

In addition to the export trade, the domestic trade for *garrulus* within Indonesia is quite substantial. Indeed, of 381 parrots of 19 species observed in markets there, 116 (30.5 percent) were *garrulus*. In the domestic market, *flavopalliatus* was the most common subspecies identified (64 percent). Taking into consideration all factors, Lambert suggested an estimate of 9,600 *L. garrulus* caught in 1991. His minimum population estimate for this species throughout its range was 43,360, thus it could be that nearly one-fifth of the entire population was trapped that year. Clearly, continued trade on this scale would bring this beautiful lory to the edge of extinction within a few years. In 1994, PHPA revised its quota to the level recommended by Lambert, according to Jepson (in *Birds to Watch 2*); but, unless the illegal trade is controlled, this step will do little to conserve the species.

Aviculture

Status: Common

Clutch size: Two

Incubation period: About 26 days.

Newly hatched chick: Down color, yellow or white. It is always yellow in *flavopalliatus* (Marc Valentine verbal comm. 1996). Krupa (1991), whose birds had an area of yellow on the mantle

which was intermediate between *garrulus* and *flavopalliatus*, described the newly hatched chicks as having white down, sparsely distributed; the beak is dark brown and the egg tooth is white; feet are dark gray. Wenner (undated) described the down as white and not very thick in the chicks from one pair of the nominate race.

Chick development: One parent-reared chick (from a pair belonging to Mr. and Mrs. R. Wallwork): at ten days a few wisps of down remained; the feet were black. At 16 days, pin feathers were appearing; at 23 days, eyes were fully open. At 38 days black, green and red quills were apparent; at 53 days tail feathers were 1 cm long and at 60 days, 2 cm long. It left the nest at 76 days (Low 1986b). At the age of three weeks the eyes of a chick reared by Nauschutz (1979) were slit and pin feathers were appearing on the wings; at 30 days the pin feathers were opening. At 40 days the feathers had erupted and the black markings on the beak were fading. (At five months this young one was talking well in a clear voice and imitating the telephone.)

Young in nest: From 10 to 11 weeks; two chicks, hatched same day, both 72 days (Bertagnolio 1967); one only, 76 days (Collard 1965).

Chick growth

Relates to two chicks hand-reared by Krupa (1991) from the age of 35 days. (Both were shown as 35 days old but one was probably a day or two younger.) Rearing food consisted of the nectar fed to the parents (fruit-flavored baby cereal and glucose with a little honey and Bovril added), plus a mixture of one part each of fruit-flavored baby cereal and wheat germ cereal.

Lorius garrulus weight table

Age in days	Weight in grams	
	Chick No. 1	No. 2
35	129	119
36	147	125
37	151	128
38	153	135
39	155	140
42	181	163
45	191	177
49	207	195
50	211	192
57	202	192
66	194	188

Age at independence of hand-reared young:
82 days (Krupa 1991); four months (one only) (Nauschutz 1979).

General: The yellow-backed subspecies is better known in aviculture than the nominate race. In the breeding register of the Parrot Society for 1992, for example, members reported breeding 37 yellow-backs but only nine chattering lories. It was of interest that the combined total of 46 for this species was the fourth highest for any lory species that year. This reflects its popularity. It is well known for its murderously aggressive tendencies toward other species and, indeed, Lendon (see under Habits) also noted this toward its own kind in birds in the wild. Wenner (undated) gave examples of the aggressive nature of this species. A pigeon was found dead in a large aviary which contained a pair of chattering lories. Some days later the pair of lories was observed attacking and killing another pigeon. In a large aviary, five chattering lories were together for several months without any trouble. An attempt was made to remove the unpaired bird but the wrong one was removed. The bird which was then without its partner was immediately killed. The following day the single one was killed, leaving only the pair. (It seems extraordinarily negligent of the person responsible to have left an unpaired bird in the aviary after the first one was killed.)

It is surprising that at the Natal Zoological Gardens in South Africa, a pair reared a youngster in an aviary occupied by another chattering lory, seven purple-crested touracos, a pair of Nicobar pigeons, a pair of fruit pigeons and one blue jay!—"all of which agree except if they perch on the tree near the breeding lories' nest box" (Collard 1965). The aviary measured 11 m (36 ft.) long, 6 m (20 ft.) wide and 3.3 m (11 ft.) high. Another breeding of interest occurred with Paolo Bertagnolio in Rome. Two pairs reared young in the same aviary; however, three of the four were offspring of two pairs. Perhaps they had been together from an early age, and were thus more tolerant than normal. In one year, one pair reared two young in two nests and the other pair produced four young in two nests (Bertagnolio 1974). These pairs were double-brooded. In South Africa, Luch Luzzatto owned three breeding pairs of chattering lories which normally nested once annually, about October. However, if the weather was favorable and spring came early, they would nest twice (Barnicoat 1983).

The display of this species was described by Krupa (1991): "Pre-nuptial displays consisting of stationary wing fluttering, eye 'blazing' and slow motion walking and hopping were observed regularly, making this one of the most enjoyable periods to observe them."

In view of the threats to this species from over-trapping, the captive population in mainstream avicultural countries is important. In the U.S.A. it is reared primarily for the pet market. In the U.K. the number of breeders is declining; the price has dropped to the degree that young are sold to dealers for export. We need to be aware of the decreasing numbers kept by breeders, lest this species is suddenly found to be as rare in aviculture as it will one day be in the wild.

Genus: *Chalcopsitta*

This genus contains four species of large, long-tailed lories—three from New Guinea and one from the islands to the east. Diamond (1972) commented on their evolutionary history, describing three of them as "a superspecies ring in the New Guinea lowlands, with contact between populations having been broken and reestablished a number of times. The genus consists of three virtually allopatric semispecies, whose distributions now form an incomplete circle (broken in the east between Astrolabe Bay and the Kemp Welch River) around the periphery of New Guinea." (He apparently overlooked the existence of the fourth species, or perhaps considered that *cardinalis* did not belong in this genus; at least one early taxonomist preferred to classify it in the genus *Eos*.) Note that Diamond describes *atra*, *scintillata* and *duyvenbodei* as superspecies. The possibility exists, however, that *scintillata* and *atra* are the same species and that *C. atra insignis* evolved from a melanistic form of *scintillata*. Bell (1984) recorded watching a group of *scintillata* which included a bird which was "purplish black but with the exact red patches of *scintillata*. This bird could have been a melanistic *scintil-*

lata or a stray *C. atra*." If it was a "stray" it would have been *insignis*, hundreds of miles outside of its range. The melanistic form is common in one other New Guinea lory: *Charmosyna papou*. It seems very likely that melanistic examples also occur in *scintillata*.

The genus *Chalcopsitta* is notable for the fact that each member has the basic plumage coloration different: one black, one brown, one green and one red.

In the field, *Chalcopsitta* can be distinguished from other lories by the larger size combined with distinctive flight. According to Beehler, Pratt and Zimmerman (1986), their wing beats are rapid and shallow but their progress is slow "making it seem as if the birds are working hard for little return."

There is a scarcity of information about the status or biology of the three mainland New Guinea species. Most knowledge comes from the avicultural literature. In captivity, the care and breeding of each species differs little except in regard to diet. Black and Duyvenbode's are mainly nectar feeders who also accept fruit, while the yellow-streaked and the cardinal take other items, such as fresh corn and other vegetables.

Black Lory *Chalcopsitta atra* (Scopoli 1786)

Description: All black, except the tail; most of the body feathers carry a vinous purple gloss which is only apparent in sunlight. The neck feathers are distinctively shaped, narrow and pointed. The rump feathers are tinged with blue and the underside of the tail is red and yellow. The iris is brown, with a yellow ring in some birds; P. Clear (pers. comm. 1995) noted that his male had the iris reddish while that of the two females was brown. The beak and feet are black and the facial skin is dark gray.

Length: 31 cm (12 in.)

Weight: 230 g to 280 g, males heavier (captive birds).

Immatures: The birds have a few scattered red feathers on ear coverts and body plumage. The bare facial skin is white and the eyes are grayish.

Note: The presence of scattered red feathers and the fact that replacement feathers of young which are plucked in the nest come in red, not black, may indicate that this species evolved from a red bird. Atavistic features can be seen in immature birds of many species. Perhaps there were two color phases, red and melanistic (black). Eventually the melanistic phase might have predominated because it was more advantageous to survival.

Sexual dimorphism: In the nominate race only, the male is larger, with larger head and beak.

Natural history

Range: The nominate race occurs in Irian Jaya (western part of New Guinea), in the western Vogelkop, also on the islands of Batanta and Salawati (see Gazetteer). *C. a. bernsteini* is from the island of Misool (Misol or Mysol), Irian Jaya, and *C. a. insignis* is from the eastern part of the Vogelkop, the adjacent island of Amberpon, and the Onin and Bomberai peninsulas. It seems likely that intergrades occur between the nominate race and *insignis* in the Vogelkop, or that there is variation in birds from Misool; two wild-caught birds at Palmitos Park resembled *bernsteini* except for the pinkish red mark on the primaries and some red on the outer lesser underwing coverts. One might question whether *bernsteini* is a valid subspecies. Ten skins of *bernsteini* examined in the natural history museum in Leiden, Netherlands, by J. Hubers, were extremely variable; five had some red marks on the primaries (Hubers pers. comm. 1996)—and this is not a feature which is acquired with maturity.

Habits: Black lories frequent coconut palms and flowering trees. They are found in open habitats such as coastal areas and savannah, also in lowland forest edge. On Salawati, Bishop (in Forshaw 1989) found them to be more common in tall mature mangroves and coastal swamps than in the drier inland forest. From late January to late February, 1986, on the east coast, they occurred in pairs and small groups and were often seen and heard in tall *Rhizophora* mangroves. They were less plentiful upstream in more complex vegetation communities.

Nesting: No information. One egg from a wild bird measured 31.0 x 25.7 mm (Ripley 1964).

Status/Conservation: Probably stable, but no information available.

Aviculture

Status: Uncommon

Clutch size: Two

Incubation period: About 25 days; 24 days recorded for second egg.

Newly hatched chick: Weight, about 6 g; whitish or pale gray down, dark brown beak, dark feet.

Chick development: See below under rajah lory, disregarding the reference to day 54.

Young in nest: Ten to 11 weeks is normal; about 90 days (Clear pers. comm. 1995).

Age at independence of hand-reared young: From seven weeks.

Black lory weight table: See below under rajah lory, whose weights are slightly lower.

General: When the black lory was commercially imported into Europe for the first time in 1970, it was not much sought after. Only the nominate race was available, which was considered to be a dull-colored bird. In 1969, however, it was such a rarity that a U.K. dealer who specialized in unusual birds advertised a pair for £300 (a large sum in those days)—and sold the pair overseas. Most wild-caught birds were, for some months, or even several years, very nervous. When they became steadier it was easier to appreciate their personalities—intelligent, inquisitive and quite often aggressive. Although I first kept black lories in 1971, it was not until 1975 that I had a true pair. One could only guess at their sex in those days. The pair was kept outdoors in a small aviary measuring 1.8 m (6 ft.) long, 91 cm (3 ft.) wide and 1.8 m high. A feeding hatch and a wire cover placed over the food bowl were essential to maintain one's hand in its original condition! The male was very aggressive!

The female laid her first clutch during the second week of May, 1976. On June 8, inspection revealed an empty nest. The first egg of the second nest was laid on or about July 9. On August 5, a chick was heard which could have hatched one or two days previously. Soon after, a second chick was heard. In those days, the wisdom of locating the nest where it could be inspected without entering the aviary was just becoming apparent, but had not been practiced! After the chicks hatched the adults became much quieter; at last there was relief from the deafening screeches which met the ears of anyone who passed the aviary. During the last week in September, the young ones were seen looking out of the nest entrance; they seemed calm and unafraid—but had been plucked. The first left the nest on October 17, aged 10½ weeks. I checked the aviary in the dark that evening and saw one of the adults roosting on top of the box with it. Next day the young one looked well and confident, but about one quarter of its feathers had been plucked. The following morning I experienced the kind of disappointment that every aviculturist knows sooner or later and recalls ever more with a sinking heart. The young one was dead beneath the nestbox. It may have been disturbed by a cat or a fox as several birds in nearby aviaries were clinging to the wire at first light—evidence of night disturbance. Autopsy revealed no sign of disease.

The black lory family was at once transferred indoors to a cage with a nestbox attached. Within minutes the surviving youngster had found its way into the box. Four days later it was seen to feed itself. Five days after that I returned the parents to their aviary. The young one continued to feed itself on nectar, sponge cake and nectar, and on ripe pear and grapes (Low 1978b).

This was only the second occasion on which the black lory had been bred in Britain. The first had occurred many years previously, in 1909, by E. J. Brook of Hoddam Castle, Scotland. His pair, collected by Walter Goodfellow, had been preceded by only one other pair, collected by Goodfellow in 1904. Goodfellow obtained birds from many parts of the tropics for wealthy private bird keepers. One of his most interesting experiences concerned a black lory. The bird in question was obtained in the Maclure Gulf, northeastern New Guinea; it succeeded in opening the door of its cage and flew off to join a flock of the same species in the neighborhood. "This was not near a village but on a hillside with virgin jungle all around, and the cages were in a makeshift shed with an open front. One morning 4 days later when feeding my birds as usual, a noisy flock of Black Lories flew overhead, and one detached itself from the others and without any hesitation flew straight down to its cage and tried to get in, which of course I helped it to do" (Goodfellow 1933).

Tame black lories are among the most delightful of all parrots. The young bird whose rearing I described above was one of the most affectionate I had known at that time. In December of the year in which she was hatched, she changed from a very shy bird. By early January she was adventurous

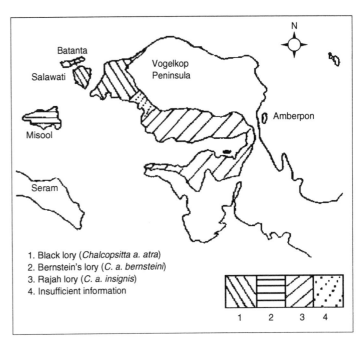

1. Black lory (*Chalcopsitta a. atra*)
2. Bernstein's lory (*C. a. bernsteini*)
3. Rajah lory (*C. a. insignis*)
4. Insufficient information

Range of *Chalcopsitta atra*.

enough to nibble my fingers. Three or four days later she was friendly and playful and would stand on my hand and allow me to stroke her. By the end of the month she would jump on to my hand and run up my arm, wanting to be taken out. If I could have carried her around all day she would have been the happiest bird alive. I have never forgotten her sweet and affectionate nature. It often happens that parent-reared birds which become tame are totally trustworthy in temperament, lacking the sometimes aggressive moments of those which are hand-reared. Such was the case with her.

I am not alone in cherishing the memory of a black lory. A member of the Avicultural Society who had lived in Singapore had kept many lories. Two were outstanding in her memory. "One was a Black Lory, a soft bronzy black washed underneath the body with a delicate sapphire-blue, a bird which followed me like a dog, and would clamber on to my shoulder and 'chuss' around my neck, keeping up a running commentary of soft mutterings. You could give the bird a piece of paper rolled up into a ball, and it would play for hours like a kitten, lying on its side and 'back-pedalling' it till quite exhausted" (O. St. A. S. 1938).

Black lories also provide me with an amusing memory. During the time I was curator at Loro Parque, Tenerife, two pairs of blacks were breeding, one pair on exhibit. One on occasion they had two newly fledged young, one of which was clinging to the front of the aviary. It often made me angry to see the way some visitors treated the birds; on this day the blacks triumphed over their unsuspecting tormentor. A man was prodding the young one on the wire. The male parent flew to the front of the aviary and squirted liquid feces all over the man. He jumped back in surprise, dripping. I observed this scene with satisfaction. Revenge had been exacted!

The other pair at Loro Parque, whose young had to be removed soon after hatching, laid four clutches annually in 1987 and 1988. In 1987, six chicks hatched and five were reared. This shows that this species could easily have been established within aviculture had enough breeders been interested in it. A few successes have been achieved in widely scattered collections but I doubt that many have had consistent success. My fear is that the black lory will die out in aviculture.

Bernstein's Black Lory *Chalcopsitta atra bernsteini* (von Rosenberg 1861)

Description: Adult birds differ from the nominate race in having dark red feathers on the inside of the thighs and, in some birds, reddish purple markings on the forehead. The lores are dark red. Feathers of the upper breast have reddish margins and centers and a vinous bloom. Underwing coverts are tinged with vinous and a small area on the bend of the wing is red. The rump is bluer than that of the nominate race, the feathers being broadly margined with blue. The tail is dark reddish at the base, the lower half being dark gray with a yellow tinge to the centers and margins of the feathers. Undertail coverts are greenish gray. Without opening the wing, it is difficult to distinguish an immature *insignis* from an adult *bernsteini*. Also, as mentioned above, some wild-caught birds (origin unknown) are intermediate between *bernsteini* and *insignis*, showing red markings on three or four primaries and some red on outer greater underwing coverts.

Length: 31 cm (12 in.)

Weight: Two females weighed 228 and 245 g.

Natural history

Habits: On Misool this lory seemed to prefer trees on the forest edge or small groves of eucalyptus scattered over the plains (Mayr and de Schauensee 1939). Bishop (in Forshaw 1989) reported that in January, 1986, it was seen in tall mangroves and in forest along the south coast. Near the Begam River a party of six or eight was observed feeding at flowering *Schefflera* in scrubby, patchy lowland swamp forest.

Aviculture: A few birds of this subspecies were imported into Europe in the 1980s in consignments of *atra* and *insignis*. Probably most were not properly identified at the time and this led to some of them later being paired with *insignis*. At the time of writing, *bernsteini* is rare in aviculture.

Rajah Lory *Chalcopsitta atra insignis* (Oustalet 1878)

Synonym: Red-quilled lory

Description: Mature birds are striking, with scarlet forehead, lores, underwing coverts, bend of wing and alula, and thighs. The contrast of scarlet against purplish vinous-glossed black pro-

duces a remarkably beautiful bird. Under tail coverts are bluish black and the underside of the tail is red near the base, yellow elsewhere. It takes about five years to acquire deep scarlet markings. An unusual specimen, either very mature or aberrant, had almost the whole side of the face red, and broad red margins to the feathers of the upper breast, with some breast feathers entirely red. Patrick Tay of Singapore owned this bird, which I saw in 1989. (Its photograph appears on page 42 of *A Guide to Lories and Lorikeets* by Peter Odekerken with the caption "Rajah Lory— cockbird"—but this was a very unusual male.) As the iris of the eye was also red, the possibility exists that the excess of red in the plumage was caused by malfunction of the liver. The rajah lory could be described as a sister species to Duyvenbode's. It is identical in areas of color (also in behavior); only the actual colors are different, plus the fact that in Duyvenbode's the color from the lores (yellow) extends all the way around the lower mandible, whereas in *insignis* there is little or no color on the throat, except perhaps in very mature birds. In *insignis* the red bases to the breast feathers are just apparent, producing a tinge of red, whereas in Duyvenbode's there is a yellow shading on brown. It differs from nominate *atra* in that the bare skin surrounding the eye is gray—not black. Its build is slightly smaller and more slender, with little or no difference in size between male and female.

Length: 30 cm (12 in.)

Weight: About 220 g; four young at Palmitos Park were recorded as follows: 200 g at eight months, 209 g at ten months, 233 g at 11 months, and 209 g at 30 months.

Immatures: They lack red on the head except perhaps for a tinge on the lores; the thighs are partly dull red; underwing coverts are red and primaries are marked with red. The skin surrounding the eye and the lower mandible is white, changing to light gray by about five months. The iris is dark gray and has changed to light gray, with an outer circle of dull brown, by about five months. I examined a wild-caught bird believed to be about two years old and noted that "the lores were partly red, but not the forehead. There was a red patch on the first nine primaries, this area being largest in primaries four to seven, where the red was touched with yellow at the top of the patch. The greater underwing coverts were red in the center. The bend of the wing was vinous black and red and the thighs were red on the inside. The under tail coverts were mauvish black.

The bases of the feathers of the throat and upper breast were red" (Low 1993).

Nesting: Nothing recorded. Eggs laid by two females at Palmitos Park measured 31.0 x 23.0 mm, 30.5 x 24.5 mm, 30.2 x 24.8 mm, 30.1 x 25.5 mm, 30.0 x 25.2 mm, 30.0 x 24.0 mm and 29.5 x 25.5 mm.

Aviculture

Status: Rare

Clutch size: Two

Incubation period   24 or 25 days—possibly 23 days for the second egg on one or more occasions.

Newly hatched chick: Weight, about 6 g; 5 g and 7 g also recorded; long white down or gray down—if white on the body, grayer on the head; down sparser on crown; gray or grayish pink feet; dark brown beak with white egg tooth.

Chick development: Day 7, little down remains; day 8, feet tinged with gray; day 10, ears open, down feather tracts visible under the skin; day 11, eyes slit, feet gray, legs partially gray; day 19, second down erupting on underparts in little tufts; day 21, wing quills about to erupt, feet mainly black; day 26, feathers of wings, mantle and scapulars starting to erupt; day 54, red underwing coverts starting to erupt, lores reddish.

Young in nest: Ten to eleven weeks.

Age at independence of hand-reared young:
45 days up to nine weeks.

Chick growth

Shows average weights, in grams, of three parent-reared young at Palmitos Park, Gran Canaria, in 1993 and 1994, and one which was reared by Duyvenbode's. The young were from two pairs. This table demonstrates the weight differences that can occur in different pairs kept under identical circumstances, but one also has to take into account how much food was in the crop. For example, one 35-day-old chick weighed 114 g with the crop nearly empty but next day weighed 142 g with a very full crop.

Rajah lory weight table

Age in days	Weight range (gm)	Average weight	No. of chicks weighed
day hatched		6	1
1		9	2
2		10	2
3		12	1
4		15	1
5	16–18	17	2
6	19–21	20	3
7		21	1
8	27–29	28	2
9	28–31	30	3
10	34–35	35	2
11	38–45	43	3
12	33–50	40	3
13	45–60	52	4
14	54–56	55	2
15	50–68	63	3
16	54–64	61	3
17	52–84	72	3
18	81–82	82	2
19	73–77	75	2
20	89	89	2
21	87–92	89	3
22	70–102	86	2
23	75–112	87	3
24	95–98	97	2
25	94–98	96	2
26	84–107	95	2
27	97–131	112	3
28	94–121	108	2
29	95–111	105	3
30	92–117	111	3
31	100–150	126	3
32	100–137	126	3
33	109–164	143	3
34	114–163	139	3
35	139–156	146	3
36	117–160	141	3
37	124–162	148	3
38	140–168	154	2
39		146	1
40		128	1
41		132	1
42		162	1
43		164	1
49		162	1
54		173	1
55		203	1
62		198	1

General: Until the late 1980s the rajah lory was virtually unknown. One or perhaps two were exhibited at Birdland, Bourton-on-the-Water, U.K., in the 1970s. Most of the few rajahs imported into Europe were immature birds and no breeding successes occurred until about 1990. In 1992, both the breeding pairs in the breeding center at Palmitos Park, Gran Canaria, started to lay. One female produced her first clutch in January of that year, but it was not until her fifth clutch, in May, 1993, that a fertile egg was

produced (and a chick reared). In contrast, the other pair was very prolific. Their first ten eggs (laid April, 1992, to April, 1993) produced eight chicks, of which seven were reared; one died at five days.

The breeding behavior of these two pairs is in contrast to that of black lories I have had in my care. They are a pleasure to work with because they are neither aggressive nor extremely noisy. Their threat display, given when, for example, I have removed the chick to renew the nest litter, consists of wing-whirring, moving the head jerkily up and down, all the time screeching and sometimes flaring the tail. The males never attack and the females quietly leave the nest when I tap gently on the door.

For various reasons (the death of one of two chicks at an early age, a deserted egg and a badly plucked chick), young have been hand-reared, as well as parent-reared. Their individuality is very apparent. For example, one reared with only an Amazon parrot as a cage companion, was first offered the warm rearing food in a container at 53 days; it filled its crop at once and never again needed to be spoon-fed. I had reared it from the egg. Another, removed for hand-rearing at six days and kept with a yellow-streaked lory, was handled as little as possible but did not perch or feed itself until it was 67 days old. Thenceforth it readily consumed the rearing food, and spoon-feeding ceased. In contrast, two chicks hand-reared from the ages of 11 and 12 days were totally independent at only 45 days. After 60 days they were not handled at all, and it was with some difficulty that I weighed them at 70 days. They were unfriendly—lunging and biting. Two black-capped lories reared with them did not behave in this way.

Most of the young rajahs were placed on exhibit at Palmitos Park in an aviary containing other young lories, mainly yellow-streaked, Duyvenbode's and black-caps. (However, they can be kept in this way only for about three years; when they mature, fighting is likely to occur.) The *Chalcopsitta* species usually remained together, not associating with the black-caps and green-napes. Their playful behavior and glorious colors attracted large crowds. Groups of *insignis* could often be seen wrestling together on their sides or on their backs, on a small shelf set into the aviary wall. They appeared as an indistinguishable black mass as they rolled around. Or they would cling to the welded mesh front of the aviary, exploring proffered fingers with their moist little tongues and showing off the colors which, month by month, were becoming more evident. In the sunlight, the purple gloss on the plumage shone like no other color I have ever seen, except the "black" of melanistic Stella's lorikeets. The rajah lory will always remain one of my favorite birds.

There are few in aviculture and few breeding successes. In Germany, Dr. Robert Peters has reared a number of young and in the U.K. there are one or possibly two breeding pairs. This subspecies is non-existent in the U.S.A. and there was but a single bird in South Africa in 1996.

Duyvenbode's Lory *Chalcopsitta duyvenbodei* (Dubois 1884)

Synonyms: Brown lory, Duivenbode's lory (This species was originally named *duyvenbodei* after C. W. K. van Duyvenbode; the use of common and scientific name with an "i" instead of a "y" may have originated from Peters, 1937.)

Description: No other parrot has a color scheme of rich brown, yellow and violet. The predominant color is dark brown which contrasts with the striking crescent of bright yellow extending from the lores and encircling the lower mandible. Forehead, underwing coverts, bend of the wing and thighs are also bright yellow. The narrow, elongated feathers of the nape and the sides of the neck are dull yellow. Lower back, rump and under tail coverts are violet or deep blue. Breast feathers have the concealed part yellow. Primaries are black with a large yellow patch on the inner web, also on the adjoining two secondaries. Beak, cere, legs and the skin surrounding the lower mandible are black. The iris is dark brown with an inner circle of pale yellow.

Length: 31 cm (12 in.)

Weight: 200 to 230 g; up to 250 g recorded in captive-bred birds.

Immatures: They have duller and usually less extensive yellow on the face, with the lores dull yellow. The upper breast is more brightly colored in many young birds because the feathers are broadly but faintly margined with yellow. The elongated nape feathers are brown, with a lighter shaft-streak. Underwing coverts are yellow, brown at the base; greater underwing coverts are faintly tipped with brown; primaries and secondaries colored as in adults. Thighs are bronze-yellow and the rump is yellow-brown, tinged with blue. Naked skin surrounding the eye and lower mandible is white, becoming gray by about eight months, thence gradually darkening to black at well over one year old.

Sexual dimorphism: Males are generally larger and bolder in appearance, with the head and beak larger.

Mutation: In 1989, Patrick Tay of Singapore obtained a Duyvenbode's which was green where the normal bird is brown, only part of the mantle and wings being brownish. The rump and under tail coverts were iridescent blue and most of the lower half of the tail was blue. The yellow areas were normally colored. Jim Hayward (pers. comm. 1989) suggested that the factor which prevents reflection of blue (to create green) had been altered so that blue and yellow could be reflected together.

Subspecies: *C. d. syringanuchalis* is said to be darker on the head and back, some specimens having a violet sheen; doubtfully distinct.

Natural history

Range: The northern part of New Guinea, between Geelvink Bay in the west and Astrolabe Bay (Gogol River) in the east. Beehler, Pratt and Zimmerman (1986) state that the species is patchily distributed through the northwestern and Sepik-Ramu regions at low elevations. The subspecies *syringanuchalis* has been described from the Aitape area east to Astrolabe Bay.

Habits: Found in forest, tall secondary growth, lowlands and hills up to 150 m (490 ft.). It congregates at flowering trees with other lories and with honeyeaters and moves along the forest edge and other open habitats, according to Beehler, Pratt and Zimmerman, who also state that these birds roost socially. In Papua New Guinea, Peter Them has seen it around Vanimo, near the May River and Bonahoi village (Them pers. comm. 1996).

Nesting: Undescribed (Coates 1985). Eggs laid by captive birds at Vogelpark Walsrode measured 31.6 x 23.1 mm and 32.3 x 20.0 mm (the latter a strange shape), according to Robiller (1992); eggs laid at Palmitos Park measured 30.5 x 24.5 mm, 31.5 x 24.5 mm and 28.9 x 24.4 mm. Two fresh eggs weighed 9 g each.

Status/Conservation: Not common—usually observed in pairs or groups of up to ten birds. Rand and Gilliard (1967) described it as irregularly distributed, being common in some districts but uncommon to rare in others. After a six-week survey, Pearson (1975a) described it as an uncommon resident of the midstory. However, very little is known of the status of this species, or whether any conservation measures are needed.

Aviculture

Status: Uncommon

Clutch size: Two

Incubation period: 24 days, a female in U.K.; 24 to 26 days, a female in Gran Canaria. Of the known incubation periods for this female, three first eggs laid hatched after 26 days and one after 25 days, and one second egg hatched after 24 days, another after 25 days and another after a minimum of 24 (i.e., 24 or 25) days. This no doubt reflects different incubation habits of the two females, as when two chicks hatched in Gran Canaria, they hatched on consecutive days except in one instance when the interval was two days, indicating perhaps that full incubation did not commence on the day of laying. Tiskens (pers. comm. 1994) gives the incubation period as 25 to 27 days.

Newly hatched chick: Weight, 7–8 g; long white down except on crown where down is sparse; feet gray-pink; beak brown-black with white egg-tooth.

Chick development: At four days the feet are tinged with gray; at eight days the second down feather tracts are just visible beneath the skin; at 12 days the eyes may start to slit but are not fully open until about three weeks; at 17 days the dark gray second down has started to erupt, first on the wings, back, thighs and underparts, as little black tufts. At

about 24 days many of the wing feathers have erupted and the yellow-brown feathers of the thighs are erupting; the head feathers appear soon after, starting with the ear coverts; at five weeks most of the body feathers have erupted except under the wings; the tail is starting to erupt. Chicks are ringed with 7.5 mm rings at 14 to 15 days.

Young in nest: 70 to 82 days; young from pair in Gran Canaria average 73 days.

Age at independence of hand-reared young: 40 to about 60 days. For example, one first fed itself at 48 days when it weighed 160 g and filled its crop to 175 g. Next day it weighed 170 g, filled its crop to 181 g and then took 10 g more from a spoon.

Chick growth

Shows the weight of a chick removed from the nest at the age of 24 days (because a bird died in the next cage and a disease risk was feared). It was hatched at Palmitos Park in June 1993.

Duyvenbode's lory weight table #1

Age in days hatched		Weight in grams
2		12 fic *
4		17 nf
6		24 fc
14		64 vfc
19		73 fic
21		98 fc
24 am		91 e
24 late am		119 fc
		removed for hand-rearing
25		91/104 **
26		98/110
27		102/113
28		109/122
29		114/126
30		121/152
32		132/142
34		134/146
36		147/165
38		153/169
40		160/177
42		171/190
44	(moved to cage	180/198
46	from brooder)	178/195
48		184/202
49		183/195 self-feeding
50		182 weaned

* fic = food in crop; vfc = very full; nf = crop nearly full; e = empty; fc = full crop; ** weights before and after first feed of the day

This table shows the weights of two parent-reared chicks hatched at Palmitos Park on 9/2/95 and 10/2/95.

Duyvenbode's lory weight table #2

Age in days	Weight in grams Chick No. 1	No. 2
hatched	7 e	7 e
1	11 fic	8 fic
2	12 fic	11 ne
3	15 fic	13 nf
4	18 fic	14 fic
5	22 fc	17 vfc
6	22 e	20 nf
7	27 nf	24 fc
8	31 ne	31 fc
9	40 fc	34 fic
10	46 fc	36 fc
11	50 fc	40 vfc
12	56 nf	41 fc
13	69 fc	48 fc
14	70 fc	46 fic
15	74 fic	53 fc
16	72 ne	58 fc
17	74 fic	77 vfc
18	90 nf	65 fc
19	93 fc	72 fc
20	95 fic	73 nf
21	115 fc	82 nf
22	119 fc	90 fc
23	119 ne	93 fc
24	126 nf	101 fc
25	128 nf	–
26	123 ne	118 vfc
27	148 vfc	108 fic
28	138 ne	123 vfc
29	159 vfc	110 e
30	136 e	–
31	176 fc	124 fic
32	154 e	132 fic
33	159 e	–
34	–	133 e
35	163 e	138 ne
36	165 ne	–
37	–	173 fc
38	214 fc	–
40	–	158 e
41	210 fc	–
47	193 e	169 e
55	–	199 e
56	216 e	–
73	–	left nest
74	left nest	

e = empty; fic = food in crop; fc = full crop; vfc = very full crop; nf = nearly full; ne = nearly empty

After 33 days, the female became increasingly aggressive when chicks were removed to be weighed.

General: In 1884, Maarten Dirk van Renesse van Duyvenbode presented a magnificent collection of skins of birds and animals from New Guinea to the Royal Natural History Museum in Belgium. Among them was a previously unknown lory which was dedicated to the donor (Prestwich 1963).

Commercial importation of live birds of this species into Europe commenced in 1973, in which year I acquired my first pair. Duyvenbode's were regularly imported from the wild in small numbers for the next five years or so. I greatly admired this species and my aim was to set up several pairs for breeding. To this end, I acquired seven more birds up to 1978. Sad to say, only two survived for more than five years and some died within a few months of importation. Cause of death was diagnosed as *Trichomonas* in two, bacterial infections in others and compaction of the gizzard caused by a lump of peat (peat was used in the nestbox) in another. Other people also found the mortality rate high, especially in newly imported birds. I recorded: "By the early 1980s no one in Britain had bred this species (which had been reared in the U.S.A. and Germany) and I doubt whether there were 20 birds in the U.K." (Low 1984).

In 1983, chicks were reared from my pair which had been together since November 1978. This was the first successful breeding recorded (Low 1984) in the U.K., and almost certainly the first breeding. Their aviary measured 2.1 m (7 ft.) long, 91 cm (3 ft.) wide and 1.8 m (6 ft.) high. The nestbox measured 23 cm (9 in.) square and 54 cm (21 in.) high. Several inches of pet litter (similar to wood shavings, bought in compressed packs) lined the base. The female had previously been kept with another female and it was several months before the new pair were compatible. The first breeding attempt occurred in 1982; two eggs were laid and one hatched, but the chick died on the first day. The next clutch was laid in April, 1983. Both eggs hatched and the chicks were removed for hand-rearing on the day of hatching. At five weeks old they weighed 130 g and 140 g, and the temperature in the brooder had been gradually reduced to 28°C (82°F). At six weeks old their tails measured 1.2 cm (½ in.) long. At seven weeks they were moved from the brooder, in which the temperature had been reduced to 21°C (79°F) to an unheated weaning cage. By then they were fully feathered, except under the wings and for the tail, which was about 4 cm (less than 2 in.) long.

The pair nested again and chicks hatched on July 21 and 23 (both after 24 days). The youngest chick had to be removed for hand-rearing, as its weight gains were poor. The parent-reared youngster left the nest on September 25 (after 11 weeks) and the hand-fed chick was also successfully reared. Since that time young from three more pairs have been produced in my care—two pairs at Loro Parque, Tenerife, and one pair at Palmitos Park, Gran Canaria. I have always retained a great affection for the young of this beautiful lory. I recall at Loro Parque a 17 m (50 ft.) long, off-exhibit aviary for young lories which was a riot of color and activity. At one time it contained 42 young of various species, but it was invariably three or four young Duyvenbode's who would land on my head and shoulders when I entered. They were the friendliest and most affectionate. Other species included black, violet-necked, green-naped, red, dusky, chattering and yellow-bibbed.

Characteristic behavior of this species, observed by Tiskens and by the author, occurs during the breeding season. They are nervous; but not bold enough to be truly aggressive, they lean forward (usually with the wings held slightly away from the body), growling like a nervous African gray parrot.

The breeding history of the pair at Palmitos Park is worth relating in some detail. They reared their first chick while on exhibit in the park. It left the nest in August, 1991. What happened in 1992, when the female surely laid one or more clutches, I was unable to ascertain. They were therefore moved, toward the end of 1992, to the breeding center, where I could closely monitor their breeding attempts (Low 1995). By December 31, the female had laid two eggs. Both hatched and the chicks were removed for hand-rearing at the ages of 10 and 19 days because their growth rates were poor. The second chick was removed on January 24. The female laid the first egg of the next clutch on February 22. Again both chicks hatched, and were removed for hand-rearing, on April 22. The female laid the first egg of the next clutch on May 20. One egg was infertile and the single chick was removed on July 11. The female laid the first egg of yet another clutch on August 9. The resulting chick, and all subsequent young, were parent-reared. In 1994, the female laid on January 25 and 27 but, for some reason unknown, these eggs were destroyed by February 6. On March 9, the female laid the first egg of the next clutch. The time between loss of eggs (or removal of young) and relaying was therefore 28 to 31, but usually 29, days.

The two young which left the nest in October, 1994, (both males) were not removed until January, 1995, when the female was already incubating two more eggs, both of which hatched. These two young left the nest toward the end of April but were not removed until the end of June, due to lack of cage space. The female laid two more eggs at the end of May which were broken, almost certainly by the young from the previous clutch, who often entered the nest. They must have been a source of annoyance on occasion, but the parents were extremely tolerant, never showing any aggression toward them. This was a perfect breeding pair.

Yellow-streaked Lory *Chalcopsitta scintillata* (Temminck 1835)

Note: The use of "*sintillata*" is an error which has been perpetuated by many authors. In common with most species, the scientific name has been changed a number of times since the nineteenth century. Temminck originally named the species *Psittacus scintillatus* in 1835. From that time until 1877, when Salvadori's *Catalogue of Birds in the British Museum* was published, the name of the species was variously rendered as *scintillans*, *scintillatus* and *scintillata*. By the time Peters' *Check-list of the Birds of the World* was published in 1937, the misspelling "*sintillata*" had appeared, and was used in this work. As this source is followed by many authors, the error continues to appear. I may be pedantic, but the use of "*sintillata*" always annoys me when the appellation obviously refers to the "scintillating" plumage of this bird. The change of gender in the current spelling was of course necessitated by the placement of the species in a different genus than that in which it originally appeared; the gender of the specific name must always correspond to the gender of the generic name.

Description: The only member of the genus

with green wings, it has the forehead scarlet, much of the head black, and the underparts streaked with orange and/or yellow on a green or brownish green and green background. The thighs are scarlet, as are the bend of the wing, the underside of the wings, and the basal part of the tail, the distal part (furthest from body) being olive yellow. There is heavy scarlet spotting on neck and throat. The outstanding feature of this species is the yellow shaft streaking of the head, neck and breast feathers; on the neck, these feathers stand away in little spikes. Beak and feet are black, the iris is brown with an inner circle of yellow, and the skin surrounding the eye is gray.

Length: 29 cm (11½ in.)

Weight: 190–220 g, but up to 240 g recorded in captive birds.

Immature: Forehead and crown black or dull dark red; generally duller throughout, especially in red areas, including underside of tail, but plumage variable. (Young from my own pair, which may be *rubrifrons*, have bright orange streaking on breast and neck.) There are brown-yellow markings at the base of the upper mandible, the iris is brownish and the bare skin surrounding the eye and the upper mandible is white. The red on the forehead becomes apparent at five or six months. This species takes about five years to become fully colored.

Sexual dimorphism: Male generally larger, with larger head and beak and usually more scarlet on forehead; but in view of subspecific variation, these features are not a reliable guide.

Subspecies: *C. s. chloroptera* has the underwing coverts, which are red in the nominate race, green or green with red markings. The body color is said to be lighter and the streaking narrower. This race is sometimes known as the green-streaked lory. Whether the subspecies *rubrifrons* should be recognized is debatable. It is said to differ in having the upper breast browner and the streaks orange-yellow; the streaking on the abdomen is said to be yellower and more extended. Rothschild and Hartert (1896) stated that "the majority of the Aru specimens have the breast more washed with brown and have very dark orange stripes along the shafts of the feathers on the breast as well as the hind-neck, but the British Museum possesses specimens from New Guinea which are just like our Aru skins." This

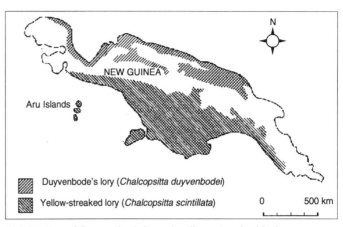

Distribution of Duyvenbode's and yellow-streaked lories.

race is sometimes known as the carmine-fronted lory, suggesting that the forehead is redder; in captive birds (wild caught) it is seldom easy to recognize this subspecies.

Natural history

Range: Southern New Guinea and the Aru Islands (see Gazetteer). The nominate race is found from Triton Bay and the head of Geelvink Bay, Irian Jaya, east to the lower Fly River, Papua New Guinea. According to Diamond (1972), "*chloroptera*, originally described from the southeastern coast, may extend in the foothills far west of the easternmost coastal population of *C. s. scintillata* (originally described from the southwest)." *C. s. chloroptera* is found between longitudes 146°E and 148°E, apparently extending westward in the foothills and the upper reaches of the south coast rivers to Soliabeda (upper Purari drainage, longitude 145°E), upper Fly River, upper Eilanden River and upper Noord River. Diamond comments that both races show much individual variation (such as the amount of yellow in the primaries of *chloroptera*, of red in the predominantly green underwing coverts of *chloroptera*, and of green in the predominantly red underwing coverts of nominate *scintillata*). He suggested that hybridization of the two subspecies was occurring after a prolonged break in contact at the Fly River. *C. s. rubrifrons* is found in the Aru Islands, southeastern Moluccas. Milton and Marhadi (1987) observed it on the island of Pulau Warmar.

Habits: There are few references to this species

199

in the literature searched, the most recent being that of Milton and Marhadi. In 1985, they observed this lory on the island of Warmar, but few sightings were recorded in comparison with the black-throated lorikeet (36 sightings). A single yellow-streaked was seen on one occasion, two birds were seen on two occasions and groups of four and nine birds were seen on one occasion. In New Guinea it is found in lowland savannahs and adjacent forest, in secondary growth and in coconut plantations. Bell (1984) noted that although Forshaw (1973) had stated that this species is often seen in parties of 30 or more birds, at Brown River he had recorded the following group sizes (number of occurrences in brackets): 1 (28), 2 (39), 4 (4), 5 (1), 6 (2) and 7 (1). Bell (1984) described this species as "more frugivorous than nectarivorous." On November 29, 1970, at Rouka village, Trans-Fly, the lories had apparently consumed fermented coconut juice. The local people were producing an alcoholic drink called "tuba," made by tapping a live coconut and draining the fluid into a bamboo tube where it ferments. A number of lories and honeyeaters were seen to fly to these tubes and drink the fluid. Five yellow-streaks were so intoxicated they were rolling helplessly on the ground below. (Intoxication of lorikeets is also known in South Australian vineyards.) Coates (1985) states that in New Guinea the yellow-streaked lory frequents forests, forest edges, partly cleared areas and heavy savannah; it visits gallery forest and sago swamps. It occurs in lowlands and in hills up to 800 m (2,620 ft.). Flocks are seen flying above the forest, at times quite high, in the early part of the day. In flight the neck is extended and the tail appears moderately long and broad, almost square-ended. The flight is direct but appears relatively slow, almost hesitant, with rapid, shallow, flickering wing beats. It feeds in the tops of flowering trees, including sago palms and umbrella trees (*Schefflera*). It sometimes feeds in company with green-naped and red-flanked lorikeets. In September, 1996, it was common at Kiunga, Papua New Guinea, where over 200 could be watched. Flock size was generally two to 20 birds (Them pers. comm. 1996).

Nesting: Almost nothing is known; 2 birds were seen investigating a dead palm stump in October on the lower Fly River (Mayr and Rand 1937). A nest in a tree hollow was found about 24 m (79 ft.) above ground at Veimauri River, southeastern Papua New Guinea (Mackay 1971). Bell (1984) saw two birds feeding another with a short tail (and therefore presumably a young one) on February 7, 1976, at Brown River. Both adults fed fruits of *Brassaia* to the young bird. Harrison and Holyoak (1970) describe a single egg as measuring 31.4 x 24.1 mm; Hubers (1995a) describes eggs of captive birds as measuring about 30 x 25 mm.

Status/Conservation: Unknown in New Guinea. It is rarely traded or held in captivity in the Aru Islands, according to Milton and Marhadi (1987); they state that its survival there is not threatened, "due to the islands extensive forest cover and small human population."

Aviculture

Status: Fairly common

Clutch size: Two

Incubation period: 25 or 26 days (Kyme 1979b; King, in Low 1986a).

Newly hatched chick: Weight, 7–8 g; long whitish or pale grayish down, sparser on head; beak dark brown with white egg tooth, feet pink.

Chick development: Similar to Duyvenbode's, except that pigment in feet slower to appear—at 30 days feet are not entirely gray (still partly pink).

Young in nest: 76–100 days recorded. Gran Canaria, 76–82 days; U.K., 81–87 days; Germany, 90-100 days (Brockner 1989).

Age at independence of hand-reared young:

2 young only, 40 and 44 days; another, stunted by early parental neglect, 88 days (weighing only 129 g, compared with 168 g and 174 g for the other two at the same age).

Chick growth

Shows weights of parent-reared young: Nos. 1 and 2 hatched by author's birds in 1990, No. 3 reared by Armin Brockner's pair (Brockner 1989).

Yellow-streaked lory weight table #1

Age in days	Weight in grams Chick No. 1	No. 2	No. 3
day hatched	8 e *	8 e	
3	17 fc		
4	17 e	20 fic	
5	23 fic		
11	36 e		
24	94 fc	75	
25	98 fic		
28	116 fc		
29	114 fic	116 fic	
31		91	
34	136 fc		
35	136 fc		
38	138 fc		
39	142 fc	130	
45		146	
46	152		139
51		148	
52		152	
53			150
58	166		
59		162	
61			180
76			156
85			164
93			168

* e = crop empty; fic = food in crop; fc = full crop

Weight of one chick parent-reared by author's birds in 1992.

Yellow-streaked lory weight table #2

Age in days	Weight in grams
day hatched	not weighed
1	8 e*
2	12 e
3	14 nf
4	16 e
5	16 e
6	22 f
7	20 e
8	24 fc
11	34 fc
12	36 nf
13	42 fc
14	40 fic
15	44 fic
17 (ringed)	50 e
18	64 fic
19	70 fc
20	70 fic
21	80 fc
22	78 e
24	82 e
25	92 e
27	96 e
28	106 fc
29	125 vfc
32	120 fc
33	122 fc
34	136 fc
36	126 e
39	138 fic
41	149 nf
46	158 e
60	179 ne
77	left nest

* e = empty; fic = food in crop; nf = nearly full; fc = full crop; vfc = very full crop

General: Because this lory comes from the same region (Irian Jaya) as the two preceding species, it too was not exported commercially until the early 1970s. The first time I saw it was at West Berlin Zoo in 1971. Although it is not among the first ten species in terms of abundance (see Popularity, Avicultural) it is fairly common and widely bred, its beautiful coloration ensuring its popularity. However, like other members of the genus, its loud voice deters some aviculturists from including it in their collections. Kyme (1979a) referred to its voice when describing the behavior of his pair: "the male is more aggressive. When the pair are fed it is the cock who attacks physically although it is both sexes which utter the highpitched, deafeningly loud, mobbing calls. While they shriek they keep the tail flared right out, the wings outstretched and the long, rather narrow, feathers to the head and neck ferociously bristled."

In 1976, this species was bred by Holger and Inge Roth of Farum, Denmark. Young were hatched at the end of January. The interesting observation was made that the parents took twigs and grass into the nest to keep it dry, because of the wet droppings of the young. This is worthy of note because lories have very seldom been observed to take such items into the nest. Also, one wonders, where would they have found twigs and grass in Denmark in February? The two young ones left the nest on April 29 and May 1 (Roth 1976).

In my experience, and that of others, yellow-streaks will accept a more varied diet than black and Duyvenbode's lories. They relish corn on the cob, especially when rearing young, whereas the other two species ignore it. They will eat a wider range of fruits and green foods and are extremely fond of orange.

In common with other *Chalcopsitta* lories, some birds become very tame. One wild-caught male has been with me for 20 years. I took him to the Canary Islands, where he lived in suspended cages for 8½ years. By the time I brought him back to the U.K., he was so tame that he would attempt to copulate with my hand—and I had never tried to tame him. His female, a captive-bred, parent-reared bird, exhibits none of this tameness. In the U.K., my aviaries have an indoor section where the birds sleep at night. On the first evening in the new aviaries, I anticipated problems in persuading the birds to go inside. But I need not have worried; they understood immediately what was expected of them. There was just one problem bird, the male yellow-streaked. When I entered the aviary he saw the opportunity for a new kind of game. He jumped on my shoulder and started to "preen" my hair. He had no intention of going inside. Finally I went for the net. I knew I would only have to show it for him to get the message! He took one look and went through the pophole!

Cardinal Lory *Chalcopsitta cardinalis* (G. R. Gray 1849)

Description: Red, except for wings and back which are brown-red; feathers of underparts are red edged with yellowish buff; primaries are bronze-orange and the tail is rusty red. The beak is orange with black at base of upper mandible in most birds; the iris is orange. The skin surrounding the eye is gray, that below the lower mandible being gray and yellow.

Length: 31 cm (12 in.)

Weight: About 220 g (none weighed)

Immatures: Duller throughout, with pale yellowish on the ear coverts. The iris is light-colored; beak black in nestlings, also the naked skin (Loman and Loman 1992). The skin around the eye is whitish. The beak is changing to dark brown by the time the young fledge and soon becomes orange.

Natural History

Range: Solomon Islands (see Gazetteer), including Guadalcanal, Makira, Isabel and Kolombangara and the islands of Bougainville and Buka which belong to the Territory of New Guinea. Most sources state that it is found on Lihir, near New Ireland; Peter Odekerken spent three weeks on Lihir in 1995 and confirmed its absence there (Odekerken 1995b). The cardinal lory is said to occur on other islands to the east of New Ireland, such as Feni, Tanga and Tabar, and also New Hanover.

Habits: It occurs in primary and secondary forest, in the lowlands and in the hills, and in coastal coconut plantations and mangroves. It feeds in forest canopy trees and in coconut palms, on blossoms and pollen and other vegetable matter. Cain and Galbraith (1956) found small berries in the crop. Jonathan Newman visited the Solomon Islands from July to September, 1990. He observed cardinal lories on Kolombangara, on the south-west coast; in the Selwyn College area of Guadalcanal; and on the north coast of Makira (San Cristobal). In the lowlands it was feeding on coconut flowers, taking nectar and pollen. In higher areas, it favored red flowering trees and shrubs, including what was probably *Erythrina*. It fed alongside Massena's lorikeet (*Trichoglossus h. massena*) and Meek's lorikeet (*Charmosyna meeki*) and was dominant over both, but displaced by eclectus parrots. The cardinal's voice is deeper than that of the *Trichoglossus*, and more harsh; clipped notes were given in flight (J. Newman pers. comm. 1994).

During the period of 1985 to 1988, habitat preferences of birds on New Georgia, Western Province, Solomon Islands, were noted (Blaber 1990). The cardinal was found to be common in primary forest, secondary forest, villages, gardens and plantations and in mangroves. Extensive logging had occurred there in the previous 20 years and industrial development was beginning.

Has the cardinal lory declined in recent years? Stevens (1969) described seeing them in "their shrieking scarlet hundreds" and Stott (1975), writing of the year 1943 when he was in the U.S. Navy, recalled how, when he was interviewing personnel who had served in the Solomons, he "was impressed by the fact that one man after another referred to the enormous numbers of parrots, red ones especially, in the Solomons. If during the rage of battle and the subsequent counter-attacks and bombings, birdlife was sufficiently abundant to command the attention of war-weary marines and sailors, it must have been prolific."

Eclectus parrots, in which the females are red, are also found there, but they do not occur in large numbers, therefore it must have been the cardinal lories which caught the men's attention. They certainly left an indelible impression on Ken Stott, for their "overwhelming abundance and beauty." He considered that this species

> ...offers one of the most amazing ornithological events on earth, which though a daily affair can become no more common-place than a view of the Grand Canyon. Just before dusk in the coastal coconut groves, clouds of Cardinal Lories swarm through the palms, their coloring enriched by the golden rays of the late afternoon sun. At this hour, a drive along the road that passes Henderson Field and the Red Beach, where so many Americans lost their lives, now as then provides the sight of thousands upon thousands of Cardinal Lories. Being large (12 inches) and long-tailed, a single bird in flight would be a noteworthy encounter, its coloration becoming more or less intense, depending on the angle of the sun on the plumage of the bird as it banked this way or that among the palms. But to see it in such vast numbers, flock after flock, rivals the masses of flamingos in East Africa's Rift Valley for sheer beauty. Sometimes Cardinal Lories may be seen flying from island to island, resembling flocks of crimson terns against the azure of Solomon skies and the turquoise of its inter-island seas (Stott 1975).

Jan van Oosten saw a group of about 25 cardinal lories in Honiara, the capital of Guadalcanal in the Solomons. They could be heard about three-quarters of a mile down the road. "They would fly across the road to venture in the grove of trees between the road and the ocean." It was about 500 yards in length. The birds would search the flowers of the coconut palm and, when satisfied, they would return to the other side of the road (van Oosten 1993). In Isabel Province, the provincial capital of Buala, van Oosten found cardinal lories and Massena's lorikeets to be very common. He observed them on a coastal strip adjacent to plantations of copra and to primary forest. On the journey back from a man-grove swamp north of Buala, he saw over 300 lories of each species fly across from the mainland two miles distant, to the island at the north end of the lagoon. A Ducorp's cockatoo (*Cacatua ducorpsi*) was seen to open a green coconut. When the cockatoo departed a cardinal lory arrived "and spent over 30 minutes drinking the sweet fresh water inside and eating quite a lot of the white gelatin material." No apparently young birds were observed.

Newman (pers. comm. 1994) reported regularly seeing groups flying over the sea, between islands; small islands are visited in order to feed on coconut palms, and for roosting purposes. One flock of 120 birds was seen at a feeding site on Guadalcanal. Larger flocks —several of over 100—were seen on their way to roosting sites in mangroves and sago swamps.

Nesting: Newman (pers. comm. 1994) saw a pair enter a hollow in a near-vertical branch stump on two occasions. It was in a 10 m (33 ft.) ridge-top tree in closed forest at 1,150 m (3,770 ft.) on Kolombangara in the Solomon Islands. The birds scolded the observer on both occasions (August 17 and 18) and were assumed to be nesting. Stevens (1969) saw a young one being fed in secondary forest in July in the Shortlands Islands, northwest Solomons.

Status: Common to abundant throughout most of the range; usually seen in groups of up to about 20 birds. Newman describes this lory as widespread and abundant in the Solomon Islands, from sea level upward, and up to 1,100 m on Kolombangara. He found it most abundant in coconut plantations. There were lower densities in primary forest and it was especially scarce in lowland forest. Even where total numbers were large, such as in plantations, groups of from five to 20 birds were evident.

Trade: It was not until about 1989 that the Solomon Islands first permitted commercial export of its avifauna, when it limited export to four species. It was decided to set the quotas every two or three years and to export the birds through the seven members of the Solomon Islands Bird Dealers' Association. In 1993 the following annual quotas were set: cardinal lories, 1,400; yellow-bibbed lories, 350; Ducorp's cockatoos, 1,400; and Solomon eclectus parrots, 1050. These appear to be realistic quotas; however, by 1996 there was evidence that these quotas were being exceeded.

When dealers receive an order, they trap only the numbers required and care for them for several weeks before export. One dealer stated that losses were one in 40 birds (J. van Oosten unpublished report 1993). In 1993, the government of the Solomon Islands requested Mr. van Oosten's advice on how to set up a captive-breeding program, indicating they were considering the idea of breeding for export rather than trapping.

Aviculture

Status: Uncommon, but numbers are increasing.

Clutch size: Two

Incubation period: 24 days (D. Brockhoff pers. comm. 1993); "around 24 days" (Sweeney 1994a).

Newly hatched chick: Weight, 6.7 g, 6.4 g and 5.9 g; dense long white down, bill mainly black but pale toward base; nails black with pale tips, crop capacity first feed, 0.8 ml (Sweeney 1994a and pers. comm. 1996); long white down (Loman and Loman 1992).

Chick development: Of chick described by Sweeney which was reared at Birds International Inc. in the Philippines in 1993: eyes had begun to slit by day 12 and were fully open by day 16; day 20, secondary down follicles apparent over much of body and by day 23 most of the gray second down had emerged; about day 28 pin feather on crown and those of flights were appearing; day 35, feathers erupting; day 60, fully feathered.

Young in nest: "Around 65–70 days" (Sweeney 1994a); 70 days, one chick only (Müller and Neumann 1996).

Age at independence of hand-reared young: 70 days, one only (Sweeney 1994a); one young cardinal lory took "9 weeks to take food by itself" (Loman and Loman 1992).

Chick growth

Shows the weights of two chicks from the same clutch which were hand-reared at Loro Parque, Tenerife, on Pretty Bird hand-rearing food (19 percent protein and 8 percent fat)(Sweeney pers. comm. 1996).

Cardinal lory weight table

Age in days	Weight in grams	
	Chick No. 1	No. 2
hatch	6.4	5.9
1	6.9	7.2
2	7.4	7.2
3	8.6	8.0
4	10.0	9.0
5	11.9	10.0
6	13.7	12.0
7	15.6	14.8
8	17.9	16.8
9	20.3	19.5
10	23.7	23.0
11	27.7	27.0
12	32	31
13	36	36
14	42	41
15	47	45
16	53	52
17	60	57
18	68	61
19	71	68
20	78	75
21	84	83
22	90	89
23	96	95
24	103	105
25	109	110
26	115	118
27	121	122
28	126	126
29	127	134
30	136	143
31	143	145
32	151	149
33	155	163
34	165	169
35	169	175
36	177	181
37	182	186
38	189	190
39	195	190
40	192	191
41	191	193
42	190	192
43	191	190
44	196	198
45	197	196

General: In the Solomon Islands, cardinal lories kept as pets are sometimes given their liberty during the day. Kenning (1993–94) described how "they fly where ever they like then late in the afternoon they come back to sleep in the house. If you take a lory out of its cage, it likes to rub its body against your fingers very gently. Also, when it lifts its wings, you get the impression that the bird is very happy. In the wild sometimes up to 5 birds can be seen together on one branch, lifting their wings, screaming all the time, dancing up and down, putting their head under the wings of their neighbors."

Newman also noted that in the Solomon Islands cardinal lories are often kept as pets; they are fed on papaya, banana, mango, etc. Adults are sometimes captured by the inhumane method of hitting flying birds with a catapult.

To date, little has been recorded about this lory as an avicultural subject. The Swiss aviculturist Dr. R. Burkard pointed out that both this species and the yellow-bibbed demonstrate close pair bonding and show little inclination to form larger groups:

> Both are extremely aggressive toward a third member of the same species. This occurs even before breeding maturity. Both species transfer this aggression to the curator during the hatching time by attacking vehemently when one enters the aviary. And these are birds which during non-breeding season are tame. Even during the hatch they will take a delicacy off the hand—though only if one stands outside the enclosure. To these paired individuals it must be added that voice contact between one pair releases a "contact-cry" of neighboring pairs. It seems that these aggressive territory-defenders are stimulating other pairs to mark their own territories through cries (Burkard 1983).

In some aspects of behavior the cardinal lory does not resemble members of any other genus. It has a peculiar display, which I have never seen in any other lory. When excited, it slaps its tail hard against the perch in a series of exaggerated movements. Perhaps this is a method of warning intruders out of its territory.

The first and probably only U.K. breeding to occur before the Solomons allowed this lory to be commercially exported was recorded by T. Morford. A pair was received in January, 1979, and male and female were quarantined in adjoining aviaries. On April 1 they were reunited in an aviary measuring 8.3 m (27 ft.) long and 4.5 m (15 ft.) wide, planted with conifers and honeysuckle. By the middle of May the female was spending long periods in the nestbox and the yellow margins to the feathers

of the male's underparts had become vivid. Two eggs had been laid by May 27, and the male spent nearly all day in the nest. From this period until June 12 no fruit was eaten. A fine mist spray was used daily from June 1–14 to increase the humidity within the nest. On June 20 the male was seen to remove two pieces of shell from the nestbox. (This is interesting; I have never observed this behavior in any parrot.) Inspection did not occur until July 14 when one chick was present. It was removed for hand-feeding on August 22 and seven days later it started to feed itself (Morford 1980).

The personality of this species is extrovert, not unlike the yellow-bibbed. In display, a male cardinal at Loro Parque would run along the perch with drooping wings, then he would flap his wings. The aggressive tendencies of this species were also apparent in the pair owned by Benno and Hilda Loman; one of this pair decapitated the first chick soon after it hatched (Loman and Loman 1992). In Denmark, Stig Lundsgaard from Frederikshavn imported a pair of cardinal lories from the Netherlands in 1996. When the female laid two eggs in the same year, the pair became very aggressive and attacked the keeper. The male had to be removed, for fear he would kill the female. The eggs hatched after 26 days and the young left the nest at 11 weeks old (Them pers. comm. 1996).

At Vogelpark Walsrode, in Germany, there was an egg-laying female, plus two males, in the collection in March, 1995. The female was placed with one of the males and they were provided with a nest log with an inside diameter of 25 cm (10 in.) and a height of 30 cm (12 in.). The nest entrance was 8 cm (3 in.) in diameter. On May 17, four weeks after the pair was put together, the female laid her first egg. On June 11 there was a dead chick in the nest. The second egg, which was pipping, was given to a pair of rajah lories with eggs. It hatched on July 13, and an *insignis* hatched on the following day. Both were reared, and the cardinal left the nest on August 22. Meanwhile, the female had laid again, the first egg on July 7 (i.e., 26 days after the previous egg and dead chick were removed). The two eggs in the clutch were exchanged with infertile eggs from a pair of black-winged lories, but one of the cardinal's eggs was infertile and, in the other, the embryo died. On August 11 the cardinal laid the first of two more

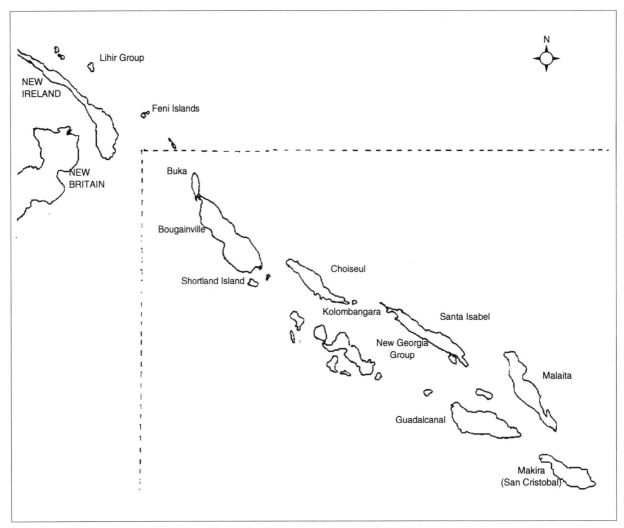

Distribution of the cardinal lory (*Chalcopsitta cardinalis*). Named islands within the dotted box.

eggs. They were first placed in an incubator, and then, on September 1, given to incubating dusky lories. One egg hatched on September 7. The chick was removed for hand-rearing when the foster parents started to pluck it. The hand-reared young one was subsequently sexed, as was the one fostered by rajahs; they proved to be a male and a female (Müller and Neumann 1996).

In South Africa the cardinal lory is being bred in a number of collections, and several pairs are kept in most locations. There is no doubt that it will be quickly established there.

Genus: *Eos*

This genus contains six species (or seven if taxonomists decide to separate *obiensis*) of medium to large-sized lories, all of which are red with black and blue or mauve markings. All are Indonesian island species, thus are vulnerable to habitat destruction and overtrapping for trade. Their status ranges from fairly common to threatened to endangered or critically endangered. Their avicultural status varies from common to rare to almost unknown. They nest readily in captivity and every member of the genus could be maintained indefinitely within aviculture in the hands of competent aviculturists. In aviaries, they accept a wide range of foods and are less dependent on a nectar substitute than most other lories.

In terms of evolutionary relationship, they are close to *Pseudeos*, *Lorius* and *Chalcopsitta*. In the streaking of nape and mantle, and in its profile, which is more streamlined than other members of the genus, the blue-streaked (*E. reticulata*) seems close to the genus *Chalcopsitta*. Smith (unpublished manuscript) states that "*Eos* was solely determined because members had red on their wings" and argues that it cannot be sustained against *Lorius* except for the comparative differences in tail and wing length, in consequence of the limited nomadism of *Eos* against the more sedentary nature of *Lorius*.

I have compared young chicks (under about seven days of age) of *Eos bornea* and *Lorius lory*; apart from the gray forehead down which is a feature of the latter species, I can find absolutely no difference between them.

Black-winged Lory *Eos cyanogenia* (Bonaparte 1850)

Synonyms: Biak red lory, blue-cheeked lory

Description: No other member of the genus has such a large area of solid black in the plumage, or such bold areas of color. The black-winged is a handsome bird, with the ear coverts and much of the side of the head violet-blue, these feathers being slightly elongated. The remainder of the lower part of the face is red, of a brighter shade than the body which is tinged with black in some areas. Upper wing coverts and scapulars are black, also most of the hidden part of primaries and secondaries, which are otherwise red with a golden yellow spot on the fifth primary. Part of the abdomen, the thighs and a spot on the flanks are also black. Central tail feathers are black with red inner webs; under tail coverts are red. The beak is orange, the cere and feet are black and the iris is dark red with a faint inner ring of white.

Length: 28 cm (11 in.)

Weight: The male of one pair was 180 g, the female 155 g (Campbell, San Diego Zoo, pers. comm. 1995).

Immatures: They have the red feathers irregularly margined or flecked with black. The beak is black and the iris is dark. Wright (1977) stated that the young had much black flecking on the back of the head, forehead and chest.

Sexual dimorphism: German aviculturist Walter Brasseler stated that in all pairs known to him, the male is compact and stocky and the female is long and thin. He also used weight as a guide to sex (Brasseler 1994).

Natural history

Range: The island of Biak (see Gazetteer) in Geelvink Bay, Irian Jaya, and the satellite islands of Numfor, Manim, Meos Num (see Gazetteer) and Supiori.

Habits: A gregarious species of coastal habitats such as coconut plantations, it apparently avoids the forests of the interior. It often flies below the level of the canopy (Beehler, Pratt and Zimmerman 1986). According to Bishop (in Forshaw 1989) it occurs in coastal forests, secondary woodland and scrubby fallow land with remnant tall trees, almost invariably in pairs. On Supiori it is also seen inland, up to about 460 m (1,500 ft.), where groups of four to six birds were regularly observed in the forest canopy. It is very conspicuous; the brilliant red and black plumage is spectacular as birds pass swiftly overhead with whirring wing beats.

While on Biak, F. Baur, an aviculturist, asked a guide to cut some inflorescences from a palm tree on which these lories fed; the numerous blossoms were very tiny. He commented that in his experience in captivity this species needs only as much food as the much smaller Josephine's lorikeet. The inflorescences were covered in tiny yellowish green insects which perhaps are also eaten by this species. Baur asked local people if the lories damaged crops—but no damage had been noted. In any case, cultivated fruit trees were so laden with fruit that not all of it was harvested (Baur 1991).

Nesting: No information. Courtship display was frequently seen during June and July, 1992, on the islands of Biak and Supiori (Bishop, in Forshaw 1989). A pair was seen at a hole in a tree on a ridge-top deep in primary forest. Schonwetter (1964) described six eggs, the average size of which was 29.0 x 23.8 mm. Eggs laid by one female at San Diego Zoo weighed between 8.6 g and 9.6 g.

Status/Conservation: In 1986, Bishop found it was common and widespread on Biak and Supiori, but virtually unknown on Numfor (where clearance of primary forest is almost complete) and on Manim. On Biak in 1991, Thomas Arndt described it as easily observed in coconut palms along the coast (Arndt 1992). Elsewhere Arndt (*Lexicon of Parrots*, p. 5) stated that the total population numbered only 500 to 1,000 birds. At the same time, Lambert, Wirth, et al., in the first draft of Parrots, *An Action Plan for their Conservation and Management*, estimated the population as less than 5,000 and described its status as endangered. They suggested that "an assessment of the reserve system in the region should be made and a management plan drawn up especially for the important Supiori reserve, which is thought to be the stronghold of the species."

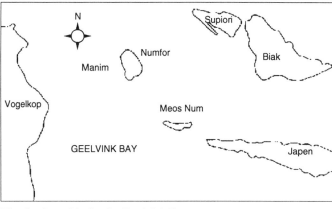

Distribution of lories in Geelvink Bay.

BIAK
 black-winged lory (*E. cyanogenia*)
 Rosenberg's lorikeet (*T. h. rosenbergii*)
 black-capped lory (*L. l. cyanuchen*)
 red-spotted lorikeet (*C. r. kordoana*)
SUPIORI, NUMFOR, MANIM and **MEOS NUM**
 black-winged lory (*E. cyanogenia*)
MEOS NUM
 Jobi lory (*L. l. jobiensis*)
JAPEN
 Jobi lory (*L. l. jobiensis*), dusky lory (*P. fuscata*)

Trade: This lory was rare in aviculture until the early 1970s when it was occasionally available, but not in large numbers. Then, during the mid- to late 1980s, very large numbers were exported from Indonesia, and trapping became a threat to its survival. Clearly, it should then have been protected from trapping or killing, at least until surveys had determined its status. According to Tabak (pers. comm. 1995), black-winged lories and Rosenberg's lorikeets are not traded locally, although quite common when he visited Supiori in January and February, 1995. The local people are interested only in black-capped lories and sulphur-crested cockatoos.

Aviculture

Status: Not common

Clutch size: Two

Incubation period: 25 to 26 days with the parents, 25 to 26 days in an incubator (Rimlinger).

Newly hatched chick: Average weight of chicks hatched at San Diego Zoo was 5.7 g; the down is white, sparse and fairly long, and the beak is black (Rimlinger pers. comm. 1995).

Chick development: At six weeks covered in midgray second down, with head feathers and primaries starting to erupt; at eight weeks nearly all the head feathers have erupted and the underparts are half covered in feathers; at ten weeks almost fully feathered except on upper breast, but tail not full length; beak black and bare skin of cere and eye gray (from photographs by Brockner 1994).

Young in nest: 75 to 87 days recorded; 75 days for two young in same nest and 87 days for single young from same pair (Wright 1977).

Chick growth

Shows approximate weights (taken from a graph) of one hand-reared at Birdworld, Surrey, U.K., in 1991. Rearing food consisted of five parts of Aviplus hand-rearing diet and one part Birdquest lory nectar mixed with cooled boiled water (Sweeney 1992a).

Black-winged lory weight table

Age in days	Weight in grams
4	7
8	11
12	17
16	22
20	38
24	55
28	70
32	90
36	105
40	115
44	130
48	135
52	135
56	123
60	120
64	115
68	120
72	120
76	130
80	125

Age at independence of hand-reared young:
The young one whose weights are given above started to feed itself at 50 days; by day 76 it was no longer spoon-fed but fresh food was given three times daily. Young hand-reared at San Diego Zoo were independent at about 75 days. Average weight at 8 weeks was 111 g; at 9 weeks, 129 g; at 10 weeks, 147 g; and at 11 weeks, 131 g. One parent-reared chick weighed 66 g at 20 days (Campbell pers. comm. 1995).

General: Where did all the black-wings go? It is unfortunate that during the 1980s there were a few importations into Europe of large numbers which dealers probably wanted to sell as quickly as possible. Many birds must have been obtained by those who were not lory specialists and the result was that few

were bred. There was little information available about the status of this species and the specialists who should have been breeding it were largely unaware that it was a threatened species. It was a wasted opportunity: large numbers exported for no other reason than that money was to be made from them yet almost certainly not sufficient demand overseas. Many must have died in export centers such as Singapore while dealers awaited orders. Few breeders appear to have extensive or sustained success with this beautiful lory. A glance at the breeding register of the Parrot Society (in the U.K.) for the year 1992, for example, shows that six members reported breeding 12 birds (not all members report their results); in comparison, 10 members bred 36 red lories of the nominate race. A decade earlier, one lory specialist, Andrew Blyth, had realized the importance of maintaining captive populations and acquired several pairs. His first young were produced in 1984, and by 1991 a substantial number had been reared. That patience pays dividends was proved by Patricia King of Cornwall, U.K. During the early 1980s her pair of black-wings produced infertile eggs for five years. Then, at last, in 1985, one egg was fertile and a chick was reared. In the next clutch the same result was obtained (Low 1986a).

Two German breeders have methods which are not dissimilar. Brasseler (1994) described the unit in which his breeding pair was kept. Each aviary had an inside flight 1.5 m (5 ft.) long, 1 m (3 ft. 3 in.) wide and 2 m (6 ft. 6 in.) high, with tiled walls and a concrete floor. The 2 m long outside flight contained growing plants such as elderberry and wild rose. On hot days a misting system sprays the outside flights and the lories hang upside down from the roof, bathing until they are completely soaked. They are fed on a commercially produced nectar (Lorinectar by Avesproduct). In the late afternoon, they receive a puree of fruit such as banana and apple. When young are in the nest, the commercially produced Loristart (Avesproduct) is given.

Armin Brockner feeds his lories three times daily, twice with nectar and then in the evening with fruits and berries. However, the black-wings consume less than the other lories, thus their second feed is not always necessary. Willow branches placed regularly in the outdoor aviary are relished by these birds. Green food such as dandelion and lettuce is readily accepted. When chicks are being reared the commercial rearing food CeDe is offered but the blackwings, unlike the other lories, show little interest in it. They consume more nectar in the evening when they have chicks, and apples and pears are also eaten. Because his two pairs were so nervous, a 3 m (10 ft.) long indoor aviary proved unsuitable, so they were moved to an aviary with an outside flight 4 m (13 ft.) long, 1 m wide and 2 m high, with a shelter 2 m long. In the larger aviary, the pair became much calmer. They were provided with an oblong nestbox 40 cm (16 in.) long, 22 cm (9 in.) wide and 30 cm (12 in.) high. In the summer of 1990, two eggs were laid. The embryos died at an early stage, apparently due to erratic incubation. At the suggestion of a successful breeder of this species, a different type of nestbox was provided, one with a "separate brooding chamber."

In 1992, both females laid early in the year, within a week of each other. A policy of not inspecting the nest was adopted—but once again the embryos died at an early stage. No fault could be found with the birdroom ventilation or the supply of vitamins and minerals. One of the pairs produced two more clutches, with the same result. When this pair laid a fourth clutch, they were permitted to enter the outdoor flight, whereas in the previous clutches they did not have access to the flight during incubation. On June 18, after 26 days, a chick hatched; a second inspection revealed another chick. In order to avoid

disturbing them, the nest was not inspected again for about 14 days. Then the youngest chick was found dead, perhaps due to lack of food. The other chick was reared by the parents and left the nest aged between 11 and 12 weeks. Surprisingly, it was very tame (Brockner 1994).

There is a prolific breeding pair in the collection of San Diego Zoo in California. Obtained as an adult in November, 1988, the wild-caught male was paired with a captive-bred female, hatched in 1984 and obtained in September, 1986. Their first chick hatched on June 13, 1989, and the last one listed on the zoo's taxon report was for March 21, 1994. Between these dates the pair produced 25 young, six of which died before independence. This is an average of five per year; from the hatch dates it is apparent that most of the chicks were hand-reared. In 8 cases, two chicks hatched in the same clutch. The intervals between the hatching of the first and second chicks were one day in 3 cases, 2 days in one case and 3 days in 2 cases; in 2 instances the chicks hatched on the same day. However, this may not be significant if eggs were removed to an incubator soon after laying. This pair was rearing chicks when I visited the off-exhibit breeding center where they are housed, in August, 1996.

The black-winged lory deserves more attention from breeders, both from the conservation aspect and because it is a most attractive bird.

Blue-streaked Lory *Eos reticulata* (S. Muller 1841)

Synonym: Blue-necked lory

Description: The distinguishing feature is the narrow blue feathering of the hind neck and mantle, creating a streaked effect not unlike that of the yellow-streaked lory. The blue streaks are almost luminous in their intensity. There is also streaking on the ear coverts, which are a darker shade of blue; the rest of the head is red, as are the underparts. The wings are red marked with black, the greater wing coverts, primaries and secondaries being partly black—the primaries appearing almost wholly black. The thighs are red and black. Upper and lower tail coverts are red and the tail, which is

slightly longer than in other *Eos* species, is black above and red below. The rump is inconspicuously marked with dull blue. The beak is orange; cere, eye skin and feet are gray and the iris is orange.

Length: 31 cm (12 in.). The build is slimmer than that of *cyanogenia*.

Weight: About 160 g. Jay Kapac from California weighed the five males in his collection. They ranged from 196 g for the oldest (minimum ten years) to 154 g for the youngest, aged five months; average weight was 172 g. His four females ranged from 200 g for the oldest (minimum eight years) to 120 g, while a four-month-old bird weighed 142 g;

average weight was 156 g (Kapac pers. comm. 1995). It was interesting to note the weight range in the mature birds, from only 120 g to exactly 200 g.

Immatures: They have the feathers of the underparts broadly edged with blue-black to give a dark and mottled appearance. The wing markings are grayer and less clearly defined. Beak and iris are dark brown. Birds in nest feather have the crown blue and the beak black. One hand-reared by the author was described as follows: "forehead, crown, cheeks, throat and part of the upper breast are red. The electric blue streaks on nape and mantle, also ear coverts, are as prominent and rich as in the adult. Wings are red and black, more wing feathers being tipped with black than in adults. Seen in sunlight, some of the black feathers have a green sheen. The primaries are black, red on the outer margins of the center of each feather. The 2 central tail feathers are black, the rest being red on the inner web and black on the outer web. Feathers of the underparts are red with bluish black tips. Feet and skin around the eye are black" (Low 1986b). However, it should be noted that hand-reared birds are often more brightly colored than those reared by the parents.

Sexual dimorphism: Slight. Hayward (1984) recorded that a consignment of wild-caught birds "were easily sexed, the cocks being larger in body, head and bill as well as being a little brighter in color than the hens with larger cheek patches and more extensive cobalt streaking."

In the U.S.A., Carl McCullough noted that "the females of my blue-streaked lories have irregular lines of black feathers between the legs and also covering the entire pelvic area. In my 3 males these feathers are totally absent. All my birds have been surgically sexed, so there is no mistake as to their identity" (McCullough 1985).

Natural history

Range: Tanimbar (Tenimber) Islands, Indonesia (see Gazetteer), of which Yamdena is the principal island and is said to account for 55 percent of the range of this species (Cahyadin et al. 1994). It was introduced to the Kai Islands (see Gazetteer) to the north of Tanimbar, where it probably no longer occurs, and to the island of Damar in the Southwest Islands. Finsch (1900) recorded it from the island of

Babar in the same group. This is the most southerly representative of the genus.

Habits: This species was observed on Tanimbar by N. Brickle. He noted that it was recorded in most habitats there—forest, forest edge, agricultural land and on plantations of coconut and sago. Outside of forest and plantations, most records were of birds in flight. They were occasionally observed feeding in fruiting trees and in sago and coconut palms. Highest densities were recorded in low, quite open monsoonal forest. The presence of this lory was easily determined by its call, "a series of starling-like chatters and whistles." Flying birds called continuously. Most records were of pairs but single birds and groups of up to eight were also seen (Brickle pers. comm. 1994).

During the April–May, 1993, PHPA/Birdlife survey on Yamdena, encounter rates were noted for this lory on two cross-island transects. They were as follows: agricultural land 2.4 km and 4.7/hr; monsoon forest 1.7 km and 5.0/hr; semi-evergreen forest 1.3 km and 3.3/hr; mangrove forest 7.9 km and 12.4/hr.

At the end of the rainy season the blue-streaked lory occurred in disturbed and undisturbed monsoon, semi-evergreen and mangrove forest and occasionally in dry-land agricultural areas. It was not found in coconut plantations but was attracted to fruiting and flowering sago palms (Cahyadin et al. 1994).

Nesting: No information. Sweeney (1993b) gave the size of four eggs laid at Birdworld, Farnham, U.K., as 29.0 x 23.9 mm, 31.1 x 26.0 mm, 28.6 x 23.6 mm and 29.2 x 24.2 mm. When newly laid, the last three weighed 9.66 g, 8.56 g and 9.04 g, respectively.

Status/Conservation: After the 1993 survey on Yamdena it was described as a "common forest bird"; previously it was described as vulnerable/endangered by Lambert, Wirth et al. (undated) after heavy trapping for export during the 1970s and 1980s. In 1992, they suggested that the population was in the region of 10,000 to 50,000 birds. However, after the 1993 survey (Cahyadin et al. 1994), its population on Yamdena was estimated at 220,000 birds, plus or minus 52,000. In 1982, a comprehensive system of protected areas was proposed for Maluku Province, of which the Tanimbar and Kai islands are a part. However, up until 1992 nothing had been done. One proposed reserve on Yamdena

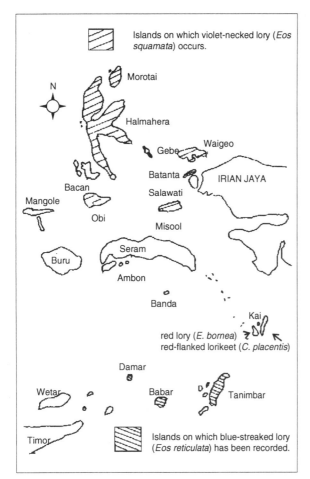

Distribution of the blue-streaked lory and the violet-necked lory.

are as follows, and are compared with those of the red lory (*Eos bornea*) and the violet-necked lory (*Eos squamata*).

Indonesian Lory Exports

	1981	1982	1983	1984
reticulata	370	534	4,252	2,785
bornea	2,735	4,853	4,968	10,022
squamata	400	—	780	275

At the 1992 CITES meeting the proposal to transfer the blue-streaked lory from Appendix II to Appendix I was withdrawn, following the government's agreement to set a zero quota for 12 months. In 1996 the nil quota was still in force. As can be seen from the official catch quotas shown below, these had been set far too high:

Blue-streaked Lory Catch Quotas

Year	Export quota	Net CITES trade
1984	10,000	4,730
1985	7,000	1,399
1986	unknown	1,452
1987	1,000	unknown
1988	1,000	2,738
1989	1,500	unknown
1990	2,000	unknown

Note that the 1984 CITES trade figure (from the 1988 annual report to CITES) is nearly double that of the numbers known to have been legally exported in the same year (Cahyadin et al. 1994).

Blue-streaked lories were normally trapped outside their breeding habitat (i.e., on fruiting sago palms in agricultural areas); however, this does not necessarily mean that they were non-breeding individuals.

After the 1993 survey on Yamdena, when there was a zero catch quota, it was suggested that, if catch quotas are resumed, consideration should be given to introducing a bird catcher registration scheme. This would provide a more reliable way to estimate numbers trapped. There should also be regular monitoring of the wild population. See Trade.

Aviculture

Status: Common (see **Popularity, Avicultural**)

Clutch size: Two

Incubation period: 25 to 27 days (Tiskens pers. comm. 1994); 25 or 26 days (Low 1986a). Gibson (1984) kept records of three clutches; the eggs hatched after 28 days except the second egg in

would protect a large part of the island and would be of great importance for the endemic avifauna. The single reserve proposed for the Kai Islands comprises about 37,000 hectares of forested hills in the north of Kai Besar (ICBP unpubl. report 1992).

Trade: Commercial export commenced about 1970 and continued for nearly two decades. However, in 1988 this species was listed among those which were not permitted to be imported into countries of the European Union. This was because it was endangered by trade. Even if Indonesia had been a member of CITES at that time, that country would almost certainly not have favored adding this and other lories threatened by trade to Appendix I, because valuable income was gained from their export. Some figures are available from PHPA (Indonesian) government statistics to show numbers of blue-streaked lories exported legally from 1981 to 1984; however, it is well known that huge numbers of parrots are exported from Indonesia illegally, due in part to the impossibility of controlling all the ports from which they might leave. The figures, quoted by Milton and Marhadi (1987), the first clutch which hatched after 27 days. A newly laid egg placed in an incubator hatched after 25 days (Kapac pers. comm. 1995).

Newly hatched chick: 8 g; fairly dense white down.

Chick development: At two weeks old the developing second down is just visible beneath the skin; at three weeks the down is starting to erupt; at four weeks young are covered in gray down; at five weeks the head feathers and the feathers of the wings and scapulars are starting to erupt; at six weeks the head and wings are half feathered. At eight weeks young are fully feathered except for the shorter tail. Tiskens reported that chicks were ringed at 14 days with 7 mm rings.

Young in nest: About 75 days; 56 days for two young in a nest, but unable to fly well (Rucker 1980).

Chick growth

Shows the weight of one chick hand-reared from the age of eight days by the author in 1986. Rearing food consisted of Milupa baby cereal, wheatgerm cereal and tinned Heinz Fruit Dessert baby food. Weights are shown before and after the first feed of the day.

Blue-streaked lory weight table #1

Age in days	Weight in grams
8	21/22.5
15	36/40
18	46/53
22	59/71
29	82/93
36	96/106
43	121/130
80	139
92	166
106	136

The weights below are of four chicks hand-reared from the egg at Birdworld, Surrey, U.K., in 1991. The rearing food consisted of 5 parts of Avi-plus hand-rearing food and 1 part Birdquest lory nectar mixed with water. The weights are approximate as they are taken from a graph (Sweeney 1992b).

Blue-streaked lory weight table #2

Age in days	Weight in grams			
	Chick No.1	No.2	No. 3	No. 4
day hatched	8	8	8	8
3	8	8	9	9
6	9	9	11	12
9	10	10	15	15
12	12	12	18	20
15	15	15	20	25
18	17	18	28	30
21	20	20	30	33
24	25	25	38	45
27	30	30	48	60
30	38	40	70	75
33	48	50	80	80
36	60	60	80	80
39	70	72	87	100
42	78	78	108	112
45	78	82	110	110
48	80	82	110	112
51	100	110	110	110
54	105	105	105	108
57	105	105	105	110
60	105	105	105	115

Sweeney (1992b) comments on the "different patterns of weight gain between the two sets of chicks, from different parentage. All four chicks were hatched within a short time of each other and were reared together, and all chicks were subjected to exactly the same rearing method, diet and feeding."

Age at independence of hand-reared birds: 56 days for one hand-reared by the author.

General: As long ago as 1822, Lord Stanley (afterward Earl of Derby) had a live specimen in his collection in England. But it was another 150 years before this species was exported on a commercial and regular basis. Since the mid-1970s it has been bred regularly in Europe and in the U.S.A. In California, Jay Kapac has been breeding this species for nearly 20 years. All his lory chicks have been incubator-hatched and hand-reared from the egg since 1976 (J. Kapac pers. comm. 1995). In the U.K., the first-recorded success was achieved in 1972, by R. Phipps. The pair had been imported the previous year by Jim Hayward—who observed, to his surprise, that they were mating a few hours after he collected them from Heathrow Airport. They had just been imported from Singapore! With Mr. Phipps they reared a single chick on a diet which included maggot pupae, corn on the cob and nectar with wholemeal breadcrumbs. This pair was then sent to a breeder in South Africa, and thence to Rhodesia where they continued to breed for Mr. D. Edwards, in 1973. These much traveled birds then went back to South Africa (Hayward 1984).

Another breeder in the U.K., John Gibson, first reared this species in 1983. For the previous seven years he had unknowingly kept together two males. (This was a common problem before the era of surgical sexing.) The female laid her first egg on May 9, 1983, and the second on May 12. Both hatched and young were hand-reared. Two eggs laid on September 29 and October 1 hatched at the end of October. Ten days later, after a severe frost, the chicks were removed "stiff with cold" and apparently dead. They were revived over a closed fire and were successfully reared. On December 11 the female laid an egg on the floor; this was followed by two more in the nest on the 14th and 16th. Again, both eggs hatched and chicks were reared (Gibson 1984).

In the U.S.A., W. and E. Cressman's pair shared their dining room at complete liberty and hatched a chick on top of the china cabinet! The male was obtained in January, 1992, and kept in a large cage. The female was obtained in November of 1992. After a 40-day quarantine period, her cage was placed opposite the male's. Three days later both cage doors were opened. The male at once established his dominance by posturing and approaching the female "with romantic noises." Bonding behavior and "head-nuzzling" were soon observed. Thirty-three days after the introduced, she laid her single egg. It hatched 27 days later in an L-shaped box (on the china cabinet). The birds accepted the presence of people in the room provided they did not go too close to the nestbox. This breeding was remarkable not only for its location but in the fact that the pair hatched a chick only 60 days after being introduced (Cressman and Cressman 1993).

Red and Blue Lory *Eos histrio* (Muller 1776)

Synonym: Blue-tailed lory

Native names: Sangihe, *Burung Luring* or *Sumpihi*; Talaud, *Sampiri*.

Description: Distinguished by a broad band of dark violet-blue on the breast in the nominate race, plumage is entirely red and blue, with black on the wings. The red head is marked with a broad violet-blue band across the crown. A streak of the same color, which meets the violet-blue of the hindcrown and mantle, appears above and below the eye. The abdomen is red and thighs are blue and black. Primaries are red margined with black, with a broad patch of red near the tip; this forms a black band when the wing is stretched. The scapulars are black, and wing coverts are tipped with black. The tail is red and mauve above and red below; under tail coverts are tinged with blue. The beak is orange, cere is white or gray and feet are gray. The bird's iris is orange-red.

Length: 30 cm (12 in.)

Weight: Of subspecies *talautensis*: males 170–195 g; females 150–184 g (de Dios pers. comm. 1995).

Immatures: Red areas variably marked with dusky blue; in *E. h. talautensis*, the plumage is duller than adult's; beak gray (soon turning orange), one only (Barnicoat 1995). Plumage appears to be variable. A *talautensis* hatched at Loro Parque in 1996 had almost the entire top of the head rich blue.

Natural history

Range: The Talaud Islands, and possibly the Nenusa Islands; the nominate race is from the islands of Sangihe (Sangir) and Siao (see Gazetteer), province of North Sulawesi, Indonesia (to the northwest of New Guinea). White and Bruce (1986), using alternative spelling, give its distribution as Sangihe, Siau and Ruang.

Habits: The nominate race is little known. The University of York/University Sam Ratulangi Expedition of 1995 encountered the lory only in the north of Sangihe, on the lower slopes of Gunung Awu. Small groups of two to six birds were seen on nine occasions during ten days in the Kendahe district. Villagers reported the species to be present also in the Tabukan Utara district. According to local sources, approximately 30 lories inhabit the area. The population had been at the same level for

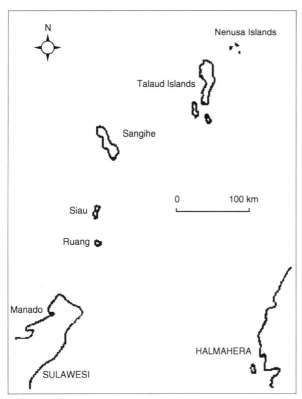

Distribution of the red and blue lory.

Eos h. histrio—Sangihe and Siao (Siau)
Eos. h. talautensis—Talaud Islands
Eos h. challengeri—reputedly Nenusa Islands

the previous five years. Trapping no longer occurred because the number of surviving birds was too small to make this worthwhile. Villagers in Kedang also reported that the lory was present in small numbers. Elsewhere, in apparently suitable habitat, reports were negative. The area is cultivated on ridges and upper valley slopes; on the steepest slopes, original forest still occurs (Anon. undated.; Riley preliminary assessment 1995).

Status/Conservation: The nominate race is believed to be critically endangered, perhaps near extinction. At the end of the nineteenth century it was no longer common and had retreated to the

mountainous interior, apparently because of the spread of coastal coconut plantations. By 1986, only a few hectares of mountain forest remained and D. Bishop was unable to find this lory. White and Bruce made the statement that in 1978 one of them had found that its status on Sangihe and in the Talauds was "much as it was last century," without indicating whether he had seen it on Sangihe. Siau and Ruang were not visited, but Bishop was informed that there were very small remnant areas of suitable habitat there. Taylor (1992) failed to find the red and blue lory on Sangihe. However, the joint University of York Exploration Society/University Sam Ratulangi, Manado, research team visited Sangihe in the summer of 1995 and they were more successful. The subjective evaluation of the team suggested a population of between 30 and 100 birds (later revised to 30), dependent on isolated patches of natural vegetation, which total less than 200 hectares. Within this population, escaped *talautensis* were seen. They were recognized by plumage, and by the presence of wooden leg rings. Pairs consisting of one member of each subspecies were seen displaying, suggesting that breeding could occur between them. The escaped birds are a problem which needs to be addressed—or the genetic distinctiveness of the nominate race will be lost.

Aviculture: Only two specimens of the nominate race have been recorded in aviculture. The longest lived, that at Vogelpark Walsrode, was believed to be 20 years old when I saw it there in 1991. It had produced young when paired with a Forsten's lorikeet. In Denmark in 1965, Carl Aage Jensen of Koge apparently reared a youngster from a male *histrio* paired to a female *challengeri*. No breeding report was published, and some doubt attaches to this breeding.

Challenger Lory *Eos histrio challengeri*

The validity and origin of this subspecies is doubtful.

Description: It is said to differ from the nominate race in that the blue line from the eye does not meet the blue of the mantle, the blue band on the

breast does not extend so far down, and the area below it is said to be mottled with red and blue.

Length: Described as smaller than the nominate race, about 25 cm (10 in.).

Immatures believed to be of this race were seen in captivity on Salebabu. Their plumage was mottled with gray, especially on breast and head (Nash 1993). However, there is no evidence that they were of this race.

Range: Said to be Miangas Islands and Nenusa (Nanusa) Islands, Indonesia. Riley (*in litt.* 1996) questions the validity of this subspecies. It was named by naturalists who visited the area on the Challenger in 1879. Riley looked at a map of the boat's voyage; it did not visit the Miangas but laid anchor at the Nenusas; however, the Nenusas were formerly known as the Meangis Islands. From the boat they could see the vegetation of Karakelong; this would not have been possible from the Miangas. Riley infers that there is no evidence these birds occurred on the Miangas Islands. As there was trade between the islands, those described from Nenusa might have been immature specimens of *talautensis*.

Status: Said to be endangered, critically endangered or extinct. According to Riley (1995) these islands are not known to have been visited by ornithologists this century; they are small and difficult to reach. Nenusa and Miangas islanders interviewed on Talaud in 1995 had no knowledge of this species.

Trade: In January, 1993, seven lories tentatively identified as this subspecies were seen in an aviary in Lirung, Salebabu Island, by personnel of TRAFFIC, Southeast Asia. They were familiar with *E. h. talautensis*. The birds were apparently a "bycatch" of fruit bat (*Pteropus* sp.) trapping. (Fruit bats are eaten throughout their range.) The bats are caught with large nets, presumably when they are feeding on fruit at the same time as the lories. In 1995, Jan van Oosten saw two captive birds on Salebabu which were noticeably smaller than the others, with different areas of blue than *talautensis*. However, in Toabatu he saw many captive birds with similar plumage and believed that this was "either a morph or intermediate adult plumage" (van Oosten pers. comm. 1995).

Talaud Red and Blue Lory *Eos histrio talautensis*

Description: It differs from the nominate race in having less black on the wing coverts and flight feathers.

Natural history

Range: Talaud Islands, Indonesia, to the northeast of Sangihe. It occurs on the island of Karakelong. A 1986 survey failed to find it on Salebabu Island; but, according to PHPA officers, it still occurs there in small numbers. It is extinct on Kabaruang (Riley 1995). A report issued the following year (Anon. [1996]) states that the PHPA in Beo also indicated it no longer occurs on Kabaruang.

Habits: A century ago Dr. Hickson visited the Saha (or Saka) Islands, islets three-quarters of a mile and less in diameter, which lie three or four miles from the coast of Salebabu. He wrote: "My attention was called to these islands by a flock of lories, consisting of many hundred individuals which flew from the main island to the larger of them as the sun was setting..." (in Meyer and Wiglesworth 1898).

On Salebabu (or Salibabu), Hickson stayed at the village of Lirung. He related (in Mivart 1896) how "the natives had sold and presented to members of our staff and crew at least 3 score of these pretty little creatures; but the mortality among them on our voyage was tremendous, and I believe that not half a dozen reached Monada alive. Three specimens were presented to me, but they all died, apparently of cramp, before we reached Sangir."

In May, 1968, David Bishop (in Forshaw 1989) described this lory as commonly seen in small groups on Karakelong. Pairs were observed feeding in flowering coconut palms within 100 km of a village. Local people told Bishop that the species was then common throughout the island, where it was often kept as a pet.

In 1995, Jan van Oosten accompanied University of York expedition members to Salebabu. He was told that from the beginning of November until March every year between 100 and 150 *talautensis* visit the mountains behind Lirung, which is now the largest town in the Talaud Islands. It is believed that they fly from Karakelong. Other *talautensis* seen flying free there were probably escaped pets.

On September 15, after a hike to the Lobo River area, the team camped on a large sand bar which gave a good view of the surrounding area. Next morning van Oosten counted 95 *talautensis* flying overhead between 5:45 A.M. and dusk—and probably missed between 10 and 25 birds which he heard depart before he arose. However, about half of these might have been counted twice. The next morning he got up before the lories left their roost between dawn and 6:15 A.M. They headed toward the coast. After an hour to an hour and a half, some of the lories flew back toward their roosting tree. Some dropped down into the forest, while others flew into the plantations. At about 4:30 P.M. they flew toward the roosting tree area; the last one was seen at about 5:45 P.M. They probably spent most of their time in the plantations, which are joined by corridors of forest. The plantations contain coconut palms, papaya, nutmeg, cloves and banana.

On September 16, Jan van Oosten and two team members headed up the Lobo River. On the way they counted 47 *talautensis*. Some were seen in an *Albizia valcataria* tree; perhaps they had been feeding on the small berries, some of which were on the river bed. In Beo, van Oosten met a trapper who regularly catches about 30 lories. First he sells to the local people, then to those living in Salebabu and Sangihe; the rest he takes to Manado. Another trapper regularly sold lories to people from Mindanao in the Philippines (van Oosten unpubl. report 1995). Trapping is, of course, illegal, as the species was placed on Appendix I of CITES in 1994.

The University of York team members were able to observe that the red and blue lory roosts communally on Karakelong. One site in the Rainis District held approximately 400 birds between September 29 and October 3. Another roosting tree, identified by villagers in Bengel, was said to be used by about 500 lories (Riley 1995).

Extensive forested areas still remain on Karakelong, and potentially suitable habitat for the red and blue lory appears to be widespread. But in the south of the island, villagers in Pulutan reported that the lories were scarce. The birds were observed feeding on trees of *Ficus*, *Canarium* and *Lansium domesticum* (Anon. [1996]).

Nesting: In the 1890s native hunters employed by Dr. Meyer informed him that on Sangihe this species lays one or two eggs (Meyer and Wiglesworth 1898). In 1995, van Oosten was told by the head trapper in Bengel, Karakelong, that the breeding season is April and May, but that nesting may also occur in November and December. He was shown a nest tree near Bengel used by this species. It was a *Canarium* situated on a slope of about 45°; the nest hole was about 10 m (33 ft.) from the ground. The surrounding area was forested and cultivated.

In captivity, newly laid eggs at Birds International in the Philippines weighed 7.64 g to 8.02 g (de Dios pers. comm. 1995). In the Netherlands, eggs laid in the collection of Gert van Dooren measured 28.2 x 22.5 mm and 28.8 x 23.0 mm (pair 1); 28.1 x 23.3 mm and 27.5 x 23.3 mm (pair 3); and 28.6 x 23.8 mm and 28.9 x 23.9 mm (pair 4).

Status/Conservation: Endangered. In 1992, the population was estimated at fewer than 2,000 birds. At that time the main threat to its existence was habitat destruction. Only one area of forest survives in the Talaud Islands; most formerly forested regions have been converted to coconut, nutmeg and clove plantations. In that year possibly 1,000 birds were captured, as much as 50 percent of the estimated population. However, the total population may have been underestimated.

Two reserves have been proposed to protect the relatively undisturbed forests (said to total about 22,000 hectares) in the center of Karakelong. At the time of writing, trade presents a serious threat to this species. Although the red and blue lory was placed on Appendix 1 of CITES in 1994, a year later it was still being trapped for the local and export trade. Hopefully, an educational campaign will bring an end to trapping in the near future.

Riley (1995) suggested the following conservation measures: (1) closer monitoring of trade by PHPA officers; in particular, the illegal export to the Philippines should be stopped; (2) a strict quota of lories trapped annually should be established; a quota of 100 was suggested; (3) trappers should be persuaded to release any adult birds caught; there should be a ban on trapping during the breeding season; (4) owners of captive birds should be taught how to care for them properly; a general education program on the islands regarding the threats faced by the red and blue lory should be initiated; and (5) a pilot captive-breeding scheme should be set up. If this is successful the objective would be to replace trapped birds with those bred by the villagers.

One aim of the 1996 expedition to Sangihe and Talaud was to publicize the threats to the lory in the provincial capital, Manado, Sulawesi. This is the major market for this species within Indonesia. Publicity was sought through television, newspapers and magazines.

Trade: 1992 was a catastrophic year for this subspecies. Although the export trade probably commenced in the late 1980s, it must have been on a small scale. In 1992, however, it seems likely that as many as 1,000 birds were captured, at least 700 of which are known to have been exported. It is surprising that a capture quota of 1,000 birds was allocated by the Directorate General of Forest Protection and Nature Conservation (PHPA). Nash (1993) was the first to investigate this trade. He described how every week people from outlying villages went to Lirung to sell up to ten lories. Trade was on a house-to-house basis—not in a market. The islanders acknowledged that the lories had been trapped to the point where it was no longer worth the effort to catch more specimens on Salebabu. Most trapping took place in the protected forest on Karakelong. This occurred despite the fact that community and church leaders were opposed to the capture of the island's wildlife on the grounds that it could upset the ecological balance. Yet trapping had probably been occurring for at least a century. When Dr. Platen visited Sangihe in the late nineteenth century he believed that the caged examples seen there were derived from the Talaud Islands (Meyer and Wiglesworth 1898).

In March, 1992, I saw 40 or more red and blue lories in the small shop of a dealer in Singapore; many were immature and probably all were *talautensis*. They were crammed together in three or four wire cages in the front of his shop which opened on to a busy street. By December, 1992, it was known that at least five importers in Singapore were involved in this trade. At least two had more than 100 lories each and the remaining three had 130 or more between them. Between 435 and 485 were traced by

TRAFFIC Southeast Asia in Singapore, all apparently *talautensis*. In addition to the numbers already mentioned, at least 200 birds died in Jakarta from disease and poor handling at the premises of one exporter (Nash 1993). Alas, this is typical of what happens when a species, even a rare one, is traded in large numbers with the birds crammed together. The very numbers seem to diminish their value in the eyes of the dealers. To them, they are merely items of commerce; the fate of individual birds is of no interest. While one would not expect them to be interested in that aspect, every death means less profits, so their attitude is hard to understand. The dealers could easily have prevented some of the deaths by caging the birds in pairs rather than large groups. This is the very worst face of trade. Serious lory breeders throughout the world would have treasured every single bird, but they did not know of their existence. At that time, two people in Singapore (not dealers) were attempting to monopolize the market in red and blue lories by purchasing large numbers with the intention of breeding from them. One lost a large number of birds due to his inability to care for so many lories in the proper manner. Finally, in 1993, the TRAFFIC Secretariat contacted the Indonesian management authority to recommend a total ban on export and on internal trade by transferring *Eos histrio* to Appendix 1 of CITES. This was carried out at the 1994 CITES conference. Regrettably, two years later, the trapping of birds on central and northern Karakelong continued.

Riley (1995), leader of the 1995 University of York expedition, identified two villages with trapping cooperatives—Bengel and Toabatu. Trappers catch independently, but trade is conducted jointly, with the total catch for the village being sold to one buyer. Potential trappers from other villages pay to learn catching techniques from the head trapper in Toabatu. The method is to remove some of the canopy from a tree near a roosting site and replace the branches with flexible ones from a certain tree; these are coated with a sticky glue made from sap. The branches are camouflaged with sprigs of vegetation, then decoy birds (up to eight) are tied to the tree.

The main trapping period is during the dry season, April to August, which, unfortunately, coincides with breeding season. This is because the glue used for trapping is often washed off during the wet season. In 1994, 240 *talautensis* were caught at Toabatu and 150–200 at Bengel. Trappers at Toabatu stated that totals were formerly higher, a single man being able to take 100 birds each year.

Trappers received the equivalent of about $2 for each bird. Buyers from the Philippines, who arrived in August each year, paid the equivalent of $10 per bird.

Aviculture: Rare, but likely to become established in the Philippines, in South Africa, and perhaps in a few European countries within the next decade.

Clutch size: Two

Incubation period: About 24 to 25 days (Barnicoat 1995); 25 to 26 days (Sweeney 1994c).

Newly hatched chick: Weight, 5.76–6.13 g; light gray down and black beak (de Dios pers. comm. 1995); "covered in thick white down" (Barnicoat 1995); "long pale natal down over entire body" (Sweeney 1994c).

Chick development: Two chicks hatched at Birds International Inc. in the Philippines were described as follows: "The eyes first began slitting on Day 10 and they were fully open in both chicks by Day 16. Secondary down follicles started to appear through the skin at around Day 10, and by Day 23 the secondary down was covering most of the chicks' body. The bill and toe nails remained black throughout the chicks' development; the bill later gains its orangish adult coloration after weaning. Pin feathers started to appear on the head on Day 28. Pin feathers on the wings became evident on Day 30. By Day 32 the pin feather quills on the head were already breaking to reveal their feathers and red coloration. Flight feather pins first started to break through their quilling on Day 34; also a few pin quills were breaking on the forehead, crop and the abdomen at this time. The chicks were fully feathered in both cases by Day 77" (Sweeney 1994c).

Young in nest: 70 days (de Dios pers. comm. 1995); one only "just over 2.5 months" (Barnicoat 1995); 71 days for two young (van Dooren pers. comm. 1995).

Age at independence of parent-reared young: about 90 days (de Dios); of hand-reared young: 80 days (Sweeney 1994c).

Chick growth

Two young hatched at Birds International Inc., whose development is described above, were weighed daily, usually before the first feed of the day. Their weights were recorded as follows:

Talaud red and blue lory weight table #1

Age in days	Weight in grams	
	Chick No. 1	No. 2
hatch	5.78	6.13
day 1	5.78	6.12
2	6.17	7.05
3	8.42	8.88
4	9.01	9.32
5	10.11	11.00
6	11.24	12.61
8	14	16
10	19	21
12	24	27
14	31	33
16	37	39
18	40	41
20	46	47
22	49	52
24	53	56
26	58	64
28	66	73
30	73	79
32	74	80
34	82	84
38	97	95
40	110	107

(subsequent weights not available)

Weights of parent-reared chicks, in the collection of G. van Dooren, hatched Aug. 14 and 16, 1995.

Talaud red and blue lory weight table #2

Age in days	Weight in grams	
	Chick No. 1	No. 2
hatch	5	6
1	6 fc	8 fc
2	8 fc	9 fc
3	10 fc	11 fc
4	12 fc	13 fic
5	14 fc	16 fc
6	15 fic	18 fc
7	19 fc	21 fic
8	23 fc	27 fc
9	24 fic	31 fc
10	29 fc	27 fic
11	34 fc	32 fc
12	37 fc	35 fic
13	35 fc	42 fc
14	47 fc (R)	48 fc (R)
15	45 fic	61 fic
16	53 fc	60 fc
17	66 fc	63 e
18	72 fc	69 fc
19	67 fic	73 fc
20	79 fc	72 fc
21	76 fc	74 fic
22	81 fc	77 fc
23	89 fc	84 fc
24	84 fic	93 fic
25	99 fc	84 e
26	101 fc	88 fic
27	93 fic	–
28	108 fic	–

fc = full crop; fic = food in crop; e = empty; (R) = day ringed (6 mm).

General: For a long time, probably many decades, red and blue lories have been trapped to meet the local demand for the pet trade. With the occasional, rare exception, none were exported to countries outside Indonesia until 1990. It may be that the first recipients were Japan and other countries where there is little interest in breeding, and where lories would be sold singly as pets. There was a period of only about one year when these beautiful lories, which would have been so highly prized by aviculturists, were imported into countries where they would have had the opportunity to breed—notably South Africa; a small number reached the U.S.A., likewise Europe. Thenceforth their importation into South Africa and Europe was prohibited. The first South African consignment arrived in February, 1993; it contained 66 birds which were advertised in an avicultural magazine for the equivalent of US $2,600—presumably per pair. In 1994, 75 males and 71 females were thriving in South Africa.

The 120 birds in the first two consignments "were sold with extraordinary rapidity to only 5 owners of large and well established lorikeet collections" (Barnicoat 1995). Losses were practically nil and, after surgical sexing, an equal number of both sexes was known to be present. It was an ideal situation considering that the birds had not been trapped as a result of avicultural demand.

Barnicoat commented on the position in South Africa: "It is fortunate that stocks in considerable numerical depth have been concentrated in the hands of a few highly experienced aviculturists who will be working from very large gene pools and should never have to resort to inbreeding." Indeed, South African breeders will be able to export captive-bred young to other countries. As the importer registered the arrival of the birds, the Transvaal Department of Nature Conservation has stated it will issue licenses for closed ringed or microchipped young to be exported. "In this way," wrote Barnicoat, South African "breeders will be making a definite contribution to the conservation of this rare species by eliminating totally the demand for smuggled wild-caught birds." However, by the time the South Africans are exporting young in significant quantities, the export of wild-caught birds from their native islands will, hopefully, have ceased, due either to the implementation of CITES or an education program, or both.

During the period 1993–94, several captive-breeding successes occurred. In 1993, the red and blue lory was bred at Birds International. In 1994, one was reared by the single pair belonging to Gavin Zietsman in South Africa. They reared young before any of the 50-plus pairs in the five collections. Barnicoat commented: "They are a relatively tame and wonderfully cheerful pair of birds, who revel in displaying to each other in a playful manner and showing off to all and sundry. Whenever I have watched them, I have felt how well they merit their Latin name 'histrio'—'an actor.'" The female laid 2 eggs in a medium-sized log in the shelter of their 3.5 m (12 ft.) x 1.2 m (4 ft.) x 1.8 m (6 ft.) high aviary.

At Jurong Bird Park, Singapore, there are six pairs, two pairs of which had bred by the end of 1994, producing five young. By the end of 1996 the total hatched was ten (Buay pers. comm. 1996). In 1994, young were reared in a private collection in Florida (J. van Oosten pers. comm.). By 1996, red and blue lories had even found their way to Australia; there were at least nine birds. In the Netherlands there were about 30 birds by 1995. Pairs belonging to Gert van Dooren and Jos Hubers both hatched chicks in that year. One pair belonging to Gert van Dooren became very aggressive when he started to weigh the youngster, and started to pluck it. They obviously resented the interference. In 1994 Loro Parque, Tenerife, received three pairs of captive-bred *talautensis* from the Philippines. They were still in immature plumage. They began breeding in 1996. Seven young had been reared from two pairs by October of that year. All were hand-reared because early attempts at parent-rearing resulted in the death of chicks within the first 48 hours. It was intended that any future young should be parent-reared (Sweeney pers. comm. 1996).

Placing this species on Appendix 1 of CITES had not (at the time of writing) been effective in ending trapping activities. Captive birds in the hands of experienced aviculturists therefore assume great importance as founder stock of future generations. The large number exported from their native islands does not guarantee their survival in captivity because so few of these birds were permitted to enter countries where serious aviculture is practiced. The founder stock will therefore be relatively small. Everyone who owns this species is urged to register their birds with the studbook, where relevant. The EEP Parrot TAG studbook is operated by Roger Sweeney at Loro Parque, Tenerife.

Red Lory   *Eos bornea* (Linnaeus 1758)

Synonym: Moluccan lory

Description: Predominantly red with small areas of black and blue on the wings. The primaries are black with the speculum red; secondaries are red, broadly tipped with black. The greater wing coverts are blue, also the scapulars in some specimens. The area surrounding the vent and the undertail coverts are cobalt blue. The upper side of the tail is darker than the body plumage; the under side is brownish red. The beak is orange; the cere, the skin surrounding the eye, and the feet are gray. The iris is orange-red. Note that the shade of red of the subspecies *cyanonothus* is darker than that of other red lories.

Length: About 28 cm (11 in.)

Weight: About 170 g; some birds fed a high-protein diet (see below) weighed as much as 240 g.

Immatures: Their plumage is variable; in some birds the red feathers of the underparts are margined with blue and the ear coverts are blue. The beak and the iris are brown. The beak gradually becomes paler at the tip and the brown coloration is lost, replaced by dull yellow-orange. Gerischer (1991) described the plumage of the young of his pair as having "red and blue markings on the back stronger than in the parents. Chest and belly were red crossed by wavy black lines. The blue patches behind the ears disappeared in time."

Specialist breeders of red lories described their young as follows: "The juveniles have far more black and less blue on their wings than adults, and the black is irregular...After each molt, the blue wing band becomes more pronounced: There is less black, and they lose the blue ear patches. The blue wing bands reach their maximum blue coloration after 5 years. Some begin to develop orange irises at about a year, although we have noticed adults that never develop the orange iris ring" (Collins and Sefton 1994).

At nine weeks old, one youngster hand-reared by the author did not differ substantially from its parents. The underparts were pure red but there were only one or two blue feathers on the thighs; a mere tinge of blue could be seen on the ear coverts. The primaries were tinged with gold on the outer edge. The iris was dull brown and the beak was yellow-orange, except for brown tips to both mandibles.

Natural history

Range and subspecies: Ambon (Amboina) (see Gazetteer) and Saparua, islands of the Moluccas (Maluku), Indonesia. (Saparua is one of the small Lease Islands to the east of Ambon.) The birds from Seram are considered to belong to the subspecies *rothschildi* and apparently differ only in their smaller size. Those from the Kai Islands are said to be slightly larger than the nominate race and have been assigned to the subspecies *bernsteini*. Birds from Goram, Ceramlaut and the Watubela Islands (to the southeast of Ceram) are considered to be intermediate between *bernsteini* and *rothschildi*. The only subspecies recognizable outside its natural habitat is *cyanonothus*, from Buru.

Habits: The bright colors and loud calls of this lory have attracted the attention of naturalists ever since Alfred Russel Wallace first set foot in the Kai Islands in 1857. He wrote: "Here my eyes were feasted for the first time with splendid scarlet lories on the wing as well as the sight of that most imperial butterfly, the 'Priamus' of collectors...Almost the only sounds in these primeval woods proceeded from 2 birds, the red lories, who utter shrill screams like most of the parrot tribe, and the large green nutmeg-pigeon, whose voice is either a loud and deep boom, like 2 notes struck upon a very large gong, or sometimes a harsh toadlike croak..." Two years later Wallace visited Seram, where he noted the red lories were too high up for him to shoot. On Amboina, the "most remarkable" species he saw was "the fine crimson lory (*Eos rubra*), a brush-tongued parroquet of a vivid crimson color, which was very abundant. Large flocks of them came about the plantation, and formed a magnificent object when they settled down upon some flowering tree, on the nectar of which lories feed" (Wallace 1896).

In the early years of the century this lory was very common in the coastal zone of Seram and in montane forest up to 1,250 m (4,100 ft.), flocks of 20 or more being observed in flowering trees (Stresemann 1914). On moonlit nights the lories flew, screeching loudly. On Buru in the 1920s, it inhabited the mangroves in preference to open grassland and hilly country near the coast. Buru then became a penal settlement and was not visited again by an ornithologist until 1980. Smiet (1985) observed this species in flocks of up to 30 birds on coastal Buru, also on coastal Ambon, Seram and in the Kai Islands. In November and December, 1989, Buru was visited by the Manchester Indonesia Islands Expedition. As a result, Jepson (1993a) described *Eos bornea* as the commonest parrot on Buru. It was abundant in the selectively logged forest above Charlie Satu from November 9 to 11, when a group of 50 or more was observed in a single fruiting tree. Maximum group size in flight at this site was 18. Red lories feed on flowering trees such as *Eugenia* and *Erythrina*. Fragments of flowers and small insects have been found in the crop. Marsden, who was a member of the same expedition, recorded average group size of red lories on Buru was 2.4 (maximum seen together was 20) and average group size on Seram was 2.9. The 22 birds in one flock on Seram was the largest group recorded for any parrot during the expeditions. Marsden (1995) suggested that on Buru this species showed a preference for recently disturbed lowland forests; open forests with many dead trees may be preferred. On Seram it favored lowland forests with many large primary forest trees, away from steep ground.

Nesting: No published information on nest sites. Five eggs averaged 30.2 x 24.2 mm (range 29.8-31.2 mm and 23.0-24.6 mm) (Harrison and Holyoak 1970). Two eggs in one clutch at Palmitos Park measured 30.5 x 23.2 mm and 30.5 x 23.3 mm, and an egg in another clutch from the same female measured 30.2 x 23.2 mm. The latter egg weighed 6 g when newly laid.

Status: Still common in most areas but it has declined due to overtrapping. If the Indonesian export ban which became law in March, 1995, is implemented, some recovery might occur.

Trade: For many years this was the only member of the genus available in captivity; even in the mid-1980s when other *Eos* were exported, *bornea* quotas were the highest. PHPA government statistics detail minimum numbers of parrots exported from the Moluccas for the international market. In 1981, the total was 18,829 of which 2,735 were *Eos bornea*; 1982 total was 25,358 including 4,853 red lories; 1983 total, 41,673 including 4,968 red lories; 1984 total, 44,056 including 10,022 red lories (Milton and Marhadi 1987). By 1984, excessive numbers of red lories were being exported and populations must have been declining. However, no studies of abundance were made during that period. It is little wonder that this species was one of the cheapest lories available.

Aviculture

Status: Common (see Popularity, Avicultural)

Clutch size: Two; Dr. R. Jerome reported (1979 convention of the American Federation of Aviculture) an incidence of a female laying four eggs, all of which hatched.

Incubation period: 24 to 25 days; Collins and Sefton have recorded periods up to 29 days and state that "the eggs take a day or two longer in cold weather."

Newly hatched chick: Weight, 6 g; dense white down; beak brown with white egg tooth; feet pink.

Chick development: Feet becoming gray on day 3; both eyes starting to slit, day 9; covered in gray second down, day 17; red feathers erupting on the underparts, day 20; feathers erupting on head, day 23; blue feathers erupting on face, day 29; first red tail feather erupted, day 31 (Pszkit 1991). Collins and Sefton note how their young lose the feathers of the crown at about 8 weeks. "This bald spot is usually about the size of a nickel, and it regrows quickly." Loss of crown feathers has not occurred in young hand-reared by author. Development of one, hand-reared from day 12, was as follows: day 2, feet tinged with gray; day 4, feet gray; day 7, second down tracts visible under skin; day 9, feet dark gray; day 11, ear open (pin-prick size); day 15, tufts of dark gray second down erupting through sparser first down on head, wings and body; day 26, thickly covered in midgray down, first red feathers erupting on head and wings; day 36, primaries and tail coverts erupting, a few head and wing feathers half-free of their sheaths, feathers erupted on flanks and breast, beak becoming yellow at base and edges; day 43, wings, breast and forehead feathered, feathers erupting on rest of head and on mantle and rump, tail feathers about 1 cm long. Day 57, fully feathered; inner edge of upper part of wing blue and some blue feathers on thighs but, except for faint tinge on cheeks, plumage otherwise lacking blue tinge; some wisps of down remaining on back; beak brownish orange with brown tip on both mandibles.

Young in nest: About 10 weeks—but up to 80 days.

Age at independence of hand-reared birds: About 50 days.

Chick growth

Shows the weight of one (Chick No. 1) reared at Pearl Coast Zoo, Broome, Australia, in 1990 (Pszkit 1991). It was reared on a mixture of Heinz junior muesli, raw egg yolk and Heinz mixed vegetable baby food for the first ten10 days. This is contrasted with Chick No. 2, hatched at Palmitos Park in March, 1995, and hand-reared by the author from day 12. Rearing food consisted of Milupa baby cereal, wheat germ cereal and Nekton-Lori (Nekton Produkte, Germany) in approximately equal parts. The chick was fed four or five times daily until day 44, when it was first offered warm rearing food in a shallow dish. It was fed once daily for the following 2 days, and thereafter was completely independent.

Red lory weight table

Age in days	Weight in grams	
	Chick No. 1	No. 2
day hatched	6	–
1	–	8
2	9	10
3	–	14
4	–	15
5	–	18
6	–	21
7	–	24
8	–	27
9	24	30
10	–	33
11	–	40
12	–	44
13	–	44
14	36	50/57
15	–	50/57
16	–	55/64
17	49	61/66
18	–	64/73
19	–	69/75
20	56	75/82
21	–	80/88
22	–	88/95
23	64	92/100
25	–	99/111
27	–	102/114
29	76	110/121
31	81	120/137
33	–	125/141
35	–	134/151
37	–	137/158
39	99	146/163
41	–	155/174
43	–	161/181
45	–	162/179 self-feeding
47	–	166/177 self-feeding, independent
49	130	168/173
54	–	172 peak weight
59	150	160
67	–	161
69	140	–

(Collins and Sefton record the peak weight of their young as 180 g.)

General: Red lories were regularly imported until the 1990s. Up to this time the price had been low (as little as $45 in the U.S.). In Australia and New Zealand, where importation of birds was forbidden for over 30 years until (in Australia) 1992, the price is very high. In 1989, a New Zealand aviculturist told me they cost $10,000 per pair there! At the time they cost about £80 each in the U.K. This is the best-known member of the genus and was the only one readily available in aviculture for many years.

Reds make wonderful aviary birds. They are hardy and often free-breeding. Collins and Sefton noted their birds seemed to prefer cool weather. "They have very thick down feathers, similar to those of waterfowl. They play outside in weather as cold as the low 30s and 20s...Our lories' highest egg production occurs in the coldest weather, typically during the months of February and March...We have also noticed that northern lory breeders with cooler climates are significantly more successful than [southern] breeders." They also noted deaths occurred during unusually hot weather with high humidity.

A common problem with this species is feather-plucking; they may pluck themselves or their young. Collins and Sefton state: "We have always been able to cure this problem by switching them over to our diet. We believe that feather picking in red lories can be an indicator that their protein is too low." An interesting suggestion which might be tested by someone with two pairs of reds which pluck. One pair should be offered a diet with an increased protein level and the others should act as control birds with unchanged protein levels.

The diet of the pair at Palmitos Park consists of nectar made from Milupa baby cereal, Nekton-Lori and honey, fresh fruit (apple, pear, banana, other fruits in season) once or twice daily, a small amount of sunflower seed and, when available, pollen-laden blossoms, and small branches of *Casuarina equisetifolia* bearing small cones. They eagerly consume the sunflower and would undoubtedly take much more. The female is slightly plucked on the breast. The male can be very aggressive when breeding and I hesitate to increase the protein level of the diet. The chick in the weight table was from this pair. I had weighed the chick daily without encountering any problems until the chick was 11 days old. When I opened the nestbox, the male suddenly flew in and bit my finger. The following day he attacked the female. She had to be removed for a few days to recover from a leg injury which he had inflicted. I then removed the chick to hand-rear and discovered it had been bitten on the back,

presumably by the male. The injuries were not serious but indicated possible mutilation and resulting death, if the chick had been left in the nest. I was told a young one bred from this pair previously was so aggressive that it had killed another young lory when it was only about four months old. The one I reared had a Stella's lorikeet of the same age as a companion. When the red was only 11 weeks old I separated them. They had previously spent hours playing together, but I feared for the Stella's when I saw the red's behavior becoming slightly aggressive. In my experience with many parrot species, aggression is inherited.

Red lories are colorful and entertaining with an often flamboyant display, which precedes copulation. It has been described by Gerischer (1991) as follows: "The male circled around the hen with outspread wings, constantly jerking his head up and down. After a while the female began the same play and soon they were chasing each other and 'embracing,' uttering hissing and cheeping cries. Both puffed out the body feathers and the pupils of their eyes narrowed and widened, showing the iris quite clearly."

Buru Red Lory   *Eos bornea cyanonothus* (Vieillot 1818)

Description: Overall a darker shade of red, almost approaching maroon on the breast. The blue is more cobalt. It is smaller and slightly more slender.

Range: Buru (see Gazetteer), Indonesia.

Habits: Little known, due to restrictions against admitting foreigners, which were enforced until the late 1980s. In the 1920s it was known to be common.

Aviculture: In the late 1980s, fairly large numbers were imported into Europe (and, no doubt, into other countries), whereas previously this race was either seldom seen or seldom recognized.

Eos bornea rothschildi (Stresemann 1912)

Description: Distinguished only by its smaller size. Considered doubtfully distinct. Some red lories in aviculture measure only about 25 cm (10 in.) but their origin is unknown.

Range: Seram (Ceram) (see Gazetteer), Indonesia.

Habits: This lory is most common in the rain-forests of the lowlands. In 1987, the observation rate near Solea was 12.7 sightings per hour. Outside the lowlands it is not common. Near Air Besar the observation rate was 0.8 per hour and at Kanekeh 1.1 per hour. It visits areas of secondary vegetation and cultivation but prefers the rainforest. When searching for food, it gathers in flocks, often in company with green-naped lorikeets. In August, 1987, several pairs were observed near holes in trees, probably searching for nesting sites (Taylor 1991b).

Trade: This had increased, according to Taylor. Birds could be bought in Seram for a low price (4,000 rupiahs). As yet, it does not appear to be endangered by trade.

Eos bornea bernsteini (Rosenberg 1863)

Description: Distinguished only by its slightly larger size. Considered doubtfully distinct.

Immatures: The birds are said to have a blue band behind the eye, extending to the ear coverts, also a few blue feathers above the eye. The feathers of the throat are narrowly margined with pale blue (Forshaw 1989).

Range: Kai Islands (see Gazetteer), Indonesia.

Aviculture: This subspecies is not usually distinguished. However, several birds believed to be *bernsteini* were imported into Singapore in 1989. One of them was orange in color, with green upper-tail coverts (Tay pers. comm. 1989).

Blue-eared Lory *Eos semilarvata* (Bonaparte 1850)

Synonyms: Ceram lory, half-masked lory

Description: Mainly bright red, distinguished by the violet-blue area on the upper cheeks, as well as on the ear coverts; the blue of the cheeks extends downward on the side of the face. Abdomen and undertail coverts are also violet-blue. Primaries are black with a red speculum, secondaries are tipped with black, and wing feathers nearest the rump (tertials) are black suffused with blue. The tail is brownish red above, dull red below and dusky near the tip. The beak is orange; the iris, orange; and the feet, gray.

Length: 25 cm (10 in.)

Weight: 170 g (estimated)

Immatures: They are paler throughout; blue on head restricted to ear coverts and area below eyes; some feathers of abdomen margined with blue; scapulars brownish gray edged with pale blue; bases of body feathers gray-brown (Forshaw 1989).

Natural history

Range: Seram, Indonesia (see Gazetteer).

Habits: It occurs in the mountains of central Seram (replacing the lowland-dwelling *Eos bornea* at higher altitudes). On Mount Binaia, Bowler and Taylor found it to be common above 1,200 m (3,900 ft.) during their study period of July to September. It occurred as high as the bare peak, where it fed on flowers of the highest tree heathers. On nearby Mount Kobipoto it was locally more abundant. It was observed as low as 800 m (2,600 ft.), attracted by certain flowering trees. In contrast to the purple-naped lory (*Lorius domicellus*), Taylor (1991b) found the blue-eared lory to be common during two short periods in Seram in 1987 and 1990. It is found at altitudes between 900 m and 3,030 m (2,950 ft. and 9,940 ft.) at the peak of Mount Binaia, but is most common between 1,600 m and 2,400 m (5,250 ft. and 7,870 ft.). At an altitude of 2,100 m (6,900 ft.) the observation rate was 3.0 times per hour. On Mount Kobipoto it occurs at an altitude of 1,100 m (3,600 ft.) but is more common above 1,300 m (4260 ft.). Bamboos and flowering trees and shrubs are its food sources. Taylor usually saw this species in pairs. On August 31, he observed copulation by a pair. During the display which preceded it, the birds bowed and whistled quietly.

Status/Conservation: Common. No conservation measures are known to be necessary. It is not kept as a pet by locals and little habitat disturbance has taken place. However, about 1995 the first birds were exported. Almost simultaneously, in March, 1995, Indonesia banned the export of several lory species, including the blue-eared. Steps should be taken to ensure this ban is implemented, or those exportations might represent the start of a trade that could threaten its existence.

Aviculture: Pagel (1985) stated that in 1965 there was a single blue-eared lory at a bird show in Krefeld. Assuming that it was not confused with an immature *Eos bornea*, this is the only bird recorded until 1996. Then a very small number of birds went into South Africa and were bought by Gavin Zietsman. In June, 1996, I saw five in his aviaries. One extremely beautiful and compatible pair was showing an interest in breeding. At the same time there were several blue-eared lories in a private collection in Singapore.

Eos squamata (Boddaert 1783)

Note: The small size and variable markings (except in *obiensis*) characterize this species. It has given rise to much confusion over the years. For example, Salvadori (1891) listed three species—*wallacei*, *insularis* and *riciniata*, all of which are now considered synonymous with *squamata* (which was lacking from his catalogue). But what exactly is the nominate race? No one seems to know. I believe that the species is still evolving, with much plumage variation within each population. The subspecies *obiensis* is distinctive and easily recognized. But how does one distinguish *squamata* from *riciniata*? According to Hubers (1996), who examined a number of museum skins, in the nominate race the crown is red, the collar is violet-blue and can be complete or absent, wide or narrow; the shoulder feathers [scapulars] are black-blue. In *riciniata* there is a wide violet-blue collar with variable gray shading. In most birds the collar extends to the middle of the head [crown]. The shoulders are red. Forshaw (1989) is uncharacteristically vague, remarking on the variable plumage coloration and the violet-blue collar around the neck which is "broad and well-developed in some birds but almost entirely lacking in others." He describes *riciniata* as having a "prominent violet-gray neck collar, usually extending up to hindcrown; some birds have violet-gray crown but red nape."

In his descriptions there are only two features which are different: the nominate race has a violet-blue collar and dull purple scapulars tipped with black and *riciniata* has a violet-gray collar and red scapulars. Unfortunately, not all birds fit neatly into one category or the other; some have only a few dark feathers on the scapulars. In a species which exhibits such variable coloration, is there justification for separating these two races? I suspect there could be a good case for naming two species: *obiensis* from the island of Obi and *squamata* (it was named first) from all the other islands within the range. However, here I follow current nomenclature and consider them as subspecies of *squamata*.

It is worth noting that Wallace's lory (*Eos wallacei* Finsch [1864]), from the islands of Wiageu, Gebe, Batanta and a small island near Misool, is figured in Mivart (1896). It was said to differ from *riciniata* in that the purple of the collar did not extend up the nape to the head—but then Mivart admitted this character was not constant. It seems that nobody has been able to define the subspecies.

Violet-necked Lory *Eos squamata riciniata* (Bechstein 1811)

Synonym: Violet-naped lory

Description: Red on the head with violet-blue or violet-gray on lower cheeks and, in some birds, also on the nape and crown; underparts violet-blue or violet-gray broken by a red band on the lower breast. Scapulars purple or red, greater wing coverts and flight feathers red, margined and tipped with black; tail purple-red above, brownish red below; under tail coverts purple or blue. The beak is orange; the cere and skin surrounding the eye, white or pale gray; feet, gray; and iris, orange-red or yellow.

Length: 26 cm (10 in.)

Weight: 110 g

Immatures: They are variably marked; the red of the underparts is mottled with dull blue (or almost absent), and there may be bluish markings on the ear coverts. There are more black feathers than red on the wings in some immature birds. The beak and iris are brownish.

Further confusion is created by the existence of a form which lacks red on the head or breast and has only a narrow line of dark blue on the neck; it has the abdomen dark blue but differs from *obiensis* in having the upper part of the wings almost clear red, and in its larger size. In Thomas Arndt's *Lexicon of Parrots*, volume 1, what appears to be an immature bird of this form is depicted on page 6, captioned as *Eos squamata squamata*. Perhaps this is in fact the nominate race. If so, it is either extremely rare or inhabits an island from which birds are very rarely collected. In 20 years I have seen only one bird of this form, an adult with very clear red plumage, which was in the collection of Raymond Sawyer of Surrey in the mid-1970s.

Another subspecies of *squamata* is listed by some authors: *E. s. atrocaerulea*. It supposedly differs in having the underparts, including the thighs, bluish black, the mantle black washed with blue, blue-black ear coverts and a darker red rump. A photograph of a lory answering to this description, but captioned *riciniata*, also appears on page 6 of Thomas Arndt's *Lexicon of Parrots*. It is clearly an immature specimen of *riciniata*, correctly labeled, and this doubtful race was probably described from a similar specimen.

Forshaw (1989) also suggests that the description "may have been based on immature plumage." There is no doubt in my mind that this was the case because young which have been reared in my care have exactly resembled this bird.

Natural history

Range: Most authors follow Mees' (1965) ar-

rangement, confining the little-known nominate race to the Western Papuan Islands of Gebe, Waigeu, Batanta and Misool, and the Schildpad Islands; *riciniata* is thus restricted to the Weda (Widi) Islands and islands of the northern Moluccas including Halmahera, Morotai, Widi and Bacan. Salvadori gave the origin of *wallacei* as Waigeu (Waigeo). The hypothetical subspecies *atrocaerulea* supposedly originated from Maju Island in the Molucca Sea; *E. s. obiensis* (see separate entry) is found only on the island of Obi, northern Moluccas.

Habits: The violet-necked lory is described as a flocking, wide-ranging island species that congregates in flowering trees and frequents coconut plantations. It is primarily found in coastal habitats and travels between islands (Beehler, Pratt and Zimmerman 1986). One observer watched flocks crossing the channel shortly before sunset each evening to roost on the opposite coast, two miles distant (Mees 1965). On the island of Bacan, which encompasses over 3,322 km^2 (1,283 sq. mi.), the coastal plains have been cleared for cultivation (coconuts, bananas, corn, etc.), but this lory does not occur regularly in plantations. It is usually encountered in forest areas (Milton and Marhadi 1987), probably because it has been heavily trapped near human habitation. Lambert (1993a) states that, between October, 1991, and February, 1992, *riciniata* and *obiensis* were recorded in all habitat types, including coastal coconut groves, mangroves and scrubby secondary growth. *E. s. riciniata* was most frequently encountered in pairs and groups of up to 25 birds during the survey and tended to be rarer at higher elevations. In 1994 and 1995, this subspecies was the subject of population monitoring by scientists taking part in the joint PHPA/BirdLife bird surveys. These were carried out in two proposed protected areas of Halmahera—Ake Tajawe and Lalobata. *E. s. riciniata* was found to be common in primary, logged and secondary forest at Labilabi, Miaf and Kulo, common in an agricultural area 3 km inland from Buli, but absent close to the village. It was also common in the Akejailolo sago swamp. At higher altitudes at Buli flocks of up to 15 birds were seen above 1,200 m (3,900 ft.). The team was surprised to find it to be common in montane forest at Buli and up to 27 km (17 mi.) inland at Miaf, as this species is usually described as a bird of coastal areas. The largest flock seen from within lowland forest was of six birds. Flocks of six to 20 birds were seen on several occasions in cultivated areas (Poulsen in prep.).

Nesting: There are no descriptions of nest sites of this common species. Schönwetter (1964) gave the average size of four eggs laid in the wild as 26.8 x 21.7 mm. One egg laid at Palmitos Park measured 27.1 x 22.4 mm.

Status/Conservation: Common, except around human habitation and disturbed areas. However, Lambert noted that this is a species which forms relatively large flocks, thus it could be common in some areas and rare in others. Lambert made the following theoretical population assessment: Halmahera 45,226 to 306,543 birds and Bacan 8,104 to 26,267 individuals. Poulsen (in prep.) noted that there are not yet any protected areas within the range of this species. He suggested that habitat destruction did not seem to be an important threat as it does well in man-made habitats, also that its abundance in accessible areas may indicate that the present level of catching for trade is sustainable.

Trade: Since 1985, the level of international trade in this species has remained relatively constant. The capture quota for 1990 was 2,400. Official records show that 2,146 were exported—but at best such records are representative of only one-quarter to one-half of actual trade. Also, mortality can be very high.

Lambert (1993a) obtained anecdotal information from numerous interviews which suggested that about 15–20 percent of *Lorius garrulus*, 7–10 percent of *Cacatua alba* (umbrella cockatoo) and probably more than 25 percent of *Eos squamata* die prior to shipment to other destinations within Indonesia. (Surely it should be possible to impose certain standards of care and to issue licenses only to trappers and handlers who can meet these standards. Such a waste of life is to be deplored, especially in instances such as careless transporting which could easily be corrected.) Lambert suggested interim catch quotas of 5 percent of the minimum population estimate until research and education are able to reduce levels of mortality and law enforcement becomes more efficient. However, in March, 1995, Indonesia banned the export of this species; whether trapping for the domestic trade was permitted is not known.

Aviculture

Status: Formerly common, but becoming less so in the 1990s.

Clutch size: Two

Incubation period: 24 to 26 days; 27 and 26 days in a single clutch (Wright 1977).

Newly hatched chick: Weight, 6 g. White down, black beak and dark-colored legs (Wright 1977).

Chick development: At three weeks not all the second down has erupted, thus some areas are not covered by plumage; at four weeks the body is fully covered with second down and the first contour feathers, those of the crown, have started to erupt. At five weeks the breast and wing feathers are starting to open. At six weeks the chick is fully covered in contour feathers, except below the chin. Some flight feathers are still contained in their sheaths and the tail feathers are just starting to erupt.

Young in nest: About nine weeks; 64 days for both young in one nest (Wright 1977).

Chick growth

Shows weights of two *riciniata* hatched at Palmitos Park in June, 1989, and removed from the nest at 18 days. The hand-rearing food consisted of Milupa baby cereal, wheat germ cereal and papaya liquidized with bottled water. Weights shown are those before and after the first feed of the day.

Violet-necked lory weight table

Age in days	Weight in grams	
	Chick No. 1	No. 2
19	35/40	35/41
20	38/42	36/41
21	40/45	38/43
23	49/55	45/52
25	56/68	56/66
27	61/68	61/66
29	68/80	68/78
31	78/82	76/84
33	82/90	82/88
37	96/108	96/104
41	104/112	100/106
45	106/116	104/114
49	112/128*	106/120*
53	106/118	104/114
60	108/122	104/118
66	100/112	98/104
76	108/114	98/100
82	—/—**	104/—**

* peak weight ** weaned

General: *Eos squamata* and *obiensis* are the smallest of the omnivorous (rather than predominantly nectar-feeding) species and can be recommended to the beginner with lories. Hand-reared birds make excellent pets and some become good mimics (see Pets). The violet-necked is one of the most free-breeding of the lories. In Europe it has been bred regularly since the mid-1970s. The first recorded breeding in the U.K. occurred in 1976 by Mrs. L. Hutchinson of Eastbourne. The pair was housed in an aviary with a large flight measuring 4 m (13 ft.) x 3 m (10 ft.) with a shelter 2 m (6 ft. 6 in.) x 1 m (3 ft. 3 in.), and 2 m high. The female laid the first of two eggs on May 8; one egg was damaged and the other hatched on June 6 (thus the incubation period was 26 or 27 days). In 1977, the first of two eggs was laid on March 20. Two chicks hatched and were reared, but one died two days after leaving the nest. The female laid again as late as November, the first of two eggs being laid on or about the 14th. Two chicks hatched on or about December 12. In the 1970s, when hand-rearing was rarely carried out, most winter-hatched lory chicks died. When these two were two weeks old, one was found dead and the other stone cold. It was removed for hand-rearing but died a few days later. The diet of the parents consisted of sponge cake soaked in Stimulite nectar, with a separate dish of nectar made from baby cereal, wheat germ cereal and honey, with one drop of Adexolin multivitamins per bird. When rearing young, large amounts of green food were eaten, mainly the heads of milk thistles, seeding grasses, spinach and chickweed. This was offered fresh two or three times a day (Hutchinson 1978).

In 1977 C. K. Wright bred from a pair obtained in 1974. A single fertile egg laid in February did not hatch. Two more eggs were laid on May 23 and 26, and hatched on June 19 and 21. They were fully feathered by July 25 and left the nest on August 22 and 24. They were housed in an outdoor aviary measuring 2 m (6 ft. 6 in.) x 1 m (3 ft. 3 in.) x 2 m high and fed on fruit, sunflower seed and a liquid mixture of Complan, Ostermilk, various baby cereals, glucose and honey. While young were being reared, milk sweetened with honey was added to the diet (Wright 1977).

There was a very prolific pair at Loro Parque when I was curator there. Young had to be removed as soon as they started to feather because the parents plucked them. As a result of their removal, four or five nests of two were produced every year.

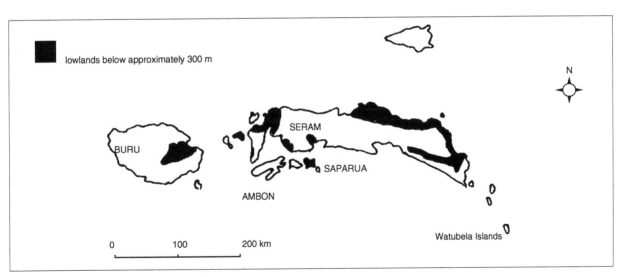

Species found on Seram, Buru and Saparua.

LOWLAND SPECIES

SERAM
red lory (*Eos bornea*)
green-naped lorikeet (*T. h. haematodus*)
red-flanked lorikeet (*C. placentis*)
purple-naped lory (*Lorius domicellus*)—and midmontane
BURU
red lory (*E. bornea cyanonothus*)
green-naped lorikeet (*T. h. haematodus*)
SAPARUA
red lory (*Eos bornea*)

MOUNTAIN SPECIES

purple-naped lory (*L. domicellus*)
blue-eared lory (*E. semilarvata*)

Buru lorikeet (*Charmosyna toxopei*)

229

Obi Violet-necked Lory *Eos squamata obiensis* (Rothschild 1899)

Description: It differs from *riciniata* in two obvious details: the head and upper breast are red and the scapulars are black. The demarcation between the red breast and the deep purple-blue of the abdomen is almost a straight line. (In *riciniata* the areas of color are more patchy, less clearly defined.) Cooke (1990) states that in females seen, and some young males, the shoulders "can have a brown appearance." She also mentions that her two females, and others she had seen, had "slightly orange feathers at the base of the upper part of the leg (near the tarsus); in older males these are scarlet. The younger males can have a mixture" of both colors. Some specimens have a little violet-gray on neck and crown, but most lack this. The beak is orange; cere and feet, gray; and the iris is brown.

Length: 22 cm (9 in.)

Weight: 100 g

Immatures: Plumage varies; some young have a blue-gray collar; throat and part of upper breast also blue-gray; lower breast mottled red and bluish black. Cooke (1990) described the first young from her pair as almost identical to the adults "except for a rather 'speckled' appearance on the breast and around the cheeks, due to some slight brown and purple 'lacing' to the feathers with just one or two of green appearance." The cere and skin around the eye are white (gray in adults), and the beak and iris are brownish.

Natual history

Range: The island of Obi in northern Moluccas.

Habits: Lambert (1993a) observed a pair on Obi "investigating various cavities in 2 tall, leafless dead trees in primary forest at 730 m [2,400 ft.] altitude. The holes that these birds were investigating were situated at points where branches had broken off major side branches at c. 25 m [80 ft.] above ground."

Status: Common but vulnerable. The Obi population was estimated at between 11,808 and 64,599 birds, based on theory (Lambert 1993a). However, Lambert, Wirth et al. (undated) suggested that the population numbers less than 7,000 birds and that hybridization (with what is not stated) is a threat. They classified it as endangered.

Trade: See under *Eos squamata riciniata*.

Aviculture

Status: Uncommon

Clutch size: Two

Incubation period: 24 to 25 days (Cooke 1993a); two eggs laid by G. van Dooren's pair hatched after 31 and 30 days, indicating that incubation did not commence with the first egg (van Dooren pers. comm. 1995).

Newly hatched chick: Weights, 4.5 g and 4.9 g; long white down, beak black, pale at base with white egg tooth; nails gray with white tip (de Dios and Sweeney 1993).

Chick development: Relates to the two chicks whose weights are given in the table below: day 10, second down follicles forming under skin; day 13, second down erupting and eyes starting to slit; day 15, eyes partially or nearly open; day 19, eyes fully open; day 25, first pin feathers appear on head; day 30, head feathers erupting; day 34, beak becoming lighter, flight feathers erupting; day 39, beak orange, red feathers appearing on breast, purple on abdomen; day 46, fully feathered except for flanks; day 60, fully feathered. One reared by G. van Dooren was ringed at 15 days with a 6 mm ring.

Young in nest: About 8 weeks (Cooke 1993a).

Chick growth

Shows the weights of two chicks hatched at Birds International Inc. in the Philippines in 1993 (de Dios and Sweeney 1993). The diet was based on the following formula: 30 ml Easy to Digest Formula, 20 ml papaya, 10 ml banana, 5 ml egg (osterized), 10 ml boiled distilled water and one tablespoonful of Apple Jungle Pellets.

Obi violet-necked lory weight table

Age in days	Weight in grams	
	Chick No.1	No. 2
day hatched	4.48	4.85
1	5.62	4.97
2	5.67	5.60
3	5.83	6.07
4	5.96	7.02
5	6.06	8.14
6	8.75	9.40
7	9.9	10.3
8	11.0	11.9
9	12.4	13.0
10	14	14
11	15	16
12	17	19
13	20	20
14	21	24
15	25	25
16	27	30
17	29	30
18	31	36
19	37	38
21	39	45
23	46	47
25	50	51
27	53	56
29	58	62
31	65	70
33	70	71
35	71	74

General: This lory was unknown in aviculture until 1987. In that year, a few were imported into the U.K. They were believed to be Wallace's lory and were incorrectly known by this name for a few months. I did not see *obiensis* until 1988 when Loro Parque received four specimens. From the description of Mees (1965), I was able to identify them as *obiensis* and published a note to this effect (Low 1989).

The Obi violet-necked lory has proved as easy to breed as *riciniata*. Enough birds—but not large numbers—were imported in the late 1980s for it to be established in aviculture. Perhaps no more importations will occur, so serious lory enthusiasts should take the opportunity to breed this distinctive lory before it is too late. In careless hands *obiensis* will soon be lost to aviculture.

The Obi lories belonging to D. and F. Cooke regularly enjoyed devouring small branches of willow about 46 cm (18 in.) long—"containing the delectable green buds they crave." They also devoured the bark. Branches of apple and hazel were also offered (Cooke 1990). Dulcie Cook noted: "Home-reared Obi Lories are delightfully tame and confiding, each one with a most engaging personality all his or her own, and different from all the others. One gets used to varying temperaments and characters of the larger Parrots but to find so much variation in a 9 in. bundle of mischief is quite an experience." Her birds sometimes attempted to breed during cold weather, but the chicks seldom survived unless removed for hand-feeding; the parents ceased to brood during the day after the chicks were two weeks old (Cooke, *Birdkeeper*, 1991).

Obi lories are rare in North American collections. In 1993, Jim and Pat Taylor were able to locate only eight pairs when they were searching for more. Their pair laid soon after they acquired them; both chicks hatched but were not fed by the parents, and died soon after. The female laid two more eggs, one of which was given to a pet blue-streaked lory who was incubating infertile eggs. The Obi lories killed their chick when it hatched; the blue-streaked fed hers until it was about four weeks old, when it was removed for hand-rearing. Soon after, the Obi and the blue-streaked laid at the same time; the blue-streaked incubated both the Obi's eggs for 25 days, until they hatched. One chick died; the other was reared to fledging by its foster parents. On a further occasion the blue-streaked repeated this success (J. and P. Taylor 1995).

Genus: *Pseudeos*

In coloration, the single species in this genus is distinctive; in behavior it is reminiscent of *Trichoglossus* and *Eos*. But its chromosomes have revealed some surprises. Simon Joshua has studied the chromosomes of many parrot species, with special emphasis on lories (more than 30 species, from every genus except *Oreopsittacus*). He revealed that the karyotype of *Pseudeos fuscata* is of considerable scientific interest. "Perhaps most importantly, it is unique among other members of the Loriinae. Its relationship with species of *Eos* is therefore not as clearly defined as has been suggested. Clearly its evolutionary path has diverged considerably from other species in genera such as *Eos* or *Chalcopsitta*. It would appear that *Pseudeos* is in fact more closely related to *Chalcopsitta* than *Eos* (although the generic name suggests closer ties with *Eos*). Secondly, with 38 chromosomes, it has the lowest number of chromosomes which have been recorded for any bird" (Joshua 1994a).

However, current impressions of karyotypes only tell part of the story. My own impression (after 20 years of keeping members of all three genera) is that *Pseudeos* has similarities and differences with both genera but is a species which is still evolving. The wide range of variation in plumage of young and adult birds suggests that this is so. I feel that the appearance of young chicks is of significance in trying to determine evolutionary relationships—and chicks of *Pseudeos* resemble *Eos* more than *Chalcopsitta*. I have recorded the juvenile food begging calls of all three genera on the same tape on the same day; to my ear, those of *fuscata*, *insignis* and *bornea* all differ slightly from each other.

If its chromosomal make-up is so different, can it hybridize with other lory species? The answer is yes, although there seem to be fewer records of hybrids, when one considers how frequently this species breeds in captivity and how readily other lories hybridize, even those of different genera. Many years ago, W. Sheffler in the U.S.A. bred a hybrid "white-rumped" (i.e., dusky) x Swainson's. (The inference is that the male was the dusky, as it is normal practice to name the male parent first.) The best evidence is that of A. Manning (1994/1995) who describes the breeding of an Obi (*Eos squamata obiensis*) x dusky lory. Photographs were taken depicting the parents and young one together, plus photographs of the young one which clearly show its plumage. At first glance it could have been mistaken for a dusky. Although the author described the young one as inheriting the red Obi coloring "as a kind of frosting on the cheeks and down the front of the bird," dusky lories in immature plumage are so variable that this does not seem to be of great significance. Of interest, however, and not mentioned by the author, is the feathered area at the sides of the mandibles which is normally bare is *Pseudeos*.

At Taronga Zoo, Sydney, two hybrids between a male collared or solitary lory (*Vini solitarius*) and a female "white-rumped" were apparently reared in 1957 and another in 1958. Taking into consideration how much smaller is the collared, this seems a bizarre happening. Even stranger was their appearance. K. Muller (pers. comm. 1976) reported that "they resembled closely the *Domicella* lories, being predominantly red and blue pattern with a short tail, and resembled closely the Black-capped Lory." The hybrids were intermediate in size between the dusky and the solitary.

Another dusky hybrid is that recorded as *P. fuscata* x *T. h. haematodus* at Munich Zoo in 1978 (*International Zoo Yearbook*, vol. 20).

Dusky Lory *Pseudeos fuscata* (Blyth 1858)

Synonym: White-rumped lory

Description: There is great variation in plumage, due partly to the presence of more than one color phase and to the variation in individuals. Without reference to skins collected throughout the range and accurately labeled, or observation of wild birds, it is impossible to resolve the question of how much this is influenced by the locality of their origin.

Most dusky lories are of the orange phase, that is, there are two bands of orange, one just below the throat and the other on the upper breast. Most of the abdomen, except the sides, is also orange, and this color varies from intense fiery orange to a more muted color. The forepart of the crown is golden orange. In the yellow phase these areas are replaced by yellow. The photograph above illustrates these two most distinctive phases. However, there is an intermediate in which the orange is duller and interspersed with yellow feathers or, in other individuals, the upper band could be mainly orange and the lower band mainly yellow, or vice versa. (Such a bird is illustrated in a photograph in Coates 1985.) Whether this is the result of pairing two birds of different color phases is unknown.

The feathers from nape to mantle and those between the two bands of color, are black and/or brown margined with orange or yellow, according to the phase, brighter on the nape and duller on the breast, where they may be buff or silvery buff. The abdomen, even in some adult birds, may be marked with black or blackish brown. Upper parts are very dark brown, browner on the wing coverts; underwing coverts are orange in the orange phase and yellow in the yellow phase. The flight feathers are broadly marked with orange or yellow. The tail is blue-bronze and orange above; the underside is pale bronze and orange. Under tail coverts are dark blue. The beak and the prominent area of bare skin surrounding the lower mandible are orange; the iris is also orange and the feet are black.

Theoretical color expectations: two orange phase birds will produce only orange phase young but two yellow phase birds can produce yellow and orange young.

Length: 25 cm (10 in.)

Weight: About 155 g

Immatures: The plumage is extremely variable, more so than in any other lory species with which I am familiar. There are basically two different types. Perhaps they vary according to locality of origin because in the 1970s when the first commercial importations of dusky lories came from New Guinea, the young produced from these birds were darker and brighter. Duskies currently in my care were imported in the 1980s when most importations probably originated from the Papuan Islands such as Salawati. Young produced from these birds are duller and lighter. Some of the young from my pair imported in 1973 were nearly as bright orange as the adults (see photograph in *Lories and Lorikeets*, opposite page 68). I described one of the young from this pair as follows: "On fledging the colors differed markedly from those of the adults. In place of the clear bands of color on the breast it had almost the entire breast bright orange with a smudge of dark brown across the upper breast where the adult has a clear band. Each feather on the wings had a bronzy, almost iridescent center, and the orange patch on the outer web of each tail feather was far brighter than that of the adults. The rump was almost golden in color, brighter than that of the male" (Low 1976).

The two pairs currently in my care (the male of one pair being the son of the other) are not such brightly colored birds of the orange phase. In their young the brown feathers of the upper breast are margined with buff or yellowish buff (margined with orange in the brighter type), and the colored areas are dull orange or yellow-orange.

In all immature birds the beak is brown on fledging, gradually becoming orange by the age of about four months. The naked skin at the side of the lower mandible is grayish or brownish yellow, gradually becoming orange, and the iris of the eye is brown.

Sexual dimorphism: Finsch (1900) was the first to record that some populations of this species are sexually dimorphic. He described the rump color of eight specimens as follows: male, yellowish white; female, silvery white; immature male, yel-

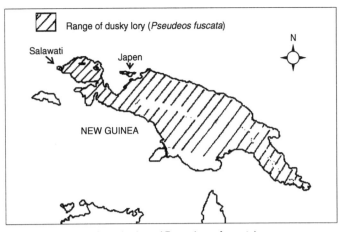

Range of dusky lory (*Pseudeos fuscata*).

lowish. These descriptions agree with the coloration of my breeding pair of the 1970s as well as three other pairs I was familiar with at that time. The rump was yellowish in all young birds. However, most of the birds now available in aviculture are not sexually dimorphic. As suggested earlier, they probably originate from a different locality.

Hybrids: These merit some attention, as already explained (see Genus: *Pseudeos*). In addition to those described, a hybrid involving three species was reared by Jay Kapac in California. The male parent was a hybrid between black-winged and blue-streaked lories and the female parent was a dusky. A photograph (back view) of the resulting hybrid before independence showed that its upper parts were black and all the primaries and secondaries were marked with a large area of orange-red. The tail was black; back and rump were mainly orange-red but some of the feathers were margined with black and/or strongly tinged with blue.

Natural history

Range: Throughout New Guinea from sea level up to 1,500 m (4,900 ft.), or more rarely, up to 1,800 m (5,900 ft.) (Beehler, Pratt and Zimmerman 1986), and sometimes even higher; also occurs on the island of Yapen (Japen) and on Salawati (see Gazetteer) in the western Papuan Islands.

Habits: Most habitats are used: forest, tall secondary growth, partly cleared areas, plantations and occasionally savannah. It flies high over forest on daily flights to and from its feeding grounds. These flights often cover long distances. Bell (in Forshaw 1979) even observed dusky lories flying over the main cordillera of New Guinea. Flocks on their way to roost can be observed flying from the mountains to the lowlands. Roosting sites can contain "thousands of individuals" (Coates 1985). At Ok Tedi in the Western Province there is a roost of several thousand birds at Ok Menga. Gregory (1995) states: "Evening roost flights over Ok Ma may total several thousand birds all heading east." In August, 1996, Peter Odekerken saw a flock of about 50–60 birds in preroosting flights. Most of the flock settled in two trees; only yellow phase birds were observed (Odekerken pers. comm. 1996).

According to Diamond (1972), "the Eastern Highlands population breeds during the rains and then leaves before the onset of the dry season, perhaps retreating to the coastal lowlands." In view of the nonsedentary nature of this species, subspecies would be less likely to evolve on the mainland—but perhaps those resident on Japen and Salawati are evolving or have already evolved into a slightly different form, thus explaining the differences in immature plumage.

Beehler, Pratt and Zimmerman describe the dusky as "most gregarious of the lories; often in large, high-flying, post breeding flocks, which are exceedingly vocal. These are often heard long before they are seen...Forages in canopy of forest, forest edge, and in plantations (especially in the *Leucaena* trees used to shade coffee)."

At Tari, in the Southern Highlands, at altitudes of 2,200 m to 2,400 m (7,200 ft. to 7,870 ft.), J. Newman observed this species in September, 1990. Several groups of five up to 20 birds were seen flying high over the grounds of Ambua Lodge in the morning (9:00 A.M.) or afternoon (4:00 P.M. to 5:00 P.M.). They were not seen at flowering trees in the vicinity, where several other lory species fed. When Peter Them visited Ambua Lodge in 1996, groups of two to eight birds were flying over. They fed in tall flowering trees, at an altitude of over 2,000 m (6,560 ft.) (Them pers. comm. 1996). When I visited the same area in December, 1994, no dusky lories were present. However, at a lower altitude, about 1,650 m (5,400 ft.), I heard dusky lories but did not see them. At 5:00 P.M. on December 7 a flock of about 20 birds was observed feeding near the Tari airport. In Karawari, Them observed several hundred dusky lories from a canoe on the river. They were flying overhead in the late afternoon, in groups of 10–50 birds. In Madang, the raucous calls of a flock of duskies attracted Peter Them to a tall tree (perhaps *Elaeocarpus sphaericus*). It was covered in purple fruit and clusters of creamy white flowers. At no time were they seen to feed on the fruit, only on the flowers. Because the bell-shaped corollas were pendant, the lories had to bend low with the body held below the horizontal to position their heads beneath a flower cluster. The head was then turned so the beak faced upward in order to ingest the nectar—and perhaps also the pollen. Each bird worked on several flowers in a cluster before proceeding to another cluster (Them pers. comm. 1996). A flock of 50 observed by Bell (1979) was feeding on the pupae of the teak moth (*Hyblaea puera*). The stomach contents of some birds taken at Manokwari consisted of small black fruit stones, light green pulp and fine vegetable matter.

Nesting:　Usually high in tall montane trees, thus little is known. One egg laid by a wild bird measured 27.8 x 24.1 mm; two laid by a captive bird were 27.5 x 23.5 mm and 28.2 x 23.8 mm.

Status/Conservation:　Common; no conservation measures known to be necessary.

Aviculture

Status:　Common (see **Popularity, Avicultural**)

Clutch size:　Two

Incubation period:　24 or 25 days.

Newly hatched chick:　Weight, 6–8 g; white down, thickest on wings and head; beak and legs pinkish (soon turning darker).

Chick development:　At seven days the down feather tracts are visible beneath the skin; by 20 days the grayish second down is erupting, in contrast to the white patch of down on the nape; the eyes are open, the beak is brown and the legs are dark gray. At 25 days the feathers of the forehead and the secondaries are starting to erupt; at 34 days the wing coverts are about 1.25 mm (½ in.) long and the tail half that length. At eight weeks the chick is fully feathered except for the tail.

Young in nest:　About 70 days, but 64 to 72 days recorded under normal circumstances. One chick which was slightly retarded by a bacterial infection left the nest at 78 days.

Age at independence of hand-reared young:
50–68 days, depending on how they are weaned.

Chick growth

Weights of two parent-reared chicks at Palmitos Park are contrasted with two reared by P. Tiskens' pair. Chick No. 1 (hatched 23/4/94) made normal weight gains. Chick No. 2 (hatched 25/5/94), from another pair, was retarded by a bacterial infection diagnosed when it was 20 days old. It was treated orally with Baytril while in the nest, but its weight was below average until the age of about 60 days. In both nests, only one chick hatched. Chicks No. 3 and No. 4 were parent-reared (Tiskens pers. comm. 1994).

Dusky lory weight table

Age in days	Weight in grams		
	Chick No. 1	No. 2	Nos. 3 and 4
day hatched			6 fc
1		9 fc	
2		10 fic	
3		11 nf	
5		14 nf	
7		18 fc	
9		23 vfc	
11		29 vfc	
13		34 fc	
14			100g (ringed)
15		35 fc	
17		41 nf	
20	88 fc	50 fc	
23	99 fc	56 fic	120g
26	106 fc	57 ne	
28			130g
30	109 fc	81 fc	
34	118 fc	94 nf	
35			150g
37		99 nf	
41		102 fc	
42			160g
44	129 fc	111 fc	
47	146 ne	111 e	
50		114 ne	
53		135 fc	first left nest and returned
55			second left nest and returned
56		155 vfc	
58		132 ne	young stayed out
62	137 e		
63			independent
70 left nest			
78		left nest	

e = crop empty; ne = nearly empty; fic = food in crop; fc = full crop; vfc = very full crop

Peak weights of two chicks hand-reared simultaneously from one of above-mentioned pairs were 150 g at 50 days, then 156 g at 66 days when it was weaned; and 149 g at 50 days and 150 g at 64 days when it was weaned.

General: The dusky lory imported into the U.K. by Mrs. S. Belford in 1972 was an avicultural rarity. I was fortunate to look after it for some while and I was entranced by its tameness and gentleness. It was the first of its species seen in Europe for probably nearly 40 years—and the forerunner of thousands; later that year commercial export from Irian Jaya commenced. In 1974, the first breeding successes were recorded and, soon after, this species was securely established in captivity. This occurred perhaps with greater speed than in any other lory species following its first influx from the wild. It has been bred in most countries where aviculture is practiced. In Japan, for example, the first success was reported in 1978 by Dr. Masahide Koji Mahara. He obtained a pair from a bird shop in 1976 and kept them in an aviary 2.1 m (7 ft.) long, 95 cm (3 ft.) wide and 2.2 m (7 ft.) high. They were fed on sunflower seed, sponge cake, biscuit and fruit (banana, apple, grapes and orange). During the breeding season they were offered bread spread with raw egg and honey. The pair nested in a box 23 cm (9 in.) square and 30 cm (12 in.) high. A chick was heard on June 24; nest inspection on July 9 revealed one chick and one infertile egg. The young one left the nest on September 2, after a minimum of 70 days (H. Shimura pers. comm. 1994).

At the time of writing, I have kept this species for 20 years—and reared it almost every year since 1975. I would rate it with the green-naped lorikeet as being one of the easiest lories to breed. However, it is not as prolific, because it is a seasonal breeder. In my experience, in the U.K. and in the Canary Islands, eggs are usually laid in April; if the clutch is successful and the young parent-reared, there will not be another clutch. If the nest fails or the chicks are removed, a second clutch may be laid, but no later than the end of August or the beginning of September—and that rarely. It has been my practice in recent years to leave the young with their parents for months after they fledge. (At the time of writing a seven-month-old youngster is still with its parents.) In colder climates, such as the U.K., even though this species is not a winter breeder, one needs to be aware of the dangers of night-exposure to newly fledged young. In 1980, I was abroad when the chicks fledged on October 10. One was picked up dead the next morning, undoubtedly because it had not roosted in the nestbox. Thereafter, I found a simple solution—to take the young indoors at night in cold weather.

The adults were very hardy birds, impervious to the worst winter weather. They could sometimes be seen jumping on the snow-covered floor of their aviary. Evidently they did not feel the cold, or they would not have exposed their feet to the snow, thereby lowering their body temperature. On many winter days it would be necessary to renew the nectar three or four times, to prevent it from freezing (Low 1985a).

Rearing food for my pair consisted of nectar to which was added wheat germ cereal (and, on occasions, Horlicks), and also milk in a separate dish. At first I gave bread and milk but the bread was ignored. Ripe pear and millet spray were given daily. The latter had never been eaten until young hatched and I saw them trying to get at the millet sprays in the adjoining aviary (Low 1976). The first three clutches laid by this pair resulted in a chick hatching each time, which died at only two days. It must have been coincidence that chicks were always reared after spray millet became part of the rearing diet, because other pairs in my care since then have not had this food; they have reared their young on the normal nectar, plus extra fruit, especially banana, also apple and pear. However, the weights of these young (see table) are lower than those reared by Paul Tiskens' pair. His lories are fed on his own liquid food in the morning and on fruit with a dry food in the evening.

Some dusky lories are aggressive when breeding. One pair in my care will not leave the nest when the chicks are young, so for the first 20 days it is difficult to view them. Others threaten, without carrying out any physical violence. Some hand-reared young can also be aggressive. Val Wittcke (pers. comm. 1994) described the behavior of two chicks she hand-reared when their mother became paralyzed: "I made such a fuss of them because they were so cute when I brought them in. I used to spend hours wasting time with them. They were fine at first and then they got a bit nippy. Now, they jump on me as soon as I go into their aviary, snuggle into my neck and then they bite as hard as they can. Not only my neck, though: they aim for my hands, face, any flesh they can get hold of."

Hiller (1987) describes how apparent aggression is expressed in terms of natural behavior. "In both sexes I have observed a 'ploughing' motion, in which the bird walks on a smooth surface, pushing the tips of both (?) mandibles along in front of it. It will then stop, and rub the beak in a forward-back movement with increasing frequency. Alternatively, a rapid 'tapping' with the beak may occur. In context, I believe this action implies aggression. If I imitate the beak-stroking or tapping movement with a finger, I can usually induce them to do the same, which shows it is a highly communicative gesture. When the pair is excited, as in during play, they will 'hop,' again along a flat surface, achieving a great deal of height."

Despite the aggressive nature of some individuals, the dusky lory often makes a delightful pet. It is one of the two or three lory species most likely to be chosen as a companion bird.

Genus: *Trichoglossus*

This genus is difficult to define, therefore it is equally difficult to limit its boundaries. The larger species, such as *ornatus*, *haematodus* and *chlorolepidotus* might be thought of as typical of the genus, i.e., they are mainly green with the breast or the entire underparts boldly barred. Some have very colorful head markings, others lack barring on the underparts and one species is entirely red (this color probably masks the barring). There is much difference of opinion regarding in which genus some of the smaller species, such as *versicolor*, *iris* and *goldiei*, really belong. The latter has little in common with other species of *Trichoglossus*. Forshaw (1989) retained it in this genus but suggested that "further investigations with fresh material could result in it being placed with *Charmosyna* or in a monotypic genus." The latter would be more appropriate as it has no affinities with *Charmosyna*. He described *Trichoglossus* as "small to medium-sized parrots with gradated tails comprising rather narrow, pointed feathers. There is no naked area surrounding the lower mandible. Sexual dimorphism is absent, and immatures resemble adults but have more sharply pointed tail feathers."

He described the closely related *Glossopsitta* thus: "In this genus there are no shaft-streaked, erectile feathers on the crown. The small bill is fine and somewhat projecting, and the central tail feathers of the wedge-shaped tail are pointed. The cere is naked. Sexual dimorphism is slight in one species—*concinna*—and absent in others."

I intend no criticism of Forshaw, only to emphasize the difficulties involved in trying to classify certain lories. In *Glossopsitta* species the bill is finer—but to my eyes it is less projecting than in *Trichoglossus*. Sexual dimorphism is not entirely absent in *Trichoglossus*, although, as in *concinna*, in some subspecies of *haematodus*, also in *meyeri*, for example, it is slight.

Modern methods of analysis, such as DNA-DNA hybridization and egg white protein analysis, have so far served to show how closely related are these and other lories—but more evidence is needed before attempting to separate or combine them, and so far no research has been carried out on certain crucial species, such as *Oreopsittacus arfaki*.

Peters (1937) favored placing five small species in the genus *Psitteuteles*: *meyeri*, *johnstoniae*, *goldiei*, *versicolor* and *iris*. Because they are comparatively recently evolved, the lories pose a puzzle to systematists which, perhaps, will never be satisfactorily resolved.

The species which are generally accepted as *Trichoglossus* range in size from 17 cm to 30 cm and in weight from 50 g to about 150 g. Their geographical range is extremely wide—most of Indonesia and New Guinea, coastal Australia, and that far outpost of lory habitation, the Caroline Islands. Their status varies from extremely common and widespread species (e.g., *haematodus*) to little known island endemics (e.g., *iris*). In aviculture, their members include some of the most widely kept and easily bred species.

Ornate Lorikeet *Trichoglossus ornatus* (Linnaeus 1758)

Description: Immediately distinguished from all other *Trichoglossus* by the red lores, cheeks and throat. Forehead, crown and ear coverts are purplish blue and the nuchal collar is yellow. Lower cheeks and upper breast feathers are red, edged with blue-black. Undertail coverts and area surrounding the vent are yellowish green. Underside of the wings is yellow; there is no colored band. The rest of the plumage is green with variable yellow markings on the abdomen. The beak is orange and the iris, dark orange.

Length: 25 cm (10 in.)

Weight: 110 g

Immatures: The birds have less pronounced barring on breast and some have more yellow on abdomen than adults. Cere and skin surrounding the eye are blue-tinged pale gray (dark blue-gray in adults). The iris is brown. The bill is black, and starts to change to orange soon after fledging.

Range: Sulawesi (Celebes) (see Gazetteer), Indonesia, and most larger offshore islands such as Manterawu, Bangka, Togian, Banggai, Muna, Butung and Tukangbesi (Kaledupa). According to van Oosten (pers. comm. 1995), the University of York Expedition observed ornate lorikeets on Sangihe in 1995.

Habits: It occurs in montane forest up to about 1,000 m (3,300 ft.). Stresemann (1940) stated that it tended to avoid dense primary forest. More recently, Watling (1983) visited northern and central Sulawesi between November, 1978, and March, 1981, and found it at the margins of forest, in secondary forest and in bush or open areas, including open upland valleys. At higher altitudes it may be found in association with Meyer's lorikeets. Dr.

Meyer described it as "the most common Parrot of Celebes; I got it at all times and everywhere in the Minahassa from January till July; at the end of March 1871 it suddenly appeared in large flocks, near Limbotto in August; near Gorontalo in September; on the Togian Islands in August; and in South Celebes in October and November." When discussing Meyer's lorikeet, Dr. Meyer commented that "it appears to be probable that the rainy season in the mountains of the Minahassa drives the birds to places where it is warmer or not so damp; at least this was the condition in 1871 near Menado, where the rainy season was very mild" (Mivart 1896).

The food of this species consists of fruit, blossoms and seeds, including those of *Casuarina* and *Tectona*, as well as pollen and nectar.

Nesting: Little information has been recorded. W. Timmis (pers. comm. 1980) wrote from his camp in central Sulawesi on February 28: "Found my first nest 2 days ago in a tall palm tree. *T. ornatus* breeds here from September to May. Many young birds are taken from nests by local people." Two eggs laid at Palmitos Park measured 25.2 x 22.0 mm

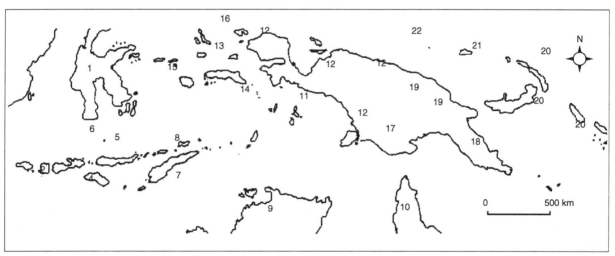

Distribution of ornate lorikeet (*Trichoglossus ornatus*) and green-naped lorikeet (*Trichoglossus h. haematodus*) and subspecies.

1. Sulawesi – *T. ornatus*
2. Lombok – *T. h. mitchellii*
3. Sumbawa – *T. h. forsteni*
4. Sumba – *T. h. fortis*
5. Flores – *T. h. weberi*
6. Tanahjampea (Djampea) – *T. h. djampeanus*
7. Timor – *T. h. capistratus*
8. Wetar and Romang – *T. h. flavotectus*
9. northern Australia – *T. h. rubritorquis*
10. eastern Australia – *T. h. moluccanus*
11. Aru Islands – *T. h. nigrogularis*
12. Irian Jaya – *T. h. haematodus*
13. Seram – *T. h. haematodus*

14. Watubela Islands – *T. h. haematodus*
15. Buru – *T. h. haematodus*
16. Biak – *T. h. rosenbergii*
17. Papua New Guinea – *T. h. caeruleiceps*
18. Papua New Guinea, Misima Island and Manam and Bagabag – *T. h. micropteryx*
19. Papua New Guinea – *T. h. intermedius*
20. New Britain, New Ireland and the Solomon Islands – *T. h. massena*
21. New Hanover and Admiralty Islands – *T. h. flavicans*
22. Ninigo – *T. h. nesophilus*
New Caledonia – *T. h. deplanchii*—see map on page 347
Vanuatu – *T. h. massena*—see map on page 347

and 25.9 x 22.0 mm. The average of 3 taken from wild nests was 26.1 x 21.7 mm (Harrison and Holyoak 1970). Wright (1981) recorded the size as 26.5 x 21.5 mm.

Aviculture

Status: Uncommon

Clutch: Two

Incubation period: Varies between 27 days for the first egg to 24 days for the second, with most recorded instances being 25 and 26 days. Two eggs laid at an interval of four days in the collection of P. Bertagnolio of Rome both hatched after 26 days.

Newly hatched chick: Weight, about 6 g. Long, dense white down (van Rooyen pers. comm. 1996).

Chick development: At an early age, when the rest of the chick is down-covered, the crown looks blackish; the last areas to feather are the yellow patches on the neck (van Rooyen). Chicks are ringed at 13 to 15 days (with 6 mm rings).

Young in nest: About 60 days, but up to 73 days recorded; also an abnormally long 80 days for two young in same nest. Two reared by Swainson's lorikeets left nest after 56 and 58 days and one reared by scaly-breasts after 60 days.

Chick weight table

Shows the weights of two parent-reared chicks (No. 1 and No. 2) from a pair belonging to Jan-Hendrik von Rooyen, in South Africa (von Rooyen pers. comm. 1997) and a single chick hand-reared from day 6 at Palmitos Park in 1990. The food for chick No. 3 consisted of Milupa baby cereal, wheat germ cereal and liquidized papaya. Weights shown for this chick are those before and after the first feed of the day.

Ornate lorikeet weight table

Age in days	Chick No. 1	No. 2	No. 3
day hatched	–	6.42 fc	
1	7.49 nfc	7.57 nfc	
2	8.87 nfc	8.92 nfc	
3	9.97 e	9.83 e	
7			11/13
8			13/14
9			14/15
10			15/16
11			17/19
12			18/21
13			19/22
14			20/23
15			22/25
16			24/26
17			25/28
18			27/32
19			29/33
20			31/36
22			33/38
24			39/44
26			44/52
28			50/57
30			55/64
35			72/82
40			82/94
45			90/98
50			96/108
55			96/104
65			90/102
70			86/94
71			86/sf
74			refused spoon

fc = full crop; nfc = nearly full crop; e = empty crop; sf = self-feeding

General: This species was bred as long ago as 1883 by Baron Cornelius in France. Another early success was that by Ruby Hood of California in 1932. After several unsuccessful nests, two young were reared. The diet consisted of cooked oatmeal mixed with milk and honey, fruit (especially seedless grapes) and sunflower seed (Hood 1932).

The ornate lorikeet has an interesting history in aviculture. Wild-caught birds were imported during the late 1960s and early 1970s, before lory breeding was popular and when this was one of the most commonly available lories in Europe. It was being bred in zoos before the first influx of various wild-caught Indonesian lorikeets, when few other species were available. It was also bred in Japan for the first time, by Akiro Sano, in 1965. When the market was suddenly flooded with species which most people had never seen before, the ornate was neglected. In 1987, soon after this fact was realized, its importation into the European Union was forbidden. Its numbers in Europe decreased rapidly. Then, in the late 1980s when its rarity in aviculture was becoming apparent, a few breeders again took an interest in the species. By the mid-1990s, its stronghold was South Africa, where several breeders were producing significant numbers.

In Denmark, Carl Nielsen bred the ornate lorikeet in 1966. His pair was housed in a flight 3.5 m (11 ft.) long, 91 cm (3 ft.) wide and 2 m (6½ ft.) high, with access to a heated shelter 1.8 m (6 ft.)

long. The first two clutches were infertile. Then the nestbox (which measured 18 cm (7 in.) square and nearly 1 m (3 ft.) high), was placed in the outside flight. It was hung at an angle of 45° and contained a mixture of wood shavings, coarse sand and a little soil. A clutch laid at the end of May produced one infertile egg and one chick. The young bird left the nest at eight weeks old. In the following year the pair nested at the end of March and, again, one young was reared. In the next clutch, two eggs were laid, both of which hatched. When the young left the nest, aged 53 and 56 days, they differed from the adults only in their dark beaks. The male fed the young for ten days after they left the nest, after which they fed themselves on fruit and nectar. They were removed to a separate aviary when 11 weeks old. These ornates were fed on a liquid food made from apple, honey, raspberry jam and bread crumbs, also grated carrot, sprouted canary seed and fruit—grapes and orange being the most popular. During the breeding season baby cereal was added to the liquid food, egg food was added to the grated carrot and liberal quantities of sprouted wheat were available (a varied and nutritious diet). The whole family of ornates was exhibited at a show and took the award for best exhibit in the breeders' classes .

In Germany, Bettina and Harald Muller were not so successful. In 1987 they purchased four young birds. They separated into couples but, on being sexed, all were known to be females. After a male was obtained, a female laid 18 months later. Two chicks hatched, and nest inspection took place once a week. When the chicks were found dead at four weeks, they were believed to have been killed by the parents as the result of inspection. Therefore, when two more eggs were laid and chicks hatched, the nest was not inspected until the chicks were four weeks old. It was then apparent that the chicks had deformed legs. They were removed for hand-rearing but died soon after (Muller and Muller 1992). This incident underlines the importance of regular nest inspection. The condition of the chicks' legs might have been corrected (see **Chicks, Foot and Leg Problems**). It seems unlikely that the pair killed their first chicks as a result of nest inspection—after four weeks. Carried out in a sensible manner on a regular basis, few birds object enough to kill their young. It is more likely that they died of some other

cause (possibly from a dietary deficiency or bacterial infection) and were then chewed up by the parents as they cleaned out the nest.

One of the breeders of this species in the U.K. during the 1970s was D. J. Hodson. The first three clutches during the first year were infertile; in the fourth clutch, one egg was fertile and hatched after 24 days. The young one left the nest at 72 days, when it was described as slightly smaller than the adults and just as brightly colored, but with a less extensive yellow collar (Hodson 1978). The display was observed as follows: "In the territorial display a bobbing action of the head and body is accompanied by flashing of the underwing and screeching. In the courtship display a grossly exaggerated vertical stance, from legs through to shoulders, is assumed with the head and neck bowed down to what appears to be a most uncomfortable position. During this painful looking procedure, a continuous hissing and flashing is maintained."

Despite its dwindling numbers, the ornate lorikeet can be very prolific, rearing three or four clutches a year. Luch Luzzatto in South Africa started lory keeping with a pair of ornates. This pair reared a total of 21 young, and died within a few days of each other at the estimated age of 18 years. Another pair in the same collection produced two or three nests of young per year. In 1996, when I visited Mr. Luzzatto, he was still breeding this species, which remained one of his favorites.

Another South African breeder is Jan-Hendrik van Rooyen. He has been breeding this species since 1988 and had 10 pairs (some of them young ones) set up when I visited him in 1996. His birds were in superb condition. In addition to nectar and fruit, they receive a mixture of cooked maize, rice and beetroot, and soaked and sprouted sunflower and sorghum. About 300 ml of food is consumed during the day, plus about 100 ml in the late afternoon. Just before the female starts to lay, food consumption increases to about 600 ml daily. During incubation, consumption decreases to only 300 ml daily. It increases as soon as the chicks hatch; by the time they leave the nest, between 800 ml and 900 ml is taken daily. The ornates were housed in suspended cages about 3 m (10 ft.) long. Each pair had a choice of two nest sites, one of which was a sisal log.

In the U.K., the fortunes of this species were

revived to a degree by Andrew Blyth and Jonathan Powell, who had several breeding pairs. In 1989, for example, they reared 11 young. However, the ornate is still among the rarer lorikeet species in the U.K. It exists in Australian aviculture as an extreme rarity.

Green-naped Lorikeet *Trichoglossus haematodus* (Linnaeus 1771)

Description: Forehead and forecrown are blue, also lores and chin; the throat is blackish blue; the area between the crown and nape is purplish brown, the feathers are shaft-streaked with dull green. The nuchal collar is greenish yellow, with a spot or two of red in most birds. The breast is red, each feather strongly margined with blue, and the red of the breast spills down under the wings. Abdomen is margined with dark green; inner part of the thighs are mostly yellow. Underwing coverts are orange-red with a broad yellow band across the underside of the flight feathers. Upper parts, including upper side of the tail, are green, bases of the feathers of the mantle are red. Underside of the tail is dull olive yellow and yellow. The bill is orange-red, the feet are gray and the iris is red.

Length: 26 cm (10 in.)

Weight: About 130 g to 140 g.

Immatures: They are slightly less brightly colored than adults. The bill is brownish black, changing to brown and then becoming orange by four or five months of age.

Natural History

Range: The nominate race occurs in western New Guinea east to the border area between Papua New Guinea and Irian Jaya in the north (Humboldt Bay), and east to the upper Fly River (Black River) in the south; Indonesian islands of Buru, Ambon, Seram (see Gazetteer for all three), Ceramlaut, and the Goram, Watubela and Western Papuan Islands, and islands in Geelvink Bay. (Its reported occurrence in the Kai [Kei] Islands is based on two specimens mentioned by E. Hartert in 1901. As the collector's specimens normally gave a definite locality but these were labeled only "Key Islands," Cain [1955] doubts that they originated from there. He states: "The known range of *haematodus* comes very close to the Kei Islands but cannot be stated definitely to include them.")

Subspecies are numerous, more so than in any other parrot species; each is covered separately. The ranges of the various subspecies are listed below, alphabetically by locality. However, some of the subspecies are so difficult to distinguish that there is not necessarily general agreement on their exact ranges. Some subspecies are considered doubtfully distinct by systematists, but for the sake of completeness, all are included here.

Trichoglossus haematodus Subspecies

Ambon (Amboina): *T. h. haematodus*
Aru Islands: *T. h. nigrogularis* and *T. h. brooki*
Australia, east and northeast: *T. h. moluccanus*
Australia, Cape York area: *T. h. septentrionalis* (if valid race)
Australia, northern (except northeast): *T. h. rubritorquis*
Bali: *T. h. mitchellii* (probably extinct there)
Biak, Geelvink Bay: *T. h. rosenbergii*
Bismarck Archipelago: *T. h. massena*
Buru: *T. h. haematodus*
Ceram: see Seram
Djampea, Flores Sea: *T. h. djampeanus*
Flores: *T. h. weberi*
Kalao-tua, Flores Sea: *T. h. stresemanni*
Lombok: *T. h. mitchellii*
New Caledonia and Loyalty Islands: *T. h. deplanchii*
New Guinea: *T. h. caeruleiceps*
 T. h. haematodus
 T. h. intermedius
 T. h. micropteryx
 T. h. berauensis (if valid race)
New Hanover and Admiralty Islands: *T. h. flavicans*
Ninigo Islands and Hermit Islands: *T. h. nesophilus*
Seram: *T. h. haematodus*
Solomon Islands: *T. h. massena*
Sumba: *T. h. fortis*
Sumbawa: *T. h. forsteni*
Timor: *T. h. capistratus*
Vanuatu: *T. h. massena*
Wetar and Romang: *T. h. flavotectus*

Habits: Coates (1985) describes this lorikeet as "perhaps the most abundant of New Guinea Parrots; it is common in most parts of its geographical distribution though locally scarce, especially in the upper levels of its altitudinal range." Common and widespread, it often occurs in large flocks, its abundance being dependent on flowering trees. It is found in rainforest and in open habitats, but tends to favor forest edges rather than closed canopy. At Ok Tedi, in the Western Province of Papua New Guinea, green-naped lorikeets (probably the nominate race) are common, although far outnumbered by the dusky lory with which they do not seem to mix. Green-naped occur in flocks of less than 20 birds, or in pairs. Occasionally they are seen at flowering trees in Tabubil town (Gregory 1995).

On Buru it occurs on the coastal lowlands up to 400 m (1,300 ft.), and was not recorded in 1989 by the Manchester Indonesia Islands Expedition at Wafawel or at Danau Rana. Groups of two to six birds were observed in selectively logged forest 1 km up the valley from Charlie Satu. A group of six was seen at this location prospecting nest holes in a tree supporting a colony of starlings, *Aplonis mysolensis*. It was moderately common in the coconut groves and a remnant band of forest behind the beach at Wanibe, and very common at Teluk Bara. Five were seen feeding on casuarina at Tifu on December 4 (Jepson 1993a). Marsden (1995), who was with the same expedition, calculated density estimates as 29 +- 5.2 in primary forest (sample size 50), 33 +- 11 in secondary forest (sample size 24) and 1.9 +- 1.3 in parkland (sample size 12). He concluded that it prefers "quite open lowland forest with dead trees. May prefer the plains rather than steep river valleys." On Seram, Taylor (1991a) found it to be common from the coast up to 800 m (2,600 ft.). In lowland rainforest it was frequently observed (9.8 observations per hour). In Air Besar there were only 1.6 sightings per hour and in Kanekeh, 1.3 per hour. In 1987 he observed a flock of 100 birds, whereas in 1990 the largest flock was of 19 birds. The decrease in numbers was due to heavy trapping. Encounter rates recorded by Marsden alone (1995) were lower: 1.8 per hour in primary forest (27 records), 1.3 per hour in secondary forest (47 records), 2.2 per hour in logged forest (68) records; these records were from altitudes up to 350 m (1,100 ft.). (None was recorded from coastal areas.)

Status/Conservation: Still common, but declining in parts of its range due to trapping and increasing deforestation. There are no conservation measures at present—but I believe that trapping should cease; it is not necessary to export this common and easily bred species. (This occurs only to provide some small income to local people, as happens in so many tropical countries worldwide).

Aviculture

Status: Common

Clutch size: Two, rarely three; Sallien (1994) recorded a clutch of three eggs. All hatched and all were reared to independence by their parents.

Incubation period: 22 to 24 days, occasionally 25 days. (From the pair studied, there was a single record of 21 days—which was either an error or the female laid after the early morning inspection—and a record of 27 days in the case of an egg found cold and transferred to an incubator.) In the first clutch of R. Flierman's pair, the eggs were laid, and hatched, at an interval of three days, with a 24-day incubation period; in the second clutch eggs were laid and hatched at an interval of two days, both hatching after 25 days; the eggs of the third clutch were recorded with a five-day interval but hatched apparently after 26 and 23 days; in the fourth clutch the interval was two days and one egg only hatched—24 days after the second egg was laid (Flierman pers. comm. 1995).

Newly hatched chick: Weight, average 6 g; 4 g, 5 g and 7 g also recorded; long, dense grayish white down on the back, sparser elsewhere; feet grayish pink.

Chick development: Feet pale gray by day four; eyes half open by day 11. At six weeks chicks are almost fully feathered, except for shorter tail and lacking feathers on nape and rump; the beak is black.

Young in nest: About 70 days.

Age at independence of hand-reared birds:
From 34 (34!) up to 59 days recorded by author, depending on how young the chick is when first offered warm food in a dish.

Chick growth

Shows weights of five chicks hatched at Palmitos Park between May and August, 1993, all from the same pair. Numbers 1 and 2 were clutch mates, as were numbers 3 and 4. The figures demonstrate two points: the difference in age at independence, influenced by the age at which they were first offered food in a dish, and the increasing expertise of the hand-feeder, who had not worked with parrot chicks until the month before the first of these chicks hatched. Weights were recorded before and after the first feed of the day. The rearing food consisted of wheat germ cereal, Milupa baby cereal, a smaller amount of Nekton Lory and liquidized papaya.

Trichoglossus haematodus weight table

Age in days	Weight in grams				
	Chick No. 1	No. 2	No. 3	No. 4	No. 5
Day hatched	–	–	–	4	5
1	4	5/5	–	5/6	5/5
2	6/6	6/6	5/5	6/6	6/7
3	6/7	7/7	5/6	7/8	8/8
4	7/9	8/9	6/8	8/9	9/10
5	9/9	9/10	8/9	9/11	11/12
7	10/11	12/12	11/12	13/14	16/17
9	13/14	14/15	14/15	17/18	20/22
11	16/17	–	18/20	21/24	24/27
13	19/20	20/23	21/24	23/26	28/33
15	23/26	25/29	24/29	27/33	33/39
17	28/31	27/30	27/30	34/40	39/47
19	28/32	33/37	31/37	41/48	50/57
21	36/40	37/42	36/43	47/56	60/65
24	40/46	41/46	45/51	61/72	71/79
27	44/51	48/53	57/65	74/84	84/96
30	53/64	60/68	68/78	86/102	96/105
33	69/77	69/79	81/94	93/102	104/116
36	79/89 sf*	83/88	93/106	107/123	111/121
39	88	90 sf*	104/122	113/127	116/128
42	98	98	112/124	116/128	120/134
45	109	107	119/131	123/138	125/135
47	114	110	124/138	124 psf*	130/144
48	113	113	128 psf*	121 psf	133/140
49	119	117	130 psf	123/132	133/151
50	121	–	129 psf	126	130/150
54	–	–	130	–	124/134
59	–	–	–	–	132 sf*

* sf=self-feeding psf=partly self-feeding

General: The green-naped has been one of the best known and most readily available of lorikeets in aviculture since lory breeding occurred on a regular basis, from the 1960s. Because it has always been common and inexpensive, the assumption has been that breeding accounts will be of little interest, thus the amount of information published on this species is scant. (I, too, failed to keep records of chick development, other than weights.) In addition, many keepers found it difficult to identify the subspecies, which again deterred them from writing about their birds. In numerous instances, even in zoos, pairs consisted of two different subspecies of *Trichoglossus haematodus*.

This is undoubtedly one of the easiest lories to breed. One generally finds that parrot species which, in the wild, are common and widespread, often occurring in large flocks, breed readily in confinement. They are successful species, without specialized requirements, and very adaptable. Not only do green-naped lorikeets breed readily, but they are prolific, laying and/or rearing clutch after clutch. However, they should be deterred from this by removing the nestbox for, say, three months of the year. This could result in the female laying eggs from the perch, and frustration at the lack of a nestbox. In this case the new-laid eggs should be substituted with infertile ones from other birds, kept for the purpose. If the egg size is not identical, the female is unlikely to object.

The frequency with which females lay eggs can be illustrated by a pair in my care at Palmitos Park. Chicks had to be removed by about five days, because the pair did not rear them beyond seven days, with a single exception. The first time they bred in my care, a chick reached 28 days, when it was found dead. On several occasions, the skin color of the newly hatched chicks was pale orange and I suspected a bacterial infection which had reached the liver. The parents were injected with antibiotics but results in future clutches were the same. If the chicks were not fostered or hand-reared, they did not survive. I gave the two chicks in one clutch a drop of the antibiotic Baytril daily from hatching for five days, but they both died on the same day, aged seven and six days. On the previous day they both weighed 14 g and appeared healthy. On more than one occasion both chicks were found dead on the same day and it appeared that the injuries they had sustained had been caused by their parents before they died. (Many parrots will chew up dead small chicks, presumably to rid the nest of decaying bodies.)

Between June, 1989, and the end of 1993 (when the eggs were broken up soon after laying), the female laid 29 clutches. In 1989 she laid two; in 1990, six; in 1991, seven; in 1992 (when the nestbox was removed from September until the following February) she laid six and in 1993, eight. After the nestbox was returned in February, the first clutch was destroyed after two weeks. The second clutch was infertile. The next two clutches were fertile but in the following clutch, only one egg was fertile. No subsequent clutches were fertile, and the eggs disappeared after shorter and shorter periods. In 1994, I suspected that the female laid on several occasions but after the first clutch in January I found only shell fragments on two widely separated occasions. Almost certainly, eggs were broken as soon as they were laid. It was almost as though they knew the eggs were infertile, although of course this was not possible. The age of the pair was unknown and it may be that they had reached the end of their reproductive lives. In 1994, I noticed that the female's beak color was not as bright as that of the male.

In the period defined, the female laid 56 eggs in 29 clutches, 31 of which hatched. The incubation period was recorded for 18 of these chicks—and the results were inconsistent. They varied from 22 to 25 days (with two exceptions). In four cases where the incubation period was known for both eggs in the clutch, in one instance the first egg hatched after 25 days and the second after 24; in another, both eggs hatched after 22 days. In the third instance, the first

hatched after a minimum of 23 and the second after a minimum of 22 days; in the final clutch recorded, both hatched after 23 days. Incubation periods for second eggs in clutches were recorded, in chronological order, as 24, 23, 24, 22 and 22 minimum, 22, 23 and 24 days.

Some of the young reared flew at liberty, in twos or threes, in Palmitos Park, and delighted countless visitors with their colors, antics and tameness. I recall one occasion when two were prancing in that exaggerated display gait, on top of one of the lory aviaries. I take this comical behavior for granted because I see it so often, but it again occurred to me why lories are so beloved by the general public. Few other birds evoke such laughter and enjoyment: lories are so different and often hilariously amusing.

Subspecies of *Trichoglossus haematodus*

There are 22 subspecies, although some of these are doubtfully distinct. The nominate subspecies, *T. h. haematodus*, has already been described. In order to assist the reader in referring to them, the remaining subspecies are described below in alphabetical order, as follows: *aberrans* (see *flavicans*); *berauensis; brooki; caeruleiceps; capistratus; deplanchii; djampeanus; flavicans; flavotectus; forsteni; fortis; intermedius; massena; micropteryx; mitchellii;*

moluccanus; nesophilus; nigrogularis; rosenbergii; rubritorquis; septentrionalis; stresemanni.

N.B.: Immature plumage is not described for individual subspecies. In each instance, it resembles the adult's except it is slightly dull and less glossy. Shaft-streaking on the head is inconspicuous until after first molt. The beak is dark brown, becoming orange by about four months; the iris is grayish at first, and almost as bright as an adult's around five months of age.

Western Green-naped Lorikeet *T. h. berauensis*

Description: It differs from the nominate race in having the green of the back, wings, tail and abdomen suffused with black, to give a less brightly colored bird. This effect is most intense on the abdomen and influences even the bars on the breast feathers. The chin, throat and nape are almost completely black and the blue of the head is restricted. The collar is pure yellow or slightly greenish; Cain contrasts this to "warm yellow" in the nominate race.

It differs from *intermedius* in being dingier, with heavier barring on the breast, a less greenish collar, less green on occiput, much more blue on the cheeks and only traces of brown on the nape.

There is local variation. Birds from Waigeu have more blue on the head. Those from Setekwa and Upper Fly rivers are small and have the blackest abdomens. Those from Misol and Arfak show the most general darkening of the plumage (Cain 1955).

Key identification features: Overall darker, with heavy breast barring.

Range: Western Papuan Islands, including Misol and Waigeu; the Vogelkop, islands in Geelvink Bay (Ron, Japen), Lobo Bay, the Setekwa River and Upper Fly River.

Aviculture: This subspecies is not recognized by most aviculturists or by some systematists. The only mention I can find of this race in the avicultural literature is that by Paolo Bertagnolio, whose

pair reared two in a single nest (Bertagnolio 1974). Eggs were laid on May 24 and 26, and hatched on June 19 and 20 (26 days for the first egg and 25 for the second). On July 18, traces of green could be seen on the abdomen, back and wings, and the white down was replaced by short, thick, very dark gray down. By August 20, the plumage differed from that of the adults only in the "yellow dotting" on the lower cheeks. The young ones left the nest at the ages of 67 and 71 days. By September 14, their beaks were reddish brown.

Brook's Lorikeet *T. h. brooki* (Ogilvie-Grant 1907)

Description: It differs from *nigrogularis* in the black patch on the abdomen being more extensive, with little or no trace of green.

Key identification features: None known. It is not clear how it is supposed to differ from *caeruleiceps*.

Range: It is reputedly known to museum work-

ers only by two skins, cage birds said to have come from Pulo Swangi (Spirit Island), off the coast of Trangan Island. one of the Aru Islands.

Aviculture: A few specimens of *caeruleiceps* imported into Singapore, Europe, and South Africa from 1989 until 1992 were incorrectly identified as *brooki*.

Pale-headed Lorikeet *T. h. caeruleiceps* (Salvadori 1879)

Synonym: Blue-headed lorikeet, Merauke lorikeet

Description: Diamond (1972) describes it as "very distinct in its much darker green belly, pronounced orange wash on the abdomen, reduced green edges to the red of the breast, and more extensive blue on the forehead and sides of the head." Cain describes it as having the crown and sides of the head pale blue, the breast paler than in *micropteryx* and with narrower barring, and the abdomen blackish green. According to Hubers and van Dooren (1994), it measures about 27 cm (nearly 11 in.). The collar is almost pure yellow, rarely marked with red; the nape feathers are green with red margins. Head color varies from pale blue (with blue-green feathers on the middle of the head in

some birds) to dark blue. The breast color varies from orange-yellow to orange-red with narrow dark green/blue margins. The abdomen varies from grass green to very dark green to almost black. The underwing coverts are orange-red and there is a broad band of yellow on the primaries.

Key identification features: Blue head, and very narrow breast barring; abdomen usually dark green. The blue of the head has been described by van Oosten (pers. comm. 1992) as "spectacular." He adds that this shade of blue rivals that of Swainson's and red-collared lorikeets.

Sexual dimorphism: Hubers states that most females have the head and the abdomen lighter colored. Most males have the hind part of the crown blue-black, and darker markings on breast and abdomen. In some males, the dark margins to the breast feathers are almost nonexistent. Young females have a blue head, light breast and greenish abdomen, whereas young males have darker head and abdomen and a redder breast. Hubers does not find it necessary to have the young surgically sexed, except those from one pair in which the female looks like a male. Sweeney (1996) described the abdomen as normally green in the female and much darker green—almost black—in the male.

Range: Coates (1985) gives its range as the southern Trans-Fly region and lower Fly River. Forshaw (1989) states that it originates from southern New Guinea between the lower Fly River and Princess Marianne Straits.

Aviculture: Hubers (Hubers and van Dooren 1994) states that this subspecies first came to Europe about 1990. Van Dooren obtained two pairs in 1992. In May of that year, the four birds were placed in an aviary which measured 2 m x 1.5 m x 2 m (6 ft. 6 in. x 5 ft. x 6 ft. 6 in.) high with an inside flight of 1.5 m x 1.4 m x 2 m (5 ft. x 4 ft. 7 in. x 6 ft. 6 in.) high. One pair hatched two chicks in July but they died at 14 days old due to disturbance. They had been ringed with 6 mm rings at the age of ten days. The pair nested again and produced two more young. These were removed for hand-rearing when

the parents started to pluck them. After the parents were moved to a new flight for the winter, a third clutch was laid. The chicks were raised by a pair of Swainson's. In the fourth clutch, three young were reared by the parents. There were also three young in the fifth and sixth clutches because the eggs were removed as laid, causing the female to lay an extra egg. The eggs were returned to the nest after the third egg was laid. In this way, all the chicks hatch at the same time. Young left the nest at the age of eight weeks. They were left with their parents until the young of the next clutch hatched.

The second pair of *caeruleiceps* was housed in an aviary which also contained ornate and Obi lorikeets—without nestboxes. The female *caeruleiceps* dug a hole in the ground 5 cm (2 in.) deep and laid two eggs there; they did not hatch.

In 1991, birds described as *brooki*, but almost certainly belonging to this subspecies, were bred in South Africa. In the U.S.A., *caeruleiceps* has been bred by J. van Oosten (six reared by 1994) and by at least 1 other breeder, who had reared 12 by the end of 1994 (van Oosten pers. comm. 1995). This subspecies has been exhibited at Loro Parque since 1994, when three captive-bred pairs were obtained. Young have been reared there. The incubation period was recorded as 25 days and newly hatched chicks weighed about 6 g (Sweeney 1996).

Edwards' Lorikeet *T. h. capistratus* (Bechstein 1811)

Description: Head green, forehead, cheeks and chin are dark blue with shaft streaks of violet-blue; throat and ear coverts are dark green, also a wide stripe above the eye which is shaft-streaked with lighter green. The nuchal band is greenish yellow or yellow. Upper breast feathers are yellow, narrowly edged with orange; the abdomen is dark green, and the sides and the thighs are dark green and yellow.

Length: About 27 cm (11 in.)

Key identification features: Forecrown and cheeks green streaked with violet-blue; yellow breast feathers (some with orange margins); yellow underwing coverts, sometimes marked with orange.

Sexual dimorphism: In some cases the yellow feathers of the breast are narrowly margined with orange in the male and with green in the female.

Range: Timor, Indonesia.

Habits: M. Bruce found it to be fairly common in lowlands below 500 m (1,640 ft.), mainly near the coast. Pairs and small flocks were seen and heard feeding in eucalyptus woodland and in gardens and villages (Forshaw 1989).

Aviculture: Edwards' lorikeet was regularly available from about 1965 until the late 1970s, when the export of birds from Timor ceased. Within a few years it was rare, indicating that despite being bred on many occasions, there were no serious efforts to establish it in aviculture. This situation changed and a small minority made an effort to breed it. In the U.K., the breeding register of the Parrot Society (to which a minority of members contribute) showed in 1977, three members reported breeding a total of five birds. In 1991, five members reported breeding 25 and the following year four members bred 21. This subspecies is also being reared in Europe and in the U.S.A. R. T. Kyme of Lincs, U.K., received a pair in 1972. Eggs were laid on June 29 and July 1, and the first hatched on July 21 (23-day incubation period). The young bird left the nest at 74 days old, with nearly bare neck, breast and back, where it had been plucked by the parents. Its beak was black; it otherwise resembled the adults. Their diet consisted of nectar, spinach beet, apple, sowthistle and white sunflower seed.

Deplanchi's Lorikeet *T. h. deplanchii* (Verreaux and Des Murs 1860)

Local name: *Perruche*

Description: It differs from *massena* in having more blue on the head and less brown on the occiput and nape. There is less yellow on the thighs and on the undertail coverts. Cain (1955) refers to "the faint powdering of light blue over the head" and comments that all the forms nearest Australia show a slight increase in the blueness of the head. (A photograph of this subspecies, taken on New Caledonia, appeared in *Gefiederte Welt* 117 (9): 311, September 1993).

Key identification features: More extensive blue and less brown on head than *massena*. Briefly, it can be described as a large *massena*.

Mutation in wild: A bird described as "part lutino" was caught near Tao and kept by a plantation owner. A photograph in *Gefiederte Welt*, September, 1993, depicts it as having the head white instead of blue; yet the throat is orange, the wings are green and yellow, and some of the breast feathers are margined with yellow. The flanks are green and the abdomen and the underside of the tail yellow and green. Eye color is not discernible.

Range: New Caledonia and the Loyalty Islands, in the western Pacific (French Community).

Status: Common on New Caledonia where it is the most numerous parrot species; probably occurs only irregularly on the Loyalty Islands (Bregulla 1993).

Habits: In New Caledonia it is found in forest, mangroves, savannah, plantations and gardens. It has a preference for coconut and banana plantations, and groves of oranges, mandarins and papaya. It visits higher regions when certain trees and lianas are flowering. During April and May, and December and January, groups frequent the flowering *Niaouli* trees in the savannah. In flight their silhouette is unmistakable—a long, thin, wedge-shaped body with a long, pointed tail and narrow wings. They fly very fast and their wings make a hissing sound. When going any distance they fly high, otherwise they fly just above the trees. If there are no trees, they fly just above the ground. When the sun shines on them, the orange-red of their plumage has an extraordinary sheen. They are normally very noisy, especially when searching for food. Near the nest, or in gardens when they are aware of being observed, they are quieter (Bregulla 1993).

According to Forshaw (1989), they apparently fly from New Caledonia to Lifu and Ouvea in the Loyalty Islands—therefore crossing at least 100 km (62 mi.) of ocean. Flocks have been seen out at sea.

Bregulla states that their food consists mainly of nectar from trees, palms and lianas. They also take pollen and small insects. Juicy fruits are eaten, papaya being preferred; the seeds are not eaten. Crops such as maize, oranges and figs, which are allowed to ripen, suffer damage by the lorikeets. Blossoms or fruits of the following trees are preferred: *Erythrina* species, *Spathodea campanulata*, *Casuarina* species, *Metrosideros quinqueneria* and *Mutingia callabura*. The breeding season is believed to extend from June or July to December or January.

Aviculture: This subspecies is rare in aviculture as New Caledonia does not normally export its fauna. Dick Schroeder keeps this subspecies in California, but it has not proved easy to breed. The young have to be hand-reared, as the chicks are killed by the parents at about two weeks of age. Three birds were seen in Mr. Schroeder's collection in 1996. Bregulla described three pairs in a large aviary on New Caledonia, at Parc Forestier, on exhibit to the public. One pair nested after only four months. Two eggs were laid at the end of June. The young left the nest after 50 to 55 days. The pairs tolerated each other but only one pair nested successfully. In France, Françoise Schrenk had bred this subspecies to the second or third generation by 1996.

Djampea Lorikeet *T. h. djampeanus* (Hartert 1898)

Description: According to Cain (1955) there is some blue on the forehead but most of the head is black with a magnificent deep purple gloss. The abdomen is the same and there is a strong dark purple color in the post-torqual region (behind the collar) and the interscapular region. This is one of the most handsome of the *haematodus* subspecies.

Length: 25 cm (10 in.); wing length is 140–146 mm, compared with 132–139 mm in *forsteni* and 129–136 mm in *mitchellii*.

Key identification features: Scarlet, unbarred breast, blue forehead and glossy black head. In each area of the body the coloration is uniform and intense, resembling *Lorius* species in this respect. White and Bruce (1986) state that the head is darker and more strongly streaked with violet-blue, and the purple band behind the yellowish nuchal collar is always present; thus it is very close to *forsteni*.

Range: Tanahjampea (Djampea) Island, Flores Sea; Lesser Sunda region of Indonesia.

Aviculture: This strikingly beautiful bird was imported into Europe in the early 1990s. Two were reared at Vogelpark Walsrode in 1990, and a pair was exhibited at Loro Parque, Tenerife, in 1994.

Olive-green Lorikeet *T. h. flavicans* (Cabanis and Reichenow 1876)

Description: Its plumage is variable, more so than in any other subspecies. Cain (1955) commented that some specimens closely resemble *nesophilus* but that in others all the green areas are strikingly yellowed, so that they appear a dull bronzed gold. Every shade between the usual midgreen of *haematodus* and olive green are found. Newman (pers. comm. 1994) observed wild flocks and noted that individuals were "very variable in the degree of yellow suffusion of the plumage, especially the underparts. Several very yellow birds were seen; the most extreme had completely yellow underparts and yellow-green uppers. This bird was with three 'normal' individuals."

Cain described *flavicans* as being much like *micropteryx* in head and collar coloration, but "with the addition of a very narrow red-brown pre-torqual band, and a strong purple gloss on the black parts. There is no blackening of the green parts, and very little or no breast-barring. The red of the breast is very brilliant and pure."

Key identification features: *T. h. flavicans* that are not bronze or yellow can be distinguished from *nesophilus* in the following respects: (1) a strong tendency to show red splashes on the edges of the collar (also seen in *rosenbergii*); (2) the abdomen is blackish green; (3) darkening of the feathers just below the mid-dorsal part of the collar (forming a post-torqual patch much as in *forsteni*); *nesophilus* has only a slight indication of this patch, a much reduced tendency to red tinges on the collar, and the abdomen slightly less blackened. Cain believed the differences to be so slight that they hardly warranted subspecific separation.

N.B.: *T. h. aberrans*, named by Reichenow in 1918, should be treated as synonymous with *flavicans*, according to Cain, who believed that the type was a small individual of the latter.

Range: New Hanover and the Admiralty Islands, Bismarck Archipelago, northeast of New Guinea.

Habits: In the Admiralty Islands, olive-green lorikeets nest and roost on the ground on some islets because there are few trees large enough to provide cavities (see **Nest Sites**). They fly back and forth to Manus Island, 0.5 km away. LeCroy, Peckover and Kisokau (1992) described numbers there as high in the morning, judging by the noise. They commented: "It would be interesting to know whether nests and roosts are confined to trees on the large islands where cats and pigs are present, whether the nesting is 'colonial' there, or whether ground nesting and roosting occur only on offshore islets where predators are absent and is 'colonial' only because of the limited number of suitable islets."

Aviculture: Chester Zoo, U.K., recorded breeding this subspecies, under the name of *aberrans*, in 1969 and 1970.

Wetar Lorikeet *T. h. flavotectus* (Hellmayr 1914)

Description: It differs from *capistratus* only in lacking orange on the breast and underwing coverts—not all *capistratus* have orange on underwing coverts. The breast and underwings are clear yellow.

Range: The islands of Wetar and Romang, near Timor, Lesser Sundas, Indonesia.

Aviculture: A few birds have been imported into Europe. Perhaps the first time was in 1958 when Wilfrid Frost collected a few specimens, at least one of which went to Wassenaar Zoo in the Netherlands. It has bred at Vogelpark Walsrode, Germany, in 1979 and on other occasions.

Forsten's Lorikeet *T. h. forsteni* (Bonaparte 1850)

Description (and key identification features): The head is very dark brown, except the forehead, which is dark blue; forehead and cheeks are shaft-streaked with violet-blue; the breast is scarlet without any barring and, according to White and Bruce (1986), it is darker than that of *mitchellii*, although the difference is slight. For further details see under *T. h. djampeanus*. According to Tiskens (1995), in comparison with Mitchell's, Forsten's have overall a much more blue-black head and generally have a blue nape. The breast is an intense red and the thighs are more yellow than those of Mitchell's.

Length: 23 cm (9 in.); fractionally larger than *mitchellii*.

Weight: About 95 g.

Range: Sumbawa, Lesser Sundas, Indonesia (14,800 km^2, or 5,700 sq. mi., in area).

Status: Vulnerable and possibly declining, but little current information available.

Habits: Rensch (1931) found it common, and moving extensively to coincide with flowering trees. He believed that it occurred in any place only for a few weeks when trees were in flower, and that this explained its varied altitudinal range. In 1993, Sumbawa was visited by ornithologists from the Cambridge University Flores Conservation Project. Encounter rates (birds encountered per 100 man hours) on southwestern Sumbawa were as follows: in semievergreen forest 200 m to 350 m (650 ft. to

1,150 ft.), 23; moist deciduous monsoon forest, dry thorn scrub 20 m to 100 m (65 ft. to 330 ft.), 8.1.

Aviculture: This lorikeet was well known for many years, but not long after importation ceased, it declined quickly and is now rare. It was described as the most common lory in the Calcutta market in 1901 by Frank Finn, was exhibited at London Zoo as long ago as 1896, and was bred in Britain in 1905 (by Mrs. Mitchell). It has been bred regularly in many countries worldwide since the 1960s; in Japan, for example, the first breeding occurred in 1968, by Mr. Akira Sano (H. Shimura pers. comm. 1994). It can be equally as prolific as the nominate race. In 1987, when I moved to Tenerife to become curator at Loro Parque, I took with me a female Forsten's to pair with the single male there. She nested only three weeks after being placed with the male, after four weeks in quarantine. By May 3, the pair had two chicks. These and the many subsequent chicks were removed for hand-rearing because they were plucked. In the next 18 months two chicks hatched on July 15, October 5, mid-November, February 3, 1988, March 27, June 6 (only 1), August 12 and November 12. One newly hatched chick weighed 5.6 g and was sparsely covered with longish white down. Two chicks started to feed themselves at 48 days.

Sumba Lorikeet   *T. h. fortis* (Hartert 1898)

Description: Forehead and cheeks are dark bluish, streaked with violet-blue; lores and throat are green, sometimes tinged with black, and the green of the nape extends as a line over the eye. The abdomen is blackish green. In Hartert's original description in *Novitates Zoologicae* (vol. 5), he stated that there was a great deal of variation in the color of breast and underwing coverts. These areas were strongly washed with deep orange in some specimens, chiefly mature males, and sulphur yellow without an orange wash in others.

Key identification features: Green line over eye and dark face; unbarred yellow breast (or tinged with orange); slightly larger size than *capistratus*, wing of males averaging 153.3 mm as against 146.1 mm in *capistratus*.

Range: Sumba Island, Indonesia, due south of Sumbawa and Flores.

Habits: It is abundant near primary and old secondary forest; it is especially common where flowering *Erythrina* trees exist (Riffel and Dwi Bekti 1991). Until Sumba was visited in 1989 by the Manchester Indonesian Islands Expedition and in 1992 by the Manchester Metropolitan Expedition, very little had been recorded about the Sumba lorikeet. Duration of these visits was from October 10–25 in 1989 and from July 18 to September 30 in 1992. Studies were made at six forest sites. The findings are summarized here (Marsden 1995; Marsden and Jones in prep.). Encounter rates (number of individuals divided by the number of hours) in primary forest were 2.8 per hour (185 records); in secondary forest 2.6 per hour (114 records) and in parkland 1.9 per hour (57 records). The lowest altitude at which they were encountered was 20 m (65 ft.) and the highest 1,060 m (3,480 ft.). Density estimates (number of individuals per km^2) was 29 +- 5.2 in primary forest (sample size 50) and 33 +- 11 in secondary forest (sample size 24). McKnight et al. (in prep.) calculated the total area of closed canopy forest on Sumba to be 1,591 km^2 (614 sq. mi.); this total was made up of 26 forest blocks. Of the total forest cover, 98 percent was in 18 blocks, each of which was over 5 km^2 (2 sq. mi.). Marsden's total population estimate for forest sites was 44,178 (minimum of 30,539 birds, maximum of 57,817). The mean group size was 2.5 and the most seen in one group was 14 birds. Marsden concluded that its preferred habitat was fairly recently disturbed higher altitude forests with many fruiting trees.

Nesting: Of the 132 parrot nests located in Sumba by the two expeditions, 21 nests were those of *fortis,* and nine were occupied at the time. Three of these nests were located "in the rooty undersides of 3 large arboreal epiphytes at Porunumbu"; the remainder were in tree cavities. Of the latter 18, four were in the main trunks of trees, four were above the first branch, two were in the first major branch, four were in the second or successive branches and four were in a side branch. The average length of the nest hole was 15.8 cm (6 in.) but sizes ranged from 11 cm to 32 cm (about 4 in. to 12 in.). The average width of the nest hole was 12.5 cm, but sizes ranged from 9 cm to 17 cm (about 3½ to 7 in.). The average height was 19 m (62 ft.), with a range of from 9 m to 29 m (30 ft. to 95 ft.). Seven nests, i.e., 24 percent, were in trees which contained another nest of the same species. No nests were in dead trees; 90 percent were in deciduous trees and 10 percent were in evergreen trees. The average size of the tree girth was 2.4 m (8 ft.); the range was from 1.1 m to 4 m (3 ft. 7 in. to 13 ft.). The average height of the nesting tree was 36 m (118 ft.), and the range was from 24 m to 46 m (80 ft. to 150 ft.)(Marsden and Jones in prep.).

Status/Conservation: Still fairly common but vulnerable. Jepson (1993b) stated that by 1993 less

than 11 percent of the island was forested, as against 50 percent in 1927. Much of the 400,000 hectares of lost forest had become unproductive grasslands. BirdLife International's Indonesia Program is collaborating with Indonesian Directorate General of Forest Protection (PHPA) on a project to develop and implement a forest conservation plan. It recognized that this must include the needs of the community. Fire had been the principal cause of deforestation. All Sumba's birds are now protected by decree and parrot smuggling has been reduced. Public awareness of conservation issues is promoted by talks in schools and villages.

Trade: "Regularly caught in considerable numbers" according to Riffel and Dwi Bekti (1991), who stated that the lorikeet (unlike the citron-crested cockatoo) had "not yet shown a comparable dramatic decline which could be attributed to over-trapping." Local catchers were paid a price which rarely exceeded the equivalent of US $2 per bird. The lorikeets were trapped with the use of nylon or horse hair nooses placed in the tops of *Erythrina* trees.

Blue-faced Lorikeet *T. h. intermedius* (Rothschild and Hartert 1901)

Description: It can briefly be described as resembling the nominate race except that the blue of the forehead does not extend to the crown; the nape is brownish, shaft-streaked with olive; and the abdomen is darker green. Cain (1955) states that it varies considerably in size and color "but in nearly all respects it is intermediate between *berauensis* on the west and *micropteryx* on the east. In comparison with *berauensis*, it is less blackened throughout and with the breast bars narrower. The collar is more greenish yellow, the occiput shows a tinge of green, and there is a brownish hue on the nape."

Length: 25 cm (10 in.)

Weight: About 115–168 g recorded by Diamond and Lecroy (1979); birds from the northern coastal range about 145–162 g; males from the Eastern Highlands about 125–141 g, and females from same area about 115–125 g; one male each from Sepik and Madang, 163 g and 168 g, respectively.

Key identification features: Blue   forehead (not crown) and brownish nape.

Range: Northern New Guinea from the Sepik River east to Astrolabe Bay (Forshaw 1989). Cain (1955) gives its range as Astrolabe Bay, lower Sepik Valley and the Maeanderberg (upper Sepik). The exact boundary was uncertain but was probably near Torricelli Mountains. Coates (1985) describes its distribution as northern and east-central New Guinea from the Torricelli Mountains and Sepik River to Astrolabe Bay in the north and Wahgi Valley and Karimui—upper Purari River in the south.

Habits: Diamond (1972) encountered it up to 1,400 m to 1,500 m (4,600 ft. to 4,900 ft.) in primary forest and up to 2,000 m (6,500 ft.) in the casuarina groves and trees of villages and open country. It usually occurred in small, noisy groups of up to half a dozen individuals perched in the middle story and canopy of trees or calling loudly in flight. Unlike the dusky lory, to which it is otherwise similar in habits, it does not leave the Eastern Highlands during the dry season.

Aviculture: It is unlikely to be distinguished; nothing recorded.

Massena's Lorikeet   *T. h. massena* (Bonaparte 1854)

Synonym: Coconut lorikeet

Native name: Bislama name on Vanuatu, *nasiviru*; on Karkar, *siril*; on Bagabag, *sir*.

Description: Forehead, forecrown, lores and chin bluish mauve; remainder of head dark brownish purple, collar yellow-green; throat purplish black. The breast feathers are red, narrowly edged with black; abdomen green; lower abdomen, thighs and undertail coverts green strongly marked with yellow. Tail green above, olive yellow below. Some individuals are variably marked with deep violet-blue on flanks and lower abdomen (Bregulla 1992).

Key identification features: Red breast feathers narrowly margined with black; shaft-streaking on head changes from blue on crown to light brown on occiput; abdomen green.

Weight: 85–120 g; males from Karkar about 95–105 g, females from Karkar 85–105 g; males and females from western New Britain 110–120 g, males from New Britain about 97–105 g (Diamond and Lecroy 1979). A size cline has been noted. Cain (1955) recorded wing lengths in millimeters as follows:

Massena's Wing Lengths

	male	female
Bougainville and northward	136.8	135.4
Islands south of Bougainville	139.3	134.6
T. h. deplanchii from New Caledonia	146.0	141.2

Diamond and Lecroy (1979) state that the birds they collected on Karkar Island (off the northern coast of New Guinea) definitely belonged to *massena*, on the basis of their smaller size, less heavy bill and paler red breast, compared to *intermedius* and *micropteryx*. They were smaller and paler than the birds from nearby Bagabag Island. These authors also made the interesting observation that in

the American Museum of Natural History there are two specimens of *massena* from Witu Island which are bronze; two similarly colored birds are also known from Bagabag, thus "bronzing is not restricted to *flavicans* and may in fact occur more frequently than collections indicate in populations from small islands."

Range: Bismarck Archipelago, New Guinea, Solomon Islands and throughout Vanuata (formerly called New Hebrides), except islands of Lopevi and Futuna (Bregulla 1992). (Republic of Vanuatu is situated southeast of the Solomon Islands and northeast of New Caledonia). This is one of the most extensive ranges of any *haematodus* subspecies.

Habits: There are several published observations from the Solomon Islands. Donaghho (1950) saw a flock of about 100 birds feeding on blossoms of scarlet bottle-brush (Myrtaceae) on Guadalcanal. In contrast, Jan van Oosten visited Honiara, capital of Guadalcanal, in April, 1992. He recorded that "it was surprising to see that with all the flowering trees throughout the city and environs that the Massena's were not visiting these trees, but only the *Casuarina*. It is further interesting to note that during the seven days [thirty hours of observation] the Massena's were never seen to visit any of the flowering trees, at least at this location. I obtained several seed pods from the *Casuarina oligodon* trees to check the seeds inside. The outer skin is not thick enough to prevent the birds with their sharp hook beaks from penetrating it. I cut the pods in half to find the seeds in a 'milky' stage. The seeds were very soft and moist. The Massena's that lived in and around the city while I was there, ate nothing but these seeds" (van Oosten 1993). Kenning (1993–4) recorded of the same city: "Walking around Honiara you can see many Massena's (Coconut) Lories which are very common in this area. They often sit in large casuarina trees and feed on the flowers."

On Karkar, Massena's lorikeets were common in flowering trees from sea level up to 1,400 m (4,600 ft.) in May and June, 1969, (Diamond and LeCroy 1979). At 1,000–1,200 m (3,300–4,000 ft.) on the outer rim of Karkar crater, they were noisy and abundant. Trees in this area were full of displaying pairs opening and rapidly fluttering their wings at each other to show the colored bands under the wings. In Vanuatu, they may be encountered in most types of wooded habitat, primarily in the lowland, although they are not uncommon in the mountains when favorite shrubs or trees are flowering. They are more often seen in disturbed habitat than in the true forest or mangroves; savannah-woodland, trees bordering watercourses or surrounding paddocks, suburban gardens and in nearly every coconut plantation. They are predominantly nomadic, their presence being governed by the availability of nectar and pollen from flowers of native and introduced trees. They particularly feed on coconut palms, also Indian coral-tree (*Erythrina variegata*), African tulip-tree (*Spathodea campanulata*) and sago palm (*Metroxylon rumphii*). They are important pollinators of coconut palms. Soft juicy fruits, also berries, are eaten including those of the introduced Panama cherry tree (*Mutingia callabura*) and cultivated fruits such as oranges, mangoes and papayas. They cause some damage to ripening fruits in orchards and may also raid crops of maize and sorghum to feed on unripe milky grains. Half-ripe seeds of *Casuarina* trees are occasionally eaten (Bregulla 1992).

Armin Brockner visited New Ireland in 1994 and observed that this lorikeet was common there. He saw it in primary and secondary forest, mangroves, coconut plantations and around small settlements. During his stay in Kavieng, the capital, a pair flew by his hotel every morning between 6:00 A.M. and 7:00 A.M. and every evening between 5:00 P.M. and 6:00 P.M.; the hotel was situated on the coast. In the mornings the lorikeets came from the coconut plantations and flew to the mangroves. On one occasion he saw a group of up to ten birds. They were also observed with purple-bellied lories in thick foliage (Brockner pers. comm. 1994).

Nesting: In Vanuatu the breeding season is mainly between August and January. The nest is a hollow limb or hole in a tree. Eggs measure 28 x 23 mm (Bregulla 1992).

Aviculture: Imported during the 1960s and 1970s, and inexpensive, it was mainly available before there was much interest in lory breeding. During that period, it was common to see "coconut lorikeets" advertised—but they were not necessarily of this subspecies as literature assisting identification was not widely available. In 1966, W. H. Brown of Kent was the first to record breeding Massena's. By March 19, the female was incubating two eggs which were abandoned when the weather became cold. By May 21, the female had laid again and one egg hatched, probably on June 10. The pair was offered a wide variety of foods but took only their usual nectar mixture (one teaspoonful each of honey, Horlick's and condensed milk in 6 oz. of water) and a little canary seed. The young one left the nest at the age of 60 days on August 9. By September 27, the female was incubating two more eggs (Brown 1966).

Southern Green-naped Lorikeet *T. h. micropteryx* (Stresemann 1922)

Description: Cain (1955) describes it as "a small *intermedius* with the green areas even more free from black, the breast barring even narrower, the collar and the occiput slightly more greenish and the red of the breast slightly paler." Diamond and Lecroy (1979) mention two specimens from 330 m (1,000 ft.) on Bagabag that show "the bronzing of the green parts which is so prevalent in *T. h. flavi-* cans." They compared these birds with the latter and found that they had the breast more deeply marked with black and very slightly more salmon colored, and lacked the narrow pre-torqual band of reddish brown, as do *intermedius* and *micropteryx*.

Length: About 25 cm (10 in.)

Weight: About 110–142 g. Males weighed from

the Huon Peninsula were about 118–121 g; males from Bagabag, about 110–142 g and females from Bagabag, about 115–125 g (Diamond and Lecroy).

Key identification features: Narrowly barred orange-red breast and greenish collar. Blackish back of head distinguishes it from *massena*.

Range: New Guinea east of the Huon Peninsula, the Waghi Range and Hall Sound; also Misima Island in the Louisiade Archipelago (Forshaw 1989). Coates (1985) gives its range as "eastern New Guinea west to the Huon Peninsula in the north and Lake Kutubu in the south; also Manam, Bagabag and Misima Islands." But some systematists assign the Bagabag birds to the subspecies *massena*. However, Diamond and Lecroy commented that "Bagabag birds conform most closely in their measurements with birds from the Huon Peninsula (*micropteryx*). Birds from the Madang area and the Sepik river seem large, despite the small sample size; particularly are they heavy. Surprisingly, birds from Manam Island agree more closely with *micropteryx* in measurements (except for tail measurements of males), although Manam is closer to Madang and the Sepik river than to the Huon Peninsula." When measurements of birds from the East-

ern Highlands, from another source, but identified as *intermedius* were included, there was a broad overlap of *micropteryx* and *intermedius*. Diamond and Lecroy suggested that these two subspecies may not be separable but that, until more material was available from the area between the Sepik River and Astrolabe Bay, it was preferable to consider the populations from Bagabag and Manam as *micropteryx*, with *intermedius* confined to the mainland. Weight may be a useful character in determining to which subspecies they belong.

Habits: Coates (1985) describes the roosting habits of this subspecies in the Port Moresby suburb of Gordon Estate. Despite the presence of other large trees, it roosts only in tall eucalypts. "If disturbed from their roost during the night they settle in other eucalypts in the vicinity, sometimes ones quite low and obviously unsuited to their needs, rather than resort to other types of trees. However, when resting during the day and in pre-roosting assembly places, they commonly frequent the big, heavily-foliaged rain trees."

Aviculture: This subspecies is unlikely to be identified by aviculturists. It was bred at Stuttgart Zoo, Germany, in the 1970s.

Mitchell's Lorikeet *T. h. mitchellii* (G. R. Gray 1859)

Native name: *Kasturi*

Description: Head and nape rich dark brown, with dark olive green or green-brown feathers on the forehead; grayish shaft-streaking on head and cheeks; cheeks and throat are also dark brown. The breast is scarlet and only a few feathers have very narrow margins of bluish, green or yellow. The abdomen is dense purplish black.

Length: 23 cm (9 in.); smaller and slimmer than Forsten's.

Weight: 100 g

Key identification features: Dark brown nape, cheeks and throat, greenish yellow collar, almost unbarred red breast, very dark abdomen and small size.

Paul Tiskens, a breeder of this subspecies, gave

the following description: "The face and the small area on top of the head are dark to black brown, marked with gray green streaks. The forehead is marked with numerous gray feathers. Many birds have a few blue feathers below the collar; in others it is sometimes more like a band" (Tiskens 1995).

Sexual dimorphism: The male's breast is usually a purer red, the female's breast feathers being margined with green or yellow, or with some feathers entirely yellow. There are exceptions, however. Tiskens (pers. comm. 1994) stated that one of his females has the breast entirely red.

Immature birds have blackish edges to some of the breast feathers. Buckell (1992) states that, in his experience, "some immatures appear with a green abdomen up until the first molt." Tiskens notes that "as a rule, young birds are more washed out in color; the head, breast and stomach plumage are marked with greenish feathers."

Range: Bali and Lombok in Lesser Sunda Islands, Indonesia. (See Gazetteer.)

Habits: On Lombok it has been recorded up to 900 m (2,950 ft.). On Bali, where it is now believed to be extinct, it was known from Mount Bratan (1,333 m), the peak of Singaraya and from Lake Bratan, in montane rainforest, forest clearings and *Erythrina* stands in coffee plantations. MacKinnon (1990) described it as "a flocking bird, flying over the forest in noisy screeching parties." He believed that "escapees are also occasionally seen in Bali and Java and may sometimes breed." On Bali, forest has survived only in the west and surrounding the mountains in the center and the east.

Nesting: Wild nests and eggs are undescribed (MacKinnon 1990). Two eggs laid in the collection of G. van Dooren weighed between 5.5 g and 6.0 g and measured 26.2 x 21.7 mm and 26.2 x 22.2 mm (van Dooren pers. comm 1995).

Status/Conservation: It survives on at least one site on Lombok (Lambert, Wirth et al. undated). Victor Mason, coauthor of *Birds of Bali*, informed me in 1994: "I have not seen *T. h. mitchellii* in the wild in Bali, where I fear it is extinct. Jarvis and I tried, without success (in 1982) to find it; however, we are convinced that birds we saw caged were of local provenance."

Mason and Jarvis (1989) wrote: "It was formerly present in some numbers and, being a blossom feeder, may have invaded Bali on the introduction of coffee cultivation and the intensive planting of coral-bean (*Erythrina*) shade-trees which flower profusely. The unpalatable fact is, however, that there has been no official sighting of this bird since World War II. As recently as 3 or 4 years ago, I used to see these Lorikeets, often in pairs, offered for sale in the Candikuning market-place at Bedugul. I was assured that they were of local provenance and were trapped high in the mountains which was doubtless true, for why would anyone bother to import birds to sell in Bedugul?... Now the cages are empty."

No conservation measures are known on Lombok.

Aviculture: This subspecies was imported into Europe in small numbers between the early years of the century and the 1970s when in the U.K., for example, it was bred in at least seven collections between 1968 and 1974. In *Lories and Lorikeets* (Low 1977a) I recorded that it was "fairly frequently imported." However, this was true only for this period. At that time there was no serious interest in lory breeding and knowledge of the status of parrots in the wild was nonexistent. It must by then already have been seriously threatened, either by overtrapping, deforestation or both. In the 1980s it was realized that its captive numbers were low and breeders in Holland, Germany and Belgium started a cooperative breeding program. In Issue 4, 1992, of *Lori Journaal Internationaal* (LJI), an appeal was made for information on existing numbers. It resulted in 45 birds being recorded (LJI, Issue 1, 1993) with 14 keepers in the following countries:

Belgium, 5 adults and 2 young; Denmark, 2 adults; Germany, 9 adults and 3 young; Great Britain, 4 adults; Netherlands, 12 adults and 6 young; Switzerland: 2 adults.

At the same time there were at least eight breeding pairs in South Africa (G. Zietsman pers. comm. 1994), where they were bred by Gavin Zietsman in 1994. In the same year only three were known to survive in the U.S.A., a pair and one unsexed bird.

In October, 1987, a pair of Mitchell's was presented to Loro Parque, Tenerife, by Jan van Oosten of the U.S.A. The female laid an egg in January, 1988, which was infertile. A clutch of two eggs was laid by March 22 and two chicks had hatched by April 13, one of which died when very young. No more eggs were laid until November; on the 18th there were two in the nest, the first of which hatched on December 11; date of hatching of the second chick was unknown. The young were plucked in the nest and were removed for hand-rearing on January 18. Two eggs in the clutch laid by a pair belonging to G. van Dooren (see Nesting) hatched after 24 and 23 days. The chicks were hand-reared from the age of 23 days. However, when they were almost ready to fly, they lost the flight and tail feathers. In the previous year, six young Mitchell's had died for the same reason. This problem was not experienced with any other lories in the collection. Mr. van

Dooren's opinion was that it is doubtful if Mitchell's will survive in captivity because other breeders have had similar problems to that which he described.

One of them was Paul Tiskens of Germany. In 1993, three pairs produced 13 clutches which resulted in 11 young birds, only three of which survived. This is no reflection on Mr. Tisken's ability as a breeder; I believe that it demonstrates what I discovered for myself: species endemic to small islands, which may have been in-bred for centuries, are difficult to breed in captivity (see **Small Island Endemics**). During the middle of 1993, young birds lost their flight and tail feathers soon after leaving the nest. With subsequent broods of young, this problem became worse, resulting in totally naked birds or abnormally formed feathers. The problem was diagnosed as psittacine beak and feather disease—but Mr. Tiskens was doubtful that this was the case. It was necessary to cull the affected young, as well as two adults which had never been in good feather. In 1994 a new male was obtained. The two eggs laid by the female in December were infertile. Both pairs were then given three drops of Baytril in the food daily for two weeks. In January, 1995, both females laid. The eggs of the pair with the new male were again infertile, but the other female hatched two eggs after exactly 20 days (Tiskens 1995; Tiskens pers. comm. 1995).

Chick growth

Shows weights of two parent-reared young hatched on December 5 and 6, 1994, in the collection of G. van Dooren in the Netherlands. They were removed for hand-rearing at 23 days because the adults started to pluck them. The parents were fed on Aves Lorifood until the chicks were aged eight and nine days when the food was changed to Aves Loristart. The chicks were ringed on the 12th day with 6 mm rings.

Mitchell's lorikeet weight table #1

Age in days	Weight in grams	
	Chick No. 1	No. 2
hatched	5	5
1	–	6
2	7	–
3	–	8
4	10	–
5	–	11
6	13	–
7	–	13
8	16	–
9	–	21
10	23	–
11	–	23
12	29	–
13	–	29
14	36	–
15	–	36
16	45	–
17	–	40
18	49	–
19	–	54
20	57	–
21	–	60
22	75	–

Shows weights of two chicks hatched at Loro Parque in December, 1988, and removed for hand-rearing when the eldest was 38 days old and the youngest was assumed to be 36 days old. Weights shown are those before and after the first feed of the day. Initial weight loss is usual in chicks taken at an advanced age as they adapt less readily to a new environment.

Mitchell's lorikeet weight table #2

Age in days	Weight in grams	
	Chick No. 1	No. 2
36	–	57/63
38	80/85	57/67
40	79/88	63/73
42	82/92	68/78
44	85/97	74/86
46	86/96	77/91
48	90/100	79/88
50	85/94	82/91
55	89/99	–
60	89/99	86/97
65	82 s/f	81/92
70	81	82/85
80	75	84 s/f

sf = self-feeding

According to the *International Zoo Yearbook*, three more were reared at Loro Parque in 1990 and two in 1991.

In Issue 1, 1994, of *Lori Journaal Internationaal*, it was reported that the numbers of known birds had decreased to forty. The comment was made that "enthusiasts tend to be more interested in 'expensive' species threatened with extinction, such as the Purple-naped Lory, *Lorius domicellus* than in their 'cheap' colleagues. However, the group of Mitchell Lory keepers (which is far too small) is determined to help this species survive"!

Rainbow Lorikeet (Australia) *T. h. moluccanus* (Gmelin 1788)

Swainson's lorikeet (aviculture elsewhere)

Synonym: Blue Mountain lorikeet

Description: The head is rich blue, almost violet, shaft-streaked with lighter blue; abdomen is deep purple-blue. Nuchal collar is yellowish green. The breast is orange, usually without any barring but marked with yellow at the sides to a varying degree; underwing coverts are orange, tinged with yellow.

Length: 30 cm (12 in.)

Weight: About 130 g; Forshaw (1981) gives the weight range for males as 115 g to 157 g and for females as 107 to 143 g. Salisbury (1985) records the weights of a pair at Currumbin Sanctuary as 190 g and 146 g (male apparently overweight). The average weight of 191 wild birds visiting Currumbin Sanctuary was 148.6 g (range 108–183 g) (Cannon 1984); however, it is possible these weights are slightly above average due to the twice-daily feeding sessions.

Key identification features: Blue head, almost unbarred orange breast and yellow-green collar.

Mutations: More mutations have appeared in this subspecies than in any other lorikeet. A photograph of a **pied** appeared in the *Magazine of the Parrot Society* for April, 1984. The normally green areas were replaced by vivid yellow except on the thighs and some of the wing coverts. There was a brief comment by Dr. A. J. Wright that this mutation had appeared in collections in South Africa during the previous 10 years. He stated it was recessive, delicate and infertile! Not surprisingly, the mutation was not established. Sindel (1987) was informed that a **cinnamon** mutation had been developed in South Africa. There are now a few in Australia. The green areas are replaced by greenish yellow; the head is lilac and whitish lilac with some orange streaks. The underparts are entirely red. In Australia there is another form which has been called the **Dutch cinnamon**; all its colors (including those of the soft parts) are paler. Stan Sindel developed an **olive** mutation in the rainbow lorikeet by hybridizing it with an olive scaly-breast. Two interesting mutations in Australia are the **blue-breasted**, which lacks red and yellow in the plumage, and the **blue**, which is various shades of blue with white breast and white nuchal collar.

Range: Eastern Australia, from Cape York Peninsula southward along the coast to Eyre Peninsula, South Australia; also Kangaroo Island. It is a rare visitor to Tasmania; there were occasional records from "Tasmanian region" last century. In 1969, a flock of about 16 birds was recorded on King Island where it remained for several months. The population in the region of Perth, Western Australia, originated in the late 1960s from escaped captive birds (see **Feral Population**). Stan Sindel (1987) pointed out how the range of this lorikeet has extended in response to urbanization: "It has been a common resident of the northern suburbs of Sydney for many years and more recently has extended its range into the southern and south-western suburbs. I often see small flocks in the southwestern suburbs where five years ago they were unknown." By 1995, or before, they were in the very heart of Sydney, for example, at Ashfield, where they visit the balcony of one apartment (Courtney pers. comm. 1995).

Status: Common and widespread except at southern limit of range.

Habits: It is found in a multitude of habitats, in fact in any locality where there is sufficient food—in mallee scrub, open eucalyptus forest, rainforest, gardens and parks. It frequents coconut palms on offshore islands along the Queensland coast. Found in mountainous areas in Queensland and northern New South Wales, it is otherwise primarily a lowland species which often moves in large flocks, in search of large trees. Observations of contributors to *The Atlas of Australian Birds* did not suggest any large scale seasonal movement.

Forshaw (1981) stated: "In the north, there are some birds present in most districts throughout the year, but in the south, they travel more widely in search of blossoms and at times completely vacate a district. Movements are usually erratic though are regular and seasonal in some places."

At Swan Vale, northern New South Wales, they regularly cross the Great Dividing Range. They come from the coast in the east each spring when white box or yellow box trees are in heavy flower. The visiting rainbows used to nest here. They ceased to do so about 1927 when most big gum trees along Swanbrook Creek were finally killed, depriving them of nest sites. John Henry Courtney, John Courtney's father, found the last known rainbow nest in this area in 1927 (Courtney pers. comm. 1995).

Rainbow lorikeets are easily attracted in large numbers to artificial food sites. The most famous of these sites is Currumbin Sanctuary, Palm Beach, Queensland, where twice daily feeding by visitors has been an attraction for many years. They can hold out plates of bread soaked in honey and enjoy (or otherwise!) the sensation of lorikeets perching on their hands and heads in their eagerness to partake. The flocks consists mainly of rainbows with a few scaly-breasts—more of the latter at the morning feed and only about 3 percent at 4:00 P.M. Total numbers vary; there were only about 200 on the December day of my visit. Des Spittall, curator for many years, told me that numbers currently fluctuate between about 50 and 1,200, but formerly as many as 1,400 might come. Numbers are influenced by abundance of flowering eucalypts and the presence of predators; on the day I was there a goshawk was in evidence, thus numbers were lower. The long-term decline in numbers is due to the Gold Coast becoming increasingly developed, and probably also to the fact that more householders put out food to attract lorikeets to their gardens. There were quite a few young ones (some of which had probably only recently left the nest) and I was told that they had bred early that year. Usually young are not seen until January.

Most residents of areas where this species is common take it very much for granted. Yet it must be one of the world's most common birds which is colorful and easily seen; a feeding flock, or one coming in to roost, is a spectacle which would be difficult to surpass. "Rainbow" denotes colorful but the intense hues of this species are not present in a rainbow. My own first view of these lorikeets in the wild, along with musks, took place in an Adelaide park on a fine April day. I was totally spellbound. It amazed me that local people were walking by without giving the feeding birds as much as a glance! I was speaking at a convention being held in the nearby university, and it was with great reluctance that I tore myself away from this wonderful scene! I still find flocks of this subspecies totally mesmerizing from the point of view of color, noise and activity.

The diet of pollen and nectar is varied with seeds, fruits, berries and insects. Forshaw (1981) states that the staple diet is pollen and nectar from the flowers of *Eucalyptus*, *Angophora*, *Melaleuca*, *Banksia* and other native and introduced trees and shrubs. They also feed on fruits of *Ficus*, mangoes which have been broken open by fruit bats, *Casuarina* flower heads, beetles and small grubs. Orchard fruits such as apples and pears are eaten, causing some damage. Maize and sorghum crops may also be attacked before they are ripe, as lorikeets like to feed on the milky grains.

Nesting usually occurs from August to January, also at other times in the north of the range. The most favored nest sites are hollow limbs in *Eucalyptus* or *Melaleuca*, generally at a height of about 16 m (50 ft.); nests as low as 3 m (10 ft.) have been recorded.

According to Courtney, the nest entrance can be a symmetrical round hole in the side of the trunk of a living river red gum tree (into which they can barely squeeze), or it can be a short dead spout leading into a living limb or trunk. Both types of nest entrance have been observed by John Courtney at Bonshaw, west of the Great Dividing Range, on the NSW/Queensland border. A clutch of three eggs from Grafton, New South Wales, averaged 27.6 x 23.0 mm (Forshaw 1981).

Aviculture

Status: Common in Australia, less so elsewhere.

Clutch: Two, rarely three. The female of one pair in Victoria, breeding since 1982, laid a three-egg clutch in 1991—the first of four such clutches. In two of the nests, three young were reared; the fourth was still being incubated (Dear pers. comm. 1994).

Incubation period: Usually 23 days, but 22–25 days recorded; an unincubated egg which I placed in an incubator hatched after 22½ days.

Newly hatched chick: Weight, 5 g or 6 g; white or silvery down, longest on head.

Chick development: Second down appears at 18 days, pin feathers at 22 days, usually on the head first; tail and flight feathers at 28 and chest and abdomen about 30 days; almost fully feathered at 45 days.

Young in nest: About 60 days, but as short as 51 days recorded.

Chick growth

Shows the weights of parent-reared chicks of the pair belonging to Alan Dear of Dromana, Victoria (chicks No. 1, No. 2 and No.3), contrasted with a chick reared from the egg by Inge and Nancy Forsberg in Sweden (chick No. 4). The basic diet of Alan Dear's birds was nectar made from brown sugar and a vitamin mixture, with wholemeal bread, plus apples, silver beet and, when available, flowering plants. The parents were at least 13 years old. Chicks were weighed at 3:00 P.M. daily, thus the amount of food in the crop varied; this accounts for an apparent but not actual weight loss on some days. The three eggs in the clutch hatched on August 17, 19 and 24 (Dear pers. comm. 1994).

Rainbow lorikeet weight table

Age in days	Weights in grams			
	Chick No. 1	No. 2	No. 3	No. 4
day hatched	–	6	4	4.8
1	8	7	6	
2	12	9	died	5.3
3	12	11		
4	14	12		7.4
5	18	14		
6	18	18		10.0
7	18	17		
8	25	22		13.4
9	28	20		
10	32	24		18.0
11	30	26		
12	34	28		22.8
13	36	29		
14	41	30		26.9
15	41	32		
16	51	ringed 42		32.4
17	ringed 54	44		
18	58	40		40.3
19	62	44		
20	58	55		50.7
21	60	54		
23	78	54		58.8
25	88	65		63.6
27	88	76		68.6
29	96	86		76.4
31	110	95		92.7
33	112	105		
35	125	109		120.9
37	126	110		
39	145	128		
41	140	120		126.4
44	145	130		
47	148	140		
50	158	150		132.0
51	170			
52	165			
70	left nest			

General: Due mainly to its wonderful plumage, this has always been a very popular lorikeet with aviculturists. Outside Australia, demand probably exceeds supply and females have become harder to find. (Buyers should note that in the U.K. birds of the nominate race are often advertised as "rainbow lorikeets." This may happen quite innocently when reference has been made to *Parrots of the World*, where the name green-naped lorikeet is not used.) Up until 1961, when Australia ceased to export her fauna, this was one of the few lorikeets which was freely available in Europe.

In Australian aviaries it breeds throughout the year. Peter Philp, a speaker at the Fifth National Avicultural Convention in Adelaide in 1989, stated that his pair had fledged young in every month except May and September. They had reared 33 young in ten years. Australian breeder Barbara O'Brien, who kept every species of Australian lorikeet, commented that her rainbows were the only ones which would eat seed and "only when they have chicks and then only hulled oats." These were limited to a handful daily.

In *Lories and Lorikeets* I recorded several instances of aviary birds burrowing into the ground in order to nest.

Hutchins and Lovell (1985) give a description of the display: "The courtship dance of the male rainbow lorikeet is rather acrobatic; he resorts to dancing up and down, then hanging upside down from the perch or hopping along the perch sideways and twisting from side to side. Often this is done in silence but at other times loud screeches are uttered and added to these antics a fair amount of bobbing and bowing is done. During this display the male's eyes dilate, this is very pronounced and very striking to the observer." It should be noted that both sexes blaze the eyes, arch the neck, hiss and bob in typical *Trichoglossus* manner.

Ninigo Lorikeet *T. h. nesophilus* (Neumann 1929)

Description: The *nesophilus* resembles *flavicans* except that the green areas are always midgreen; the color variations shown in the latter do not exist.

Range: Ninigo and Hermit groups of islands, north of eastern New Guinea. (A remote group west of Manus, the chief island of the Admiralty group.)

Habits: According to Bell (in Forshaw 1989) in 1970 it was "commonly distributed, even on very small atolls." Two to five birds were observed in most habitats, including coastal mangroves, and were seen flying from island to island.

Black-throated Lorikeet *T. h. nigrogularis* (G. R. Gray 1858)

Description: Crown and sides of the head are light blue and throat is black. Cain (1955) states it has "a strong tendency to the development on the breast feathers of a yellow band just proximal to the dark bars and is further characterized by reduced barring."

Key identification features: Black throat, dark sides of head, narrowly barred orange breast tinged with yellow, and purplish or purplish black abdomen with some green feathers.

Length: 28 cm (11 in.). An excellent photograph of this subspecies appears in *Lexicon of Parrots*, vol. 1, page 13/5.

Range: Aru Islands, eastern Kei Islands, to the southwest of New Guinea.

Habits: Flocks have been observed on several occasions crossing the two-kilometer stretch of water between the islands of Warmar and Wokam (Milton and Marhadi 1987).

Aviculture: Walter Goodfellow collected the *nigrogularis* subspecies in the beginning of this century, and it was first kept by Mrs. Johnstone in 1904. Since then it either has rarely been exported or, more likely, has rarely been recognized. It was bred at Kelling Park Aviaries, Norfolk, in 1966, and on several occasions at San Diego Zoo from 1969.

Rosenberg's Lorikeet *T. h. rosenbergii* (Schlegel 1871)

Description: An immediately recognizable subspecies, its head is like that of the nominate race, but even bluer, with no greenish streaks on the occiput. There is a strong purple gloss on nape, throat, chin and chest bars. Notable is the extremely heavy blue-black barring and the wide greenish yellow collar—much wider than in any other race—which extends down the mantle. The primaries and secondaries lack yellow, which is replaced by red, and there is a corresponding extension of red down the sides and abdomen toward the tail. Between the yellow collar and the nape is a narrow dull reddish band.

Length: 28 cm (11 in.)

Key identification features: Enormous greenish yellow collar, very heavy breast barring.

Range: Biak Island (previously called Mysore), Geelvink Bay, Irian Jaya. Hubers (1994) states that it also occurs on the island of Korrido.

Habits: There is little recent information. In 1939, Mayr and de Schauensee recorded that it moved about in deep forest more than *Eos cyanogenia*. "Small screaming parties of five or six traveled rapidly back and forth over the jungle from one flowering tree to another. Their shrill, constantly repeated notes in flight, changed to a softer but still continuous conversation while feeding."

Thomas Arndt visited Biak in 1991. After inquiring where he might find this lorikeet, he was taken to the village of Korem, northwest of the city of Biak. He had a brief glimpse of one bird, near a river lined with coconut palms. He suggested that it might be found only in this area; if so, the population might be very small (Arndt 1992).

Nesting: Nests in the wild are undescribed. Two eggs of one clutch laid by a female belonging to G.

van Dooren measured 27.5 x 21.6 mm and 28.0 x 22.0 mm, and from another clutch, 28.0 x 21.8 mm and 27.7 x 22.3 mm; one egg from another clutch weighed 5 g (van Dooren pers. comm. 1995).

Status/Conservation: Because of its small range and the destruction of much primary lowland forest, it is considered to be a vulnerable form.

Aviculture: Rarely available. P. H. Maxwell (1952) acquired a young one in the U.K. from the well-known collector Wilfrid Frost. Maxwell had formerly kept another one that died at an old age, in the London Zoo Parrot House in 1943. According to the *International Zoo Yearbook* (vol.16), three Rosenberg's were bred at Barcelona Zoo, in Spain, in 1974; at least one of the parents was bred there. Vogelpark Walsrode exhibited a pair in 1979. In the late 1980s and early 1990s a few were imported into Europe. Jos Hubers (1994), publisher of *Lori Journaal Internationaal*, could trace Rosenberg's with five breeders. In 1990 he obtained seven females and three males, all young. Through exchange and purchase, he was able to make up five true pairs. The first pair started to breed at the end of 1991. Two eggs were laid, the first of which hatched after 25 days, and the second on the following day. At 12 days the chicks were ringed with 6 mm rings. When feathers started to appear at about four weeks, the parents plucked them. By the end of 1993 this pair had produced 10 young which were hand-reared or placed with foster parents. Pair No. 2 started to breed in April, 1992; two infertile eggs were produced. Two young were reared in the next two nests, but one died at five months. From May, 1993, until near the end of 1994, nine young were reared from this pair. When they plucked their young, the nest-box was placed on the floor with the lid open, and plucking ceased immediately. One youngster had been reared from pair 3 and pair 4 had produced their first clutch which was infertile. Hubers' birds were fed on Aves Lorinectar, with Loristart (higher protein content) when young were in the nest.

Only a few Rosenberg's reached the U.S. In the U.K. there were several males only until 1996 when Trevor Buckell acquired a female and two young unrelated pairs from the Netherlands. In that country, two eggs from a pair owned by G. van Dooren were laid on October 31 and November 4, 1994. The first egg was infertile and the second hatched after 24 days. The chick weighed 6 g on hatching, 10 g on day 2 and 15 g on day 4; it died on day 11. In 1995 the female laid the first of two eggs on July 14. They hatched on August 8 and 10 and the chicks' weights (weighed at 6:00 P.M. daily) are given below.

Rosenberg's lorikeet weight table

Age in days	Weight in grams	
	Chick No. 1	No. 2
hatch	6 e	6 e
1	8 fic	9 fc
2	11 fc	11 fc
3	12 fc	12 fc
4	15 fc	16 fc
5	16 fc	19 fc
6	20 fc	22 fc
7	22 fc	25 fc
8	27 fc	30 fc
9	35 fc	33 fc (R)
10	36 fc	40 fc
11	40 fc (R)	45 fc
12	45 fc	50 fc
13	51 fc	51 fic
14	56 fc	58 fc
15	54 fic	58 fic
16	60 fc	68 fc
17	65 fc	59 e
18	71 fc	67 fc
19	75 fic	63 e
20	84 fc	73 fc
21	75 fc	78 fc
22	84 fc	83 fc
23	89 fc	81 fic
24	94 fc	80 fc
25	91 fic	91 fc
26	93 fc	87 fic
27	101 fc	91 fic
28	100 fic	97 fic
29	104 fic	94 e
30	107 fic	104 fic
31	104 e	105 fic
32	113 fic	109 fic
33	114 fic	—
34	117 fic	—

fc = full crop, fic = food in crop, e = empty, R = ringed (6 mm)

Red-collared Lorikeet *T. h. rubritorquis* (Vigors and Horsfield 1827)

Description: The whole head is bright blue with light blue shaft-streaking. The unbarred orange of the upper breast continues upward to form an orange nuchal band; scapulars are blue, variably marked with orange toward mantle. The abdomen is bottle green, appearing almost black; compared with *moluccanus* there is a broader yellow band on the underside of the flight feathers and thighs, and undertail coverts are more heavily marked with yellow.

Length: 30 cm (12 in.)

Weight: About 130 g. Salisbury (1985) gave the weights of the males of two captive breeding pairs as 136 g and 131 g, and the females of the same pairs as 125 g and 128 g.

Key identification feature: Orange collar.

Sexual dimorphism: According to Odekerken (1995a) the abdomen viewed in strong sunlight is bluish black in males and green-black in females.

Immature birds often show a few scattered medium blue feathers among the greenish black ones on the abdomen, which are conspicuous (Courtney pers.comm. 1995). Courtney believes that this suggests that *rubritorquis* evolved from blue-bellied ancestral stock, like *moluccanus*.

Range: Northern Australia, from Western Australia (vicinity of Broome) northward and eastward approximately to the border of Northern Territory/Queensland in the Gulf of Carpentaria, extending inland for about 200 km. *The Atlas of Australian Birds*, which charts the distribution of all species in 10' blocks suggests "that the gap between the Red-collared and the Rainbow at the eastern end of the Gulf of Carpentaria is narrow, if it occurs at all. There are no records in 10' blocks between 139° 30'E and 140° 50'E, an area where difficulty of access has precluded many visits..."

Status: Nomadic; numerous where food is available. *The Atlas* states that this is "one of the most common parrots of the Top End but in parts of the Kimberley Region is less common than the Varied."

Habits: This lorikeet will move long distances, according to *The Atlas*. No large-scale, seasonal movements are suggested but there are local shifts in population which may correspond with the flowering of trees. Such movements are most conspicuous at isolated sites such as Groote Eylandt; some birds stay throughout the year but large numbers visit during April and May when the banksias are flowering. Red-collared lorikeets usually live along watercourses in eucalypt forest, moving into eucalypt and paperbark woodland when the trees flower. Like rainbow lorikeets, they can be observed in towns. Darwin is a good place to look for them, also the Howard Springs Nature Park, 30 km (19 mi.) east of Darwin. In addition to pollen and nectar, they feed on fruit crops, especially mangoes. Sindel (1987) recorded: "They would gorge themselves for weeks and as the last of the fruit became over-ripe and fermented in the hot sun, the lorikeets would become so intoxicated that they would be unable to fly and would wander around the trees in a stupor." He also observed them feeding on the orange blossoms of *Eucalyptus miniata*. Nesting usually occurs during spring and summer but autumn nests have also been recorded. They usually use holes in eucalypts, at heights varying from 5 m to 30 m (16 ft. to 98 ft.).

Aviculture

Status: Fairly common in Australia; rare elsewhere.

Clutch size: Two, occasionally three.

Incubation period: 22–25 days, average 23 days. In two clutches incubation began when the second egg was laid, thus chicks hatched on the same day, 24 days after first egg and 21 days after second egg (Salisbury 1985). But in another pair breeding at the same location, the incubation period was 25–27 days.

Newly hatched chick: Weight, 6 g; covered in silvery white down.

Young in nest: Average 61 days (Sindel 1987), but as short as 52 days recorded by others.

Chick development: Simon Harvey of Adelaide described the growth of two chicks to Neville Cayley (Lendon 1979): feathers were erupting when they were 15 days old and by 22 days the believed male chick "had a bright blue head." Twelve days later both chicks were well feathered and the red collar was showing distinctly on the believed male. Both had black beaks.

General: Red-collared lorikeets are hardy and free-breeding if well cared for. A breeder in Victoria achieved excellent results on a diet of nectar (basically brown sugar and water) and wholemeal bread moistened with milk, plus apple and sultanas daily. They had thrived on this for ten years and not one adult or fledged youngster had died during this period—a remarkable record. Alan Dear had 40 red-collared at the time (Dear 1986).

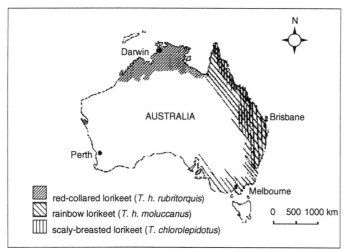

Distribution of red-collared, rainbow and scaly-breasted lorikeets.

Wilson (1988) described the display of the male of his pair: "As the male approached the hen he was seen to bob and duck, jump sideways along the perch, stretch his neck and back almost parallel with the perch while uttering a short whistle and displaying brief bursts of fluttering wings, hypnotic eyes dilating and contracting all the while."

Salisbury (1985) stated that the display of *rubritorquis* is characteristic of *Trichoglossus* in some respects, but diverges from *moluccanus* and *chlorolepidotus* in its complexity. "The display is much more aesthetically pleasing and consists of a series of reptile-like movements, including erratic head bobs and sways, feather fluffing, screeching and wing flapping.... Much of the ritual is done face to face, with both birds moving their heads in unison and then in opposite directions. Each pose is held momentarily, making it reminiscent of a traditional Hindu dance." Before copulation, one male was observed jumping completely over the female from the perch, landing neatly on the other side, and back again, while the female quivered her wings. It appeared that both birds made the same kind of movements, and that no part of the display was typically male or female.

In the 1940s and 1950s the red-collared lorikeet was freely available to European aviculturists. "It was, perhaps, the most popular of all lories and received glowing accolades from the writers of the time" (Low 1977a). In the 1960s, when Australia no longer permitted the export of fauna, except to certain zoos, it soon became rare in aviculture. San Diego Zoo bred it with great success between 1969 and 1974 (over 30 were reared); most of the young were sent to zoos and private aviculturists who squandered the opportunity to establish it in the U.S. Within a few years, very few were left. In Europe, it did not fare much better. By the 1980s only a few remained, some of which had been hybridized with Swainson's lorikeets. Wassenaar Zoo in Holland received six *rubritorquis* from Australia in 1953. They were released into the Louise Hall, a very large, planted, heated indoor aviary. One female laid on October 18 and 21, and chicks hatched on November 14. During incubation, the male entered the nestbox whenever the female left. The young are reported to have left the nest on December 30, which would give a short fledgling period of only 46 days. By 1954 this colony had grown to number 14 birds which went on to breed at least into the 1960s. As early as 1916, E. J. Brook, a very successful aviculturist, had been breeding red-collared lorikeets for eight years. He described them as being "as hardy and prolific as Budgerigars." At Keston Foreign Bird Farm in the early 1930s, they bred "steadily the whole year round."

However, in the 1940s and 1950s there were no serious breeders of lorikeets. Australian parrakeets took priority in virtually all parrot collections; the lorikeets would have been nothing more than a diversion, with probably no more than one pair kept in each collection. Given these circumstances, and the fact that the surviving birds had become inbred, it is not surprising that they failed to become established.

Northern Blue-bellied Lorikeet *T. h. septentrionalis* (Robinson 1900)

Description: It differs from *moluccanus* in the brighter and purer blue of the head and abdomen. Its size is slightly smaller. It is no longer considered a valid subspecies by most systematists.

Range: Northeastern Australia in the Cape York region. Cain (1955) states that it intergrades with *moluccanus* and forms a cline, increasing in size to the south.

Stresemann's Lorikeet *T. h. stresemanni* (Meise 1929)

Description: The forehead is blue with blue shaft-streaking and the back of the head is green with green shaft-streaking. The abdomen is dark green.

Key identification features: The most orange breast of any subspecies except *rubritorquis*, although some feathers are yellow at the tip in some birds. The bases of the feathers of the mantle (interscapular) are yellow. According to Cain (1955), the greenish yellow collar is intermediate between that of *forsteni* and *mitchellii*, and birds of this race are larger than either of those subspecies. The abdomen is dark green. An excellent photograph appears in *Lexicon of Parrots*, vol. 1, page 13/10.

Range: The island of Kalao-tua, Indonesia, 104 km (65 mi.) north of Flores.

Aviculture: This subspecies was reported to have bred at Natal Lion Park, South Africa, in 1976, when three were reared. (It appeared in the *International Zoo Yearbook* under the name of "Yellow Forsten's," with the scientific name *stresemanni*.) It was, perhaps, otherwise unknown in aviculture until the mid-1980s when a few were imported into Europe and the U.S. It has been reared at Loro Parque since 1991.

Weber's Lorikeet *T. h. weberi* (Buttikofer 1894)

Taxonomic note: Mivart (1896) considered *weberi* to be synonymous with *T. euteles*, quoting Salvadori (1891) that specimens from Flores [*weberi*] are darker, with "greener and darker heads" and noted that "the underside of the body is a darker green." Mivart mentioned that Buttikofer "has erected forms found in Flores into a distinct species, *P.* [=*Psitteuteles*] *weberi* . . . But a careful examination of the numerous skins of *P. euteles* preserved in the British Museum has led us to believe that *P. weberi* can nevertheless be but a marked variety." He goes on to mention that some of the distinctive features assigned to *P. weberi* were not apparent in the skins examined.

Smith (1975a) incorrectly claimed that Mivart had "made Weber's a subspecies of *T. euteles*." He then described how "in 3 adjacent aviaries, ideal for making comparisons," Ray Kyme once housed perfect, Weber's and rainbow (*T.h. rubritorquis*) (*sic*) lorikeets. After spending several hours comparing the three forms, Smith felt that, in his opinion, Weber's "seemed to have more in common" with the perfect lorikeet than with the rainbow. However, he noted that the perfect lorikeet "has an extensive range and Lomblen, one of the islands in this distribution, is close by the island of Flores" where "the sole *Trichoglossus* ...is Weber's." He further notes that some of the other islands inhabited by the perfect lorikeet are shared with one or two different subspecies of *T. haematodus*, and concludes that *T. euteles* and *T. haematodus* therefore do not hybridize. He is left wondering to which species, "unless one uncritically accepts Peters," does the subspecies *weberi* actually belong? This is a very good question and will not be resolved without further study.

Description: The head, sides of neck and throat are midgreen with brighter green shaft-streaks; feathers of the forehead are faintly tipped with blue. The nuchal collar is greenish yellow, also the upper breast, some feathers of which are margined with dark green. Abdomen and under tail coverts are dark green with some light yellowish green markings, especially on thighs and area surrounding the vent. The remainder of the plumage is dark green except for the greenish yellow underwing coverts and blackish primaries with a large yellow patch on inner webs. Underside of the tail is olive green.

Length: 23 cm (9 in.)

Weight: 85 g

Key distinguishing features: The only haematodus marked entirely in green and yellow, but liable to confusion with the perfect lorikeet (*T. euteles*) from which it is distinguished by forehead color: green in weberi and olive yellow or yellow in *euteles*.

Range: Flores, Lesser Sunda Islands, Indonesia (17,160 km^2, or 6,600 sq. mi., in area).

Status: It is still reasonably common but has declined considerably as a result of high levels of trapping for trade (Butchart pers. comm. 1994).

Habits: It occurs in rainforest, also in stands of *Casuarina* up to 1,400 m (4,600 ft.). As a result of the Cambridge University Flores Conservation Project of 1993, expedition leader Stuart Butchart obtained the following information (Butchart pers. comm. 1994). Encounter rates for Weber's lorikeets, per 100 man hours, were as follows: in semievergreen rainforest in southwest Flores, at 350 to 1,100 m (1,150 to 3,600 ft.), 170; in secondary scrub around village in southwest, at 350 m (1,150 ft.), 640; in upper montane forest in western Flores, at 1,150 to 1,700 m (3,770 to 5,570 ft.), 100; in moist deciduous monsoon forest in western Flores, from sea level to 500 m (1,640 ft.), 52. It was found to be most common in fruiting and flowering trees in village scrub around Kampung Langka in the southeastern part of the island, and was also found from lowland scrub to upper montane forest.

Nesting: No information. Eggs laid in the collection of G. van Do oren in the Netherlands measured 25.0 x 20.3 mm, 25.3 x 21.4 mm, 27.0 x 21.5 mm and 25.8 x 21.2 mm (van Dooren pers. comm. 1995).

Avicultural status: Rare.

General: Because of its comparatively dull coloration and sporadic availability, this subspecies has not excited much interest. Lory enthusiast Ray Kyme of Kirton, Lincs, was the first to breed it in the U.K. In 1970, two chicks hatched in February which lived for only three days. The third bird in the aviary was then removed and the pair nested again, unsuccessfully. The third clutch was laid on May 20 and 22 and chicks hatched on June 15 and 17 (26 days for each egg). The young left the nest (badly plucked) on August 13 and 14 (after 59 and 58 days); the beak of one was reddish brown, that of the other blackish red. They were removed from the aviary on September 25 when they (the young) were behaving aggressively toward their parents! The pair continued to breed until at least 1972 and was sold, with their young, in the following year. Weber's are now rare in the U.K. and in Europe, where there may be only two or three bloodlines. In Germany they were reared by Theo Pagel and then the same pair bred at Cologne Zoo. In 1994, the pair and three young (one from each of three nests) were kept together in an aviary 8 m (26 ft.) long.

In the U.K., Trevor Buckell keeps two pairs; the male of one pair and the female of another originated from Jan van Oosten in the U.S.A., who has a special interest in this subspecies. He obtained three birds in 1983, two males and a female. A pair hatched two chicks in May, 1984, and one more in each of the following months: April and June, 1985; April, 1986; and August, 1987. The female hatched in May, 1984, was paired with the other wild-caught male and produced her first chick in April, 1985. The young one was a female; her first youngster to be reared hatched in July, 1987. Unfortunately, by 1994 very few weberi still survived in the U.S.A. and females were scarce. Jan van Oosten obtained a hybrid female reared from a male Weber's and a female Edwards' which, in appearance, resembled the Weber's 100 percent. He intended to pair this bird to a male Weber's (van Oosten pers. comm. 1994). A pair of Weber's which Mr. van Oosten presented to Loro Parque, Tenerife, in 1988, reared three young in 1990 and two in 1991 and have continued to breed. In South Africa there are pairs in several collections. In 1996, Rex Duke of Melmoth had three pairs, two of which were breeding.

Perfect Lorikeet   *Trichoglossus euteles* (Temminck 1835)

Synonyms: Olive-headed lorikeet, plain lorikeet and yellow-headed lorikeet

Description: The head is olive yellow; the green nuchal band merges into the green of the wings and mantle. Abdomen is yellowish green, very faintly barred with dark green. There is a yellow band on the underside of the flight feathers. Tail is green above, dull yellow-green below; undertail coverts are yellow-tinged green. The beak and the iris are orange; the cere is pale gray.

Length: 25 cm (10 in.)

Van Dooren (1992) examined skins at the British Museum (Natural History) and in Leiden Museum in the Netherlands. He found that males collected in the western parts of the range were brighter. The green of the abdomen of birds from Kisser, in the east, was more intense, but generally coloration was quite variable.

Weight: 80 g to 85 g

Immatures: They are duller on the head; the tail is slightly shorter. On fledging, the beak is dark brown in some young, nearly orange in others. The iris is dark. In the nineteenth century, the Alor lorikeet (*T. alorensis*) was named, but proved to be nothing more than an immature *euteles*.

Sexual dimorphism: Head color is slightly brighter in the male.

Range: The Lesser Sunda Islands of Timor, Pantar, Lomblen, Alor, Wetar (see Gazetteer), Romang, Kisar, Moa, Leti, Damar, Babar, Luang, Teun and Nila (White and Bruce 1986). There is no evidence that it occurred on Flores. Skins in the British Museum (Natural History) collected there in 1861 were probably captive birds taken from nearby Timor. Van Dooren reached this logical conclusion because their plumage was dirty and damaged.

Natural History

Habits: It apparently mainly inhabits arid islands with little forest and sometimes much cultivation, according to White and Bruce (1986). They believe that it replaces *T. haematodus* over most of

its range and may be derived from the same ancestor. Bruce believed that it was more abundant on Timor than *T. haematodus*, and generally replaced the latter at higher altitudes (Forshaw 1989). In the Ramelau Range, Bruce observed small flocks in primary forest, secondary forest and savannah woodland, up to about 2,300 m (7,550 ft.). The birds were flying back and forth above the treetops. Only once was this species encountered below 1,000 m (3,300 ft.)—between Same and Betano on the south coast. However, as it was the dry season, small flocks may have dispersed widely throughout the lowlands, making them less conspicuous. Arndt, who visited Timor in the early 1990s, described the perfect lorikeet as local and not common, although flocks of up to 100 are said to congregate when the breeding season is over.

Nesting: No information available. Two eggs laid in a clutch measured 24 x 20.5 mm and 25 x 21 mm; their weight was 5 g (van Dooren 1992). One egg laid at Palmitos Park measured 23.2 x 20.0 mm.

Status/Conservation: Unknown; insufficient information. Visits to Timor by ornithologists have been few and fairly brief in recent years.

Aviculture

Status: Common

Clutch: Normally three, rarely four (in Denmark, J. Aarestrup's female laid four eggs, as did G. van Dooren's female in the Netherlands). A female belonging to Alan and Lesley Brown of Tring, Herts, U.K., laid six eggs, two of which were removed; the remaining four hatched and all the young were reared (Low 1979b). This is the only non-Australian lorikeet with a normal clutch size larger than two. Why is this so? Timor is near to Australia so the Australian influence is strong. This species is an inhabitant of dry localities and must breed quickly when the rains come to take advantage of the sudden increase in food supplies. However, the other two lorikeet species of Timor lay two-egg clutches.

Incubation period: Usually 23 days. In a clutch placed in the incubator four days after the third egg was laid, the first and second hatched after 23 days and the third after 24 (Sweeney 1993a). In another clutch, the first egg (marked) was infertile; the second hatched after 23 days (van Dooren 1992).

Newly hatched chick: Weight, 4 g; longish white down. Sweeney (1993a) noted that the beak and nails were black, the upper mandible with a paler base. Initial crop capacity of one chick was 0.2 ml.

Chick development: Day 4, bill paler; day 12, pale base extended half way up bill, leaving only the tip black. By day 33 the beak was orange with a black tip and by day 45 it was completely orange. The eyes began to slit at day 12 and were fully open by day 17. Pin feathers started to appear at about day 19 and by day 27 they had emerged from the skin over much of the body and had erupted on the head. By day 36 the feathers of the underparts and tail were erupting and by day 40 the head, wings and tail feathers were free from quills. By day 45 only the flanks and the area under the wings remained unfeathered. By day 55 young were fully feathered and the bill was rich orange (Sweeney 1993a).

Chick growth

Shows the weights of two chicks hatched at Loro Parque, Tenerife, in 1987, and hand-reared from the ages of 24 and 26 days. Rearing food consisted of baby cereal, wheat germ cereal, jars of baby food and papaya blended with water. The weights are those before and after the first feed of the day.

Perfect lorikeet weight table

| Age in days | Weight in grams | |
	Chick No. 1	No. 2
25	29/32	–
27	30/33	24/27
29	31/34	25/28
31	33/37	28/31
33	35/40	29/35
35	37/45	31/36
37	40/46	34/44
39	42/49	37/45
41	47/53	41/43
43	60 weaned	– weaned
45	61	58
47	68	63
49	71	68
51	75	72
53	77	76
55	78*	79*
60	78	76
69	–	68
76	–	72
100	–	70

*peak weight

Young in nest: 57–63 days.

General: This is an ideal species for the beginner with lories, being free-breeding and low-priced. I know of owners of pairs so prolific that they did not always find it easy to sell the young, although the price was reasonable. Those who know the species appreciate it but others tend to favor more brightly colored lories. Because little is known about its status in the wild, breeders should ensure that viable stocks are maintained in aviculture. Experienced lory breeders should find a place for this possibly vulnerable species. One breeder described it as an attractive contrast to the more gaudily colored lories—and this is true.

Its small size and tolerable voice make it a good choice for apartment dwellers who wish to keep lories. I know of more than one pair which has bred in a cage (by no means large) in a spare bedroom!

G. van Dooren of the Netherlands described his experiences with this species (van Dooren 1992; van Dooren pers. comm. 1994). His female laid her first clutch when she was 15 months old, on January 11 and 13. Only the second of the two eggs was fertile. The chick hatched after 23 days and left the nest after 57 days. It was ringed at the age of seven days with a 5.4 mm ring. (Some breeders use 6 mm rings and ring at a slightly later age.) In May the clutch consisted of three eggs. In August, four were laid! The pair was housed in an indoor aviary in which the temperature did not fall below 10°C (50°F). It measured 1.5 m x 1.3 m x 2 m (approx. 5 ft. x 4 ft. 3 in. x 6 ft. 6 in.) high. The nestbox used was 18 cm (7 in.) square and 40 cm (16 in.) high, with wood shavings in the bottom. The basic diet was Aves Lorinectar (a Dutch product), with Aves Loristart when there were young chicks in the nest. This pair was very hardy; when the female laid in January, 1992, the temperature was -12°C (10°F). They bathed daily in standing water but had never been seen to rain-bathe (presumably because they are from very dry localities).

Most perfect lorikeets will eat some seed, which is best limited to the small seeds such as canary and spray millet, or a little soaked sunflower when they are rearing young. They will eat wild green foods such as seeding grasses, chickweed and dandelion. The pair belonging to Mr. and Mrs. Brown, of Tring, U.K. (mentioned under Clutch), reared young in a large aviary overgrown with stinging nettles and other weeds. They fed on the

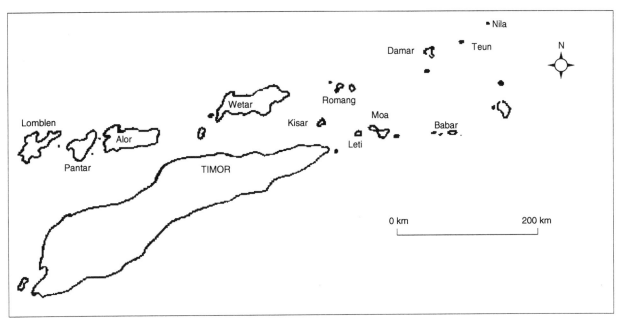

Islands inhabited by perfect lorikeet (*Trichoglossus euteles*).

pollen of the stinging nettles, on chickweed and other weeds (Low 1979b).

The European Union has banned the import of this lorikeet since 1989. Although it is free-breeding and has many attractive qualities, it is not colorful. It is considered as the ideal beginner's bird; however, beginners tend to want to move on to other species after they have achieved two or three breeding successes. The status of this species on Timor is unknown—but war and political instability will have done nothing to aid its survival. If the news one day emerges that it is endangered in the wild, it would be sad indeed if by then it had gone from aviculture.

Ponapé Lorikeet *Trichoglossus rubiginosus* (Bonaparte 1850)

Synonym: Cherry-red lorikeet

Native name: *Se-rehd*

Description: This is one of the most distinctive (and isolated) members of the genus. The easiest way to visualize it is as a maroon *haematodus*; Kenning (1995) describes it as dark chestnut (appearing chocolate brown in flight). Feathers of the neck and underparts are margined with a darker shade, to suggest indistinct barring. Flight and tail feathers are dull yellow olive, the undersides being yellowish. The beak is orange in the male, more yellowish in the female. The iris is dark orange (Kenning) but according to Baker (1951), it is yellow-orange in the male, grayish white in the female. If this is so, it would be unique among members of the genus. Legs, cere and skin surrounding the eye are dark gray.

Length: 24 cm (9½ in.)

Weight 70 g to 85 g

Immatures: The birds resemble adults except for beak and eye coloration. This has not been described but one would expect it to be the same as in other *Trichoglossus* species—the beak brown and the iris brownish. This is borne out by Kenning's photograph (*Australian Birdkeeper*, April/May 1995, p392) of a chick apparently aged about seven weeks. The cutting edge and base of upper mandible are yellow at that age; legs, cere and eye skin are light gray.

Plumage: Cain (1955) describes the Ponapé lorikeet as dull dark red except for a honey-colored tail and indistinct black bars on the underside and on

269

the back. The wings are black on the underside. He attempts to answer the questions: Why does it seem to have abandoned a color scheme which viewed dorsally is probably semicryptic? Are there no avian predators? The only resident bird of prey is the osprey, thus *rubiginosus* may be the only form in the superspecies not resident in an area containing bird-eating hawks. He comments: "We may conclude therefore that in *T. rubiginosus*, because one restraint (the necessity for a semi-cryptic color-pattern) has been removed, a pattern more compatible with other demands has been achieved. This suggestion is strengthened by the lax and downy texture of the contour feathers, which will prevent the bird from flying as fast (in times of crisis) as the other forms."

Evolutionary origin: According to Segal (1994), some ornithologists believe that this species evolved from Meyer's lorikeets which were blown from Sulawesi, and that over a long period they evolved their distinctive coloration—and much larger size. This seems totally illogical! The ornate lorikeet also occurs on Sulawesi and it is only fractionally larger; Meyer's is only half the size. The ornate also has a strongly barred breast. On the other hand, as the nearest point on Sulawesi to the Caroline Islands is 2,000 km (1250 mi.), I have great difficulty in believing that more than one lorikeet could be blown so far and survive, let alone survive to breed. Surely it is more likely that one (or more) subspecies of *haematodus* reached Pohnpei by means of trade, escaped and colonized the island, and evolved into a lorikeet of very different appearance. (Aitutaki in the Cook Islands has been colonized from escaped or liberated birds after the endemic lory became extinct.)

On the other hand, as suggested to me by John Courtney, there were probably populations of *Trichoglossus* on islands much nearer to the Caroline Islands; if so, these populations have long been extinct.

Range: Pohnpei (Ponapé) in the eastern Caroline Islands, Pacific Ocean. It is reported that it also occurs on the atoll of Mokil, 160 km (100 mi.) southeast of Pohnpei.

Habits: This lorikeet is found over almost the entire island but is especially numerous in the lowlands, in coconut plantations and mangroves. Often they are heard before seen. They are noisy and quarrelsome and their calls cannot be mistaken for those of any other species. They usually occur in small groups of from three to seven birds in their search for pollen, nectar and fruit. A specimen collected apparently had mostly vegetable matter in its stomach, gravel in the gizzard and fly larvae in the crop. During two weeks spent on Pohnpei, Kenning observed the lorikeets flying down from the hills to the mangroves every morning, in groups of four to six. They were present in the gardens where breadfruit and mango trees and banana palms are grown. Nectar from coconut and banana flowers are important items in their diet. They also feed on the flowers of the coral tree (*Erythrina fusca*) which blooms in February and is grown between the mangroves and the mainland. Mangoes are their favorite fruits and they do much damage to them.

Kenning recorded that "one morning I saw two lories playing, flying from branch to branch, hanging upside down and gradually moving closer toward me. When they were about ten meters away one bird with its tail fanned and wings spread far apart flew toward me and then flew in a tight circle above me 2 or 3 times calling all the time before disappearing into the forest."

Kenning states that the lorikeets' only enemies are rats, which move about the trees with great ease. They can balance on the smallest branches. He recalled how one evening, at dusk, he saw three rats running toward two banana flowers to feed on them. Turning 90 degrees, he saw another two rats on a coconut flower. Thus rats are food competitors as well as predators (Kenning 1995).

Nesting: Ponapé lorikeets nest in high, smooth-trunked native palms and in old mangrove trees. Neither offer food to attract rats. Kenning believed that the large numbers of lorikeets on the island show that they are adapting to the threat posed by rats and select holes "out of reach" of them. The few reports state that only one egg is laid. If so, this species is unique among lorikeets. But the eggs remain undescribed. The lorikeets breed from December until March, April or even May, according to Kenning. A chick was taken from a nest in a mangrove swamp in the middle of January, when it was about one week old.

Status/Conservation: It is a common bird, or very common where there are abundant food sources. Fortunately, hunting and killing it are prohibited, as it is the official state bird of Pohnpei. Taking young from nests is allowed but export permits are not issued, notes Kenning, who believes that its future is secure.

Aviculture: The Ponapé lorikeet is virtually unknown. However, two or more were imported into England in the 1960s. They were exhibited at Kelling Park Aviaries, Holt, Norfolk, which specialized in lories. One was bred there in 1967. Also, John Wilson of Norwich imported one in 1967 or 1968. According to the *International Zoo Yearbook* (vols. 12 and 13), Los Angeles Zoo reared one Ponapé in 1970 (from two hatched) and another in 1971. No other records are known. On Pohnpei, this species is commonly kept and Segal describes it as a much loved pet. Tame birds learn to "talk." Kenning cared for a chick for two weeks which was at least four weeks old when he arrived. It was in good

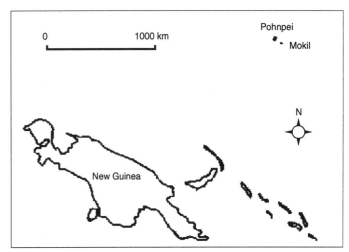

Distribution of Ponapé lorikeet (*Trichoglossus rubiginosus*): Pohnpei and Mokil only.

condition, despite being fed only cooked rice from its owner's mouth. Kenning continued to feed it in this manner—but on a variety of foods.

Scaly-breasted Lorikeet *Trichoglossus chlorolepidotus* (Kuhl 1820)

Description: A simple but pleasing color scheme: it is midgreen with the feathers of the underparts and upper part of the mantle yellow, broadly margined with green; underwing coverts and a band across the undersides of the flight feathers are orange-red. The crown is tinged with blue (more prominently in some males). The tail is green above, yellow olive below; the bases of the outer feathers are broadly edged with orange-red on the inner webs. The bill is orange and the iris is orange with an inner ring of white. The feet are gray.

Length: 23 cm (9 in.)

Weight: 70 g to 89 g (Forshaw 1989); male and female of a captive pair weighed 98 g and 90 g, respectively (Salisbury 1985). Wild birds (31) weighed at Currumbin Sanctuary averaged 89.2 g (range 74–100 g) (Cannon 1984); however, as food was provided twice daily, these weights could be slightly higher than average.

Immatures: They are slightly duller; there is less yellow on the upper mantle. The beak is dark brown to black or, in some individuals, suffused with

orange. The iris is dark brown. By the age of three months the eye color is similar to that of an adult and the beak is orange (Sindel 1987). Matthews and Matthews (pers. comm. 1995) describe the beak as dark brown to black on fledging, changing to orange over three to four months. They have had a success rate of about 95 percent in sexing young birds, basing their opinion on head size and color.

Mutations: Blue, jade (laurel green), cinnamon and lutino mutations have appeared but, as yet, have not been established. The only established mutation is the olive. The first chick of this species hatched in the collection of Antonio de Dios in the Philippines (1993) was a cinnamon! In each of the first three nests of chicks produced by the pair, one of the young was cinnamon and the other was normal. The cinnamon could be identified at hatching by the light-colored eyes visible under the skin (Sweeney pers. comm. 1995). In November, 1972, a yellow scaly-breasted lorikeet was seen at a feeder, along with normally colored birds and rainbow lorikeets. It was bright yellow with faint bars of dark green (appearing blackish). The beak coloration appeared less bright than in a normal (Dawson 1973).

Natural history

Range: Extends further west than the other lorikeets of the eastern seaboard. It is found over most of eastern Australia from Cooktown in the north to approximately 100 km (62 mi.) south of Sydney, in the Illawarra region. During the charting of the ranges of all Australian species for *The Atlas of Australian Birds*, an irruption apparently took place and the species was recorded throughout most of the Cape York Peninsula. It has not been recorded north of 15°S. The range is otherwise unchanged from the historical periods except for the small breeding colony around Melbourne that became established in the 1970s, probably from aviary escapees. Peggy Mitchell first noticed single birds at Mordialloc and Keysborough, 20 km (12 mi.) from Mount Eliza, in 1969. By 1979 they had become established on the Mornington Peninsula, south of Melbourne. As many as 20 scaly-breasts visited her garden at Mount Eliza. On March 8, 1979, she saw immature birds there for the first time. At one stage the two young ones were with two adults and were adopting a begging attitude to be fed (Mitchell 1979).

Habits: According to *The Field Atlas*, no substantial north-south seasonal movements are indicated. Like most lories, they move around to find flowering trees. Inland, their distribution extends along river systems where they are rare compared with coastal sites. They will live in cultivated areas, including suburban gardens, orchards and where sorghum is grown. Though usually described as a lowland species, in the Atherton Region they are more common above 600 m (2,000 ft.) than on the coast. *The Field Atlas* suggests a difference in altitudinal distribution in northern and southern populations. At 17°S the reporting rate above 500 m (1,640 ft.) was 24 percent and below 500 m, 19 percent, whereas at 30°S the corresponding rates were 0 percent and 55 percent. Sindel states that it is mainly nomadic but that resident populations have been established, particularly in Brisbane and Gold Coast areas. Alan Lynch, a friend of Sindel's who resides 12 km (7 mi.) from Brisbane, has fed lorikeets on his property for many years. Every afternoon flocks gather, consisting of 50–80 percent scaly-breasts, with fewer rainbows. Numbers of scaly-breasts vary from more than 1,000 to only 100. He believes that the scaly-breasts are permanent residents and the others are nomads. In most sites where lorikeets are attracted to artificial food sources, rainbows far outnumber scaly-breasts.

Food: Pollen and nectar from the blossoms of *Eucalyptus* are their most important food source; these lorikeets also feed extensively on *Grevillea*, *Banksia* and *Xanthorrhoea*. Seeds, fruits, berries and insects are also eaten and they will attack ripening sorghum crops when the seed heads are at the milky stage. Cayley quoted Florence Irby (Lendon 1979) from Narrango, Casino, New South Wales. She gave a graphic account of the arrival of scaly-breasted lorikeets.

> In the Richmond River district this is by far the most common species of the Psittaci; every flowering eucalypt, bean-tree, grevillea or other blossom-bearer attracting its quota of joyously screaming green birds. Although nectar-feeders, they come in their hundreds during the autumn months to feast on the sorghum or saccaline heads as they ripen in the fields, often three or four clinging to one head. And they are not easily frightened away; if shot at they will merely fly a little distance and return directly they think the danger past.
>
> In their swiftness and direct manner of flight they remind one of the beautiful Swift Parrakeets. They often fly very high when going to their feeding-trees. I have seen large flocks, looking no larger than mosquitos as they sped across the sky, their gladsome screech making them easily recognizable. Often too they may be heard passing at night. How do they know when the trees are flowering? Waking one morning you find they have arrived in their hundreds where there was not even a sign of 1 the evening before. In the early spring of 1926 the Scaly-breasted Lorikeets came in thousands to the flowering gums in the Casino district. They were accompanied by many Rainbow Lorikeets. The noise they made as they scrambled among the blossoms was almost deafening; and the ground was carpeted with the flowers they had pulled to pieces.

(See also **Honeyeaters, Association with.**)

Breeding usually occurs in spring and summer; there are some records of winter nesting. A hollow tree limb is the preferred site, at heights which vary from as high as 20 m (66 ft.) to as low as 3 m (10 ft.) from the ground. Nest holes may be as deep as 2 m (6 ft. 6 in.) (Sindel 1987).

Aviculture

Status: Common in Australia, uncommon elsewhere.

Clutch: Two, rarely three. Matthews and Matthews (pers. comm. 1995) have had three-egg clutches which resulted in three young fledged. Neff recorded clutches of one egg and three eggs on rare occasions.

Incubation period: 22 or 23 days, up to 28 days; in the cooler months (June to September) the incubation period is longer, according to Shannon Johnson (pers. comm. 1995) of South Australia. Incubation usually commences with the first egg (Matthews and Matthews) or with the second egg (Neff 1994).

Newly hatched chicks: Weight, 6 g (Dear pers. comm. 1996); an incubator-hatched chick weighed 4 g (Sweeney pers. comm. 1996). The gray-white down is 1–2 cm long (Neff 1994). Matthews and Matthews describe the first down as white, fairly long and sparse.

Chick development: The gray second down starts to appear at about 12 days, the eyes are open by 14 days and pin feather development is visible by about 20 days. Neff recorded slightly more rapid development: second down erupts at 9–12 days when the eyes start to slit; contour feathers start to develop at 2½ weeks. Chicks are ringed at 10–12 days with 6 mm rings.

Young in nest: 50–60 days. Young reared by Neff's two pairs left the nest at about 57 days and were independent 14 days later (Neff 1994).

Age at independence of hand-reared young: About 56 days.

Chick growth

Shows the weights of young reared by a pair belonging to Alan Dear of Dromana, Victoria; diet as for rainbow lorikeets (see page 260). Chicks were weighed at 3:00 P.M. daily, thus weights show an apparent but not actual loss on some days, according to crop contents. Both hatched on August 18, 1994.

Scaly-breasted lorikeet weight table #1

Age in days	Weight in grams	
	Chick No. 1	No. 2
day hatched	6	6
1	7	9
2	9	11
3	10	14
4	11	15
5	13	14
6	14	18
7	16	20
8	20	21
9	22	26
10	20	24
11	26	30
12	26	30
13	29	32
14	30	36
15	34	34
17	40	40
19	40	44
21	50	50
23	52	52
25	56	56
27	62	62
29	60	68
32	66	70
35	70	72
38	75	80
41	80	80
44	86	86
47	82	82
50	84	86
52	82	82
57	left nest	left nest

The weights below refer to one chick hand-reared at Loro Parque (Sweeney pers. comm. 1996).

Scaly-breasted lorikeet weight table #2

Age in days	Weight in grams
hatched	4.0
1	4.6
2	5.0
4	6.6
6	9.1
8	11.0
10	13.6
12	17.7
14	22
16	28
18	34
19	39
20	41
22	47
23	52
25	58
27	67
29	78
32	80
35	88
39	93
41	95

General: The scaly-breast has lived very much in the shadow of the rainbow lorikeet; in Australia it is seen in fewer collections. Matthews and Matthews state that it is just as hardy and as easy to maintain as the rainbow lorikeet. On one occasion they had to place two eggs from another pair under an incubating female. All four eggs hatched and all the chicks were reared to maturity by the pair. Some young have been hand-reared. They nest all year round and make excellent parents. Overseas there has been some interest in scaly-breasts but numbers are not high. In South Africa this species is well established. Luch Luzzatto has bred it for many years. One of his females has been without three-quarters of her upper mandible since she was young, yet the pair had reared 12 young in three years by 1996. Sometimes three eggs are laid but only two hatch. In Germany, Ruediger Neff obtained two pairs in 1991. They were kept in a room where the temperature was maintained at 18–24°C during the winter. Flight size was 3.5 m (11 ft.) long, 1.5 m (5 ft.) wide and 2.5 m (8 ft.) high. During a period of 25 months, the pairs produced 17 eggs. Each pair broke the first clutch and two eggs were infertile. All the other 11 eggs were fertile and produced chicks which were reared to independence. The diet consisted of nectar, dry food (see page 47), fruit and green food. The young reared were sexually mature at 12 to 15 months old (Neff 1994).

The display of this species is typical of the genus, the male moving toward the female stretched to his full height. He arches his neck, dilates his pupils, then hops along the perch excitedly, sometimes turning a full circle.

Trichoglossus flavoviridis (Wallace 1863)

This species is comprised of two very distinct forms, with the subspecies *meyeri* being much better known than the nominate race.

T. f. flavoviridis is 4 cm (15/8 in.) longer in overall length and weighs approximately twice as much as *T. f. meyeri*.

Yellow and Green Lorikeet *T. f. flavoviridis*

Synonym: Sula lorikeet

Description: The head is yellow with a blackish appearance, due to the dark blackish or greenish margins of some feathers, the extent of which varies in individuals. This color scheme is accentuated on the cheeks, throat and entire underparts, the feathers being yellow, broadly margined with dark green. The lores are dark, nearly black and there is a narrow brownish nuchal collar. Feathers of vent and undertail coverts are yellowish green with darker margins and the tail is green above and dull yellow below. Underwing coverts are yellowish green. The rest of the upper parts are green. A conspicuous feature of this lorikeet is the area of bare skin around the eye which is orange-yellow (paler in birds not exposed to sunlight). The beak is orange and the cere and feet are gray.

Length: 21 cm (8½ in.)

Weight: Two captive males at Loro Parque, 100 g and 95 g; 2 females, 91 g and 81 g (Sweeney pers. comm. 1996).

Immatures: The birds have less well defined markings; the head feathers are speckled or margined with gray; the iris and beak are brown, and the cere and the skin surrounding the eye are white. Buckell (1996) stated that the shade of head and body was lemon yellow (one bird only).

Range: Sula Islands, to the east of Sulawesi, Indonesia. This lorikeet occurs on the islands of Taliabu and Mangole, the two largest islands of the group. Whether it occurs on any of the many small islands in the same group has not been recorded.

Habits: The Sula Islands probably had not been visited by ornithologists for many years until 1988 to 1990 when Taliabu, Mangole and Sanana were visited by BirdLife International teams. They recorded that the yellow and green lorikeet was common in selectively logged lowland forest and heavily degraded lowland forest around Menanga base-camp, Taliabu, in flocks of up to 12 birds. "Numbers seemed to decline thereafter with birds in smaller groups and pairs, perhaps suggesting the onset of breeding," (Davidson et al. 1994). Density estimates per km^2 on Taliabu were 159 (64 = standard error of density estimate) in montane forest and 53 (30 standard error) in selectively logged lowland forest. (See also Bean 1996).

Status: Common to very common. Its relative abundance in different habitats on Taliabu was as follows: in montane forest, very common; primary lowland forest, common; selectively logged lowland forest, very common; heavily degraded lowland forest, common; agricultural land, common; it was not recorded in inland swamps. Of 18 species whose density was estimated in selectively logged lowland forest, the lorikeet had the thirteenth highest estimate and of the 10 species estimated for montane forest, it was the fourth highest.

It seems its status has changed little since it was first obtained by Alfred Russel Wallace's assistant, probably early in the 1860s. According to Meyer and Wiglesworth (1898), Wallace's assistant Allen "appears to have found the species abundant. The islands in which he collected were the southern and eastern ones of the group, i.e., Sula Mangoli and Sula Besi."

Aviculture

Status: Rare

Clutch: Two

Incubation period: Not recorded; assumed to be 23 or 24 days.

Newly hatched young: Not recorded; assumed to be similar to *T. f. meyeri*.

Chick development: Not recorded; probably as for *T. f. meyeri*.

Young in nest: 58–60 days (Robiller 1992); one only, 55 days (Buckell 1996).

General: I can still recall the first time I saw this lorikeet, at Vogelpark Walsrode, in 1978. The unusual orange skin surrounding the eye and the larger size surprised me; it looked so different from *mey-*

eri. The first recorded captive breeding had occurred at Walsrode during the previous year. Two pairs were housed in an aviary measuring 2.7 m (9 ft.) long x 2 m (6 ft. 6 in.) wide x 2.2 m (7 ft.) high in the Lori-Atrium. They shared the enclosure with a pair of eclectus parrots. The lorikeets were provided with logs 40 cm (16 in.) high and 25 cm (10 in.) in diameter in the inner part. The breeding pair defended a territory of 50 cm (20 in.) radius around the nest log and would attack the eclectus (much larger parrots) in this area. After clutches of two infertile eggs from both pairs, a chick hatched on July 1, 1976, from the second pair. It was removed from the nest on August 2 when the parents stopped feeding it, but it died six days later. On January 12, 1977, pair 1 had two more eggs, which proved to be infertile. In February, pair 2 laid; the first chick hatched on March 8 and the second on March 10. Both left the nest on May 7. Pair 1 hatched two chicks by May 27 which left the nest after 58 and 60 days. In 1977, a total of four young fledged, all after 58 to 60 days. However, by the end of the year, one breeding male and all the young had died, most deaths being due to diseases of the liver. Lory breeder Karl Bruch received two of these young, one of which died from a liver condition. One of the original females continued to lay until 1984, when she died due to an accident (Robiller 1992).

A few years passed before the yellow and green lorikeet was imported into Europe again. At the end of the 1980s a few arrived in Germany, and Hubers (1993b) reported that "they are now bred quite frequently." In 1992, Dutch aviculturist Jos Hubers obtained his first pair from a German breeder. He found that males could be quite aggressive and had to be separated from the females on occasions. The first clutches produced by his birds were infertile.

In South Africa, this species is well represented in specialist collections. Gavin Zietsman of Johannesburg and Chris Kingsley acquired pairs in 1993. In 1996, I saw young in the former's collection, also one being hand-reared by Rex Duke. During the first six months of 1996, Dr. Kingsley had reared 11 from two pairs; the young from one of the pairs were parent-reared and those from the other were hand-reared, due to the parents' habit of plucking.

In the U.K., Trevor Buckell was the first person to hatch (and perhaps the first person to keep) the yellow and green lorikeet. Unfortunately, his first two chicks, hatched at the end of December, 1995, died at 29 days, with the parents, after a very cold night. However, his second pair hatched a chick on January 28, 1996, which left the nest on March 24. Two more chicks were being reared at the beginning of May. Trevor Buckell commented that yellow and green lorikeets are much noisier than Meyer's or perfects, and are more reminiscent, vocally, of the ornate lorikeet, especially "the scolding and chatter when annoyed" (Buckell 1996; Buckell verbal comm. 1996). DNA analysis would be of interest in an attempt to ascertain whether the two forms presently given subspecific status are in reality two different species.

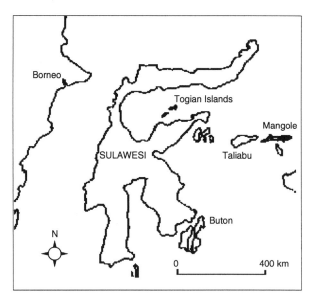

Distribution of *Trichoglossus flavoviridis*.
Sulawesi and Buton—*T. f. meyeri*
Mangole and Taliabu—*T. f. flavoviridis*

Meyer's Lorikeet *Trichoglossus f. meyeri* (Walden 1871)

Description: Distinguished from the nominate race by the broader green margins to the breast feathers, different facial coloration and smaller size. Meyer's lorikeet has the crown olive brown, the ear coverts yellow and the feathers of the lores, cheeks and throat scalloped with brownish green and yellow. The feathers of the underparts are yellow, broadly margined with green on the breast, and more greenish yellow and more narrowly margined with green on the abdomen and undertail coverts. Underwing coverts are yellowish green, underside of the tail is mainly yellowish and the rest of the plumage is dark green, except the forehead. The beak is orange and the iris is orange-red.

Length: 17 cm (nearly 7 in.)

Weight: Males 44–51 g; females, 40–43 g.

Immatures: They have yellow extending to the margins of some of the feathers of the mantle; the breast markings are less well defined with narrower and duller green margins. Head coloration is variable, but usually lighter and yellower, with quite extensive yellow in some young males, reaching to the crown. The beak and the iris are brown. They

are often easier to sex in nest feather. At one month old, the first two which I reared showed the typical dimorphism of head shape, the elder (male) having a noticeably broader upper mandible, head and body. As they feathered, the male's forehead and most of the crown became golden yellow, whereas the female's forehead was lemon yellow and the rest of the crown was green (Low 1977a). However, 20 years later, it appeared that some of the Meyer's available in the U.K. were from a different population. Immature birds from more recently wild-caught pairs had darker head coloration; sexual dimorphism is not apparent in immature or adult birds except in that the males have the slightly larger head and beak.

Sexual dimorphism: Slight; in the male the crown is olive brown and the forehead is golden or strongly tinged with gold. In the female the forehead is olive brown; males usually have brighter yellow and slightly more extensive ear coverts.

Range: Sulawesi (Celebes) (see Gazetteer), Indonesia, in mountain forests between 500 m and 2,000 m (1,640 ft. and 6,560 ft.), also Buton Island, southeast Sulawesi.

Habits: This is a true forest species, common and widespread from the lowlands to upper montane rainforest (Watling 1983). It occurs with the ornate lorikeet, especially at forest edges, but the latter is not found above about 1,000 m (3,300 ft.). Meyer's is usually seen in small groups, sometimes in open country, especially when Erythrina trees are flowering. It tends to be shy, keeping to dense foliage where it is well camouflaged. Holmes and Phillipps (1996) state that Meyer's replaces the ornate in the hills, although there is a considerable zone of overlap.

Nesting: Very little information is available. On February 28, 1980, W. Timmis reported from central Sulawesi: "*T. flavoviridis* is common and is about to start breeding, according to the local people" (Timmis pers. comm. 1980). Two eggs laid in the author's collection measured 21.6 x 18.6 mm and 23.5 x 19.2 mm. Michi and Michi (1983) recorded egg sizes of 21.6 x 17.8 mm and 21.5 x 19.0 mm.

Status/Conservation: It is common, probably with a stable population where forest survives. About 60 percent of the island was under some form of forest cover, most of it in the hills and mountains (Whitten and Whitten 1992). However, much of the lowland areas of south Sulawesi have been converted to agriculture (Cahyadin et al. 1994).

Aviculture

Status: Formerly common but rapidly declining.

Clutch: Two

Incubation period: 23 or 24 days.

Newly hatched chick: Weight, 4 g; dense, longish white down on upper parts; dark brown beak with white egg tooth.

Chick development: Hand-reared: 16 days, eyes starting to open; 18 days, eyes almost fully open, underparts covered in light gray second down; 23 days, feathers erupting on underparts; 29 days, head covered in feathers in quill; 32 days, tail rectrices about 6 mm long; 33 days, primaries 2.5 cm long, feathers of abdomen 80 mm long; 40 days, feathering complete except under wings and on flanks, tail 2.5 cm long, total length about 13 cm; 46 days, fully feathered, beak dark brown; 62 days, flying; about four months, first molt commences.

Parent-reared (one only): 5 days, tracts of developing second down visible under skin of underparts; 7 days, still thickly downed; 13 days, eyes slitting; 17 days, eyes wide open; 19 days, green feathers of ear coverts and tail erupting; 54 days, left nest.

Young in nest: About 52 to 60 days; one chick not well fed, 68 days; one chick only, 64 days (Michi and Michi 1983); two young, 50 to 53 days (Michorius and Wierda 1994).

Age at independence of hand-reared young: 54–65 days. Age at independence of parent-reared young: about 65 days.

Chick growth

Shows weight of two reared by author's pair in a cage in an outdoor birdroom, hatched March and June, 1996. In the second nest, an iris lorikeet was also reared. Rearing food consisted of Aves Lorinectar, apple, pear, millet spray and sunflower seed.

Meyer's lorikeet weight table

Age in days	Weight in grams Chick No. 1	Chick No. 2
hatched	4.5	–
1	5.5 fic	–
2	7 e	–
3	9 fic	–
4	9.5 e	–
5	12.5 fic	–
6	15 vfc	
7	16 fic	–
8	18 fic	–
9	20 fic	19 fic
10	22 e	19 fic
11	26 e	20 e
12	29 e	22 fic
13	30 e	24 e
14	33 fic	–
15	35 fc	28 e
16	38 nf	30 e
17	39 nf	30 e
18	38 fic	31 e
20	40 ne	35 e
22	–	38 e
26	–	42 e
28	–	43 e
33	51 ne	–
36	56 fc	–
40	55 nf	–
44	55 ne	–
54	left nest	left nest
73	–	49

fc = full crop; nf = nearly full; fic = food in crop; ne = nearly empty; e = empty

Weight at independence of parent-reared young: About 48 g.

General: The first commercial importations to the U.K. arrived in 1973. That year the late Mrs. S. Belford brought in nine birds, most of which were in immature plumage. I was entranced by them and bought all nine. When I left the U.K. in 1987, I had been breeding the species for 11 years. It was with great reluctance that I parted with them. Most of my young were hand-reared from the age of about six days and their cheeky, friendly personalities were a source of great joy to me. I ensured they were reared with others of their own species, thus they did not become imprinted and were suitable for breeding. I found them to be continuous nesters, hatching young in any month of the year, indoors or outdoors, in cage or aviary. This is an ideal species for apartment birdrooms, being quiet, nondestructive and very willing to breed in small cages. To me they have tremendous charm and appeal. Indeed, their individuality resulted in me giving many of them names, unlike nearly all the other breeding birds in my collection. When I returned to the U.K. to live, the only lorikeets which I added to my collection were a breeding pair of Meyer's. They arrived on February 15 and I placed them in a heated, outdoor birdroom. Their cage was made entirely of welded mesh, except for the aluminum trays. It measured 2 m (6 ft. 6 in.) long, 43 cm (17 in.) wide and 61 cm (2 ft.) high. It was the top cage of a double unit on a 60 cm (2 ft.) high base, on castors. The pair had been in a slightly smaller cage but settled down immediately. Two days later they were using the nestbox. It was not long before the male started displaying and trying to mate with the female. On nearly every occasion I witnessed he was rebuffed by the female.

She laid her first egg on March 2 (16 days after arrival). The second egg was laid two days later and hatched on March 27. The first egg was infertile. The chick weighed 4.5 g on hatching. It grew rapidly and was ringed at only nine days with a 5 mm ring. The nest litter had to be changed every three or four days, then every two days, as it soon became wet. Despite my attempts to keep the nest dry, the parents plucked the chick's underparts and removed some of its flight feathers. When it left the nest it was completely naked below. On the first day it sat quietly in the cage, occasionally begging for food from its father. Its mother was never seen to feed it. On the next day it knew its way around the cage and entered the nestbox that night. Thereafter it usually

retired at about 5:00 P.M. In the early morning of June 1 the female laid the first egg of the second clutch. I was reluctant to remove the young one because it had left the nest only 12 days previously. However, by 1:00 P.M. the unfortunate youngster had been attacked; several toes had been bitten and one nail was gone. As I prepared a cage for it, I saw the male lunge aggressively at it. A short while later, he fed it. Soon after being placed in the new cage, it was drinking nectar and nibbling at chickweed.

On the same day that the Meyer's laid, an iris who sometimes broke her eggs also laid. I placed the iris's egg in an incubator and next day gave it to the Meyer's. This may have prevented her from laying a second egg. Her egg hatched on June 24 and the iris's egg on the same day or the next day. Both young were slightly plucked—but not as severely as the young one from the previous nest. As the iris has a less placid nature than the Meyer's, I wondered whether this might have been because the iris objected strongly to the indignity of being plucked! A week after the iris left the nest I removed the male Meyer's. He was returned a week later when the two young were removed. The female laid again 13 days later but the eggs were infertile. In the subsequent clutch both eggs were fertile.

My Meyer's are fed on nectar, spray millet, apple, and green foods such as lettuce, chickweed and seeding grasses; also a little sunflower and, when in season, pomegranates. The pair described above eat quite large amounts of dry sunflower; soaked seed is ignored.

Surprisingly little has been written about this species in aviculture. Perhaps it has always lived in the shadow of the equally prolific and more colorful Goldie's lorikeet. But it has its devotees, such as Michorius and Wierda (1994), who describe it as "a lory with much to recommend it!" They mention that it is robust, does not have a harsh voice and can be kept by those who want planted aviaries. On the subject of rearing they report: "Meyer's lorikeets are not tied to any seasons for their breeding. Our breeding pair has been particularly productive. Inside the 10 years they have reared 60 young. Since the last half year the breeding activities have become less intense which may be due to their age. As far as we can ascertain, they both are about 12 years old."

One of my males was still breeding at 11 years old and perhaps continued to do so for some years. At about this age I would expect fertility to decrease, although breeding (or attempts) should continue for another five years or so. Another advantage of this species is that it breeds well in indoor cages thus is ideal for those who have limited space in which to keep birds. I recall how some years ago I placed a pair in a cage in an indoor birdroom. Previously they had always occupied outdoor aviaries. The male had bred but the female, who had not laid during the previous three years, laid within two weeks of being placed in the cage.

Meyer's lorikeet is reminiscent of the budgerigar in its color scheme, warbling song and the way in which the courting male flicks his head; however, head-flicking is more rapid in the budgerigar.

Mount Apo Lorikeet *Trichoglossus johnstoniae* (Hartert 1903)

Synonym: Mindanao lorikeet

Native Bagobo name: *Lish lish*

Description: Distinguished by the pink (dark brick pink) of the forehead (usually brighter and more extensive in the male), throat and forepart of the cheeks and by the band of brown-maroon which extends from the lores through the eye and across the nape. The ear coverts are greenish yellow and the feathers of the underparts are yellow, margined with green, darker green on the upper breast.

Underwing coverts are yellowish green, greener toward the edge of the wing; the first primary is black and all but the first three have a large yellow patch on the middle of the inner webs. Undertail coverts are yellowish green, as is the underside of the tail; upper parts are green. The beak is orange with the base of the upper mandible gray; feet and skin around the eye are gray. The iris is brown.

Length: 16 cm (6½ in.)

Weight: Male, 43–50 g; female, 40–50 g (de Dios pers. comm. 1995).

Immatures: The birds are delicately hued with soft salmon pink facial coloration. (Of two young observed at Loro Parque, this color was more extensive in one—perhaps a male—than even in the adults.) The green margins of the breast feathers are of a softer shade and less extensive. The skin surrounding the eye is whitish. Sweeney (1994b) states that immature birds have black beaks which start to become orange soon after fledging but that "the paler coloration of the cere and periorbital ring remains noticeable for several months after fledging."

Range: Island of Mindanao (37,000 sq. mi. [96,200 km^2] in area), southern Philippines. It occurs on Mount Apo, near Lake Lanao and Mount Katanglad, also Mount Piapayungan and Mount Matutum. The subspecies *pistra* is from Mount Malindang.

Habits: Found in montane forest in central and southeastern Mindanao from 1,000 m (3,300 ft.) to about 2,500 m (8,200 ft.) on Mount Apo and about 1,700 m (5,570 ft.) on Mount Malindang. This species was discovered by Walter Goodfellow in early 1903 on the Apo volcano, the highest peak of the Philippines, whose altitude was then estimated at a little over 3,000 m (10,000 ft.). Goodfellow (1906) described the terrain inhabited by this lorikeet as follows:

The whole mountain is covered with dense jungle up to about 8,500 ft., beyond which comes a broken, white stony slope and crumbling cliffs intersected by many burning fissures from which proceed an incessant noise as of colossal machinery at work underground. The white slope viewed from a distance conveys the impression of a snow-capped summit and it is difficult to believe that it is not so. At sunrise and sunset it glows with all the beautiful tints of a snowy peak.

The upper forests are dark and gloomy, and the thick, hanging—and often black looking—mosses which cover every trunk and branch, give a funereal appearance to the whole. One seems to come upon this depressing region with a strange suddenness, for a little below are many deep arms in the mountain containing hot springs from which steam, always rising, causes a rank growth of the most verdant tropical vegetation imaginable to spring up. Giant ferns and beautiful orchids here struggle with each other for supremacy.

On a bright early morning I first came across *T. johnstoniae*. Ascending from my camp to the upper forest limits, I stood for a while looking up the white slopes to the crater above, when I was surprised to hear the unmistakable sound of Lorikeets chattering nearby. I had no sooner located the noise in an isolated tree which stood a lonely giant amidst all the stunted vegetation around, than a flock of thirty or more flew out, and after circling around at a great height again entered the thick tree top. Beyond noticing the yellow undersides of their wings flash as they turned in the sunlight, it was impossible to gather any idea of their appearance. I was convinced that this must be a new species, but there seemed little chance of securing any specimen there, as the tree was quite inaccessible from where I stood with no means of getting round to it. The next day I was again on my way to the same place, but before I reached the forest limit my Bagobo guide stopped and pointed to some birds he had seen in the thick bushes overhead. Personally, I could see nothing; neither was there a sound to be heard, but at length I noticed the leaves moving, although it was impossible to distinguish any bird so gloomy was it all round. I fired and nothing fell but the air was suddenly alive with the screaming of Lories beyond the tree tops. Even then I could get no sight of them. My boy assured me I had killed one, and at length I could see it hanging by its feet and only after much trouble did we succeed in dislodging it, and then for the first time I held *T. johnstoniae* in my hand.

The above is the only account I know of that describes how a lory species became known to science. Goodfellow was a remarkable collector who obtained many rare parrots for aviculturists, including Mrs. Johnstone, after whom he named this beautiful bird. On that visit Goodfellow's time in the area was cut short by the start of the wet season. On his return journey by boat, one of the passengers was a small boy with a tame Mount Apo lorikeet fastened by the leg to a stick. Goodfellow tried in vain to buy it and the little boy cried in anticipation that it would be taken from him. But Goodfellow, understanding how the child cherished his pet, did not wish to deprive him of it. The bird had come with the Moro people from the Lake Lando region at a considerable altitude.

Goodfellow returned in 1905. Every evening from his camp on Mount Apo he would hear the lorikeets passing overhead in small flocks to their roosting sites lower down the mountain. They returned to the higher forests at daybreak. In the highest village, Tandaya, all day long they frequented trees bearing masses of beautiful scarlet flowers. Goodfellow was told that this village, at 1,200 m (4,000 ft.), was the lowest altitude at which they were seen. Goodfellow kept temperature records at his camp on Mount Apo during February and March. At 6:00 A.M. it was often as low as 1°C (34°F). The highest noon temperature in the shade was 17°C (62°F) but more usually it reached only about 14°C (57°F). Later in the year the temperature would be higher (Goodfellow 1906).

In 1994, Craig Robson observed this lorikeet on Mount Katanglad and at Sitio Siete near Lake Sebu in the southwest of Mindanao. He saw them on five occasions—twice in pairs, twice in a group of four, and once a feeding flock of six to ten birds. The altitude was between 800 m (2,600 ft.) and 1,300 m (4,260 ft.). They seemed to prefer forest edge and deforested and partly cultivated areas, with scattered patches of secondary growth and scattered trees. All observations were made in late April. Robson (pers. comm. 1995) described their calls as "rather raucous, sharp, twangy calls." The flight is fast and direct and they evidently cover considerable distances each day in their search for their favorite trees.

Nesting: According to Dickinson, Kennedy and Parkes (1991) there are no records of wild nests or eggs. However, Harrison and Holyoak (in Forshaw 1989) record the average size of two eggs as 22.1 x 19.1 mm. Newly laid eggs (in captivity) weighed 4.30–5.27 g (de Dios pers. comm. 1995).

Status/Conservation: Like all the birds of the Philippines (which has more endangered species than almost any other country worldwide), it is declining as a result of deforestation. Dickinson, Kennedy and Parkes (1991) describe it as "uncommon and local." De Dios (in Neumann and Patzwahl 1992) suggested that only about 600 birds survive on Mount Apo. Lambert (1993b) assessed the situation on one side of Mount Katanglad (above km post 1,503, north of Malaybalay) in 1989–90. He gained the impression that much of this area would be devoid of trees within ten years unless drastic measures were taken to prevent further forest clearance and fires. (He pointed out, however, that the situation on the other side may be different.) The lower slopes had been nearly cleared; little forest survived below 1,200 m (3,900 ft.). Fires swept through parts of the range in 1983; large patches of dead burned trees could be seen at surprisingly high elevations, 1,500 m (4,920 ft.) to the tops of the ridges. Forest in deep valleys extends down to about 1,100 m to 1,200 m (3,600 ft. to 3,900 ft.), but between these valleys the forest has been extensively cleared for agriculture or has been burned. Impoverished farmers were growing vegetables on the poor soils, outside the valleys, up to 1,450 m (4,750 ft.). Chainsaws were heard daily. Lambert commented that "the outlook for forests on this side of the Katanglad is extremely bleak."

Several other observers in 1993 and 1994 (see *Birds to Watch 2*) suggest that this species is now "very uncommon." According to Sweeney (1994b) the main range of the species is protected by a national park—but "it must be viewed with concern that increasing domestic pressures seem to be forcing industrial development infringement into areas which were formerly covered by the national park. The most high profile of these developments has been the proposed building of a new electricity power plant upon what has been national park land."

Many endemic Philippine birds are already seriously endangered (and some are extinct), so conservation resources have so far been limited to a few critically endangered species. It seems unlikely that much attention will be paid to the lorikeet until or unless it falls into this sad category.

Aviculture

Status: Rare, but increasing.

Clutch: Two

Incubation period: 25 days

Newly hatched chick: Two chicks weighed 3.55 g and 3.65 g; long grayish white down; nails black; bill black with white egg tooth (Sweeney 1994b). Crop capacity (first feed) of newly hatched chick, 0.3 ml (de Dios pers. comm. 1995).

Chick development: The growth of a hand-reared chick was described by Sweeney: both eyes fully open by day 16 when the second down was starting to emerge (much of the natal down had been lost by day 9); second down covered most of body by day 20. By day 25, there were pin feathers on head, flights and tail, and feathers covering the crop were just erupting. By day 28 the head feathers were open-

ing, as were those of flights and tail. Except for the flanks, the bird was nearly fully feathered by day 40.

Photographs (Neumann and Patzwahl 1992) depict two parent-reared chicks at seven and nine days with longish white down, densest on the back; at 24 and 26 days, with the first head feathers just starting to erupt and the feet pink tinged with gray; at 35 and 37 days they were covered in grayish white second down except for the head, where green and pink feathers had erupted in the eldest, and wing feathers were apparent in both; at 43 and 45 days they were almost fully feathered except for upper breast and rump and shorter tail. The black beak of one was visible. My reaction on seeing these photographs was the marked resemblance of these chicks to those of iris lorikeets. They differ in the shade of red: cherry in Mount Apo, more scarlet in iris.

Young in nest: 55 days for young described above; 50 to 60 days at Birds International Inc. (BII) in the Philippines.

Chick growth

Shows the weights of two chicks hand-reared at BII in the Philippines (No. 1 and No. 2), contrasted with one parent-reared (No. 3) (de Dios pers. comm. 1995).

Mount Apo lorikeet weight table

Age in days	Weight in grams		
	Chick No. 1	No. 2	No. 3
hatched	3.5	3.6	–
1	4.1	3.9	–
2	4.4	4.9	–
3	5.1	5.7	–
4	5.8	6.2	–
5	7.0	6.9	–
6	7.7	7.8	–
7	9.8	8.9	–
8	11.1	10.4	–
9	12.3	12.2	–
10	14.8	14.0	–
11	15.6	15.9	–
12	17.3	19.0	–
13	18.6	22.0	–
14	21.7	24.0	20.0
15	22.2	27.0	22.2
16	24.3	31.0	23.7
17	28.0	33.0	25.5
18	29	36	26.0
19	29	39	–
21	31	43	–
22	33	44	–
23	34	45	–
24	35	47	–
25	38	51	–
26	40	50	–
27	44	53	–
28	44	56	–
maximum weight	59	64	–

Age at independence: 70 days for parent-reared young at BII; 56 days for one hand-reared youngster.

Aviculture: Walter Goodfellow was the first to attempt to keep this species—in 1905. He eventually persuaded the inhabitants of Tandaya to catch some. He took the lorikeets down to the coast and on to Davao, but there they suffered from the heat, and three died after several months. He recalled the following:

> It was remarkable how with one exception they quickly got tame; several climbing on to my hand together while I was feeding them. Coming down from the mountains I fed them on wild honey diluted with water, but as soon as I reached the coast I substituted Swiss [condensed?] milk for it: they took to this readily as I have found all Lories will, in fact their supply required to be regulated or they filled their crops so full that it ran from their beaks. This was the only food I could get them to take for a long time, until one after another they began to nibble a little banana which they finally got to like very much. On the voyage home they also readily ate grapes and oranges. They were very fond of bathing and every day plunged fearlessly into the water one after the other until they were completely drenched.

At night they slept together crowded into one corner at the bottom of the cage, with their heads to the wall. After some months they paired off and started to fight. One quarrelsome male had to be removed from the cage.

> At this time they added a sweet warbling love song to their usual calls, and their love dance was really amusing if somewhat absurd. At these times the males swayed backwards and forwards on the perches with all feathers ruffled, and uttering a blowing noise.

The dominant pair ruled the cage. This pair and a third bird (the fourth died on arrival at the docks) went to Mrs. Johnstone in November 1905 as soon as they landed in England. The third bird, a female, was bullied in February so the pair was moved to the aviary next door measuring 3 m (10 ft.) square and 2.4 m (8 ft.) high with a small outside flight. They roosted in a coconut husk and scratched out most of the fiber. After about three weeks the male started to take an interest in the female next door, the "hen sitting inside the aviary looking the picture

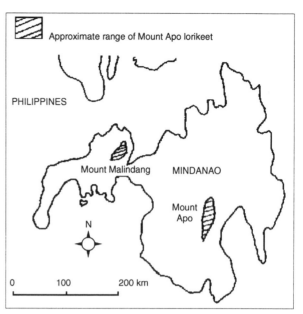

Approximate range of Mount Apo lorikeet (*Trichoglossus johnstoniae*).

of misery." Mrs. Johnstone then placed the single female in with the pair—a very unwise action under the circumstances. (The females should have been changed over.) Not surprisingly, the original female was "almost scalped" and had to be removed. The newly formed pair took possession of a nestbox measuring about 18 cm (7 in.) square and 25 cm (10 in.) high. They spent much time inside and it was believed that eggs had been laid and perhaps hatched. On inspection the nest was empty. The pair nested again and this time two chicks were hatched and reared. The immature plumage was accurately described (Johnstone 1906). While rearing young, the pair fed mainly on spray millet, sweetened bread and milk and half an orange daily.

This species may not have been seen in aviculture again until a pair was obtained by San Diego Zoo. They hatched two young in 1941. It was 30 years exactly before this lorikeet bred there again; during the intervening period it was rare in collections. Strangely, it made a brief appearance in the U.K. (probably due to someone with contacts in the Philippines) in the 1960s. Trevor Flatt from Norfolk caused a stir among parrot enthusiasts by exhibiting a pair at the 1966 National Exhibition of Cage and Aviary Birds. In 1968, it was exhibited at Kelling Park Aviaries in Norfolk. To return to San Diego Zoo, between December, 1971, and July, 1972, 10 chicks were hatched, most of which were hand-reared by an excellent bird keeper, Roby Hewitt. By October, 1973, 18 had hatched, eight of which lived for more than six months. Sadly, the species was not established at that time. It might have vanished forever from the avicultural scene were it not for Antonio de Dios. During the 1980s, in his native Philippines, he was building up what is almost certainly the largest collection of parrots ever assembled. By 1992, it totalled 6,000 birds, including about 30 pairs of Mount Apo lorikeets, plus a number of unpaired males (van Oosten pers. comm. 1992).

In that year de Dios sent 11 pairs to Germany (shared between Vogelpark Walsrode and two breeders), four pairs to Loro Parque and two pairs to Jan van Oosten in the U.S.A. Those at Walsrode were housed together in an aviary in Walsrode's famous Lori-Atrium. The first clutch was laid on May 30 and June 1, 1992. Two chicks were reared. In 1994, Loro Parque was successful in breeding Mount Apos, which have since proved prolific there. The first two young left the nest on July 27, and the parents continued to feed them for approximately ten days more (Sweeney 1995). Jan van Oosten's birds also proved to be free-breeding.

Each pair was housed in a suspended cage measuring 1.2 m (4 ft.) x 61 cm (2 ft.) with a small L-shaped nestbox. They were fed on nectar, apple and a cup of cut fruit (papaya, kiwi, pear, grapes and apple) with cooked brown rice, corn, chopped beans and vegetables, to which were added vitamins and Spirulina. Both females laid their first clutch in November, 1992; thenceforth they usually laid within two weeks of each other. Their first and second clutches were infertile. In March, 1993, both females laid their third clutches—this time both fertile. Between that time and early 1995, 20 fertile eggs were laid, all of which hatched. Six of the young did not survive; four died in the nest during very cold weather and one died during surgical sexing. All the young were parent-reared and some of these formed three unrelated pairs which were set up for breeding (van Oosten 1995).

In May, 1994, Mr. de Dios informed me that there were 58 pairs in his breeding aviaries and that young were no longer removed for hand-rearing. In 1993, 50 were reared; in the first four months of 1994, 34 had hatched. At last this beautiful bird was established in captivity!

Iris Lorikeet *Trichoglossus iris* (Temminck 1835)

Taxonomic note: The genera in which the iris lorikeet has been placed by modern systematists include *Psitteuteles*, by Peters (originally placed there by Bonaparte in 1854), and *Trichoglossus*, by Forshaw (1989) (after Muller and Schlegel 1839-44). In *Lories and Lorikeets*, I placed it with *Glossopsitta* on the grounds of head coloration and epigamic display. Lendon (1973) wrote of the genus *Psitteuteles*: "It seems likely that the Iris Lorikeet, *iris*, of Timor, strangely omitted from Mivart's monograph, fits properly into this genus, as placed by Peters, although it would be interesting to know whether this species exhibits in life the obvious naked periorbital skin so characteristic of *versicolor*." (It does not.) Lendon stated that food is not held in the foot by members of the genus *Psitteuteles* and that head scratching is under the wing. But no lorikeet, except the two species of *Neopsittacus*, is more adept at holding food in the foot than is the iris. After writing *Lories and Lorikeets* I had the

opportunity to keep a female varied lorikeet, probably the only varied outside Australia. It was presented to me by Chester Zoo after the death of its mate. I kept it with an iris, although I never felt it was comfortable with this boisterous male companion. It was so different in temperament, much quieter and more gentle. It was some months before the two birds were compatible. A clutch of eggs laid about one year later was infertile, but I suspected that mating had not occurred. Keeping these birds together as a pair, two species which have similar mating displays, unfortunately, did not provide me with any new insight about their relationship. The varied, iris and Goldie's have all been classified as *Trichoglossus*, and the latter two have also been placed in *Psitteuteles*, which appears to have been a convenient label for small lories that did not neatly fit into other genera. Placing *iris* in *Glossopsitta*, along with the varied lorikeet and the musk lorikeet, might be seen as logical until one examines the other two members of the genus, the little lorikeet and the purple-crowned lorikeet, which are small black-billed species. The little might have been classified as a *Charmosyna* if it had an orange bill! The only answer to this conundrum is to synonymize *Glossopsitta* (and *Psitteuteles*) with *Trichoglossus*. Here I do not recognize *Psitteuteles*; I now believe that the iris is most closely allied with *Trichoglossus*. (See **Systematics**.)

Description: Its most distinctive features are the scarlet forehead and the mauve ear coverts; it also has the feathers of the throat and lower cheeks salmon pink. The crown is partly scarlet but the color varies (see below), extending to the yellow nuchal collar in some individuals—and birds from the same parents vary in this detail. In other respects it resembles members of the genus *Trichoglossus*, with barred breast (greenish yellow feathers with green margins) and yellow nuchal band. The underside of the tail is yellow-green. It is otherwise green (of a particularly beautiful shade), more yellowish on abdomen, thighs and underwing coverts. The iris and beak are orange and the feet and skin surrounding the eye are blue-gray.

Length: 18 cm (7 in.)

Weight: 65 g to 75 g, males averaging slightly heavier.

Immatures: They are quite variable (see also Sexual Dimorphism) but all have duller colored ear coverts and less well defined markings on the breast. The beak is orange or brown, soon changing to orange, and the iris is brownish. One young male, hatched about May 8, had the crown mauve in nest feather (although the father's crown was green). On August 24, the first green feather appeared in the crown and by mid-September the crown was almost entirely green, tinged with mauve (Low 1977b.)

Sexual dimorphism: Slight; the crown is usually more colorful in males. In an earlier work (Low 1977a) I wrote: "The sexes are generally described as being alike. In the male of the four breeding pairs with which I am familiar the breast barring is much more prominent than in the female and the mauve ear coverts are larger. Also, the head is much bolder. The crown color differs in individuals and the evidence so far suggested by taxonomists for separating three races appears to be confusing. From my own observations it would seem that there are two

phases of iris, one in which there is sexual dimorphism of the crown color and one in which the crown color is identical in male and female. The birds which suggest this theory are fully adult, thus there is no possibility that immature plumage is confusing the issue."

Males have the crown scarlet and partly green, or entirely scarlet; in others, the green is replaced by mauve. The first male reared from my original pair lacked the mauve crown of his father in nest feather, when the crown markings were similar to that of the female (red and green) but not so bright. But a male bred in the 1970s from another pair, in which crown coloration was identical in male and female, had a mauve crown in nest feather. At a partial molt at four months, the mauve feathers were replaced by green ones. Later (Low 1986b) I wrote: "This species can often be sexed by the violet ear coverts which extend further downwards in the male, which usually has more pronounced barring on the upper breast." I should have noted the pink feathers of the throat were deeper in color and more extensive in the male. For many years I could sex my young by these features. Then in 1988 I received my first outcross, that is, the only bird which did not originate from those imported into the U.K. in the early or mid-1970s. Jan van Oosten of the U.S. generously sent me a female bred by him in 1987. She hatched her first chick in 1989. Her young did not exhibit pronounced sexual dimorphism. Although this female was visually no different from my females, the coloration of her young males and females was unpredictable. Thenceforth I could understand why at least one aviculturist (Fergenbauer-Kimmel 1992) insisted this species is not sexually dimorphic. The color of the side of the head (and of the ear coverts) appears the same in all individuals and I have never seen one which I could identify as *wetterensis*. It therefore seems unlikely the unpredictable results are due to pairing together two subspecies.

Subspecies: *T. i. rubripileum* from eastern Timor was described by Salvadori (1891) as having the red of the crown extending to the hind part where in some birds it was tinged with green, and in having the ear coverts violet-blue as against reddish violet in the nominate race. Forshaw (1989) incorrectly states "band on hindneck violet-blue." Skins show that in some birds the red extends as far as the nape. Whether these birds are exceptional, or represent a genuine subspecies, is currently unknown. It should be noted that the iris lorikeet is the only lory species in which more than one subspecies has been named in a population from one island (excluding New Guinea). As lories are nomadic—and on Timor there is no known barrier to prevent birds moving throughout the island—this is interesting and perhaps debatable. *T. i. wetterensis* from the island of Wetar is said to be darker, less yellowish green on the sides of the head, and slightly larger (wings of four males averaged 127.0 mm against 117.4 mm in *T. i. iris*).

Natural history

Range: Timor (see Gazetteer) and Wetar, Lesser Sunda Islands, Indonesia.

Habits: This species occurs up to 1,500 m (4,900 ft.). White and Bruce (1986) describe it as "a common lowland species"; then Bruce comments that it was "the least common of the three lorikeets found on Timor." In 1974, Bruce related to Forshaw (1989) his only encounter with this species "on the south coast, near the village of Betano, where a party of about five birds flew in above the treetops, pausing briefly in the canopy of a tall tree before disappearing from view. The flight is swift and slightly undulating."

In his classic *The Malay Archipelago* published in 1896, Alfred Russel Wallace described his travels, which took him to Timor in 1861. He was in the region of Delli, where the country was undulating and open, with small patches of forest in the hollows. His traveling companion assured him this was the "most luxuriant" part of the island he had seen. It was a misty, cloudy area, with seldom more than an hour or two of "fitful sunshine." They were climbing to Baliba, at an elevation of 650 m (2,000 ft.), over bare hills covered with small pebbles and scattered eucalypts. On the way he shot "the pretty little lorikeet (*Trichoglossus euteles*). I got a few more of these at the blossoms of the Eucalypti, and also the allied species, *Trichoglossus iris*" (Wallace 1896).

How ironic that in the literature of more than a century ago, I found the first reference to a food source of this species—yet no further references in the century which followed! It seemed strange that Wallace should refer to *euteles* as the "pretty little lorikeet" yet not comment on *iris*. Did he confuse them in his mind? After traveling continuously in the region for eight years and collecting more than 125,000 specimens, that would not be so difficult.

According to Collar et al. (1994), the species appears to have declined on Timor, "as recent observers have found it scarce; it was seen at only two localities during a nine-week survey of west Timor's remnant lowland forest in 1993." Even less is known, ornithologically, about Wetar. A small team of ornithologists who visited the island in 1990 were the first since 1911. They failed to find the iris lorikeet, although extensive forest survives there.

Nesting: No details of nests have been recorded. One egg in the British Museum measures 23.5 x 19.5 mm. Sizes of eggs laid by my females are variable. The largest recorded was 25.0 x 20.1 mm. Two eggs in one clutch varied considerably: 23.0 x 20.0 mm and 23.2 x 18.5 mm. Others have been recorded as follows: 23.8 x 20.6 mm and 22.3 x 20.3 mm; 22.8 x 20.5 mm and 23.8 x 21.0 mm. Still others, from different clutches, were 23.5 x 19.6 mm, 24.9 x 19.8 mm and 23.9 x 20.5 mm. Newly laid eggs weigh 5 g.

Status/Conservation: Vulnerable and almost certainly declining. Whitten and Whitten (1992) recorded: "East Timor has very few patches of forest left, but some of these contain the last natural stands of very important trees such as the plantation tree *Eucalyptus urophylla* and the Sandalwood. In the inland areas, the steep and relatively infertile mountain soils, combined with the long dry season and low rainfall, influence the types of forest present as well as the behavior and livelihood of the human communities. Fires, deliberate or accidental, sweep through the lowland hills with the effect of increasing fodder production for cattle, goats and sheep. The remaining natural forests are under increasing threats because the inland people need to range widely to find sufficient fodder for their animals. While this did not matter when the population density was low, the rapid increase in the population and their herds, is causing serious environmental degradation."

Among the areas of the Lesser Sundas designated as being of global conservation importance is Mount Mitis on western Timor. Let us hope that the iris occurs there so it is protected in at least one area. It seems that fieldwork is urgently needed to determine the status of this vulnerable endemic species.

Aviculture

Status: Uncommon to rare.

Clutch: Two

Egg interval: Of 24 occasions recorded in author's collection, the interval was three days on 14 occasions, two days on nine occasions and four days on one occasion.

Incubation period: Usually 24 days with the females, but 22 on one occasion; 23 and 25 days also recorded. All artificially incubated eggs hatched after 24 days. Fergenbauer-Kimmel (1992) gives the incubation period (of 12 hatched) as 23 or 24 days; Brockner (1993a), as 24 days for each of the two eggs of one clutch; and Kyme (1975), as 23 days for both eggs in one clutch. The most bizarre incubation suffered by one egg occurred when I moved home and birds, by ferry, from Tenerife to Gran Canaria. During the journey, one egg, which had been incubated by the female for six days, was placed in a tiny box taped under my armpit for about 10 hours. When the incubator was set up in the new location, the temperature fluctuated wildly for 1 week. Yet the chick pipped at 22 days and hatched after 24 days!

Newly hatched chick: Weight, 4 g; longish, whitish down; the beak is brown.

Chick development: At 6 days, sparsely covered with down; 8 days, eyes slit; 11 to 14 days, eyes and ears open; 14 days, thick gray second down erupting; 24 days, some head and flight feathers erupt; 27 days, red feathers of forehead erupt; 32 days, wings nearly fully feathered, beak dark brown, almost black; 39 days, almost fully feathered except tail and rump, beak lighter brown; 45 days, fully feathered, cere grayish. On leaving nest beak is part brown and part orange, or entirely pale orange.

Young in nest: About 56 days; as long as 62 recorded or as short as 48 days, but in the latter case the young one stayed out only briefly for ten days or so. They usually look out of the nest for a few days before actually leaving but a single young one, unusually, was looking out for 12 days before leaving. Fergenbauer-Kimmel (1992) recorded of 12 young that they left the nest after 55 to 59 days; Brockner (1993a) gave the period for two young of a clutch as 53 and 55 days. Two reared by Kyme in 1974 left the nest after 65 and 67 days.

Age at independence: Hand-reared birds, 40 to 58 days; parent-reared birds, about 65 days.

Chick growth

Shows the weights of three parent-reared chicks from different nests from the author's pairs in Gran Canaria in 1989 and 1990. (The sibling of No. 1 was removed for hand-rearing at 27 days.)

Iris lorikeet weight table #1

Age in days	Weight in grams		
	Chick No. 1	No. 2	No. 3
1	6 e		
2	8 fic		
3	10 fic		
5	14 e	11	
6	16 fc	12	
7	18 fic	14	15
8	20 fic	15	18
9	24 e	18	22
10	26 fc	19	23
11	28 fic	21	26
12	30 fic	22 nf	28
13	33 e	25 nf	32
14	35 nf	28 nf	–
15	37 fic	32 nf	–
16	40 nf	34 nf	36 nf
17	44 nf	35 nf	–
18	47 fic	42 fic	49 fic
19	52 fic	41 nf	–
20	54 fic	46	55 fc
21	57 fc	48 nf	57 nf
23	59 nf	56 fic	57 nf
25	59 e	55 nf	–
27	62 ne	57 nf	–
29	64 ne	64 nf	–
31	64 nf	–	68 nf
33	64 fic	66 nf	–
35	63 e	68 nf	70
37	66 fic	68 nf	–
39	64 e	70	–
41	66 fic	–	–
43	64 fic	–	–
45	62 ne	72 nf	–
47	61 fic	–	–
48	–	–	left nest
51	60 e	–	
53	58 e	–	
55	–	left nest	
57	left nest		

e = crop empty; ne = nearly empty; fic = food in crop; nf = nearly full; fc = full crop

Weight of one hand-reared chick

Hatched in March 1988, it was hand-reared from the age of seven days when it was found cold and unfed early in the morning. Rearing food consisted of baby cereal, wheat germ cereal, some baby food in jars and liquidized papaya.

Iris lorikeet weight table #2

Age in days	Weight in grams
7	11
9	14
11	16
12	17 ringed
14	19 eyes slit
18	23
22	32
25	38 head feathers erupt
30	46
34	49
38	55
40	53 feeding on own
44	60
46	61 peak weight
51	58

General: Iris lorikeets were first imported into Europe by dealers in 1972. Since then they have been available only sporadically and in small numbers. A few pairs are kept in Germany, the Netherlands, Britain and Denmark. In 1994, only five were known in South Africa (G. Zietsman pers. comm. 1994). In the U.S.A., numbers are not high but there are a few devotees, such as Roland Cristo in California who has bred iris lorikeets over a long period.

In 1971, Mrs. S. Belford imported a pair into England from Singapore. Soon after, I volunteered to look after them while she visited Singapore. Inevitably, I bought them—and they stayed with me until their deaths in 1992 and 1993. I became a fanatical admirer of this species. Why does it have such appeal for me? First, its beauty; the scarlet forehead and mauve ear coverts are very pleasing. Secondly, all my birds are so tame and inquisitive; I have seen aloof, wild-caught birds in other collections which did not attract me at all. And thirdly, the satisfaction of breeding a species which is not well established in aviculture which, for me, has always been a strong motivation. By working with lesser known species and recording as much information as possible, one can make a greater contribution to aviculture. And, yes, the last reason is quite simply that the species concerned is one which provides me with a lot of pleasure.

The members of my first pair of iris were among the most remarkable bird "characters" I have ever known. The female was also one of the most aggressive. Despite a crippled foot (an injury which had occurred before she came into my care), she was undoubtedly the most fearless bird in proportion to her size which I have ever encountered. Aggressive in the extreme, she had no hesitation in flying at my face or fingers in an attempt to bite me. I tolerated this quite good-naturedly for over 20 years! The problem was that young birds of her own species housed in the vicinity watched her aggressive behavior and tried to copy it—with a degree of success. When the pair was about 18 years old, I gave them new companions because the male was suffering from her increasingly dominant behavior. Although the female continued to lay until she was 18 years old, in all that time she produced only four young, two of which were reared to maturity—both males. One of them, hatched in 1977, was the founder of a new line and was still breeding in 1991. The display is the neck-arching, necks-almost-entwined and hissing typical of *Trichoglossus*, sometimes with another element which is very amusing to watch. When on a flat surface, they progress in exaggerated jumps, then the two birds jump up and down alternately, both feet leaving the ground. More often, however, they are seen bobbing up and down, with the feet on the perch. Neither male nor female dominates the display. Ray Kyme, the first breeder in the U.K., recorded: "Courtship display is mutual. They bow the head down and hiss through the open bill. During bowing the feathers over the shoulders are raised. Perhaps the male is slightly more demonstrative in the display than the female; certainly he is the dominant character and bickering fights are very frequent" (Kyme 1975).

The breeding season appears to be influenced by climate. In the U.K., eggs might be laid during any month of the year. Multiple clutching is not unusual, although not all females lay more than two clutches. In Gran Canaria, the breeding pattern was different, no doubt due to the much hotter climate. There, females usually laid the first clutch of the season in December or January, the second clutch by February and, if previous nests failed, the third clutch in March. On only one occasion were eggs laid as late as May. In other words, they laid during the cooler months of the year. In Tenerife, also in the Canary Islands, where the climate in the north was much wetter and also cooler than Gran Canaria, but much warmer than the U.K., eggs were laid in January, March, May, August, October and November. Armin Brockner (1993a) recounted the results of one pair in Germany: June, 1989, one egg (infertile); 1990, did not breed; May 21 and 24, 1991, infertile eggs laid; January 18 and 20, 1992, eggs laid which hatched on February 11 and 13.

The diet of this species in captivity differs from most other lorikeets, except Musschenbroek's, in that only a small quantity of nectar is consumed. Iris eat a wide variety of seeds, fruits and vegetables. Spray millet and the small seeds of soaked figs are favorite items which suggests that in the wild, seeds form an important part of their diet. It is also interesting to note that iris are more adept at holding food in the foot than any other lorikeet with which I am familiar, with the exception of Musschenbroek's. Again, this suggests that pollen and nectar are not a staple food source. Most iris from the importations of the 1970s did not survive long and I suspect this was because they were fed like typical nectar-feeders—which they are not.

Ray Kyme noted: "Iris Lorikeets are very fond of soft fruit and the youngsters were probably mostly reared on raspberries, indeed they ate these at such a rate that my wife had trouble salvaging enough for the house and jam-making." I have not found them to be especially fond of soft fruit *per se*; what they evidently relished were the tiny seeds in the raspberries. They are very fond of guavas for the same reason. Almost nothing has been recorded about the natural diet of this species. Why does it have such a strong beak? My guess is that it feeds extensively on the small seeds of a fruit with a hard exterior. I have given them the tiny cones of *Casuarina equisetifolia* but they are interested in the branches only as objects of destruction, whereas the small *Charmosynas* to which I offered them soon removed the tiny seeds. Another interesting point is that overgrown beaks are quite common in captive iris lorikeets, as often occurs with captive fig parrots. The latter have to open the hard pericarp to reach the figs.

Kyme's pair was fed on nectar, sunflower seed, apple and spinach beet daily. In the U.K., my iris were formerly fed on nectar, sometimes with sponge cake added, spray millet, panicum millet, canary seed, niger, hemp (of which they were passionately fond), sunflower seed, dried figs which

Range of the iris lorikeet, the varied lorikeet and the musk lorikeet.

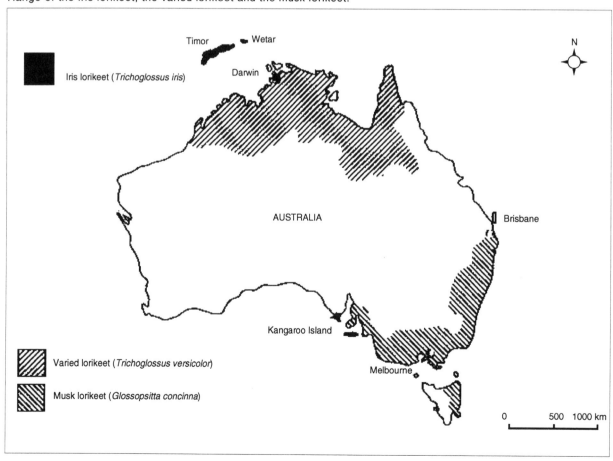

288

were soaked in water, apples, grapes and other fruits, carrot, celery and lettuce and a biscuit-based softfood. Sunflower seed should be given in moderation; however, my pair which lived for over 20 years had unrestricted access to sunflower for more than two-thirds of their lives and they never became overweight. In the Canaries they received various fruits, a mixture of small seeds and sunflower (sunflower being strictly limited in recent years because in the warmer climate some birds were overweight), plus cooked beans and maize, of which they eat little, spray millet rarely, soaked figs and fresh carrot frequently and fresh corn or tinned sweet corn on occasions. Irregularly, they received leaves of lettuce or Swiss chard. Only small amounts of nectar are consumed. Back in the U.K. again, the diet is slightly different. Nectar is always available, together with a mixture of small seeds, in which canary forms the basis. In the morning they receive a small amount of sunflower seed and apple or,

when in season, pomegranate, and in the afternoon they receive either carrot, millet spray or dried figs which have been soaked for a short while. This is a strange diet for a "brush-tongued parrot." I have yet to see the "brushes" on the tongue of an iris and assume that they are very poorly developed, as one would expect in a species which is apparently not dependent on pollen and nectar. Iris should receive twigs of apple or willow, for example, on a regular basis; they quickly remove the bark.

Due to the uncertain, possibly endangered, status of the iris lorikeet and the unstable political situation of the only (or main) island on which it occurs, aviculturists should give more attention to this species. It is of primary conservation interest, with perhaps little future opportunity to conserve it in its natural habitat. Breeders should also be aware of what an attractive bird it is to keep. Compatible pairs can be prolific, and they are suitable for those who are unable to keep birds in outdoor aviaries.

Varied Lorikeet *Trichoglossus versicolor* (Lear 1832)

Ptiloschlera versicolor (Gould 1865)

Synonym: Red-capped lorikeet

Taxonomy: The varied lorikeet was originally named *Trichoglossus versicolor* by Lear in 1832, but in the following 50 years it was placed in five different genera: in *Psitteuteles* in 1856; in *Coriphilus* (*Vini*) in 1859; in *Nanodes*, by Schlegel, in 1864; in *Ptiloschlera*, by Gould, in 1865; and in *Neopsittacus* (!), by Salvadori, in 1882. However, in 1891, when Salvadori compiled the *Catalogue of the Birds in the British Museum*, he placed *versicolor* in the genus *Ptiloschlera*, as its only member. This name has not been used in modern literature.

The appearance of this species is so distinct that the view that it belongs in a monotypic genus has to be considered. If *Ptiloschlera* is restored, however, it would perhaps be logical to reinstate other genera, resulting in a total of 14 or 15 genera. This seems too many, as all lories share a relatively recent origin. Forshaw (1981; 1989) again placed *versicolor* in *Trichoglossus*. I find this very difficult to

accept because in appearance it bears no resemblance to members of that genus. In addition, its incubation and fledging periods are shorter—and similar to those of the genus *Glossopsitta*. In *Lories and Lorikeets* I placed it with *Glossopsitta* on the basis of appearance and display. More recently, Sindel (1987) placed it as the only Australian member of the genus *Psitteuteles*, but firmly stated that it has no close affinities with other genera and should be installed in a monotypic genus. If this is so, it would seem logical to revert to the use of Gould's *Ptiloschlera*; but I am not totally convinced that it is sufficiently different from all other lories to merit this treatment. Two of Sindel's reasons for placing this species in a monotypic genus are not valid. The first is that it is sexually dimorphic; yet other parrot genera contain both dimorphic and monomorphic species, e.g., *Poicephalus*. His second untenable argument is that varied lorikeet chicks hatch with less down than any other Australian lorikeet. Sparse down, however, is an adaptation to a tropical climate, rather than a trait useful in separating a genus. Within the parrot genus *Amazona*, for example, down density varies greatly; it is thick in *tucumana*, which occurs in a temperate climate, and sparse in those from warmer climates. Note that Peters (1937) placed the following species in the genus *Psitteuteles*: *flavoviridis*, *johnstoniae*, *goldiei*, *iris* and *versicolor*.

Odekerken (1994) described the display thus: "The pair will bob up and down sitting side by side and then fluff their chest and rump feathers and stretch themselves in unison away from each other with feet placed on the perch. During the stretch both individuals will move their mandibles and arch their necks, possibly emitting a very low noise, which I have not been able to hear. Occasional more vigorous runs, bobs, and jumps occur and are usually performed by the male." He commented that the display "is one of the reasons that absolutely staggers me as why people place this bird in the *Trichoglossus* genus."

Description: A very beautiful bird, whose plumage is set off by the scarlet forehead, crown and lores. Head coloration is most unusual, the throat, cheeks and back of the head being grayish blue narrowly but boldly streaked with yellow. The entire underparts are vertically marked with yellow streaks—broadly on the vinous pink background of the upper breast and narrowly on the green abdomen. The upper parts are green streaked with greenish yellow. The tail feathers are green marked with dull yellow. Another unusual feature is the striking white skin on the cere and surrounding the eye. The beak is orange and the iris is yellow or brown.

Length: 19 cm (7½ in.)

Weight: 55 g to 62 g (Forshaw 1981); 55 g to 60 g (Odekerken 1994).

Immatures: Birds bred by Stan Sindel (who has extensive experience of this species) were described as being similar to adult females when newly fledged, but without the red cap. There is a reddish frontal band. He stated that captive-bred birds lack the reddish suffusion around the facial areas as seen in immatures in the wild. They acquire the red cap within three months of leaving the nest, i.e., when just over four months of age (Sindel 1987). Stacey Gelis provided photographic evidence that the

young of one of his pairs do not conform to this color development. At the age of 36 days a male had forehead and crown red and lores and cheeks heavily suffused with red. It also had a yellow collar—a feature absent in adult birds. A photograph taken at the same time shows that one of its nest mates, a female, had a large area of yellow on the crown, just extending to the nape. The forepart of the yellow patch was tinged with red (Gelis pers. comm. 1991). In the following year, one chick again had a yellow collar, also a red face; another had a bright red crown and pink face, while the third had only the forehead red. Two young with pastel colors were produced by a normally colored pair owned by Barbara O'Brien in the mid-1980s. They molted out to show normal coloration (Gelis pers. comm. 1996). It is apparent that there is considerable variation in head color of immature birds.

The beak is brown, tinged with orange, and changes to yellow-orange by the age of three months. Matthews and Matthews (pers. comm. 1995) state that the beak changes to orange over the period of three to four months.

Sexual dimorphism: Usually pronounced. The male is more brightly colored, the red on the head being more extensive and more intense. His yellow ear coverts and breast striations are brighter and the vinous pink of the breast is more extensive. In his classic book *Foreign Bird Keeping*, Edward Boosey, who was probably the first person to breed this species outside Australia, stated: "They are not by any means easy birds to sex, though colors in some hens are slightly duller than in cocks, and their red caps are a trifle smaller." As most Australian aviculturists find them quite easy to sex, and as Boosey was a renowned breeder of his era, this is hard to explain. Sindel (1987) states that immature aviary-bred birds are easily sexed, the males being brighter, especially on forehead and ear coverts. He notes that immature wild birds can be easily sexed by the brightness of the red face and ear coverts. David and Elaine Judd hand-reared six young from three pairs in July, 1994. When they were just starting to decline food, David Judd (pers. comm. 1994) noted: "Hens and cocks were easily discernible by the much larger amount of color on the cocks, and by the brighter colors of the head and shoulders."

Mutations: A red-eyed cinnamon varied

lorikeet has been reported in Sydney (Dosser in litt. 1996). A bird incorrectly described as a pied varied lorikeet is illustrated in Peter Odekerken's 1995 book *A Guide to Lories and Lorikeets*. The breeder confirmed that it is a cinnamon purple-crowned.

Natural History

Range: Northern Australia, from the Kimberley region (Western Australia) to the east coast of northern Queensland. Blakers, Davies and Reilly (1984) mention unconfirmed reports from the Atherton region (Cairns area) and as far south as Gin-Gin to the west of Bundaberg (latitude 25°). Forshaw (1981) wrote: "I have searched for it in north-eastern Queensland, even questioning local farmers about its occurrence, and my conclusion is that it is a decidedly uncommon and highly irregular visitor to the eastern extremity of its range." In Queensland, I was told by a dealer in reef fish, who is also a birdkeeper, and who frequently visits the area, that this species is found further north than distribution maps indicate, almost to the tip of Cape York. (He also told me that the birds there lack the pink breast; this would indicate a resident population.) I have been unable to substantiate his information, mainly because so few people go to this area.

Habits: This species is associated with tropical lowland forest and is usually more numerous in dense woodland. It occurs in most habitats with trees, except mangroves. Blakers, Davies and Reilly found that movements of the varied lorikeet "are unpredictable. Sometimes large flocks gather on flowering paperbarks, eucalypts and other species." They state that it is an irregular visitor to Darwin and is sometimes abundant there when trees are flowering, especially in April and May and September and October. However, John McKean, a renowned bird observer, informed Sindel that he had seen varieds in every month of the year in the Darwin area but had never recorded them breeding there. Sindel (1987) believes that the Cape York population is more scattered and flock size is smaller (10–20 birds) than in the west of the range. He commented: "The varied lorikeet could be termed migratory in the Gulf Country or more precisely in the Argylla Ranges which lie between Cloncurry and Mt. Isa. Their arrival in this district, usually in June, coincides with the flowering of the

bloodwood trees *Eucalyptus terminalis* and *Eucalyptus polycarpa* followed by the *Bauhina* (*sic*) and the *Melaleucas*. An old friend, Don Payne, from Cloncurry, assures me this has happened every year for as long as he can remember, a period spanning 50 years. The birds then breed, laying their first clutch of eggs in early July; the youngsters fledge in late August or early September. In a good season when the blossom is plentiful a second brood is often reared. I have seen flocks of thousands of these lorikeets, made up of 75% immatures, screeching and zooming through the river gums along the dry watercourses. When in flight, this species of bird gives the impression of being much larger than it actually is."

Sindel (1987) went on to relate that when the bloodwood trees flowered a month earlier than usual, the lorikeets arrived and bred a month earlier. The belief of the local residents is that the varieds migrate from the Northern Territory, but this has not been proved. How they coordinated their arrival with the early flowering bloodwoods is not understood.

Odekerken (1994) observed a pair and their two offspring feeding in a flowering tree. They were hanging upside down, batlike, for a full 15 minutes. The adults were hanging on both sides of their youngsters and were preening them. A short siesta followed with all birds maintaining this position. It was the only time he had seen this habit.

Queenslander Allan Briggs, who regularly visits the Cape York region, told me that varieds there are inadvertently killed in huge numbers. They die in troughs of molasses which the farmers provide for their cattle. Surely this practice should be investigated if it is truly responsible for many deaths.

Phillipps (1903) quoted Fred L. Berney, from northern Queensland, who reported in November an interesting incident with this species, on p. 218 (1903) of the Australian journal *Emu*: "I examined 3 specimens recently which suicided in a well. They were all females, and, like the one I skinned, contained in their ovaries only very minute eggs. The bird sent fell into the sheep water-trough. I rescued it (only to make a specimen), when it squealed so vigorously that in an instant I was standing in a cloud of the Parrots, which settled on my arms, hands, shoulders and hat till they weighed down the broad felt brim of the latter, almost to shut out my sight. There must have been 2 or 3 dozen on me. It was a wonderfully pretty sight."

In his *Handbook of the Birds of Australia*, published in the nineteenth century, the famous naturalist John Gould quoted what he had been told by Gilbert, one of his collectors. It "congregates in immense numbers; and when a flock is on the wing their movements are so regular and simultaneous it might easily be mistaken for a cloud passing rapidly along, were it not for the utterance of the usual piercing scream, which is frequently so loud as to be almost deafening."

Massive aggregations of the varied lorikeet are, fortunately, not a phenomenon of the past. Don Franklin of the Wildlife Research Unit, Palmerston, states that in the tropical wet season, varieds concentrate in the region between Pine Creek and Katherine. They are known to feed on the nectar of the bridal tree (*Xanthostemon paradoxus*). From June, 1995, Franklin visited Yinberrie Hills and the surrounding lowlands in every month. He first noted varied lorikeets on August 12, and continued to see them on every trip thereafter until May, 1996. Initially numbers were small. In December, varieds became abundant and were seen feeding in the hills and in the lowlands, on the flowers of the bridal tree. He was impressed by the lorikeets' abundance, which was far greater than he had observed in two years in the Top End. By January 29, the majority of the flowers had wilted, but the round-leaved bloodwood (*Eucalyptus latifolia*) had burst into a mass of blossom. This is the dominant tree over perhaps 100 km^2 (39 sq. mi.) of rolling granite country and almost every tree of the species was flowering. It was in this tree that the varieds were then concentrated. In his diary, Franklin noted: "Literally everywhere you go you see flock upon flock upon flock. They flush across the road in front of you, and I even saw a flock from the radio-tracking plane."

At every place it was the same: a constant cacophony of hundreds of varied lorikeets, regardless of the time of day. During a visit to the area between February 16 and 18, the varied lorikeets were still very abundant. On February 27 they were scarce. A few were present in March and April but none were recorded in May. Franklin believed that "the numbers could only be described as 'immense,' a description that may not have been applied to the species since the observations of John Gilbert and G. A. Keartland last century" (Franklin 1996).

Nesting: Although breeding might occur at any time, normally the northern dry months, April to August, are favored (Forshaw 1981). The preferred site is a eucalypt or paperbark near water. Sindel (1987) makes the interesting observation that most nests which he has observed have been in limbs angled from horizontal to 45 degrees. They are prepared by male and female and are often more than 1 m deep. The two to four eggs are laid on wood dust or on eucalyptus leaves. A set of four eggs from Cape York Peninsula averaged 24.0 mm x 20.0 mm.

Status: Common—more or less so than the red-collared lorikeet, according to locality and season. Forshaw (1981) concluded that it is an uncommon and highly irregular visitor to the eastern extremity of its range. Its occurrence is unpredictable; even during widespread flowering of eucalyptus they may be rare in some districts but abundant in others. Odekerken (1988) pointed out how irregular is its appearance in the northern part of Queensland: "During my two-year stay in the 'Top End,' I did not see many varieds on the Adelaide River flood plains and in mid October 1986 large flocks were found feeding beside the Arnhem highway for approximately a week. Even though I traveled this road regularly, it was the first time in two years that I met with this species feeding in the area."

Aviculture

Status: In Australia, formerly rare, now uncommon but increasing; elsewhere, was rare, now unknown.

Clutch size: Normally three or four.

Incubation period: Usually 22 days; 21 days for two eggs of three when female commenced incubation with third egg, both hatching on same day (Lewis pers. comm 1996). Incubation usually commences with second or third egg (Matthews and Matthews pers. comm. 1995).

Newly hatched chicks: One weighed 2.71 g (D. Judd pers. comm. 1994); two in same clutch 2.6 g and 2.8 g (Lewis)—photographs of these chicks show the down as white, sparse on the head, very sparse on the middle part of the back and longer but still sparse on the lower back and thighs; beak brown.

Chick development: A parent-hatched chick (with some food in the crop) weighed 3.48 g at approximately 24 hours and another 3.59 g at approximately 48 hours (Judd pers. comm. 1994). From photographs taken by Rachel Lewis, development was seen as follows: six days, almost naked, except on back where tracts of developing feathers are visible below skin and on wings; eyes slitting. (Eyes fully open at 17 days.) Three weeks, white second down visible on nape and body; contour feathers erupting over most of body except nape; legs grayish, beak orange-brown. Six weeks, fully feathered except for shorter tail, with plumage coloration very muted; feet gray, beak brown.

Young in nest: From 37–41 days (various breeders).

Age of hand-reared young at independence: About 52 days.

Weight of hand-reared young at independence: 3 males, 47 g, 49 g, and 49 g; 3 females, 41 g, 46 g and 51 g. All of them were reared together by David and Elaine Judd.

Chick growth

Shows the weights before first feed of the day of one hand-reared from 14 days by Rachel Lewis.

Varied lorikeet weight table

Age in days	Weight in grams
hatch	2.8
15	15
16	15
17	16
18	17
19	18
20	20
22	22
24	25
26	28
27	29
28	31
29	33
30	35
32	38
34	40
36	43
38	47
40	47
42	48
44	48
46	46
50	47
55	49
79	50

Two chicks which probably hatched on August 5 and 6, and one which hatched on August 9, were removed for hand-rearing on August 30. They then weighed 49.8, 47.8 and 37.9 g; the first two were feathering well and the third had feathers just erupting (Judd).

General: Boosey (1956) described this lorikeet as "charming and beautiful" which perfectly sums it up. Its beauty has a special quality which is difficult to describe and which even some of the most talented bird artists have not quite been able to capture. The first live varied lorikeets were not brought to Europe until 1902. Reginald Phillipps (1903) lamented the fact that it was "humiliating" that the honor "of first bringing live examples of this interesting species from Australia does not fall to the Britisher, and that only the 'remnants' should have come to us." Apparently, they were imported by an Italian dealer and about nine pairs of these were taken to London. Mr. Phillipps obtained two of the birds, which he believed to be a pair. He described the male's habit of standing "stiff and straight as a bit of wood, head down and tail up at an angle of about 45 degrees. Sometimes he varies this position by sitting (quite naturally and easily) as it were on the side of the perch instead of on the top, the head being straight down, the tail pointing to the skies, in which position he will remain perfectly stiff and still for 40 to 60 seconds." Stacey Gelis noticed that when a pair, sitting together on a perch, becomes alarmed, male and female stretch their bodies and necks at a 60-degree angle away from each other to form a V (Gelis pers. comm. 1996). Peter Odekerken (Low 1977a) described another form of behavior in his pair: "They liked to lie along a branch, stretched out with one wing on one side of the branch and their head on the other; once I noticed them like this for ten minutes. From this stretched-out position they propelled themselves along the branch with their feet, keeping their breast on the perch and moving the head from side to side."

They were fed on a milk food, also grapes, dry crumbled biscuit (sweet but plain), and sultanas, of which they were especially fond. Peter Odekerken (1988) offers his a dry mix (as fed by Stan Sindel) and a liquid food consisting of 2 parts each of Heinz High Protein baby cereal and finely crushed Nice biscuits (Arnott's brand) and 1 part of wheat germ cereal, plus small quantities of Sustagen Sport Mineral and multi-vitamin powder and 2 heaped table-spoonfuls of Nestle's malt powder. This was mixed with brown sugar or honey and warm water. The varieds also received apple, pear, celery, silver beet, carrot, corn and lettuce. Sponge cake moistened with sugar water was "greedily consumed" while young were in the nest.

It is apparent that the most successful breeders of this species pay special attention to the diet. Sindel warns that varieds are prone to obesity. David and Elaine Judd, of Mildura, Victoria, feed their own nectar mixed with 1 part of Wombaroo Lorikeet Wet, also orange cake. This is a homemade pound cake to which is added fresh oranges, calcium carbonate and salt. Rolled, dried raisins (not soaked) are a favorite food. Fruit such as rock melon, pear, orange and apple is also eaten.

Some Australian breeders who specialize in native lorikeets have found the varied harder to breed than the other species. Matthews and Matthews suggest a logical reason for this, and noted the following: "In the early days they often went to nest very early in the season while the weather was still very wintry. This often caused problems with the hens not brooding well enough and the young succumbing to frost-bite. We related this to the species' northern origins where the weather is somewhat warmer, even at that time of year. We suspect that our early pairs were very close to wild stock. It appears that second and third generation females are not behaving in this manner and are good brooders."

Other possible reasons why difficulties are experienced in breeding this species are that the diet is incorrect, the pair is not compatible or the climate is too dry (they may need the stimulus of rain). The fourth reason might be that the nest site is not to their liking: note Sindel's observations (see above under Nesting) regarding the angle of wild nests. The Judds obtain excellent results from three pairs in a planted aviary which contains no other species. The nests are hung at an angle of about 45 degrees. Eleven birds are housed in this aviary but only three pairs breed. This may be related to the size of the enclosure, which is about 8 m (24 ft.) long and 2.1 m (7 ft.) wide. Possibly in a larger aviary all would nest. However, not all those who have attempted to breed this species on the colony system have found this to be a good idea. Placing all the birds in the aviary at the same time may

be crucial for success, but is no guarantee. J. L. Mitchell placed 12 birds in an aviary measuring 4.9 m (16 ft.) long x 3 m (10 ft.) wide with a shelter attached which was 1.8 m (6 ft.) long. A dozen logs were provided but no breeding attempt was made until the third year, when two females produced clutches of four eggs. All were fertile but the only chick to hatch died after a few hours in the bleak July winter weather. There were numerous instances of interference and once or twice daily all the birds would start screeching, causing the incubating females to leave the nest and join in. One bird was killed at once when it entered the log of an established pair. The second clutches were equally unsuccessful, so Mr. Mitchell removed all the birds but one pair. They nested in a log 1 m (3 ft. 3 in.) long and 23 cm (9 in.) in diameter. The log was hung at an angle of approximately 45 degrees. A chick hatched on September 29 and left the log after 43 days, on November 10. It was completely independent one week later. The rearing food consisted of apple, bread and milk sweetened with sugar or occasionally with honey, and flowering eucalyptus branches when available. During the rearing period more bread and milk and more apples (one per bird per day) were consumed (Lendon 1973).

Varieds are not normally aggressive toward other species. Some breeders have kept them in planted aviaries with finches without any harm to the smaller birds. The breeding of a hybrid was an unexpected event that occurred in the aviary of a friend of Harald Mexsenaar, John McKean. He had eight lorikeets in an aviary—five purple-crowned, two little and a male varied. When he caught the lorikeets to transfer them to Mr. Mexsenaar's aviary while he went overseas, he found a chick in a nestbox, aged about three weeks. The nestbox was transferred to the new location and the parents continued to feed the chick. The parents were the male varied and a female purple-crowned! Mr. Mexsenaar photographed the chick just before it was fully feathered. It was a darker green than a varied, with a dull red frontal band and bluish streaks, reminiscent of the varied, on the upper breast. When fully feathered it had yellowish green ear coverts; the cheeks, throat and upper breast were strongly suffused with pale blue. It had characteristics of both species, but the eyes had a grayish blue cast to them and the bird seemed to be short-sighted.

Goldie's Lorikeet *Trichoglossus goldiei* (Sharpe 1882)

Synonym: Red-capped streaked lory

Taxonomy: At the time of writing, it is not clear in which genus Goldie's lorikeet should be placed. Salvadori (1891) favored *Glossopsitta* and Peters (1937) placed it in *Psitteuteles*. In appearance it seems to me to have some similarity to *Oreopsittacus*, in head coloration and bill color and in general shape. Joshua (1993a) states that recent studies of chromosomes have shown that "it has differences to the species within *Trichoglossus*." Its behavior is also different from members of that genus. Here it is retained in *Trichoglossus* because as yet there is not sufficient evidence to place it elsewhere.

Description: This is a distinctive species, with its scarlet crown and forehead and vertically streaked underparts. The area bordering the crown and the eyes is mauve and the cheeks and facial area are pink, with the dark green streaks of the underparts extending upward to the lower part of the cheeks. The entire underparts are streaked (more broadly on the sides), also the nape and the undertail coverts. Upper parts are dark green. The beak is black and the iris is dark brown.

Length: 19 cm (7 in.)

Weight: 55–65 g (Bosch 1993)

Immatures: The plumage is duller and streaking less pronounced throughout. Newly fledged young are dull plum-colored indistinctly streaked with black on the crown and cheeks and usually slightly brighter red on the forehead. The hindcrown may be green or dull purple. There may be a small area of blue below the eye. Forshaw (1989) states that the crown is green, variably marked with red—but "variably marked with red" suggests an immature bird molting into adult plumage. The cere and the skin surrounding the eye are whitish. Kapac (1985) states: "Goldie's are easier to sex upon just weaning when the male shows more color on the head."

Sexual dimorphism: Head and bill are usually slightly larger in the male. Some breeders believe that the mauve on the head is more pronounced in the male. Bosch (1993) states that the "adult male is heavier in build and has more red on the top of the head." Joshua (1993a) noted that "the margin of purple head coloration is also often straight in males, while in females the line behind the eye has the green body color cutting into the purple in a V shape. This however is not an absolute measure." The larger head and beak of the male and the slimmer, more feminine appearance of the female are perhaps the best indications of sex where several birds are available for comparison.

Natural History

Range: New Guinea, in the main central mountain ranges, also in the mountains of the Huon Peninsula.

Habits: In the Eastern Highlands, Diamond (1972) found it between 1,700 m (5,500 ft.) and 2,580 m (8,500 ft.). He mentioned that Bulmer commonly saw flocks of up to 40 in December and January but only occasionally at other times of the year, implying local migration. Coates (1985) found it to be "generally rather scarce, locally common, often rare or absent." Only in Goroka town could he be sure of finding this small lorikeet. It was always present in flowering eucalypts and silky oaks. He usually observed it in small parties and flocks of up to 30 or 40 birds, congregating in flowering trees. Brown (1978) also found Goldie's to be numerous in Goroka, in the Central Highlands. At the time of his visit in mid-January, the flowering eucalypts "attracted myriads of Lorikeets, presumably from the whole surrounding area, for the number of birds in perhaps 15–20 trees must have been in the region of 1,500–2,000 birds in all, with the Red-capped Streaked Lorikeet [Goldie's] most numerous.... Whilst the feeding activity was at its highest perhaps an hour after dawn there were birds arriving in the trees to feed at the break of day. It was fascinating to watch the activity build up in the trees as birds arrived in small groups from all directions. After they had fed they would usually depart in the same small groups, up to 7 or 8 but sometimes in pairs." In the Ok Tedi area, Western Province, Gregory

(1995) describes Goldie's lorikeet as rather rare; it was recorded by Murray at 700 m (2,300 ft.) and by Gregory at 2,100 m (6,900 ft.). In the Baiyer River area, it is a common nomad, in small flocks. In the Port Moresby region it is sometimes locally common, observed by Coates from mid-June to early January. Sometimes seen flying high over the forest in flocks, its flight is swift and direct. Forshaw (1989) quotes Bell that at Brown River, near Port Moresby, in July, 1976, there was an influx of this species and of fairy lorikeets, both normally associated with midmontane habitat, and of red-flanked lorikeets. In the Western Highlands, Graeme George encountered flocks numbering from 40 to more than 100 in casuarinas in open country (Forshaw 1989). Beehler, Pratt and Zimmerman (1986) state that it occurs throughout the Central Ranges from 1,500 to 2,800 m (4,900 to 9,100 ft.). It is gregarious and nomadic and may make long daily flights to feeding sites. Beehler et al. (1994) observed Goldie's lorikeets in the western Central Province, in the Lakekamu-Kunamaipa Basin. They were present in large numbers in 1982. Flocks were seen flying at great altitude, morning and afternoon. Apparently they were leaving to feed and then returning to customary roosting sites each afternoon. They were present in smaller numbers in April and July (but not in August), 1992.

In December, 1994, above Ambua Lodge, at Tari (Southern Highlands Province), at about 2,300 m (7,500 ft.), I had fleeting glimpses of two or three Goldie's feeding in the top of a tall white-flowering tree along with Stella's, Josephine's and Musschenbroek's. The Stella's were comparatively easy to observe, despite the height of the tree; the Goldie's were by far the most difficult. Without binoculars, it would have been impossible to pick out the occasional speck of red which was a Goldie's head. For a few minutes a Goldie's perched in the crown of the tree, at the side, its head and shoulders just visible through the binoculars. This incident is etched on my memory, for distant as the bird was, it was clearly recognizable. In fact, at that time, there were at least 30 Goldie's in or near that tree; when a harpy eagle flew over they were revealed—scattering in all directions.

Nesting: Almost nothing is known of its nesting habits in the wild. It has been suggested that this species nests in tall pandanus. Judging by the reluc-tance with which wild-caught birds enter nestboxes, I suspect that Goldie's may also nest in "moss gardens." No eggs of wild birds have been described. Measurements of some laid in captivity are as follows: average of four eggs 23.0 x 19.8 mm (maximum size 23.5 x 20.0 mm); weight 4.5 g and 4.7 g (Bosch 1993). Average of 20 eggs 23.0 x 20.0 mm; weight 4.6–4.8 g (Regler 1994). Two eggs 23 x 20 mm; weight 4.5 g (Gerischer 1990).

Status/Conservation: The least common of the midmontane lories, according to Diamond. The general opinion is that it is generally scarce and only locally common. Because it is nomadic and/or migratory it is difficult to study and therefore there is absolutely no indication of its true abundance. Lambert, Wirth et al. (undated), who tried to estimate the populations of all parrot species, suggested a total population of more than 100,000 for Goldie's. To even try to make an estimate seems unrealistic. As a midmontane species its habitat may be more secure than lories from lower altitudes. Its skins are not in demand by native peoples for decoration, thus no threats to its existence are known.

Aviculture

Status: Common, becoming less so.

Clutch size: Two

Incubation period: Usually 23 days, but 22–24 days recorded.

Newly hatched chicks: Weigh 4 g and are covered in white down.

Chick development: One hand-reared chick, whose development was slightly less rapid than that of parent-reared chicks from the same pair, had its ears open at 16 days, when the eyes were still closed; at 18 days the second down was just apparent and at 20 days it was covered with the light gray down, except on crown and cheeks; at 32 days the breast feathers and the red feathers of the forehead were erupting and the bill was black and the feet bluish; at 38 days the underparts were half feathered and the second down was thick; at 44 days it was fully feathered but for the nape, and the tail was 2.5 cm (1 in.) long. At 61 days it was independent and weighed 45 g. This bird was a female and reared young from the age of two years.

Young in nest: 57–62 days (various reports); 7–8 weeks (Regler 1994).

Chick growth

Shows weights in grams of two young, No. 1 and No. 2 reared by the author's pair, hatched in May 1984. Chick No. 3 was hand-reared at Palmitos Park, from the age of 17 days, and hatched in February 1990. Weights are those before and after the first feed of the day.

Goldie's lorikeet weight table

Age in days	Weight in grams		
	Chick No. 1	No. 2	No. 3
12		13	–
14	17	–	–
18			25/28
20	–	–	28/31
25	–	–	34/42
26		32	34/42
28	32	–	39/46
30	–	–	42/47
32		41	45/51
34	41	–	49/56
36		42	50/60
38	49	–	51/57
43		56	50/55
45	60	–	58 independent
47		60	58
49	61	–	58
50		58	54
52	58	–	54
60	left nest	left nest	50
67	–	–	51

Maximum weight after four weeks, 60 g to 70 g (Bosch); weight in fourth week, 65 g to 72 g (Regler 1994).

Age at independence of hand-reared young: 45 days (one only).

General: Perhaps no other lorikeet achieved such rapid popularity as the Goldie's. Its good qualities were soon apparent: it is quiet, nonaggressive and free-breeding. First exported to Europe on a commercial basis in 1977, by the mid-1980s it was firmly established in aviculture. The first consignment to reach the U.K. was imported by Mrs. S. Belford. I can still recall seeing the 30 birds, all together in a single cage (by no means large). But they thrived; not one was lost. I bought seven, for I was enchanted by my first sight of this species. I soon discovered that they lacked the playful and boisterous ways of *Trichoglossus* species. Often they creep about in a cautious manner, with the head held low, almost horizontal to the body. This is so unlike the energetic manner of *Trichoglossus*. Bosch (1993) perfectly summed up their nature: "Even the tamest bird always is a little bit timid and one has the impression that all their behavior (with the exception of their voice) is subdued and secretive in contrast to their more turbulent relatives."

Bosch had a trio consisting of two females and a male which repeatedly reared young successfully. The females were probably sisters and had been together from an early age. All three were very friendly when not breeding. At the onset of nesting the male always favored one female to whom he remained faithful, the other female's eggs being infertile. I had exactly the same results with a trio of Goldie's kept together for about a year. They were young; both females would usually lay at the same time but only one female produced fertile eggs. Bosch inferred that both his females laid in the same nest, whereas the females in my care used separate nestboxes a few inches apart. I know of several instances of Goldie's being successfully bred on the colony system, with up to 10 birds in an aviary (planted in one instance). However, others who have attempted this have not been so fortunate. After all, just as with humans, it only needs one aggressive individual to upset a harmonious society. An attempt to let three pairs of Goldie's choose their own partners in an aviary 2.1 m (7 ft.) long, 1.2 m (4 ft.) wide and 1.8 m (6 ft.) high failed. It contained plenty of branches and separate feeding stations. A pair was formed within weeks. Several days later a male was found dead; no disease was found on necropsy. Days later a second male was found trapped in the wire mesh and was being attacked by the other Goldie's. This male was saved and all attempts at colony breeding were abandoned (Casmier and Wisti-Peterson 1995).

Dick Schroeder of California attempted to breed from a pair of Goldie's in a large planted aviary containing eight species of softbills. They were a proven pair but made no attempt to nest in this aviary, probably being intimidated by the larger birds. When they were moved back into a cage, they nested again. Exactly the same happened at Pittsburgh Aviary (Schroeder 1995b).

Dr. Merck (1983) described unusual behavior in this species. His pair had been provided with a horizontal nestbox for roosting and, at a later date, a natural log. The female chose the new log but the male stayed in the old one. Eggs were laid on April 20 and 22, and two chicks had hatched by May 16. They left the nest on July 17 when the male showed a hostile attitude toward them. Apparently he did not acknowledge the young as his own, perhaps because they did not come from his nestbox. The female fed the young for some weeks after they fledged and eventually the whole family lived in harmony.

Regler (1994) had been breeding Goldie's since 1983. His two pairs were kept in an aviary with indoor quarters 2.5 m (8 ft.) x 2 m (6 ft. 6 in.) wide x 2 m high; the outdoor flight was 3 m (10 ft.) long, 1.5 m (5 ft.) wide and 2 m high. To stimulate them to breed in winter, artificial light was gradually increased to give 15 hours daily and indoor heating was increased to 15°C (59°F). The nestboxes provided measured 40 cm (16 in.) high, with an inner diameter of 15–20 cm (6–8 in.). Holes were drilled in the floor, which was covered to a depth of 5 cm (2 in.) with sterile sawdust. In 1990, one pair reared six young in three clutches, and another pair four in two clutches. The oldest pair, a 14-year-old male and a 12-year-old female, were still producing two clutches annually—but only one chick per clutch.

Regler, a German aviculturist, must be one of the most consistent breeders of this species, which was not available until 1977. The three pairs which I bought in that year each took a year to enter a nestbox! This species was first bred in the U.K. in 1979 and my birds first bred in 1980. One pair had the choice of a small natural log, hung horizontally, and a nestbox which measured 11 cm (4 in.) square and 23 cm (9 in.) high. They chose the box. Both chicks hatched and were reared by the parents. The

Trichoglossus goldiei, aged 38 days weighing 55 g.

first left the nest at eight weeks old. The rearing food had consisted of nectar (made from baby cereal, glucose, malt extract and condensed milk), soft pear, apple and spray millet. The young birds nibbled at the leaves of the growing elderberry which nearly filled their small aviary; the adults did not. The nest was inspected only about once a week, when the nest litter was changed.

I cannot recall these birds being especially sensitive to nest inspection, unlike those currently in my care. The problem with them and their offspring has been that they do not keep the nest dry. I recall that when I started to work with this pair they had two young chicks. I found the nest in a wet state, changed the wood shavings and replaced the chicks. A couple of days later I found both chicks dead. With sorrow I realized why. The adults were not used to nest inspection and had not returned to the nest. Six years later, almost to the day, I found a wet nest and two new laid eggs from a pair, one of which was their offspring. I had not forgotten that incident; thus, three hours later, I was not surprised to find that the female had not returned to the nest. I placed her eggs in the incubator and put two infertile eggs of fairy lorikeets (the nearest-sized eggs to their own available) in the nest. I guessed that when they went in the nest to roost at night the female might settle on the eggs again. Next morning the substitute eggs were warm so I replaced them with her own.

On a visit to Copenhagen, I watched Danish breeder Jan Eriksen offer the bushy seeding heads of canary grass (*Phalaris arundinacea*) to his Goldie's. These were consumed with great enthusiasm.

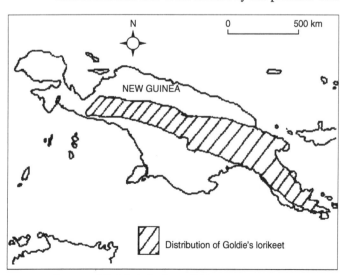

Range of Goldie's lorikeet (*Trichoglossus goldiei*).

Genus: *Glossopsitta*

It is difficult to define the characteristics of the members of this genus, except to state that they are small Australian lorikeets, with short, wedge-shaped tails; they weigh between 35 g and 65 g. Forshaw (1981) suggested that my proposed inclusion (Low 1977a) of *Trichoglossus goldiei* in this genus was "worthy of close scrutiny" but opposed the inclusion of any other species. In my earlier work, *Lories and Lorikeets*, I included the varied and iris lorikeets (*Trichoglossus versicolor* and *T. iris*) in this genus, mainly on the basis of some aspects of behavior. Then I noted: "A characteristic of the display which has not been observed in any other genus is that both birds sway away from each other in a circular fashion as though their feet were anchored to the spot."

In 1990, John Courtney wrote to Dr. Richard Schodde at the CSIRO Divisional Headquarters in Lyneham, ACT for his comment on this behavior and how significant it might be. He replied that many parallelisms in behavior may not be as indicative of phylogeny as appears at first sight. Evidence from the examination of egg white proteins (work in which he had been engaged with L. Christidis) suggested the species of *Glossopsitta*, "*Psitteuteles*," *Trichoglossus*, *Lorius*, *Neopsittacus* and *Oreopsittacus* that had been examined had genetic distances so close they could all be members of one genus! I now prefer to retain only the accepted species in *Glossopsitta*, realizing the difficulties in placing goldiei and versicolor in any existing genus. I believe the iris lorikeet is best retained in *Trichoglossus*.

Little Lorikeet   *Glossopsitta pusilla* (Shaw 1790)

Synonyms: Green keet, red-faced lorikeet, tiny lorikeet

Description: Small size, black bill and red face distinguish it from other Australian lorikeets (in flight distinguished from the purple-crowned by green underwings). It is mainly green, a lighter shade below, with forehead, lores, throat and forepart of cheeks red. Ear coverts are green, shaft-streaked with yellowish green. Nape and upper part of the mantle are bronze, tinged with green; underwing coverts are yellowish green; outer webs of primaries are narrowly margined with pale yellow. The tail is green above and yellow olive below, with orange-red on the inner webs, mainly on the basal half. The bill is black and the iris is orange-yellow; the skin surrounding the eye is dark brown, almost blackish.

Length: 15 cm (6 in.)

Weight: Three captive males (one to three years), 40 g, 47 g, 48 g; three captive females (same age) 40 g, 41 g and 44 g; three unsexed birds aged 23 weeks, 40 g (two) and 41 g (Mexsenaar pers. comm. 1996); 34–53 g (Forshaw 1981).

Immatures: They are slightly duller green and more orange-red on the face; there is only a faint trace of the brown collar of the adults. The bill is dark brown, and cere and bare skin surrounding eyes are light bluish gray. The iris is dark grayish brown; adult eye color is obtained two months after leaving the nest (Courtney pers. comm. 1995).

Sexual dimorphism: Slight. Sindel (1987) recorded that some mature males have more extensive and deeper red facial coloration and darker brown on nape and mantle. Wilson (1991–2) noted that some males "will exhibit a great deal more color in the bronzy-olive patch on the mantle and often brighter red faces, but this is never an accurate way of sexing." In some pairs, male and female are almost indistinguishable. Peter Philp of South Australia suggested holding a bird in each hand, face to face, to compare their beaks. In the male the beak is one-eighth bigger. The male's build is also larger.

Natural History

Range: Eastern coastal region of Australia, from about 100 km (62 miles) north of Cairns in northern Queensland to approximately as far south and west as Naracoorte in South Australia. In New South Wales, they do not occur further inland than the western slopes of the Great Dividing Range. Blakers, Davies and Reilly (1984) state that there are several unconfirmed sightings in the Tasmanian region but that only one appears to be reliable.

Habits: Most little lorikeets are nomadic, moving around to feed on flowering trees in woodlands and bordering watercourses. Their preference for feeding in the canopy, their small size, and their mainly green plumage make them difficult to observe. However, those who know the distinctive calls are soon alerted to their presence. Lendon (1973) described these calls as feeble and wheezy; they are thin and high-pitched, although a little harsher than those of the purple-crowned. Sindel (1987) states that when little lorikeets arrive in an area, they feed in the canopy over a period of weeks, during which time the highest blossoms are eaten first; they gradually feed lower and lower until quite close observations can be made. They often feed in company with other lorikeets. Numbers vary from small groups to 100 birds or more. Writing of nearly 50 years ago, Sindel recalls how the little was the only lorikeet to occur with any regularity in the southwestern suburbs of Sydney. Now "they are not in the numbers of the days gone by, just odd pairs that drop out of the sky when they hear my aviary birds calling. They usually land in adjacent trees, make a few low passes over the aviaries and then fly off; odd ones may linger for an hour or two, calling excitedly to my birds."

Cayley (1938) found them "still fairly abundant, in some seasons, in the neighborhood of Sydney. Often in association with the Musk and Rainbow Lorikeets, their range being the same. Usually in flocks frequenting chiefly the tall-timbered areas of flowering eucalypts, and is remarkably fearless and noisy, both while feeding and when in flight. From April until the end of June one sees large flocks of these birds twisting and turning as they fly swiftly over the tree-tops in search of flowering trees."

I well recall my first sight of this species in 1990, in the Chiltern State Forest in Victoria. Early one morning, a flock of 30 or 40 birds were flying over the canopy of eucalyptus trees. They were feeding and calling—and they looked so tiny, high above. Pollen and nectar from various species of eucalypts and *Melaleuca* form their diet. According to Forshaw (1981), they also feed on the berries of mistletoe (*Amyema cambagei* and *A. gaudichaudii*), apparently extracting juice from the flesh around the seed. In the Wallangarra district of Queensland, Haines (1946) observed that little lorikeets always fed on mistletoe when in flower, whereas the musks preferred the blossoming eucalypts. As mistletoe flowers at a different time of the year to eucalypts, the two species of lorikeets were rarely seen together. They are also known to feed on loquat fruits (*Eriobotrya japonica*).

Nesting: It usually occurs between July and January, but may start as early as May in the north of the range. The site chosen is often high in a living eucalypt near water, according to Forshaw (1981). Dosser (1990) stated: "The Littles that I have observed in the wild usually nest in small hollow limbs, very high up in the river red gum *Eucalyptus camaldulensis*, and close to some kind of waterway." My good friend John Courtney, who has spent a lifetime watching this species in the Glen Innes region of New South Wales, told me this: "You will read that these lorikeets nest at considerable heights—a myth perpetuated by ornithologists who are unable to find their nests—but it is certainly not true here." He showed me one little lorikeet nest about 4.5 m (15 ft.) from the ground. The entrance of one nest, situated not far from a main road, was

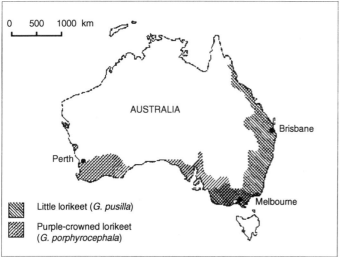

Range of the little and purple-crowned lorikeets.

in a lump in the side of a white gum tree. The hollow was horizontal and there was a long, narrow entrance passage which measured 21½ in. (55 cm) from the entrance to the back of the nest cavity. The actual entrance was round and measured 1 1/8 in. to 1¼ in. depending on the serrations in the bark.

Little lorikeets, also musks, return to their nest sites at least every two or three days, and perhaps daily, throughout the year, except in January and February. They do this to keep the nest in good repair. If they did not continually nibble around the entrance, the small hole would be closed within six months and within two years it would be completely grown over. They also attend the nest out of the breeding season to prevent it being occupied. John Courtney showed me a former nest of littles which had been taken over by introduced European bees. (This is a recurring theme. In the tropics, parrots lose their nests to introduced African bees.) He stressed that littles (and musks) are not opportunistic nesters. At least in the region where he lives, they have highly specialized nesting requirements:

This area has (or had) lots of yellow box (*Eucalyptus melliodora*) and white box trees (*E. albens*), which are the main source of food; and lots of smooth-skinned gums (white gums, *E. viminalis* and red gums such as Blakely's red gum (*E. blakelyi*) and the smoky or tumbledown (*E. dealbeata*), not a tall species), which supply the nest sites. However, unless white box (*E. albens*) is flowering in the vicinity, they will not breed. This is important because the white box is very rich in pollen whereas the yellow box produces much nectar but little pollen. The nest trees and other *Eucalyptus* trees in the area flower unpredictably and for a very short time only, e.g., Blakely's may flower for only a couple of weeks. The main food trees, yellow box and white box, flower for many months and each individual tree may have blossoms for several months, thus are a reliable, continuous food supply during the breeding season which, of course, is determined by when these trees flower. The white box starts flowering about March or April and goes right through to about September; the yellow box generally starts about August and continues until about November. The musks and littles have the majority of small young in late July, August and early September. Even in the most frightful years of drought, when human observers are unable to locate blossom, the lorikeets which occupy the best nests always seem to manage to breed (Courtney pers. comm. 1991).

In addition to the conditions already described, the little requires a nest hole with an entrance exactly 1 1/8 in. diameter.

A combination of factors make a small area just beyond the boundary of John Courtney's farm one of the most important breeding areas for both the little and the musk in New South Wales. It is dry schlerophyll forest, a narrow strip about 80 to 90 miles wide, extending down the western (inland) side of the Great Dividing Range. Unfortunately, on his 1,000 acres there is not one nest site. The two or three days during which John shared these birds with me I will always remember as my first intimate glimpse into the life of lorikeets. To observe a small population through the eyes of someone who had been watching them for 35 years was a unique privilege.

He summarized his observations of the annual cycle of the little lorikeet briefly, noting that "they arrive around April and stay permanently until early December, when the blossoms are no longer available. From mid-December (the onset of molting) they go, and only return, apparently from far away, every four or six weeks, in order to chew the hole. Even if I have not chanced to see them, this can be deduced by the bright, fresh chewing around the entrance. January, February and March sees only an occasional visit. Where they spend that time, especially when molting, from mid-December until the end of January, I do not know. I believe it to be a long way away."

On February 17, 1993, a white box tree was in heavy blossom. When a flock of eight musks flew overhead, John Courtney decided to inspect two of the littles' nests for occupancy. Both pairs were in residence. He concluded that when they are normally absent, they will occupy their nests for a while if there are a few white box in blossom in the area.

During the 35 years or so up to 1989, John had found only 45 to 50 nests of little lorikeets. Only four or five little nests had survived. Yet he knew of nests which had been active for 25 years or more. Sadly, the farmer on whose property the nest trees stand, had commenced to bulldoze the small stand of eucalypts four years previously, despite John's pleas to leave it intact. Three-quarters of the little lorikeet population of the entire area nested on this man's property; he had destroyed at least 60 percent of the nests. Where a decade previously there had been stands of white box, now there was only tree-less cultivation (Low 1990). This senseless destruction made me sad and angry. The eucalypts had been felled for no purpose; the ground was too stony to plough—yet these trees had held the key to this population of lorikeets. All the nests would have been destroyed but for John's fervent pleas as the bulldozer advanced. He marked nesting trees with a white cross and implored the driver not to fell them. One tree in which a pair of littles had nested for years, was bulldozed down. When it fell the nest entrance was only 90 cm (3 ft.) from the ground. Yet the attachment of this pair to their nest was so strong that they nested in what was then a highly unnatural site for a further three years (the roots were still in the ground), until the farmer burned the tree.

How poignant it was for me to witness the result of such pointless destruction and to realize that these two lorikeet species now exist there as remnant populations. Before the advent of the settler a couple of hundred years ago, the area was covered in euca-lyptus. Enormous flocks of littles and musks, prob-ably thousands strong, surely inhabited the area. If we could only turn back the clock of time and resurrect their habitat....

In July, 1993, John Courtney wrote to me that "a wonderful blossoming of white box tree is now on here, with all the known lorikeet nests in ac-tion. However, many of the little lorikeet nests are now gone, trees blown over, limbs snapped off, etc., and no new nests are being created. I am now down to three little lorikeet nests only. In the last month, dozens of rainbow lorikeets have flown in from the coast to take advantage of the prolific blossom. Sadly, there are few musks, and no little lorikeets that are not actually breeding." It should be pointed out that July is midwinter, when the tops of the distant mountains may be capped with snow.

John Courtney speculated that "by next cen-tury, say about 2010, they will be gone from here, an area teeming with littles when I was a boy in 1944. In that year it was not unusual to see 50 or 60 in one heavily-flowering tree. Now there are none,

except the three pairs faithfully tending their nests. Only one tree in one thousand or more may have a suitable hole, into which they can just squeeze." (In contrast, the rosellas were thriving and many had nested in hollow fence posts in the previous October or November.)

More information on the nest sites of this spe-cies and the activities of one pair is given elsewhere (see **Nest Sites**). John Courtney recorded this pair at their nest on video during August and September, 1989, during which time they were feeding young. The ritual by which they left or entered the nest is worth describing here. The bird always alighted on the lower rim of the nest, head downward, so that all or most of the entrance was obscured by its body. From this position it could depart very quickly, simply by dropping downward. However, there were three alternatives. It might turn around and run into the nest. It could run around the outside of the nest entrance up into the big limb to the left or it would fly off downward.

It is interesting to look at the distribution map for this lorikeet in *The Atlas of Australian Birds*. As for all species, the map indicates in which areas it has been reported and in which areas it is reported to breed. The reporting rate is indicated by the size of the circle. In the case of the little, one area stands out as of great importance, where maxi-mum breeding has been reported, i.e., just north of 30° latitude. This is the Glen Innes region, where John Courtney resides.

Bill Boyd (1987) describes the nest of this species as usually being located "in a knot hole or protruding hollow in the main trunk or fairly large branch." He had examined 25 to 30 nests, mainly in the late 1960s and early 1970s, in the coastal Lake Macquarie area. One nest contained five eggs—the only one found with more than four. The maximum number of young reared from any of these nests was three. He stated that incubation commenced with the laying of the third egg, and lasted about 20 days.

Egg data: Rachel Lewis (the artist whose work adorns the cover of this book), weighed five eggs, soon after they were laid, from her pair. Weights were as follows: clutch 1, 3.08 g, 3.28 g and 3.37 g; clutch 2, 3.25 g and 3.32 g. Harald Mexsenaar measured 13 eggs from three females. Five eggs from female 1 measured on average 20 x 17 mm (20 mm invariable); of five eggs in a clutch from female 2, two were 20.0 x 17 mm, and the others measured 20.5 x 16 mm, 20.0 x 16.5 mm and 20.5 x 17.0 mm; of three eggs in a clutch from female 3, two measured 20.0 x 17.0 mm and one was 20.5 x 17.0 mm (Mexsenaar pers. comm. 1996).

Status/ Conservation: Status varies according to locality. This is a common species in New South Wales, fairly common in Victoria and the least common of the lorikeets in South Australia. However, as the little is declining as a breeding bird in the center of its breeding range, it is imperative that action is taken to preserve its nest sites. Alas, it may already be too late in the area described. If preservation of nest sites does not occur, the population will crash dramatically within a very short period of time. Indeed, this rapid decline is already taking place, recorded only by Courtney, since so few people know this lorikeet as a breeding bird. He informed me:

> This season 1994–5 I glimpsed only about 3 pairs of little lorikeets zipping about in the field away from their nests (where I can usually locate them). In 1944–45 I used to see up to 30 feeding in one single tree, and of course there were lots of others in nearby trees. Their numbers were very strong in the mid-1960s, and toward the end of that decade I saw about 60 feeding in one well-blossoming tree. The decades of the 1970s and 1980s saw a gradual decline, down to the crash of the 1990s. It is sobering to realize that, at 60 years old, I have lived through the abundance, then decline and fall of the population of the little lorikeet in northern New South Wales, to bordering on extinction. (Courtney pers. comm. 1995.)

It is important to emphasize that lack of nest sites has brought about this decline. According to Courtney, some regeneration of trees is now being allowed to occur by an increasingly tree-conscious generation of people—but as the few surviving big old trees die, there will be none to replace them over the next two centuries. The extensive ring-barking (tree-killing) which occurred from about 1880 to about 1920 meant that those trees which would now be nest trees fell victim to the clearing of the pioneering era. The future looks bleak for this species.

Aviculture

Status: Not numerous in Australia; unknown elsewhere.

Clutch size: Three to five eggs, usually four; three females laid 13 clutches, of which one was of three eggs, 8 were of four eggs and 4 were of five eggs (Mexsenaar pers. comm. 1996); in the wild, usually four (Boyd 1987).

Egg interval: Two days (33 eggs) or three days (5 eggs) (Mexsenaar); usually two days, sometimes three (Matthews and Matthews pers. comm. 1995).

Incubation period: 20 to 22 days; in the wild "about 20 days" (Boyd 1987). According to Matthews and Matthews, incubation normally commences with the second egg, sometimes with the third. Harald Mexsenaar also recorded that incubation commenced after the second egg was laid; average length was 21.2 days, but the figures below (Mexsenaar pers. comm. 1996) show that females seldom start to incubate when the first egg is laid:

Days from laying to hatching (each row represents one clutch):

Little Lorikeet Incubation Periods

Eggs:	1st	2nd	3rd	4th	5th
Female No. 1	24	22	22	22	–
	22	21	20	20	21
	21	21	20	20	
Female No. 2	22	20	20	19	
Female No. 3	23	22			
	23	21	22	22	
	23	22	23	21	
	23	23	22		
	23				
	24	24	22		
	25	23	21		
	24				
range:	21–25	20–24	20–22	19–22	

Newly hatched chicks: Three chicks in one clutch weighed 2.13 g, 2.27 g and 2.35 g (Rachel Lewis pers. comm. 1995); three chicks in one clutch, 2 g, 2 g and 3 g (Mexsenaar pers. comm. 1996). The down is off-white, fairly long and sparse (Matthews and Matthews).

Chick development: The second down is gray. The feet are pale flesh colored.

Young in nest: 39–46 days, average 44 days

(Sindel); up to 50 days (Dosser 1990); average time, 40.5 days (Mexsenaar pers. comm. 1996).

Chick growth

The table shows the weights of four parent-reared chicks in a single nest, owned by Harald Mexsenaar in Queensland. The chicks hatched on May 26, 27 and 29, and June 1, 1996.

Little lorikeet weight table

| Age in days | Weight in grams | | | |
	Chick No. 1	No. 2	No. 3	No. 4
hatched	2	3	3	3
1	4	3	4	4
2	6	4	3	5
3	7	6	5	6
4	7	6	8	7
5	8	9	11	10
6	11	12	14	12
7	13	12	15	12
8	14	14	17	14
9	17	18	20	17
10	19	17	20	19
11	21	20	23	20
12	22	22	25	23
13	24	23	27	27
14	26	25	29	28
15	28	27	29	29
16	29	29	33	35
17	30	29	31	35
18	33	34	35	36
20	38	37	38	40
22	42	38	36	42
24	43	39	39	36
26	43	43	42	43
28	45	43	43	50
32	49	47	46	53
36	49	46	45	52
39	left nest	50	46	47
40	–	left nest	left nest	45
41	–	–	–	left nest

Harald Mexsenaar weighed six unsexed young aged eight weeks: 37 g, 39 g, 40 g, 42 g (2) and 43 g. At 12 weeks, three males weighed 36 g (2) and 40 g.

Age at independence: Harald Mexsenaar noted that parent-reared young were independent 11 days after leaving the nest.

John Courtney (pers. comm. 1988), who has hand-reared young, commented that "the young of larger lorikeets always beg to me, from musk upward, but out of a number of nests, never the little. They always remained wild and aloof."

General: Gentle and not destructive, little lorikeets are quite often kept in planted aviaries in company with finches. Even today, this species is not kept in large numbers but in recent years, partly due to more suitable diets, the little has been established in Australian aviculture. Unfortunately, during the 1980s there was an influx of wild-caught birds which had the effect of lowering the price, thus discouraging avicultural interest. In Victoria, in 1995, fewer littles were kept than any other native lorikeet. The Department of Natural Resources and Environment issued figures showing the numbers of Australian parrots kept in aviaries in Victoria for the 6 months up to September, 1995. At the start of the period there were 97 little lorikeets; at the end there were 98. Twelve had died and none had been bred; purchases and sales accounted for the other movements. This can be contrasted with 286 musk lorikeets at the start and 244 at the end with 14 bred and 24 deaths (Pace 1996). Figures for a 12-month period would have been more revealing but those presented suggest that the little was the least prolific species. Wilson (1991–2) suggested that "most Littles will only breed in a planted aviary" and that "anything that grows bushy and fills a portion of the aviary, so long as it isn't toxic, is suitable." Alternatively, branches of eucalyptus can be placed in a container of water and will keep fresh for a week.

From the date of the first breeding in Australia in 1948 until the early 1980s, few successes were recorded. In the 1970s, Peter Odekerken went to live in South Africa, taking with him a pair of little lorikeets. They were kept in Rustenburg for four years but did not breed, possibly because the climate there was too hot (Barnicoat 1976). In 1975, they were acquired by David Russell, a veterinarian who, at that time, had probably the largest collection of lorikeets in South Africa. The female laid four eggs in her first clutch that year, the fourth on August 19. Three chicks hatched, one of which died at about two days old. The two young left the nest when aged between 38 and 42 days. Two young were reared in the second clutch, which contained three or four eggs. The incubation period was 22 days. The diet of these littles consisted of nectar made from three teaspoonfuls of Complan and one each of honey and condensed milk with a vitamin additive, mixed with half a liter of warm water. At lunchtime a mixture of Pro Nutro breakfast cereal, honey and water was given. Apple was also eaten (Russell pers. comm.

1976). These days, littles can be offered a commercial nectar formulated for small lories or a mixture of such items as Complan, honey and baby cereal, plus apple and other fruits. Pollen and nectar from branches of flowering eucalypts, *Melaleuca* and acacias will be relished. Some breeders offer madeira cake when young are in the nest.

Graham and Glenys Matthews of Barmera, South Australia, hand-reared this species from hatching on one occasion. They commented that because of the small size, Glenys had to feed the chicks for the first week as Graham found it difficult to handle them!

Dr. Russell's pair were provided with a choice of nest sites—a natural log or a small nestbox of the type used for Australian finches. They chose the box. Stan Sindel has found that lovebird nestboxes 30 cm (12 in.) long, 15 cm (6 in.) wide and 17.5 cm (7 in.) deep have proved most suitable. There is a small entrance at one end near the top. Holes 3 mm in diameter are drilled through the bottom, which help to keep the nest dry. He recommends hanging the box at an angle of 15° from the horizontal. His pairs are housed in suspended cages. Some are double-brooded. In the Sydney area, where he lives, aviary birds lay between June and October but he suggests that those kept nearer the coast would have a longer breeding period.

Harald Mexsenaar from Brisbane, Queensland, keeps his pairs in suspended cages 2 m (6 ft. 6 in.) long, 90 cm (3 ft.) wide and 90 cm high. They lay between May and October; three clutches might be laid if one clutch is not successful. Young pairs breeding for the first time, when they are about one year old, usually lay in August. It is not unusual for the female to lay her second clutch while the young of the first nest are still being fed in the aviary. The nestbox was cleaned out every second day. When the young were temporarily removed for this reason, they would protest loudly, unlike the purple-crowned chicks which were always very quiet when removed for nest cleaning. The three pairs which he kept during 1993 to 1995 bred very successfully; indeed there was 100 percent fertility and hatchability from two pairs—17 young hatched from 17 eggs, with 16 reared. The three pairs reared 32 young in 10 clutches: pair no. 1 reared 12 in three nests, pair no. 2 had 16 in six nests and pair no. 3 reared four in one nest (Mexsenaar pers.

comm. 1996). A good diet obviously influences breeding results; details of Mr. Mexsenaar's diet are given under **Dry Diets**.

Regarding the display of this species, Sindel (1987) describes how the male approaches the female, stretched to his full height, but without arching the neck (unlike *Trichoglossus* species). He bobs and hops and wipes his beak frequently along the perch as he advances toward the female. During the display he emits a soft, low whistling note.

In 1989, when John Courtney filmed the little lorikeets at their nest, he observed their display. "They were bobbing up and down in front of their mates, then leaning away. One even behaved the same way apparently in threat behavior to an intruding pair of little lorikeets. Once, I was lucky enough to capture a few seconds of the bobbing and leaning away display of a little lorikeet to its mate, with a video camera" (Courtney pers. comm. 1990).

However, it was a captive bird which provided a real insight into the display of this species. In about 1991, a neighbor of John Courtney found a very young little lorikeet fluttering about in a car park in the shopping mall at Inverell. Its beak was still pale brown, appearing almost transparent in bright light. The neighbor reared it, but by the time it was six months old, it was no longer a welcome member of the household. He intended to release it but John Courtney told him that it would be unable to survive: there was no blossom at that time and no little lorikeets were present. John reluctantly took the bird and kept it for five years or so. Unlike other little lorikeets that he had kept, it was extremely tame. He recorded:

> When it gained maturity at about 2 years old, it became obvious that it was a female, for in the breeding season, when I placed my finger near it in the cage, it would sometimes flatten itself out on the perch in a typical mating attitude. It then began to display to me, and did so for several years until its untimely death. Whenever I went near its cage, it would sidle over on its perch, as close to me as it could get, and would then bounce vigorously up and down on the perch, by straightening and bending its legs. After about 3 or 4 vigorous bounces, it would slowly straighten right up as far as it could while leaning right away from me as far as it could lean, its feet always anchored to the same spot.

After a few seconds of leaning away, it would return to normal position and begin bouncing again, then lean away again. After 3 or 4 bouts of bouncing and leaning away, it would stop this and bite vigorously into my finger! (Apparently, somewhere in this display, I had not responded appropriately!) During the whole time that it was displaying, it erected the red feathers of the face above a line (figuratively speaking—not a visible line) from in front of the eye, into a little red crest. Over the whole of the forehead every red feather stood up vertically, and every green feather was kept tightly flat. It had total control over the red feathers of the forehead, as distinct from the green ones touching the red ones! It has always done this so consistently with every display that it ever made, that there can be no doubt that this is part of the normal display of the little lorikeet (Courtney pers. comm. 1995).

His description is interesting. It indicates that some elements of the display, the bouncing and leaning and erecting the head feathers, are shared with the display of some of the red *Charmosyna* lorikeets. In fact, through this species, the link to the genus *Charmosyna* might be quite strong. This occurred to me when I looked at a photograph I had taken of a newly fledged red-flanked lorikeet. It was a head-on study and the young one, with its brown beak and red mask, was almost identical to a little from that position, except for its green forehead. In adult red-flanked only males have red on the face and yet most immature females have as much red there as males. In many lory species, atavistic features are apparent in immature plumage.

Those like John Courtney, who have extensive knowledge of a species in the field, as well as avicultural experience, are rare. There is at least one other person who can make this claim—Bill

Boyd, also from New South Wales. He had been observing little lorikeets for 25 years before he kept them in his aviaries. He described the display as follows:

> In the early stages it consists of head bobbing and pupil dilation while uttering a soft warbling. As this becomes more serious the head bobbing becomes a complete jerking upward of the entire body similar to the display of the Gouldian finch but without the feet leaving the perch. This is also performed to a lesser degree by the hen. Excited preening also accompanies the display where the beak is thrust quickly amongst the feathers of the back or the primary feathers of one wing. Much head scratching is also done by the cock and as this all subsides the upward jerking becomes the most important feature. The male progresses along the perch in pursuit of the female who may stop and lunge at the male in a similar jerky fashion but usually without malice. All the while a soft warbling is emitted to accompany each upward jerk. The cock will also sway away and back to the hen in the same fashion as the larger lorikeets and if the birds are really serious the hen is forced into a soliciting position by the cock repeatedly placing one foot on her back in trying to copulate (Boyd 1987).

It seems to me highly regrettable that the little lorikeet is not represented outside Australia. It would be eagerly sought and greatly valued by breeders. An established population, albeit in aviaries, in other continents, would at least ensure its survival in aviculture. There are few people in its homeland at the present time who can be aware of its threatened status when it still appears to be so common.

Musk Lorikeet *Glossopsitta concinna* (Shaw 1791)

Synonyms: Green keet, red-crowned lorikeet, red-eared lorikeet

Description: The only Australian lorikeet with red frontal band and red ear coverts (the red extend-

ing to side of neck and to lores). The crown and the back of the head are blue and the mantle is bronze, tinged with green. A yellow patch on the side of the breast is mainly hidden by the wings but is more extensive in some individuals. The plumage is otherwise midgreen, more yellowish on the under parts. The tail is green above, yellow olive below; the inner webs of the lateral feathers are edged with orange-red. The upper mandible is unlike that of any other lorikeet: black at the base and at the cutting edge, and coral elsewhere; the lower mandible is black. The iris is orange. Cere and bare skin surrounding the eye are light-colored.

Length: 22 cm (9 in.)

Weight: 59 g (Sindel 1987), 52–65 g (Forshaw 1981); male of a captive pair 109 g, female 86 g (Salisbury 1985)—the male, at least, must have been very overweight.

Sexual dimorphism: Some but not all males have brighter and more extensive blue on the crown. Sindel (1987) recommends surgical sexing. He noted that "females often appear to be larger than males."

Immatures: The birds are duller, the ear patch and forehead being brick red—not bright red—or even nearer to orange. Beak and iris are dark brown. The cere and the bare skin around the eye are pale gray. At three months the upper mandible has acquired the coral coloration and the iris is brown.

Mutations: No natural mutations have been recorded. Sindel produced an olive mutation (now widely available in Australia) by hybridizing the musk with the olive scaly-breasted lorikeet.

Natural history

Range: Eastern and southeastern Australia. The northern limit of the range seems to have changed during the twentieth century (Blakers, Davies and Reilly 1984). One was collected at 19°S lat. (Cape Upstart, central Queensland) in 1843 and it was a regular visitor to 23°S (Duaringa). It no longer breeds further north than 26°S. Today it occurs in New South Wales and Victoria, and in the Spencer Gulf region of South Australia. It is also found, and breeds, in eastern Tasmania but has rarely been recorded on King Island. There is a small feral population in Perth.

Habits: At least on the fringes of its range, this lorikeet is nomadic (Blakers, Davies and Reilly 1984), but its movements are erratic, with local eruptions. For example, in 1936–42 it was a regular and apparently nonbreeding visitor to Murphy's

John Courtney at Swan Vale in April, 1989.

Creek (27°S)—and then was not seen there again until 1952. In some regions it appears only when eucalypts are in flower, whereas in Tasmania there is little difference between numbers in summer and winter. Sindel (1987) states that it is "more seasonal in its movements than other lorikeets" and thought that no resident populations had been established. Courtney (pers. comm. 1995) does not believe that this is true. His careful observations over four decades suggest "that a few pairs occupy the specialized nest-holes, and produce the vast majority of the young. This picture is in stark contrast to the popular folklore image of these birds nomadically following the blossoms and nesting anywhere the blossoms are. While the majority of the population do follow the blossom in a nomadic way, the breeders are tied to a particular area."

The musk lorikeet is found in most habitats with trees, except in some highland locations, but prefers more open areas, rather than closed forest. I saw and photographed musks feeding with rainbow lorikeets in the park by the university in the center of Adelaide in 1989. They were the first I had seen in the wild and I marvelled at how unafraid they were and how low down they fed, almost within human reach. I could not tear myself away from this wonderful sight; all the while people were passing with never a glance at this colorful vision. No doubt they saw it every day and took the birds as much for granted as

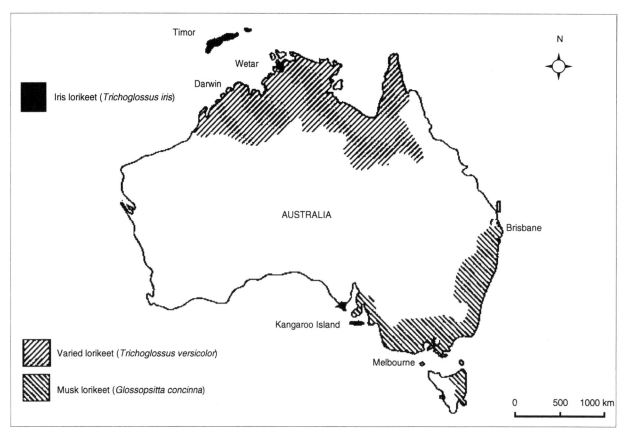

Range of the iris lorikeet, the varied lorikeet and the musk lorikeet.

a European does the house sparrow. Yet the lorikeets are infinitely more colorful and amusing.

For the three years prior to and including 1995, there was a large influx of musk lorikeets in Melbourne during January and February. Small numbers of purple-crowned and little lorikeets accompanied the musks. The lorikeets seen in the area varied according to location. Andrew Isles had seen purple-crowns near his home at Prahran, but never littles which visit the other side of the town (Isles pers. comm. 1995). In 1994, VORG (Victorian Ornithological Research Group) organized a survey of musk lorikeet sightings.

Courtney (pers. comm. 1995) has described the yearly activities of the musk lorikeets at Swan Vale, in the eastern part of the Inverell district of northern New South Wales. For those unfamiliar with the geography of this region, he points out that eastern Australia is divided into two by the north-south Great Dividing Range (GDR). A narrow coastal strip east of this range has high rainfall and tall trees; the rest of the country to the west in northern New South Wales has a lower rainfall and shorter, less luxuriant trees. Down the western (inland) side of the GDR runs a strip of country of similar width to,

and parallel to, that of the narrow coastal strip, which Courtney terms "the Western Corridor." This strip is mild in climate and has a higher rainfall than inland. It has a different combination of tree species than other areas and thus is unique. It is the main breeding habitat of the musk and little lorikeets, which do not occur any further west, and only sparsely on the east coastal strip, which is occupied by rainbow and scaly-breasted lorikeets. They seem to be highly adapted to this area and are probably there by choice, rather than as victims driven out from the coast by competition with the larger lorikeets, as suggested by G. A. Smith (1975b). (Their choice of small nest holes and their lighter weight—enabling them to exploit blossoms on the extremities of branches—would probably allow them to survive such competition.)

To understand the year of the musk lorikeet, it is important to define the seasons. In Australia, December, January and February are the summer months, March, April and May are autumn, winter is from June to August and spring occurs from September to November.

All the trees in the area described, except white box (*Eucalyptus albens*) and yellow box (*E. mellio-*

dora), are useless for food for breeding purposes as they have very short flowering periods of only ten days to one month. They usually do not overlap, so there is no continuity of blossom. In addition, they blossom at erratic, unpredictable periods. In remarkable contrast, the white box begins to flower in March or April and continues until September. The yellow box starts to flower about September and continues until early December. Individual trees of these two species may flower for months in contrast to, for example, Blakely's red gum (*E. blakelyi*), individuals of which flower synchronously for only about ten days.

Suitable nest sites are provided by the following gum trees: Blakely's, the smoky or tumbledown red (*E. dealbata*), white (*E. viminalis*) and, to a lesser extent, by the rough-barked apple (*Angophora floribunda*). Breeding areas occur in patches along the Western Corridor, where good stands of white and yellow box occur within a couple of kilometers of the suitable nest trees. The past 140 years of European settlement have seen a catastrophic reduction in the numbers of food trees and nest trees; they may remain only in narrow strips along roads. While there may always be a few stands of food trees sufficient to support a small population of lorikeets, only an extremely small percentage of trees of species which are used for nesting ever develop holes of a size acceptable to lorikeets. While many young trees (perhaps only ten years old) can, and do, blossom profusely, it may take 200 years before nest tree species reach a stage of maturity where decay begins to occur and holes form. (See also Nesting under Little Lorikeet.)

Musk lorikeets usually begin to arrive in the area about April or May. The breeding pairs take up permanent residence in their nest holes which they maintain from year to year. They chew the living bark around the rim, thus preventing the hole from closing. Thus the annual cycle starts at the beginning of autumn. During the three autumn months they presumably build up their body reserves. During the winter months the females lay and hatch their first clutch of eggs. There are usually young in the nest in August. If the yellow box then flowers strongly, a second brood is started and these young are ready to leave the nest in late November or early December. Fledged young do not return to the nest for long, if at all. In his diary for September 22,

1992, John Courtney recorded the moment of fledging of a young musk, the only one in the nest, which he inspected with the use of a ladder at 9:20 A.M. At 10:00 A.M. it was looking out, when one of the parents arrived at the nest, but did not feed it. Instead, the adult flew to a nearby tree. The young one then started to push, squirming from side to side. Suddenly it "popped out" and flew immediately. It headed south, instantly accompanied by both parents, one flying each side of it and no more than 30 cm (1 ft.) away. They turned slowly in a huge semicircle, until they arrived back about 125 m (410 ft.) from the nest tree. Height of the flight was about 9 m (30 ft.) above ground.

From about mid-December until mid-March most nesting pairs have left for parts unknown. The pairs return every few weeks for about a day, to vigorously gnaw at any new bark around their nest entrance. Their continuous maintenance of their nest leaves little doubt that the same individuals are involved. However, small nomadic flocks of nonbreeding lorikeets can arrive in the area at any time, if a few trees, such as *Angophora*, start to blossom. During this period of absence, the breeding pairs apparently molt. (Courtney's description of the annual cycle is continued under the heading of Nesting.)

Wilson (1992) described their arrival in Mudgee, New South Wales: "Around December-January hundreds of them, especially early in the morning, were noisily flitting from one flowering gum to another amongst the beautiful bushy gardens of our town. At times they were very quiet, feeding nonchalantly as though there were no observers at all. At other times their voices become very raucous and if I got too close, the alarm would be raised and an immediate sheet of green and red would lift and spread in all directions." The musks stayed in the area for about three months.

Although the musk lorikeet is protected by law, in some fruit-growing areas killing them is permitted. Elsewhere they are shot illegally. (See **Crop Pests**.) They feed on pollen, nectar, seeds, insects and cultivated fruits. Their liking for the sweet waxy secretions of the sugar lerp insect is well known. When foraging in new leaf growth of native trees, particularly eucalypts, they are feeding on lerp secretion.

Nesting: Occurs between August and January. The nest trees chosen in the Inverell District (see above) have extremely smooth bark, which may be easy to gnaw, around the nest entrance. After a couple of seasons, the entrance is "molded" to the body size of the bird because they maintain the holes at a size into which they can only just squeeze. The gnawed area dies and hardens and the living bark keeps growing. After several years the entrance tunnel is lined with hard, dead wood, which the bigger lorikeets cannot enlarge to enter.

Two eggs from one clutch measured 25.0 x 20.4 mm and 24.1 x 19.8 mm; a clutch of four averaged 24.6 x 20.0 mm (Forshaw 1981). Apparently the eggs of musk lorikeets are among the rarest of the common Australian birds in egg collections because few nests are found.

In 1989, I was fortunate to obtain nest photographs—even though the month was April (out of the breeding season). I was in the company of John Courtney, at Swan Vale. We witnessed musks throwing debris out of the nest entrance. The nest was in a Blakely's red gum at a height of about 5 m (15 ft.). A few moments later another pair flew in and landed near the nest which, perhaps, they coveted for their own. They were rapidly escorted away from the nest by the rightful owners, who made it very clear that they would not tolerate their presence. There must be so much competition for nest sites that many pairs never have the opportunity to breed. A nest site must be protected by its owners. If other lorikeets do not usurp them, introduced European bees could do so. (John Courtney knew of one nest of musk lorikeets and one of littles which had been taken in this way.)

In Tasmania, musk lorikeets are resident throughout the year in the eastern part of the island but few nests of this species have been found. On July 25, 1996, part of a white gum (*Eucalyptus viminalis*) containing a musk nest and two chicks was taken to the office of the Parks and Wildlife Service in Hobart. The tree had been cut down and when sawing it up, the owner had discovered the nest. The young were about ten days old; they were removed and hand-reared. The notable fact is that the eggs from which these chicks hatch would have been laid in midwinter. In Tasmania most parrots do not nest until September or even October (Brown 1996).

Status/Conservation: Fairly common but likely declining in the northern part of the range, if not elsewhere, no doubt due to reduction in feeding and nesting sites. There may be a marginal improvement in the situation. Courtney (pers. comm. 1995) states that in contrast to the position of the little lorikeet, he has seen a "reasonable amount of nest recruitment in the Swan Vale area in the last year or so." The key to the survival of this species may already have been lost. As mentioned above, cavities for nesting only form in very mature trees and a period is likely to occur (at least in New South Wales) when there will not be any trees with suitable cavities, due to the large numbers of trees that have been destroyed in the nesting area in recent decades. The population will appear to be stable yet will consist almost entirely of nonbreeding birds that are unable to find nest sites. If this happens, the population could crash very suddenly toward extinction.

Aviculture

Status: Fairly common in Australia, uncommon but increasing elsewhere.

Clutch: Usually two, occasionally three.

Incubation period: Average 22 days, but varies from 19 to 25 (Sindel 1987); 23 days (Wilkinson et al.); 24 and 25 days have been recorded by other breeders in Australia.

Newly hatched chick: Weight, 3–4 g (Wilkinson et al. 1993; A. Dear pers. comm. 1994); 5 g, beak brown (Digney pers. comm. 1996); they have "long silvery white down" (Sindel) or "sparse long down feathers on the body and head" (Wilkinson et al.); "off-white to light gray, fairly long and dense" down (Matthews and Matthews pers. comm. 1995).

Chick development: One chick hand-reared at Chester Zoo, covered in long dense white down at one week; eyes began to open at two weeks; feather sheaths on body began to open after four weeks but at six weeks much gray second down was still present (Wilkinson et al.).

Chick growth

No. 1 was hand-reared from the egg at Chester Zoo in 1991. It was fed on various baby foods (Wilkinson pers. comm. 1995). No. 2 is a chick hand-reared from the age of 16 days by Sharon Casmier of Washington. The figures are taken from a graph. Over 20 young had been hand-raised and their weights hardly varied from that shown.

Musk lorikeet weight table

Age in days	Weight in grams	
	No. 1	No. 2
hatched	3	
4	5.5	
8	6.5	
12	8	
15	10	
16	12	17
21	19	30
23	22	37
25	25	43
29	34	55
31	41	56
33	45	60
34	49	63
36	52	65
39	54	67
40	56	70
43	57	72
45	59	78
47	61	85
48	62 peak weight	
52	61	92
57	61	95 peak weight
62	–	90
65	–	88
68	–	87
74	–	87

One parent-reared chick weighed 8 g at three days and 17 g at ten days (Digney pers. comm. 1996). Twenty hand-reared by Philip Eglinton in New Zealand attained a peak weight of 65–70 g at between 42 and 45 days (Eglinton pers. comm. 1995).

Young in nest: 47–58 days (Wilkinson et al.); average 48 days (Sindel); 61 days, one only (Dear pers. comm. 1994).

General: In Australia, little interest was shown in the native lorikeets, except the rainbow (which has long been a favorite) until the 1980s. Since then the musk has been bred in an increasing number of collections. In the early days, dealers' shops often contained cages full of musks. They were destined to die after a very short while, because they were fed on nothing but seed. Those which died soon after purchase must have deterred many bird keepers from again venturing into lorikeets. This lorikeet is unfamiliar to most aviculturists outside Australia, as, since 1959, that country has been permitted to export only captive-bred young destined for mainstream zoos. During the 1980s and early 1990s, the only musks known in the U.K. were at Chester Zoo, the offspring of those reared at Rotterdam Zoo, and their own young. Then, in the early 1990s, musk lorikeets became available from New Zealand, which allows the export of non-native captive-bred birds. Vogelpark Walsrode and several private breeders received such birds. This resulted in the species being well established in some European countries by the mid-1990s.

Musk lorikeets should be kept and bred one pair per aviary. At the Fifth National Avicultural Convention in 1989, Peter Philp spoke on lorikeets. His attempts at breeding from more than a pair per cage had not been successful. He housed each pair of musks in aviaries 4 m (13 ft.) long x 1 m (3 ft. 3 in.) wide and 2 m (6 ft. 6 in.) high. The floors were of earth (not to be recommended), in which they would sometimes dig. Fresh eucalypt branches (when available) were placed in plastic tubs. During the breeding season, when the temperature rose to 40–45°C (104–113°C), the young had to be taken during the day, hand-fed, and returned to the nest at night. On one occasion they were removed for two days—and the parents accepted them when they were returned. Young are usually removed from the aviary two weeks after they leave the nest, unless the parents are aggressive, in which case they are taken out earlier.

Sindel (1987) has bred musks in a large mixed aviary containing doves, finches, parrakeets and purple-crowned lorikeets. No aggression occurred, even when the birds were breeding. In contrast, others have found musks too aggressive for a mixed aviary. No doubt much depends on the size of the aviary and on the temperament of the pair.

The musk is so-called because of its odor, which has been described as a "sweet, perfumy smell." Unrelated groups of parrots smell different (e.g., the subtle odor of an Amazon is unlike that of cockatoos), and within the Loriinae odors are faint or stronger and different.

The display of this species has been described by Odekerken (1992) as follows: "Courtship consists of the male arching his neck with head held over female. His head is moved to both sides of the female's head and the male's pupils dilate. During this performance he chatters, occasionally hisses, puffs out the yellow feathering on the sides of the breast and has the brownish nape raised. Female utters a begging call during this display. Male frequently moves away from female and returns to her side with an exaggerated hop. Female arches back and bows head to level of perch to receive male for copulation."

Musks are strong flyers which do well in larger aviaries. Wilson (1992) preferred to house his in aviaries at least 4 m (13 ft.) long, 1 m (3 ft. 3 in.) wide and 2.4 m (8 ft.) high. They were provided with nestboxes approximately 20 cm (8 in.) square and 30 cm (12 in.) deep, with peat moss and wood shavings to a depth of 5 cm (2 in.). One pair threw out all the nesting material and usually hatched and reared their chicks on the bare wooden base.

Outside Australia, musks were occasionally imported into Europe prior to 1959, but were not established. When export ceased, the species quickly became unknown. The first musks I ever saw, in 1974, were at San Diego Zoo, where this lorikeet reared young for several years from 1971. Another zoo which received musk lorikeets via an Australian zoo was Chester in the north of England. As previously mentioned, theirs came from Rotterdam Zoo. The first egg was laid on March 1, 1987; 11 days later it was found broken on the aviary floor, and contained a well-developed embryo. On March 26 the female spent the whole day in the nest. On May 22, when the parents had deserted, the nest was found to contain a dead chick. The nest was very damp, which could have contributed to the chick's death. During the next four years the female laid 16 clutches, 12 of two eggs, 2 of one egg, 1 of three eggs and 1 of an unknown number. The 30 or more eggs produced 14 chicks, seven of which were reared.

The nesting attempts of the pair can be summarized as follows:

No. Eggs	Date	Result	Remarks	Hatched	Fledged
1	1/3/87	broken on floor 12/3	fertile	0	0
?	end of March	chick heard 24/4	dead by 23/5	1	0
2	17/6	2 chicks by 10/7	dead by 19/7	2	0
3	18/3/88	1 chick by 24/4	first U.K. success	1	1 10/6
2	10/6	both 'addled'		0	0
2	30/6	1 chick	died soon after	1	0
2	28/1/89	1 egg & 1 dead chick by 21/2		1	0
2	15/3	1 hatched by 4/4	1 died; nestbox removed for a few days	1	0
2	8/5	2 hatched by 5/6; 2nd chick died 31/7	1 dead by 14/6 enteric infection	2	1 25/7
2	16/8	2 hatched by 3/9	parents aggressive; young removed; force-fed	2	2
2	28/3/90	2nd hatched after 23 days		2	2
2	5/9	egg gone by 22/9		0	0
1	15/10	egg gone by 27/10		0	0
2	7/1/91 9/1	1st egg, 2nd egg	embryos died	0	0
2	1/3	1 infertile, 1 pipped & died		0	0
2	10/4 13/4	1st egg; hatched in incubator 7/5		1	1
		female died 17/4	from Yersinia pseudotuberculosis; male died soon after		

The preceeding account is valuable because it documents all the problems encountered—and these are not as unusual as might be thought. Few breeders recount these in detail, preferring to describe only the results with a happy outcome. This can give a false impression to the uninitiated regarding the ease with which parrots breed. On the other hand, of course, there are pairs which are consistently successful. The above pair might have harbored the *Yersinia* infection for a long period. Viral and bacterial infections are frequently the cause of poor breeding results.

Varied diets have been described by those successful in rearing this species. At Chester Zoo the musks were fed on sunbird nectar produced by SDS (Special Diet Services), chopped fruits (apple, grapes and pears), grated carrot and a small quantity of soaked seed. When rearing young they were also provided with mealworms, crickets and chopped tomatoes. Australian breeders provide branches of eucalyptus blossom, seeding grasses and other green foods. Peter Philp reported that his eat hulled oats and sunflower as soon as chicks hatch. Careful attention needs to be paid to diet as musk lorikeets are susceptible to obesity if not housed in a large aviary.

In the U.S.A., Sharon Casmier of Washington

hand-reared 25 musk lorikeets from two pairs between 1990 and 1996. Nineteen of these were from one pair, which had been imported from New Zealand. She was also the stud book keeper for the musk lorikeet. In 1995, 49 birds were registered in the U.S.A. At least 13 breeders of this species were known, but unfortunately not all of them were willing to register their birds.

Purple-crowned Lorikeet   *Glossopsitta porphyrocephala* (Dietrichsen 1837)

Synonym: Zit parrot

Description: Distinguished by its small size and yellow-orange (or pale orange) forehead. The crown is dark purple-blue, lores and a small area in front of the eyes are orange, ear coverts are orange-tinged yellow and the rest of the head is light green. Underparts are light green, very strongly suffused with pale blue, with hidden yellow patches on sides of breast. The mantle is bronze, tinged with green. The bend of the wing is light blue; primaries are dark green, narrowly margined on outer web with pale yellow; underwing coverts are crimson. The tail feathers are dull yellow olive below with faint orange-red on the inner webs. The rest of the plumage is green, more yellowish on the thighs and flanks. The bill is black and the iris brown.

Length: 16 cm (6 in.)

Weight: Average 45 g (Sindel 1987); one-year-old male 52 g, three-year-old female 47 g, two females aged 23 weeks 50 g and 52 g (Mexsenaar pers. comm. 1996); 37–50 g (Forshaw 1981).

Immatures: They lack the purple crown, or have this color much reduced and duller, also the orange on the forehead. Mexsenaar described newly fledged parent-reared young as "duller than the parents. The red on the facial area is just coming through and the purple crown is about 50% the size of that of an adult." The iris is dark brown and the cere and skin surrounding the eye are paler gray than in adults.

Sexual dimorphism is not evident; coloration varies slightly among individuals, regardless of sex.

Mutations: A dilute or cinnamon mutation exists, in which the green areas are replaced by pale yellow. I saw one in 1992 in the aviaries of Joelle Marshall of New South Wales. The first bird she

ever bred of this species (in the previous year) was of this mutation! The parents had been obtained from two different sources. My second sighting was unexpected—a wild bird! It occurred on November 20, 1994. I was passing through Nhill, Victoria, with friends and we stopped when we heard the sound of purple-crowns from large trees lining the main street. Almost the first bird I saw (its coloration was so conspicuous) appeared mainly light yellow. From my brief glimpses and the photograph I took at the time (fortunately, my camera was already in my hand), I believe it to be the same mutation. Gordon Dosser has bred several cinnamon purple-crowns; they vary in the intensity of yellow and his females tend to be brighter. However, a bright yellow male is known in another collection (Gelis pers. comm. 1996).

Natural history

Range: Southern Australia, not generally above latitude 30°S, in Western Australia, South Australia, Victoria and New South Wales; there are occasional sightings in northern New South Wales and Queensland. The range of this species is disjunct (discontinuous); Blakers, Davies and Reilly (1984) state that there is no record between 129°E and 131°E. However, they point out that due to the mobility of the species, "the eastern and western populations may not be isolated." They could detect "no morphological differentiation" between eastern and western birds. In the Nullarbor Plain, the distance between the closest recorded localities of the respective populations is well over 160 km (100 miles). Sindel (1987) states that the species ranges across southern Australia, extending north almost to Shark Bay, Western Australia, and in the southeast as far as Eden, New South Wales. It is not found in

Tasmania. This is the only lorikeet native to southern Western Australia.

Habits: The purple-crowned lorikeet occurs in most types of timbered country which offer flowering or fruiting trees and shrubs. Forshaw (1981) describes it as essentially a bird of inland habitats, being most characteristic of the dry mallee areas of the south. It is also found in eucalyptus forests, open savannah woodland and coastal scrublands. More inclined to feed in the lower branches of trees than is the little, its habits are otherwise similar. So absorbed are purple-crowns while feeding, that they often allow a close approach. Their search for flowering eucalypts explains their nomadic life-style and the fact that they are absent from some areas for long periods. Their flight is so swift that some may be killed by flying into telegraph wires, netting fences and windows, which they cannot avoid in time. Their roosting sites may be long distances from their feeding grounds.

John Warham (1958) observed that this species "is most likely to be seen when a flock, breaking cover, dashes across a clearing or a road, a tight-knit body of birds which rips through the air and calls with sharp '*tsit, tsit*' cries. Country children often refer to them as 'Zit Parrots' because of their voice." Once one is familiar with the voice of this species, it is easily located as it feeds, even when high above.

In South Australia, Jim Pearson of Munno Para, lives in an area which was developed for housing about 1980. Parks and nature strips in the area attract purple-crowned lorikeets, but in smaller numbers than musks. Groups of purple-crowns number up to eight or ten birds, but sometimes from two to five. Occasionally they are seen flying with small flocks of musks. Purple-crowns seem to spread out more than musks when feeding in areas with several trees close together. Half a dozen birds may occupy three or four adjoining trees, whereas a dozen musks might feed in a tree then move, a few at a time, to the next tree. Sometimes a single purple-crowned gets left behind. Purple-crowned seem to feed on trees with smaller flowers, as well as on larger flowers. Musks seem to prefer medium and larger flowers, the larger flowers being 2–3 cm in diameter (Pearson pers. comm. 1995). The purple-crowned is a eucalypt specialist, extracting pollen from young flowers and nectar from older ones.

Food sources include the following *Eucalyptus*

species: *accedens, buprestium, calophylla, diversicolor, marginata, salmonophloia, sargentii, wandoo, globulus, baxteri* and *microcarpa*. Other food trees may also be utilized, such as casuarina and orchard trees. In orchards they can cause significant damage.

Nesting: It takes place in spring and summer, between August and December. A hole with a small entrance, usually in a eucalypt near water, is chosen. The tree may be dead or alive. Ornithologists have found nests whose entrances vary from only 4 m (13 ft.) above ground to as high as 40 m (130 ft.) in a giant karri (Forshaw 1981). Sometimes purple-crowns nest in colonies; at a site in Western Australia, 55 pairs were nesting in an area of about 2.5 hectares! The *Atlas of Australian Birds* indicates that breeding occurs in only about one-quarter of the usual range of this species.

Warham (1958) was shown a nest at a height of 12 m (40 ft.) in the heart of a white gum forest. The young were well grown. "Feeding was by now infrequent and sometimes other lorikeets would alight in the tree. My birds were quite tolerant of these visitors; indeed, the pair once preened each other while a third perched only two feet from them. On the last day of my watching, the chicks were fed only three times—shortly after dawn, about 2:00 P.M. and presumably at dusk, when the pair retired to the nest to roost. This was their last day in the nest."

Purple-crowns lay three or four eggs; a set of four averaged 20.3 mm (19.8–21.0 mm) x 16.7 mm (16.5–16.9 mm) (Forshaw 1981). A clutch of four laid by a captive female measured 20.0 x 18.0 mm (two), 20.0 x 17.5 mm and 19.5 x 17.5 mm; a six-month-old female laid four eggs which were smaller than the average: 19.2 x 15.7 mm, 19.5 x 17.0 mm, 19.3 x 16.2 mm and 19.2 x 17.0 mm; each weighed 3 g. Eggs of normal size laid by two females, four in each clutch, measured 21.5 x 18 mm, 21.0 x 17.5 mm and two 21.0 x 17.5 mm, and 21.0 x 17.5 mm, 21.0 x 16.5 mm and two 20.5 x 17.0 mm (Mexsenaar pers. comm. 1996). Two eggs laid by Rachel Lewis's pair weighed 3.12 g and 3.22 g soon after being laid.

Status/Conservation: Common in some areas, especially in South Australia, where it is the most widespread lorikeet; moderately common in Victoria, even plentiful in the suburbs of Adelaide. As with the other small lorikeets (see Musk Lorikeet, subheading Status/conservation), its survival is linked with the existence of eucalyptus trees old enough to form suitable small nesting hollows.

Aviculture

Status: Relatively uncommon but increasing in Australia; unknown elsewhere except possibly in New Zealand.

Clutch: Three or four (various sources); four to six, usually five (Williams pers. comm 1985).

Incubation period: 19 to 22 days. Baskerville's pair laid four eggs, two of which hatched 22 days after incubation commenced. Incubation usually commences with second egg (Matthews and Matthews pers. comm. 1994); 22 to 26 days, longer period in cooler months (S. Johnson, South Australia, pers. comm. 1995).

Harald Mexsenaar of Queensland recorded the following days from laying to hatch in two pairs:

Purple-crowned Lorikeet Incubation Periods

Egg	1st	2nd	3rd	4th
Pair 1	22	20		
	23	20	18	
	22	20	19	19
Pair 2	22	21	21	
	21	21	23	(?)—

Newly hatched chicks: Weigh 2 g and have silvery gray down (Mexsenaar pers. comm. 1996); 3 g, very sparse gray down, brown beak (Digney pers. comm. 1996).

Chick development: At about 12 days the eyes are opening and shorter second gray down is appearing. "Pin feather development is noticeable at 22 days and the chicks are feathered by 45 days" (Sindel 1987). Development was more rapid in the young reared by Mexsenaar's pairs: "At six days the skin is getting darker. A shorter grey secondary down is replacing the silvery grey. The eyes begin to open after nine days. Pin feather development starts after 16 days. The young are fully feathered by 38 days."

Young in nest: 45–60 days recorded by Sindel (1987). Two young in a clutch of W. Baskerville's

pair left the nest aged 45 days. Average time 42 days (Mexsenaar).

Chick growth

The weights of two parent-reared chicks, both hatched on November 20, 1995, were recorded as follows (Mexsenaar pers. comm. 1996):

Purple-crowned lorikeet weight table

Age in days	Weight in grams	
	Chick No. 1	No. 2
hatch	2	2
1	4	3
2	5	4
3	6	6
4	10	10
5	12	11
6	13	12
7	15	14
8	16	14
9	20	18
10	23	22
11	24	24
12	28	27
13	31	30
14	32	31
15	35	35
16	37	36
17	40	40
18	46	43
19	44	43
20	44	43
21	46	46
22	53	50
23	51	50
24	55	54
25	55	54
26	54	54
27	50	50
28	52	52
29	57	57
30	54	54
31	57	57
32	62	59
33	59	58
34	58	57
35	61	59
36	62	57
37	60	56
38	55	55
39	fledged	58
40		fledged

These weights seem high. In comparison, a parent-reared chick at Rainbow Jungle weighed 13 g at ten days (Digney pers. comm. 1996).

Age at independence of parent-reared young: 56 days. Three young aged 11 weeks weighed 47 g, 48 g and 51 g and three young aged 13 weeks weighed 48 g, 49 g and 52 g (Mexsenaar).

General: In Australia there was little interest in

this species, which was considered a difficult bird to breed (Johnson 1952), until the 1980s. In the previous 50 years there had been a few isolated successes. The first recorded breeding in New South Wales occurred in 1952. Mr. C. Lambert's pair was housed in an aviary 3.6 m (12 ft.) x 2.4 m (8 ft.) x 2 m (6 ft. 6 in.) high with other parrots. From the logs and boxes provided, the pair chose a hanging log 60 cm (2 ft.) long and 25 cm (10 in.) in diameter, with a side entrance hole. Four eggs were laid, all of which hatched. At the age of about five weeks, one youngster became trapped in the entrance, resulting in the deaths of the other three. The survivor left the nest at the age of eight weeks, when it was a dull edition of its parents, with a purple crown. In the second nest, three young hatched. Food offered consisted of a mixture of honey, milk, Farex baby cereal, breakfast cereal, sugar, crushed biscuit and plain cake. Chopped apple and an occasional mealworm were also offered (Johnson 1952).

Graham and Glenys Matthews of Barmera, South Australia, describe purple-crowns as "the most ravenous and messiest feeders of all Australian lorikeets." They are also the most in need of a constant food supply. If there is no food available they will "sit with heads in feathers looking extremely off color until fed." No doubt their metabolism is more rapid than that of the larger lorikeets. They must spend more hours feeding to supply their energy requirements. The Matthews also found this species to be more susceptible to candidiasis.

One of the first consistent breeders was Lyn Williams of Western Australia. By 1985, her colony numbered 55 birds, most of which had hatched in the aviary. All the birds were compatible, and some trios existed (males predominated). A lemon tree in the aviary was a great source of enjoyment; they nibbled the leaves and fed from the flowers. Their diet consisted of two types of nectar. The most favored contained Complan and honey mixed with water. The second type contained high-protein baby cereal, wheat germ cereal, raw sugar and a calf-raising formula called Denkavit. The females usually commenced to lay at the same time, and double-brooded. In 1984 they started to lay in August, but in some years they started to nest in April and continued throughout the winter. Fertility was good but some dead-in-shell occurred. The time young spent in the nest varied; those in deep nests could remain for two weeks longer than those which left before they were able to perch well. Losses among three-quarter-grown chicks were high; this also applied to hand-reared young if they were not given a midnight feed (Williams pers. comm. 1985). This is unusual and does not seem to apply to other lorikeets, so perhaps some unknown factor was responsible.

Harald Mexsenaar noted that in the Brisbane area captive birds lay between mid-May and the end of October. Another breeder had chicks hatch as late as early December.

Genus: *Phigys* (G. R. Gray 1870)

Synonymized with *Vini*

In the single species of this genus, the collared or solitary lory, the most obvious feature in which it differs from the *Vini* lories is the elongated feathers of the hindcrown and mantle of the adult bird (Forshaw 1989). However, to found a genus on such a characteristic would be equivalent to placing *Charmosyna papou* in a separate genus on account of its elongated tail feathers. In any case, in *kuhlii* the elongated feathers of the nape fall on to the mantle. In appearance the collared lory is otherwise similar to *Vini* species and its behavior does not differ. Steadman and Zarriello (1987) state unequivocally: "Based on osteology and plumage, the monotypic genus *Phigys* of Fiji should be merged with *Vini*, a genus that is distinct osteologically from other Polynesian parrots." Their studies led them to "recommend synonymizing the genus *Phigys* G. R. Gray, 1870, with *Vini* Lesson, 1831."

The collared lory is slightly larger than the other extant *Vini* species, but two extinct species are known which exceed it in size. Forshaw states that "the body form is stockier than in *Vini*"; but it is no stockier than that of *Vini peruviana*, for example.

There can be no justification for maintaining the genus *Phigys* and here the collared lory is placed in the genus *Vini*.

Genus: *Vini* (Lesson, 1831)

This genus contains six extant species; two others are known to have become extinct. Steadman and Zarriello (1987) believe that one or more species of Vini probably existed throughout the many islands of Polynesia that are not inhabited by parrots today. They speculate that "in light of the historic or prehistoric extirpations from certain islands of *V. kuhlii, V. ultramarina, V. australis* and *V. peruviana* and the complete losses of *V. sinotai* and *V. vidivici*, we might expect additional undescribed species of Vini to show up in future archeological or paleontological excavations in Oceania. At a minimum, we can expect new island records for the known species."

From the foregoing it is evident that the distribution and number of *Vini* species known today is but a relict of a genus once widespread and richer in species. Some survive perilously, their existence being threatened by the introduction of rats to the islands they inhabit.

Vini lories are small, short-tailed and either very colorful or unusually and exquisitely colored. Immature plumage differs from that of adults. The beauty of the feathers of adult birds has contributed to their range reductions and possibly to their extinctions. Their remote habitats and/or protection in recent times has mainly precluded them from aviculture but traditionally they are and were kept by native people.

Their breeding biology differs from that of most other lories in that males play a role in incubation and rearing of the young which is as active (or more so) than that of the female. In this respect and in others (such as appearance of chicks) they resemble the *Charmosyna* lorikeets, to which they are closely related. A slight (or pronounced, in two *Charmosyna* species) elongation of the primaries is a characteristic shared by some *Vini* and *Charmosyna* species.

Collared Lory *Vini solitarius* (Suckow 1800)

Synonyms: Solitary lory, ruffed lory

Native names: Fiji: *kula*, *kakakula*; Samoa: *sega'ula*, *segafiti*

Description: Surely one of the most beautiful birds in the world, for the brilliance and purity of its colors and its pleasing form. Forehead, crown and back of head, also lores, are deep purple, as well as thighs and lower abdomen. Its unique feature is the nuchal ruff of elongated bright green (almost iridescent) feathers. Porter (1935) described the ruff feathers as "one of the most brilliant greens in the whole of Nature"—and this is correct. Porter also observed that the ruff can be "erected from each side, forming two fans on the side of the head." The red feathers of the mantle are also elongated. Wings are dark green and the rump light green; underwing coverts, tail and undertail coverts are also green. There is an irregular spot of dull yellow-orange on the tail feathers, apparent only on the central feathers until the tail is fanned. Elsewhere the plumage is red. In adult birds, the primaries are slightly elongated, but not to the degree of *Charmosyna papou*. The bill and iris are orange and the feet are pinkish orange (salmon).

Length: 20 cm (8 in.)

Weight: Males average 85.0 g, females 74.5 g (Rimlinger pers. comm. 1995); males average 78.3 g, females 72.9 g (Campbell 1997).

Sexual dimorphism: Applies in adult and immature birds, according to Amadon (1942): immature birds "show the same sexual characters as the adults but in a less pronounced manner. Males have green in the hind-crown only near the base of the feathers, and usually concealed. In some thinly feathered male juveniles the green is more or less visible but it does not extend to the tips of the feathers as in all females." Gwendolyn Campbell noted that "upon close inspection, the female has an obvious leaf green tinge to the rear half of her crown, while the front half is indigo blue. The male has an indigo blue forecrown as well, but the rear half is deep purple...Sexes appear to be distinguishable when the first crown feathers erupt from the

sheaths, but we have only monitored this in a few clutches" (Campbell 1997).

Immatures: They have purple tips to the breast feathers, which have concealed yellowish green spots. The red and green feathers of the hind neck are short and do not form a ruff. The bill is brownish, soon becoming yellow, the iris is dark brown or black and the legs are gray. Clunie (1984) mentions that young birds "usually lack the orange patches in the tail, but so do some adult birds." However, most of the dozen or so young birds I saw at San Diego Zoo had yellow-orange in the tail. Watling (1982) incorrectly stated that "immature birds are similar to adults."

Bahr (1911) described two wild-caught young birds; when they were three months old the beak became bright yellow and at five months they underwent a partial molt of head and breast feathers; at eight months they had a complete molt. King (1940) described two captive-bred young, aged about six months, as having "eyes, feet, and beaks...black instead of orange; their crown and feet [*sic*] are darker blue and purple; the feathers on the nape and upper mandible [*sic* (mantle, not upper mandible)]] are shorter, which makes the bands of red and green look narrower. The young hen is more heavily washed with green on the head than the old hen. When flying, an arc of orange spots is shown on the upper tail coverts [*sic* (tail feathers, not coverts)]."

Patten (1941) described one young bird bred at Taronga Zoo; it had the red band on the upper mantle narrower and duller; the mandibles were slate colored, the feet gray or putty and the eye nearly black.

Natural history

Range: Fiji Islands, except the southern part of the Lau group (absent from Totoya, for example); it occurs as far south as Lakemba and Oneata.

Habits: Noisy and conspicuous, the collared lory is usually found where *Erythrina indica* and other flowering trees on which it feeds are in bloom. Watling (1982) states: "The Collared Lory is a common bird in the forests and in the wetter, windward areas of Viti Levu and Vanua Levu, but is less frequently found in the open, agricultural areas of the leeward coasts. They are a highly mobile, nomadic species and soon find any large concentrations of flowering trees. In the Sigatoka Valley of Viti Levu, Collared Lories are normally rare, but in the months of August and September they suddenly appear in large numbers as soon as the Drala, *Erythrina indica*, blooms." Forshaw (1989) mentions that they are the only rainforest birds that have invaded urban Suva (but are they true rainforest species?) and that he observed "good numbers feeding in flowering coconut palms lining the streets of Suva." He described how when feeding in these palms "the lories alight on the fronds and make their way down to the flowering stalks in a series of fluttering hops."

In the 1930s, Sydney Porter observed the collared lory on Kandavu, where it was very abundant:

Its shrill cry of "lish-lish" is heard on every hand. I have often watched them on the flowers of the tall coco-nut palms, feeding either on the pollen or the honey of the flowers; the birds seem to start at the bottom of the spray, running up licking the flowers quickly with their long brush-tipped tongues as they go. They often alight on the very ends of the palm fronds and dance along the mid-rib with the characteristic whisking movements of the Lory tribe until they get to the flowers at the base of the leaf (Porter 1935).

I had long wanted to see the collared lory in its natural habitat and, in March, 1993, spent four days on Viti Levu, en route to New Zealand. I was enchanted by this species. My first sight of it was of a pair, glittering like mobile jewels as they fed in the top of a coconut palm. No doubt they were gathering pollen from the inflorescences. The colors of brilliantly plumaged birds often disappear into drabness when seen high above, but collared lories gleam with color like no other bird I have ever seen. The intensity of the "cape" or "shawl" of vivid green impressed me; it seems to stand away from the rest of the plumage. Indelibly printed on my memory are the two mornings when I watched this species feeding low down in hibiscus. Imagine this brilliant red and green lory feeding among the scarlet flowers! Its plumage provided fairly good camouflage yet it

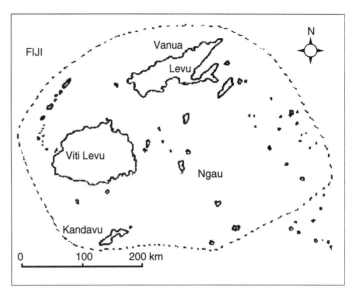

Distribution of collared lory (*Vini solitarius*): islands within dotted line.

was too noisy and active to go undetected. The birds moved rapidly through the leaves and every now and then a little head appeared and some of the bright plumage was exposed for a second. The flash of my camera, only 2 m (6 ft. 6 in.) distant, left the lories unperturbed. One stopped to survey me as I hid in the foliage. It was totally unconcerned.

Many Fijian gardens are a blaze of color from flowering shrubs, thus collared lories commonly occur in close proximity to man. It was obvious that they were not persecuted in any way—or they would not be so unafraid. Porter also commented that "the natives hardly ever interfere with the bird."

When I saw this species, I had just come from Aitutaki, from four days of intensive study of *Vini peruviana*. The behavior of *solitarius* seemed to be identical to that of its smaller cousin. They both spend much time in the crowns of coconut palms, descending lower to search for blossoms, pollen, nectar, insects and, apparently, also for fruit. They have been seen feeding on mango (*Mangifera indica*) and on soursop (*Annona muricata*). Clunie (1984) states that they eat caterpillars and other insects. He observes: "Feeding flocks are very active. Birds hang upside down, swing on whippy branch ends, chase each other about the tree tops. Two may grapple and fall to the ground, chittering. Or they may hang upside down, each by one foot, 'shaking hands' with one another, swinging wildly and nuzzling each other's neck." He states that typically they occur in small flocks but that there may be up to 50 together. They "fly quickly, with sudden changes in direction; short, whirring wings, brief glides."

It has been recorded that the peregrine falcon (*Falco peregrinus*) can catch collared lories. However, the peregrine (an endemic subspecies) is now very rare and approaching extinction. Bahr (1911) mentioned that where the introduced mongoose was found (such as on Kandavu and Taveuni), the lory was becoming rare, but I have seen no mention of this threat in recent literature.

Nesting: Bahr (1911) described nests found on Taveuni in 1910, by a planter. They were situated low down in dead stumps which had been left in a coconut plantation. In most cases the nest site was on a level with the ground, making them vulnerable to predation by mongooses. Nests were said to be hard to find because, while under observation, collared lories are wary of entering their holes. Clunie states that they breed in the second half of the year, especially after August, "but records are sparse." Two eggs are laid in a hole or a crack in a stump or a tree. They are pugnacious near their nests and have been known to attack even large barking pigeons which were perched in the top of their nest tree. Holyoak (1979) mentions a nest found on the island of Gau seen by a Fijian. It was in a hole in a tree, only 1.5 m (5 ft.) above the ground. A single egg described by Schönwetter (1964) measured 27.1 x 24.2 mm. Average weight of eggs laid at San Diego Zoo was 5.25 g (Rimlinger).

Status/Conservation: Collared lories are common and widespread. Lambert et al. (undated) suggest that the total population numbers between 10,000 and 100,000 birds. A possible threat to this species would be destruction of nest sites—but at least in coastal areas good numbers of coconut palms are likely to survive. Commercial logging is carried out on a large scale on Viti Levu and Vanua Levu.

Culture: The red feathers of the collared lory were, for centuries, of extreme importance in Fijian culture. See **Importance of Lories**.

Aviculture

Status: Rare; commercial export has never been permitted. In the U.S. it became available to a private aviculturist in 1996 due to consistent breeding successes at San Diego Zoo.

Clutch size: Two

Incubation period: 26 days with parents, as short as 24 days in an incubator (Campbell, San Diego Zoo, pers. comm. 1995). Sparse information in the literature: there are two records and in both cases the chick hatched 28 days after the first egg was laid (Low 1977a) but perhaps relate to cases (common in the closely related *Charmosyna* species) where incubation does not commence until three days or so after the first egg is laid. Rimlinger reported that incubation is carried out by the female but the male spends much time in the box (but it may be that he incubates at night).

Newly hatched chicks: Weigh on average 3.65 g; they have sparse white down, fairly long; the beak is black (Rimlinger pers. comm. 1995); incubator hatched chicks weigh 3.3–4 g (Campbell, San Diego Zoo); two hatched in 1996 weighed 5.1 g and 4.4 g (Lewins pers. comm.).

Chick development: A chick hatched at Taronga Zoo in 1940 was described as "a ball of white fluff" at the age of seven days and covered with gray down at 15 days. At 22 days the development of red and green feathers was observed and five days later the head was covered with dull green feathers. The breast and sides were the last areas to feather, by the age of 42 days (Patten 1941).

Young in nest: 61 days, one only (Patten 1941); 61–63 days (Campbell pers. comm. 1995).

Chick growth

Average weights of chicks hand-reared at San Diego Zoo (Rimlinger/Campbell pers. comm. 1995):

Collared lory weight table

Age in weeks	Weight in grams
1	7.2
2	14.6
3	25.0
4	37.2
5	48.7
6	62.8
7	75.6
8	79.3
10	71.1

Hand-reared young are independent at about 59 days and parent-reared at 77 days (Rimlinger). Also at San Diego Zoo, four young at fledging weighed 87.5 g, 90.9 g, 83.2 g and 78.2 g.

At the time of going to press more detailed weights for chicks of *Vini solitarius* were available (Campbell 1997).

Chick Weight Comparison (average)

	Parent raised (2 chicks)		Hand raised (8 chicks)	
Day	range	average	range	average
1	4.4–5.1	4.8	3.0–4.0	3.7
5	7.5–14.3	10.4	4.8–6.1	5.6
10	19.5–24.9	22.2	8.5–11.9	9.9
15	25.5–32.7	29.1	13.5–19.3	15.5
20	30.3–36.2	33.3	20.2–26.1	23.3
25	41.7–48.5	45.1	30.6–37.4	32.5
30	51.8–57.6	54.7	34.8–42.9	40.0
35	61.5–70.9	66.2	43.8–57.2	48.2
50	78.2–83.2	80.7	64.3–89.1	76.9
60		N/A	65.8–94.0	77.7
70	76.2–82.1	79.2	63.3–81.6	72.4

General: In the first half of this century, a very few privileged aviculturists were fortunate to keep this lovely bird; today it is unknown in private collections. Bahr mentioned the difficulties in keeping wild-caught birds alive. He obtained two young ones and tried to feed them on honey, without success. By accident, he discovered how to rear them:

One of them, on being placed on the breakfast table, made a bee line for the porridge and commenced feeding on it with great alacrity with his brush-like tongue. Tea with sugar and milk he absolutely could not resist, though milk and sugar alone or Nestle's milk he was not at all partial to. Henceforth the younger bird was fed by means of a spoon on sweetened tea and milk, and became very fat and grew rapidly; porridge and gruel he would not or could not take. The older bird became extraordinarily tame and familiar and never attempted to fly away.

Subsequently 5 more young birds were brought in by a Fijian; they were half-starved and were being fed on mummy apple [papaya], which did not agree with them. So infested were they with white mites that I also became covered with these creatures whose bites caused considerable inconvenience. Frequent baths with dilute lysol effectively cleansed the birds of these parasites. Though the weather was very warm they required a considerable amount of extra heat; the youngest bird especially was never so happy as when placed in the incubator at 97° Fahr. Of the new arrivals I lost one, which vomited all food and died in convulsions. The others lived in a large cage and became very

tame and familiar. They were always lively and cheery, tumbling about the floor quarrelling like monkeys and greeting others of their kind with shrill cries as they winged their way over our house (Bahr 1911).

Bahr attempted to take five of these little lories back to England by boat, via Canada. Four were seized with convulsions *en route* (three days from Honolulu, where the weather was still warm), resulting in two having permanently cramped feet. He kept them warm on the journey over the steam-heating apparatus and one recovered from the cramp. Two survived and were fed on porridge, also fruit, especially grapes, apples and cherries. They were eventually deposited at London Zoo, the first of their species to be exhibited there.

In 1937, Dr. Derscheid of Belgium presented the Duke of Bedford with a pair. They reared two young in 1939 (almost certainly the first captive breeding) and at least one during 1940–41. Unusually, he did not write about this success. However, Harold King described the first breeding in the weekly magazine *Cage Birds* (January 26, 1940). The pair was kept in an outdoor aviary. The nestbox stood in a dish of water in the aviary shelter and was partly filled with peat and rotten wood. The first of two eggs was laid on August 1 and these had hatched on or by August 29. They were successfully reared by the parents on Allinson's baby food, grapes, pears and apples.

In 1938 Taronga Zoo in Sydney, Australia, received several collared lories from Fiji. The following year a pair nested in a hollow log in a large outdoor aviary. One egg was laid but the chick died in the shell. In the following year an egg was laid on September 23 and hatched on October 21. The young one was reared by the parents, mainly on nectar, the fruit being ignored on some days. It left the nest on December 9, i.e., at the age of 61 days. It was in excellent condition, despite the fact that the nesting log was in an insanitary state. During 1943–44, two pairs nested at Taronga, each pair rearing one youngster.

In recent years, few collared lories have been seen outside Fiji. Dr. R. Burkard in Switzerland kept and bred them during the 1960s and 1970s. In his experience "they usually have 2 eggs but often only one is fertilized. If both are fertilized the parents usually raise only one of the youngsters; the other dies in the egg or shortly after leaving it" (Burkard pers. comm. 1977). His birds did not breed until their fifth year. In 1968, Dr. Burkard had seven pairs. He found keeping and breeding this species difficult. By 1975, there were 16 birds; losses had resulted from candidiasis and aspergillosis. The birds were placed with an experienced breeder of lories in Switzerland, J. Maier. In 1983 the last one with Dr. Burkard was obtained by Vogelpark Walsrode. It was a 10-year-old female and she was still alive four years later.

In the early 1990s, San Diego Zoo in California obtained some collared lories from Fiji. I saw three pairs at the zoo in October, 1992. At that time, no breeding attempt had been made. By the time of my next visit in August, 1996, flights in the off-exhibit Propagation Center were filled with young birds. It was an impressive sight! The species was breeding well and nearly all the young were parent-reared. The pair on exhibit in "Wings of Australasia" had a youngster with them. The family made an enchanting sight in their planted aviary. Exclamations regarding their beauty were heard from many passing visitors. Off-exhibit, a two-year-old bird was the focus of attention. Of normal coloration for most of its life, it had recently molted out with melanistic plumage. The feathers of the collar and the wings, green in a normal bird, were black, except for a few wing feathers (possibly the molt was not complete). The tail was also black, with red (not orange-yellow) marks. Continued success at San Diego Zoo may eventually result in this species being available to aviculturists in the U.S.A.; by 1996 the numbers bred had permitted some to be placed with a local breeder.

Blue-crowned Lory *Vini australis* (Gmelin 1789)

Synonym: Samoan lory

Native names: Tonga and Niue, *henga*; Samoa, *sega* or *segavao*

Description: Mainly green, it is conspicuously marked with red and blue on head and abdomen. The crown and occiput are blue, shaft-streaked with sky blue; lores, face and throat are red (a bib-shaped area which just extends to the upper breast). This color is repeated on the abdomen, below which is a small patch of purple; the red of the abdomen may be faintly tinged with purple. The thighs are also purple. The tail is yellow or greenish yellow below. The bill and feet are orange and the iris is brownish orange.

Length: 19 cm (7½ in.)

Weight: Males 43–49 g, females 39–44 g (Campbell, San Diego Zoo, pers. comm. 1995); 47–52 g (Forshaw 1989).

Immature birds: Red on the face and throat is less extensive and duller; the red feathers on the abdomen are mixed with green, and the red area is smaller and irregular. The purple is mainly confined to the vent and may or may not meet the red area. The thighs are green. The blue feathers on the crown are shorter. Bill and iris are dark brown and the legs are dark gray and pink.

Natural history

Range: Polynesia: Samoa, including Upolu, Savaii and the Manua Islands (but not Tutuila); Futuna and Alofi (Iles de Horn, between Fiji and Samoa); Tonga (including Niuafo'ou and Late), also Tongabatu, Hapai and Vavau, the most northerly of the Tongan islands (but not Eua); Niue (one of the Cook Islands, situated between Tonga and Rarotonga); Fiji: the southern islands of the Lau Archipelago only. (The collared lory occurs in other parts of Fiji and there appears to be competitive exclusion.)

Extinctions: Extinct on Eua; Steadman (1993) states that it was last seen there about 50 years previously. Rinke states that it became extinct on the main Tongan islands, on Tutuila in American Samoa and on East Uvea (Wallis) after the arrival of Europeans. On East Uvea and perhaps elsewhere its extinction was due to ship rats.

Habits: The blue-crowned lory became known to science during Cook's third voyage (1776–80), at Tongatabu. Mivart (1896) quotes the Rev. S. J. Whitmee that in Samoa, this species was "very abundant on all the islands during a part of the year. The natives believe this bird migrates; but all I have been able to learn on the subject is, that they are seen passing in flocks from the western to the eastern islands. A few may be found all the year round; but during several months of the cooler season the coconut trees swarm with them. They appear to feed chiefly on the nectar of the coconut flowers; but when the *Erythrina indica* (a very common tree near the coast) flowers in July and August, they may be seen about it in great numbers. Native boys are very expert at snaring the Sega on the coconut trees. I have never heard of the bird breeding in Samoa; and the natives positively affirm that it does not." According to Gibb et al. (1989), this species does breed there. In Western Samoa, forests are under threat from forestry and agriculture. In Upolu and Savai'i, blue-crowned lories were often seen feeding on coconut flowers in the agricultural plantations adjacent to the national park, in August, 1984 (Bellingham and Davis 1988). In the southern foothills of Savai'i flocks fed on *Elaeocarpus ulianus* flowers growing between newly established agricultural plantations.

Rinke (1994) states that it is a strong flyer; he saw them flying between islands, a distance of up to 20 km (12 mi.). During his stay on Niuafo'ou during March, 1984, only one juvenile bird was seen. He observed that this lory prefers open habitats such as coconut plantations, beach forests and forest edges. Coconut palms and trees with densely clustered red flowers were favored. Groups of up to seven birds were seen there. When visiting Niuafo'ou in 1990 Rinke (1991) observed that blue-crowned lories were abundant. They also occurred on the islands in the crater lake. He commented: "The two smaller islands seem to be roosting and breeding sites because they lack flowering trees to provide food (*Casuarina* dominates)." However, *Vini* lories are known to feed on *Casuarina*, e.g., *V. kuhlii* on Rimatara (Seitre pers. comm. 1996). On Niuafo'ou Rinke could approach feeding birds as closely as 1.5 m (5 ft.) but at other times they flew when approached as close as 10 m (33 ft.). He saw them feeding at unripe mangoes and at the flowers of 'iku'ikuma (*Stachyrtarpheta urticifolia*), a

woody plant which grows up to 2 m (6 ft. 6 in.) high. He suggested that these foods might have been taken due to a recent hurricane causing a food shortage. On Late, Tonga, an island of 18.5 km^2 (7 sq. mi.), there are no ship rats but the comparatively harmless Polynesian rat does occur. During seven days there in early 1990, Rinke (1991) found this species "not very common." It was rarely heard, and only two birds were seen—feeding on flowers in a high forest tree. It was recorded mostly from above the forest canopy. This species usually occurs in small groups, frequenting flowering coconut palms and other flowering trees and shrubs. Its flight is swift and direct with rapid wing beats and it calls loudly in flight. On Upolu (Samoa), Dhondt (1976) regularly saw small groups flying above the forest, between mid-1973 and late 1974. In mid-March he saw many blue-crowned lories feeding alongside wattled honeyeaters (*Foulehaio carunculata*) in a red-flowering tree. On Niuafo'ou (Tongan Islands) the flowering plants on which it feeds include *Sterculia fanaiho*, *Kleinhovia hospita*, *Musa paradisiaca*, *Alphitonia ziziphoides*, *Pueraria lobata*, *Barringtonia asiatica*, *Hibiscus tiliaceus* and *Pometia pinnata* (Rinke 1986).

Nesting: Little information exists. One egg supposedly laid by this species was described as measuring 27.1 x 24.2 mm (Schönwetter 1964)—a strangely shaped and extraordinarily large egg for such a small bird. Size and weights of four eggs laid in a private collection were as follows: 21.0 x 18.5 mm (3.86 g); 20.8 x 18.4 mm (3.80 g); 21.0 x 18.4 mm (3.82 g); and 20.4 x 17.9 mm (3.79 g) (Neff, pers. comm. 1997). Two eggs laid at San Diego Zoo weighed 4.24 g and 4.26 g 23 days before hatching and a nearly fresh egg weighed 4.3 g (Lewins pers. comm. 1996).

Status/Conservation: Although still common throughout much of its range, it is declining. On Niue, in the Cook Islands, it is not common. Forshaw (1989) quotes Kinsky and Yaldwyn to the effect that during three weeks in late August and early September, 1972, a group of scientists recorded this lory on only four occasions. Each time it was in flight across open areas between stands of mature forest. Rinke (1993) described the findings of the Brehm Fund for International Bird Conservation, whose work in the Pacific region included field studies and captive breeding of the blue-crowned lory. "Our surveys found the lorikeets uncommon (on uninhabited forested islands) to very abundant (on islands with extensive coconut plantations), but absent from islands which have been colonized by ship rats. The total number in Tonga may be about 20,000 birds."

Watling (1982) stated: "In Tonga it is apparently decreasing in numbers and is no longer found on several islands from which it has previously been recorded." This is almost certainly due to rats colonizing these particular islands. Seitre (1994) describes the blue-crowned lory as still quite common on Futuna and Alofi, where "numbers range between several hundred and several thousand individuals." Ship rats were absent. This lory, like other *Vini* species, can live close to man and benefits from the planting of ornamental shrubs—but it can offer no defense against rats. If it is to survive, it is imperative that no wharves are built on some islands within its range. A wharf brings ship rats and these rats bring extinction to small lories—possibly in less than a decade.

Aviculture

Status: Rare

Clutch size: Two

Incubation period: 25 days, with an average of 48 hours from pip to hatch (Campbell 1993); about 24 days (Rinke 1994). Incubation by male and female.

Newly hatched chick: Weight, 2.7 g; sparse white down, fairly long; beak black (Campbell pers. comm. 1995); four chicks hatched at San Diego Zoo in 1996 weighed 2.3 g, 3.0 g, 3.7 g and 3.7 g (Lewins pers. comm.).

Development of young: Eyes open at approximately ten days (Lewins 1996). A photograph of two chicks described as three days and five days old shows that the upper parts, except the nape, are covered in fairly dense down (described as yellowish white). A chick aged about 35 days was depicted by Rinke (1994). It was covered in dense light gray down; feathers had erupted on the head, wings and underparts and the red feathers below the lores were just starting to appear. Chicks photographed at San Diego Zoo at a similar stage of development, and with the tail feathers just starting

to erupt, were described as "about 4 weeks old." The feet were black.

Young in nest: "About 60 days" (Rinke 1994); 55–59 days (Rimlinger, San Diego Zoo, pers. comm. 1995).

Age at independence of hand-reared young: No information—but two started to feed themselves at 49 days (Campbell pers. comm. 1995).

Chick weights: The average weights of chicks hand-reared at San Diego Zoo were 4.6 g at one week and 7.4 g at two weeks. Parent-reared young weighed 24.2 g at 15 days, 31.3 g at 17 days and 37.1 g at 19 days (Rimlinger pers. comm. 1995). Two young weighed 45.6 g and 45.8 g at fledging and 49.7 g and 44.8 g one week later (Lewins pers. comm.). Two chicks reared by Claudia Matavalea at the

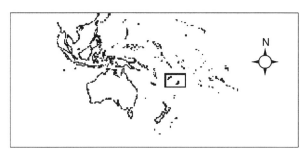

Range of blue-crowned lory (*Vini australis*): area within box. Maps below show details of range.

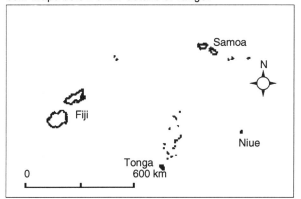

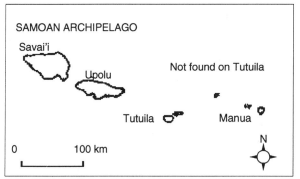

Tongan Wildlife Centre weighed 45 g and 40 g at about 46 days old (Matavalea pers. comm. 1995).

General: The Rev. Whitmee was quoted by Mivart (1896) to the effect that although the Samoans were very clever at rearing and keeping birds, they had never succeeded in keeping these lories alive for more than a few weeks. This presumably partly accounts for their scarcity in aviculture in earlier days. More recently, their export has not been permitted except in special circumstances. Swiss aviculturist Dr. R. Burkard was the first person to record keeping and breeding this species. According to Robiller (1992) the first success occurred in 1965. In 1974, Dr. Burkard wrote in *Gefiederte Freund* that feeding ant pupae played a vital part in this success. By 1976 he had bred blue-crowned lories on two occasions. The pair used a nestbox placed on the floor; it measured 91 cm (3 ft.) long and 7.7 cm (3 in.) in diameter. He found, not surprisingly, that "in a group, only the dominant pair breeds successfully. The difficulty lies in distinguishing the sex, which can only be judged by their behavior" (Burkard pers. comm. 1977). He found my query regarding their incubation habits difficult to answer because male and female spent so much time in the nest together.

The blue-crowned lory was next bred at San Diego Zoo. Five birds were received from Apia, western Samoa, in November, 1970. One pair produced three young, all of which were reared. They hatched on March 29, March 31 and July 10, 1973. The male had been in captivity for 13 years when he died in February 1981. The female outlived him by more than three years and was known to be at least 16½ years old. In April, 1991, San Diego Zoo received eight blue-crowned lories on loan from the Kingdom of Tonga. Four of these were donated to Assiniboine Park Zoo in Canada and two pairs were retained. One of these pairs proved very prolific. Between March, 1992, and December, 1994, they hatched 14 young, all of which survived to independence. These proved to be eight males and six females. Four nests produced two young and in each case they were a male and a female. The other wild-caught pair hatched eight young between March, 1993, and December, 1994, six of which were reared. Two nests produced two young: two females and a male and a female. During the first four months of 1994, a pair which were both the offspring of the most prolific pair, hatched three young, two of which were reared. The male was only 18 months old when their first chick hatched; the female was 22 months old.

By May, 1996, the two pairs received in 1991 had hatched 29 young and their offspring had hatched 10 chicks. The young obtain full adult coloration and are able to breed at one year old. Young were parent-reared, hand-reared and fostered under Goldie's lorikeets. Regardless of the rearing method, they bred successfully. The conclusion was that this species "is actually very adaptable and can be prolific in a captive environment" (Lewins 1996).

These breeding pairs were housed off-exhibit in aviaries 3.7 m (12 ft.) long, 1.2 m (4 ft.) wide and 2.4 m (8 ft.) high. One-third of the cage construction consisted of solid panels, as protection against the weather. Potted plants, such as *Ficus benjamina* and bamboo were provided for perching, cover and amusement. Campbell (1993) recorded that the lories were active practically "from dawn to dusk." In courtship display "the male approaches the female with rapid hops along the perch. Head bobbing with feathers puffed up and a curious side-to-side sway takes place and is followed by copulation."

The pairs were supplied with L-shaped nestboxes measuring 46 cm (18 in.) high, 20 cm (8 in.) square and 41 cm (16 in.) long. Pine shavings were provided as nesting material. It was recorded that in the first pair which bred successfully, the female appeared to do most of the incubating but the male was often in the nest with her. The second clutch laid by this female was fostered to Goldie's lorikeets. The eggs had been laid on February 19 and 22 and hatched on March 15 and 17. The chicks were fed by the Goldie's until March 31 when they were removed for hand-rearing. Spoon-feeding took place every three hours from 6:30 A.M. until 6:30 P.M. Four days later, feeds were reduced from five daily to four. From 32 days they were fed three times daily and from 41 days, twice daily.

As of May, 1996, 46 blue-crowned lories had been bred at San Diego Zoo; 7 from a pair between 1973 and 1979; 29 from the two pairs received in 1991, and 10 second generation from these 29. Their diet consisted of Nekton-Lory mixed with water (ratio 1:10) and fruit, usually apples, pears or papaya. A few mealworms and a small amount of chopped Romaine lettuce was also given. Gwendolyn Campbell (pers. comm. 1996) noted that they preferred nectar and mealworms, whereas the solitary lories preferred the fruit. At that time, the birds at San Diego Zoo, and some of the young on loan to a local aviculturist, were the only known specimens outside the Pacific region. As they were proving prolific, there is a good chance that more aviculturists in the U.S.A. will have the opportunity to keep them.

As part of the Brehm Fund's work in the Pacific, captive breeding of shining parrots (*Prosopeia*) and blue-crowned lories was initiated. Aviaries were built in 1990 and the first two young lories were reared in 1992 (Rinke 1994). However, because more than one pair was kept in each aviary, Rinke (1993) reported that only one pair per aviary had bred. He commented: "Lorikeets apparently need a different design of aviary, or may have to be kept in pairs." At that time the shining parrots and the lories were kept in two large aviary complexes, each with

a 110 m² colony aviary and six smaller side aviaries. After about five years, this work was taken over by the Tongan Wildlife Centre (a nongovernment, nonprofit trust). The main aim of the trust is to breed threatened species, to eventually release them on uninhabited islands, and to educate the people about their native wildlife. In March, 1995, there were ten pairs of *australis* in the center's aviaries. In 1994, eight young were reared. Some of these and some young shining parrots were sold to help finance the running of the center.

In my opinion, this is the *Vini* species which is most likely to be established in captivity, provided that the few birds available are in the care of experienced breeders. It is, perhaps, slightly easier to breed than the other *Vini* species. Valuable data can be obtained which could be applied to the rarer ones, if captive breeding is ever used as part of their conservation management.

Rimatara Lory *Vini kuhlii* (Vigors 1824)

Synonyms: Kuhl's lory, ruby lory, scarlet-breasted lorikeet

Native names: Cook Islands and Kiribati, *kura*; Rimatara, *'ura*

Common name: As suggested by Gerald McCormack, a single-island endemic should have a name which reflects its special conservation status and, in this case, the only island community to maintain a natural population. The name also emphasizes the pride in this species felt by the people of Rimatara, and will further encourage them to conserve it.

Description: Extremely colorful, it is mainly green above, tinged with olive on the mantle, and scarlet below. The feathers of the crown are elongated, green with paler green shaft-streaking; the longer, narrow feathers of the nape are rich blue. The lower back, rump, flanks, upper and undertail coverts are an unusual shade between green and yellow. The wings are mainly dark green; the primaries are blackish, with blue on the outer web. Examination of a captive bird showed that the tips

of the primaries are elongated (Schroeder pers. comm. 1996). The entire underparts, from the lores and throat to the small patch of purple on the abdomen, are scarlet. The thighs are dark purple. The tail is multicolored and the feathers are irregularly marked: green at the tip, dark purple on the outer web, red on the inner web and dull crimson below. Bill and feet are orange; the iris is gray with an inner ring of yellow.

Length: 19 cm (7½ in.)

Weight: 55 g estimated; one captive-bird (overweight): 74 g.

Immatures: They are said to have less red in the tail and the underparts marked with grayish purple. It is likely that the blue feathers of the nape are shorter than in adults. Bill and iris are brown.

Natural history

Range: Rimatara in the Austral Islands, French Polynesia (south of Tahiti); otherwise extinct throughout its natural range. It was introduced to

Teraina (Washington) and Tabuaeran (Fanning) in the Northern Line Islands (Kiribati Islands) prior to 1798. Watling (1995) points out that these two groups of islands are 3,000 km (1,860 mi.) apart.

Historical: Mivart (1896) was informed by Dr. Streets that natives from islands to the south visited Washington and Fanning islands to make coconut oil. These workers had tame Rimatara lories with them. Apparently some escaped—and lived to populate both islands. This may be a rare incidence of accidental introduction which has fortuitously extended the range of an endangered species. However, Seitre and Seitre (1991) suggest that this lory could have originated from Kiribati (Kiritimati, or Christmas Island) and was introduced to Rimatara. Watling states that the arid and unpredictable climate of Kiritimati make the long-term establishment of lories there unlikely. The few birds occasionally reported from Tubai Island (Austral Islands) are probably escaped pets. An attempted introduction to Christmas Island, of only six birds, was made in 1957; three were seen in 1959. Several more were introduced in the early 1960s, and more than one was still present in 1982. Three more were liberated in 1991 and at least two were present in 1993 (Watling 1995).

Extinctions: Kuhl's lory was widespread in the southern Cook Islands, including Aitutaki (see Gazetteer), Mangaia—385 km (239 mi.) from Aitutaki and 215 km (133 mi.) from Atiu—and Atiu, during prehistoric times. In 1987, Aitutaki and Atiu were surveyed by Steadman (1991) for the remains of prehistoric birds. Bones found, including those of *Vini kuhlii* from Atiu, were probably at least 700 years old (Steadman 1991). The first Cook Islands record of *kuhlii* was a bone found in a cave in Magaia (Steadman 1985). Numerous additional bones of this lory were found in an archaeological site on Mangaia in 1989. Steadman (1991) pointed out that "with the discovery of this species in prehistoric deposits on Atiu and Aitutaki as well, it is likely that *V. kuhlii* rather than *V. peruviana* was the indigenous small parrot of the Cook Islands. This seems particularly probable since the plumage of *V. kuhlii* is largely red while that of *V. peruviana* includes no red feathers. The Rarotongan word kura means 'red, scarlet, glowing or crimson, or ornaments with red feathers' as well as 'a native bird with

scarlet plumage, now extinct on the island of Rarotonga' (Savage 1980: 122). Savage also states that the *Kura* survived into the nineteenth century on Atiu and Ma'uke." However, Savage, in his *A dictionary of the Maori language of Rarotonga*, confused the two species (McCormack and Kunzle 1993).

Habits: On Rimatara, Bruner (1972) found that this lory preferred dense, wet rainforest in the vicinity of coconut groves. But from August 5 to 11, 1992, McCormack and Kunzle found this species to be common in the mixed horticultural woodlands which cover one-third of the island (about 2.2 birds per acre); moderately common in the villages (1.4 per acre); uncommon in the central hills (0.9) and in the coastal coconut plantations (0.8); rare in the extensive makatea (ancient raised reef) (0.6) and absent from the swamplands. The mixed horticulture habitat supported about 61 percent of the estimated population of 905 birds.

McCormack and Kunzle recorded the following food plants and the number of times they were recorded feeding on each: inga (*Inga ynga*) 33; Pacific ironwood (*Casuarina equisetifolia*) 18; falcata (*Paraserianthes (Albizia) falcataria*) 18; banana *(Musa* spp.) 9; mango (*Mangifera indica*) 8; kapok (*Ceiba pentandra*) 7; coconut palm (*Cocos nucifera*) 6; Chinese hibiscus (*Hibiscus rosasinensis*) 3?; Indian coral-tree (*Erythrina variegata*) 4; red-bead tree (*Adenanthera pavonina*) 3; rose apple (*Syzygium jambos*) 3; siris tree (*Albizia lebbeck*) 2.

All the above, with the exception of coconut palms, Pacific ironwood and tree hibiscus, are recent introductions. The lories fed on the nectar in most cases but also licked the leaf stalks of some species, especially kapok and falcata. Residents reported that Rimatara lories also feed on coffee (*Coffea arabica*), which had no blossoms at the time of McCormack and Kunzle's visit. Seitre and Seitre (1991) reported that they fed on tree hibiscus (*Hibiscus tiliaceus*), also *Barringtonia asiatica* and *Pometia pinnata*. Watling (1995) mentions that almost no native forest has survived on Teraina and Tabuaeran, where the lories are effectively confined to the coconut plantations.

Nesting: Little is known. On Rimatara suggested nest trees include *Barringtonia asiatica*, *Pisonia grandis*, falcata and coconut palm, but there is little evidence. Watling was informed that on

Tabuaeran a lory nest was found in the rotten trunk of a *Pandanus* tree. He speculated on whether the lory has survived there because it has nested away from the "normal" coconut habitat of the ship rat, adding "this should be investigated because it has important implications for the future conservation of this and other *Vini* lorikeets."

Status /Conservation: On Rimatara it was common (over one-third of the island) to uncommon in 1992. On Teraina and Tabueran it is apparently quite common, especially in coconut plantations; it is often seen near settlements (Pratt et al. 1987). A report made by M. C. Garnett in 1979 stated that on Teraina (Washington) the population was over 1,000, with 200 more on Tabueran (Fanning). However, Watling (1995) made a three-day survey in February, 1993, of Tabuaeran. He estimated the population to consist of possibly fewer than 50 birds. All were restricted to one location on the northwestern islet of the atoll. Watling wrote: "It would appear that the lorikeets have managed to survive for over 70 years in the presence of the ship rat, but have been confined to one location for the last 15 years at least, and perhaps much longer." The ship rat was first noted there in the early 1920s and again in 1937. Since plantation operations were abandoned in 1983 the coconut stands have not been cleared or thinned, thus now provide better habitat for rats. Today rat damage is a serious problem for subsistence farmers. No lories were observed or reported from the other two main islets. On Tabuaeran the lory is effectively confined to coconut plantations and is especially vulnerable to nest predation by rats, especially *Rattus rattus*. Although there have been recent introductions to Kiritimati, Watling considered its chance of future survival there to be poor not only because of the severe climate, but due to the fact that the government intends to develop this island.

This species is classified as endangered because of its small range—Rimatara is only 3 km (1.8 mi.) in diameter and 800 hectares in area—and vulnerability. McCormack and Kunzle suggested the lorikeet population on Rimatara was about 900 birds—and possibly up to 1,500. They emphasize the first conservation priority there should be to confirm the absence of the ship rat and to launch a major quarantine campaign to prevent its accidental introduction. If it is present in small numbers, steps could be taken to eradicate it.

"The second conservation priority would be to monitor the lorikeet population every two or three years, by comparing the maximum counts along the three transects...in the horticultural belt.

"The third conservation priority would be to reestablish part of the former natural range of the lorikeet by reintroducing it to one of the islands in the southern Cook Islands. On the belief that the *Kura* was lost from the Cook Islands because of prehistoric feather-collecting rather than some other environmental change, such as the introduction of the ship rat, we compared the present environments to assess the likelihood of a successful reintroduction. A comparison of the present indigenous and introduced plants of Rimatara and the makatea islands of Mangaia, 'Atiu and Ma'uke, showed great similarity, and we conclude that these islands would provide adequate food and nest sites for a successful reintroduction" (McCormack and Kunzle 1993).

Index trapping on Mangaia in November, 1992, showed that ship rats were present (probably in small numbers). A ship-rat-free island would be selected, and Atui is the most likely. During August, 1994, Roger Malcolm and Ron Dobbs carried out 376 trap-nights of trapping in five habitats on Atui. They obtained 78 *Rattus exulans* and no specimens of the dangerous ship rat (*Rattus rattus*). However, there would be one obstacle to overcome before reintroduction was carried out. The main mosquito vector of avian malaria (*Culex quiquefasciatus*) is present in the Cook Islands. The accidental introduction of the parasite could be disastrous for the indigenous birds, which presumably lack an acquired immunity. Therefore, McCormack and Kunzle state that "any lorikeets obtained for reintroduction to the Cook Islands must be adequately quarantined." This poses the questions: where and for how long? *Vini* lories are very susceptible to stress, also extremely aggressive when closely confined. Losses could be high during the quarantine period. The ideal would be to release them as soon as possible after capture—but this might not be permissible. Seitre and Seitre (1991) suggested transferring birds to the rat-free Maria-Atoll. On the Line Islands, there are plans by the Kiribati Government to resettle people on Teraina and Tabuaeran. C. Wilson informed McCormack and Kunzle that a conservation program was being developed to protect the lory there.

According to Watling (1995) the Kiribati government favors rapid development on Kiritimati. This "may entail a considerable increase in the human population and sooner or later the construction of a jetty," which would "inevitably lead to the introduction of the ship rat." Kiribati is not a signatory to any international or regional environmental conservation convention. It does have well developed wildlife legislation and Kuhl's lory is "fully protected" under the Wildlife Conservation Ordinance 1975 which prohibits the hunting, killing or capture and possession of fully protected birds. A Wildlife Unit of three men is stationed on Kiritimati but is able to make only very rare visits to Teraina and Tabuaeran. The Unit was unaware of the international significance of Teraina's population of the Rimatara lory until this was pointed out by Watling. Villagers and officials were aware that all birds are protected but did not realize that capture and possession were prohibited. There is a need for greater awareness and enforcement of the legislation.

Culture: During ancient times, the small red feathers of this lory were prized by the inhabitants of the southern Cook Islands, especially Aitutaki, Atiu and Ma'uke. They used them for personal adornment, to make caps and ceremonial headdresses and god images (McCormack and Kunzle). Almost certainly, the lories were trapped to extinction in order to obtain the valued feathers. On Rimatara, the lories were traditionally protected by the Queen. Hunting was forbidden—and this tradition has been maintained (Seitre and Seitre 1991).

Aviculture

Status: Very rare; currently unknown but for a single bird.

Clutch size: Two

Incubation period: No information, but assume 25 days.

Newly hatched chicks: No information.

Chick development: Patten (1947) provided a brief note on the development of one chick: when judged to be two or three weeks old (but surely at least four weeks) it had "just a touch of color, red and green, showing on the pin feathers."

Young in nest: Patten suggested the single youngster left the nest when aged seven weeks—but it may have been older.

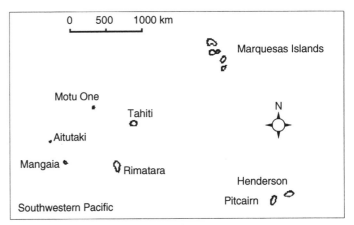

Species of the southwestern Pacific.

Rimatara or Kuhl's lory (*Vini kuhlii*) – Rimatara
Stephen's lory (*Vini stepheni*) – Henderson Island
Tahiti blue lory (*Vini peruviana*) – Aitutaki and Motu One (and other islands)
Ultramarine lory (*Vini ultramarina*) – Marquesas Islands

Chick weights: Not available.

General: The very few who have been privileged to keep—or even to see—this exquisite little lory have been impressed by its beauty and behavior. Commenting on a single bird observed at London Zoo in 1936, the renowned aviculturist and ornithologist Dr. Jean Delacour wrote "only those who have seen the specimen now in the possession of the London Zoo can have an idea of its delightful ways. Tame and gentle, it plays like a kitten; it is indeed the most wonderful pet I ever saw" (Delacour 1936). Its beauty was captured for all time in the painting by Roland Green which accompanied Delacour's notes. It was apparently an immature bird, for the beak was brownish and the skin surrounding the eye was white; it had a tinge of blue on the breast feathers. The dark blue feathers of the nape were shorter than those of an adult (assuming that it was accurately depicted by the excellent artist Green).

There was a single Rimatara lory in Germany in 1879, and a female imported into England in 1935, owned by P. H. Maxwell, lived until 1943. But because of their proximity to the Pacific, it was California aviculturists who were more likely to receive it. The first of only two recorded breeding successes occurred in California, probably in the 1930s. In 1936, Jean Delacour saw a pair in the collection of Gilbert Lee. He noted in the *Avicultural Magazine* in 1937 (page 135): "These have been nesting repeatedly for several years, but only one young one was so far reared, all the others dying after a couple of days."

In 1940, Taronga Park Zoo in Sydney, Australia, received five Kuhl's lories, one of which died shortly after arrival. There were two adults ("the envy of all who view them") and two young birds. The adults had two unsuccessful clutches before the two pairs were housed separately in 1943. Then the older pair nested in November. One egg was broken and a hatched dead chick was found on the ground beneath the nest. Two more clutches produced chicks which died soon after hatching. The successful clutch commenced in February, 1946. When the chick was at least five weeks old the log was so dirty, it had to be replaced with a new one, and the chick's feet and legs washed. Despite this, the parents went inside the new log at once. On leaving the nest the young one spent several days on the ground, indicating it was weak. However, it survived. It was described as being much duller in coloration than the adults; bill and legs were a dark slaty color and the iris dark, almost black. Like Delacour, Patten (1947) compared the behavior of this species to the play of kittens. "Their gorgeous colouring is a

delight and their playful habits, particularly when rolling on the ground, reminds one of kittens rollicking in the sunshine."

This species remains an extreme rarity. According to Neff (pers. comm. 1996), two pairs have been kept on Tahiti for some years. Also, in California, R. Schroeder had a pair in 1991. Unfortunately, the female died. I had the enormous pleasure of seeing the male in 1996. The beauty of this species has not been exaggerated; if anything, it has been understated. The array of colors is dazzling, even for a lory! And the rich blue, elongated plumes of the nape, so long they fall on to the mantle! These feathers are semierectile. Seldom has a single bird left such an indelible impression on me—the colors so rich and vibrant and perfect, and the bearing so regal! In the way he moved about, he reminded me of Stella's lorikeet. He had to put up with more than 50 flashes from my cameras but seemed quite unperturbed. Indeed, he clearly enjoyed the attention!

Stephen's Lory *Vini stepheni* (North 1908)

Synonym: Henderson Island lory

Description: The only *Vini* without blue on head or nape; the crown is green shaft-streaked with lighter green. Underparts are scarlet with a variable band of green and purple across the breast; underwing coverts are red and green and the thighs and lower abdomen are purple. Upper parts are green, more yellowish on the rump. The tail is greenish yellow and the undertail coverts are yellowish green. Soft parts are incorrectly described in the literature; correct colors are: bill, orange with dusky markings at tip of upper mandible; iris, pale to medium orange; feet and legs, orange (Graves 1992).

Length: 19 cm (7½ in.)

Weight: A single male, 55 g; four females, 42–51 g (Graves 1992).

Immatures: They are mainly green on underparts with purple and red markings on throat and abdomen; the tail is dark green. Bill and iris are dark brown and legs are orange-brown.

Natural history

Range: Henderson Island in the Pitcairn Group in the southcentral Pacific (see Gazetteer). This represents the eastern extreme of the distribution of the Loriinae. Graves suggested that it may formerly have occurred on Pitcairn island.

Habits: This is one of the least known of all lories because it occurs on only one island, 37 km^2 (14 sq. mi.) in extent. Graves observed this species during the period of May 12–22, 1987. It was conspicuous on Northwest beach where it fed on the nectar of coconut palms. He noted: "They were frequently seen in small flocks of 3–5 (possibly family groups) flying 20–40 m over the forest canopy. Despite their brilliant red, green and yellow plumage, they were difficult to locate once they alighted in foliage. Feeding birds were often detected by their soft twittering calls. Flight calls were louder and could be detected at a distance of 100 m, away from the surf. Because lorikeets made relatively large daily movements, I estimated their local abundance by recording the number of lorikeets observed along the beach and interior forest trails during a continuous census, taking into account the location of flocks and the destination of flying birds. Single census maxima of 14 and 11 individuals were observed, respectively, at Northwest beach and North beach. Estimates of total population size were difficult to make because coconut groves along the beach attract lorikeets from the island's interior. However, I speculate that lorikeet density falls between 0.2 and 0.5 individuals/ha, a total population size of 720 to 1820 individuals." The lory was not discovered there until 1907. It inhabits forest and feeds on nectar, pollen, arthropod larvae and fruits. Rosie Trevelyan studied the diet and feeding ecology between January and March 1992. Other members of the Sir Peter Scott Commemorative Expedition to the Pitcairn Islands studied it during a 15-month period. It feeds at canopy level in coconut palms to ground level, on shrubs. Precise details were noted by Trevelyan (1995) as follows: "Nectar and/or pollen was extracted from *Scaevola sericea*, *Timonius polygamus*, *Cyclophyllum* sp., *Xylosma suaveolens*, *Thespesia populnea*, *Cordia subcordata*, *Psydrax* sp., *Senecio* sp., *Cocos nucifera* and *Pandanus tectorius*. Fruits of *Eugenia rariflora*, *Nesoluma st-johnianum*, *Guettardia speciosa*, and *Timonius polygamus* were eaten. One observation of lories extracting 'juice' from *Caesalpinia* leaves was also made. Arthropod larvae, almost exclusively lepidopteran larvae from the sporangia of *Phymatosorus* ferns, were also commonly eaten. "It was not always possible to detect whether the principal component extracted from different flowers was nectar or pollen without making detailed comparisons of visited and non-visited flowers. A comparison of *Timonius* flowers which had been plucked by lories with non-visited flowers revealed that both nectar and pollen were removed. After examining visited and non-visited *Scaevola* flowers, it was assumed that nectar was the chief component removed by lories."

The foods most frequently taken by the lories on the northwest point of the island's plateau were *Scaevola* flowers (48.1 percent of all observations), insect larvae (27.8 percent), *Timonius* flowers (22.2 percent) and *Eugenia* fruits (1.9 percent of all observations). This information was obtained by spot checks which excluded the less widespread food plants such as coconut and *Cordia*. Few *Pandanus* were flowering during the time the checks were made, although lories were frequently seen feeding on the male flowers of *Pandanus* at other times. Although female *Timonius* plants were relatively common, there was only one observation of lories feeding on them.

Status/Conservation: Highly vulnerable; in 1987 its population was estimated at between 720 and 1,820 birds, while in 1992 it was estimated at about 1,200 pairs. Numbers were difficult to assess because of the lory's mobility and patchy distribution (Collar et al. 1994). The island is administered by the British government which, in 1983, rejected an application from someone wishing to settle there. The species will probably be safe as long as the island remains free of human habitation. Rosie Trevelyan noted the absence of rats.

Graves (1992) recommended that "translocation or captive propagation of Henderson's endemic species as well as many other Polynesian endemics should be implemented in the near future while populations are still high."

Aviculture: Unknown

Tahiti Blue Lory *Vini peruviana* (P. L. S. Muller 1776)

Synonym: Tahitian lory

Native name: *Kuramo'o* (Aitutaki)

Description: Glossy, deep violet blue throughout, appearing almost black in poor light. The feathers from the forehead to nape are spiky and shaft-streaked with shiny lines rather than with a contrasting color; part of the face (area below eye extending to ear coverts), the throat and the upper breast are white. Flight feathers and the underside of the tail are dark gray. The small eyes are reddish brown. Beak, cere and legs are orange-red—brighter in the male in breeding condition. The white "bib" may be slightly deeper in the male. (The coloration of this species suggests that it evolved from a blue mutation of *Vini australis*.)

Length: 15 cm (6 in.); 15–17 cm (Cappel 1988); 18 cm (Forshaw 1989). (Live young adults or freshly dead adults measured by the author were only 14 cm.)

Weight: 31–34 g (males slightly heavier); average weights of birds at San Diego Zoo, males 33.5 g, females 28.0 g.

Immature plumage: In nest feather, the upper parts are dull bluish, brighter on the head; the underparts are dull blue-gray. There are a few grayish or grayish white feathers at the side of the beak. Feet are dark gray and the bill is brownish. At about 20 weeks white feathers appear on the ear coverts; adult plumage is attained at about six months but it takes longer for feet and beak to assume adult coloration (Low 1985b).

Natural history

Range: In southeast Polynesia, recorded from more than 20 islands, but now extinct on most; in the Society Islands, it was common on Scilly and Bellingshausen, but is now plentiful only on Motu One—about 500 birds, according to Collar et al. (1994); westernmost islands of Tuamotu Group, including Rangiroa, Manuae (suggested population 300 to 400 pairs), Apataki, Tikehau, Niau, Aratua and Ahe, in the Pacific. Introduced to Aitutaki (see Gazetteer) in the Cook Islands; earliest record dates back to 1899 when 7 specimens were collected for museums. McCormack (1993) suggested that it was probably introduced around the 1820s when the first mission ships journeyed between the southern Cook Islands and the Society Islands.

Extinctions: It became extinct on Tahiti and Moorea in the Society Islands, probably in the early years of the twentieth century, due to the introduction of rats and other vermin. It is now extinct on all the main islands of the Society group, such as Meetia and Huahine. It was known to occur on Bora Bora; in 1971, Forshaw searched intensively but unsuccessfully there; rats were prevalent (Forshaw 1989).

Habits: Usually found on or near the coast, where it feeds on coconut palms; it also inhabits banana plantations and any areas with native or exotic flora which provide pollen and nectar. Such areas include gardens and other cultivated places.

Only one island within the present range of this species is easily accessible: Aitutaki. The opening of an international airport on Rarotonga made possible my long-held dream to see this species in the wild. In 1991, I was due to visit Australia and decided to travel via the somewhat circuitous route of Honolulu and Rarotonga. I was able to spend only 2 1/2 days on Aitutaki—such unforgettable days that when I went to New Zealand in the following year, I again planned my journey via Rarotonga. This gave me three more memorable days on Aitutaki. This atoll is roughly oblong, being about 3 km (2 mi.) wide at its widest point and 7 km (4 mi.) long. Its sparkling turquoise and deep blue waters are irresistible to most tourists—but during my short stays I preferred to look for a different kind of blue! One does not have to search far or long. I soon heard that sweet lisping call and saw my first bird—in the form of an apparently black bullet speeding overhead (Low 1992c). It was 10:00 A.M. and sightings are not so frequent at that hour. Tahiti blue lories are most active and vocal for about three hours after sunrise and for a couple of hours before sunset and for a while after. Unlike many parrots, they might be located in some areas at almost any hour by their calls.

No native vegetation survives on Aitutaki except for one small strip of forest—but this is not a forest-dwelling species and neither is it indigenous. Like most lories, it will feed on any introduced tree, shrub or flower which is in blossom. However, pollen and nectar from coconut palms probably account for the major part of its diet. I had many sightings of blue lories feeding high in the coconut palms. My visits were made in February and March, when little else was flowering, except for hibiscus and bougainvillea around homes. I had a brief but close observation of one bird searching the scarlet flowers in a hibiscus hedge—but rain on the previous day had deprived them of that food source. The lory flitted down the hedge, staying less than a second on each flower. The second time it came in to feed, it surprised me by flying very low over my head. It could not possibly have been unaware of my presence. In that area the lories are used to seeing people and do not regard them as a threat. I also observed them in mango trees and flame trees, and in other trees where they searched the undersides of the leaves, probably for tiny insects. I have been fortunate to observe a wide variety of parrot species throughout the tropics but none has ever given me so much pleasure as this exquisite little lory. Its speed, beauty and curiosity make every sighting a joy. Usually I heard their *zsit zsit* calls before I saw them. They call to each other in flight, feeding or resting. Sometimes I knew they were close, but could not see them. Then they would take off, flying very fast and straight, their wings beating frantically, or sailing on open wings as they came in to land. When they flew relatively low overhead, the sun lit up their brilliant coral feet and beaks, in breathtakingly beautiful contrast to their unadorned plumage. Their *joie de vivre* seemed to exceed that of the very few other bird species, even the inquisitive, chattering, snow-white fairy tern (*Gygis alba*) and the ubiquitous (also introduced) common mynah (*Acridotheres tristis*).

They were active before the sun rose, when it was possible to approach them within 3 m (10 ft.) as they fed down low. The best sightings always came soon after first light and until 8:00 A.M. The sky lightens at 6:00 A.M., with small stormy gray clouds on a backdrop of soft orange. One morning, before first light, I positioned myself near a banana palm (in a garden not sprayed with chemicals). There was a flower only 1.5 m (5 ft.) from the ground. I guessed that it might receive an early morning visit from a pair of lories which I believed roosted nearby. I was right, and enjoyed the satisfaction of close observation. There were no other opportunities to observe them so near—but the half light did not permit the close-up photograph for which I longed! From the banana palm they flew to a large deciduous tree and perched in the top. Then they would depart for their favorite feeding area, about 100 m (330 ft.) distant. It was here, on another morning, that I watched a group of eight birds feeding on the introduced *Leucaena leucocephala*, a neotropical species. It has small, white flowering "pom-poms" which were probably a rich source of pollen.

One long-term observer of this species is Aileen Blake of Aitutaki Lodges (an ideal location for someone whose main interest is observing the lo-

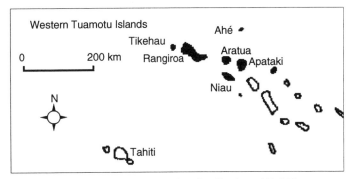

The solid black islands indicate the distribution of Tahiti blue lory (*Vini peruviana*) in the Tuamotu Archipelago.

ries). When I asked her if she thought their numbers had declined over the years, she told me that they had. She dated the start of the decline to the early 1970s when spraying of bananas (the main produce of the island) commenced. She also told me that the blue lories had learned not to feed in the plantation near her house. Steadman (1991) states that commercial citrus groves were developed on Aitutaki in the 1950s, with heavy use of pesticides.

In the heat of the day, blue lories often rest motionless in the midst of large trees where they are impossible to detect until they call. They probably also rest in the tall coconut palms, up to 30 m (100 ft.) high, where they hide in crevices at the bases of the fronds. The long, yellow inflorescences of these palms, which appear to be heavily laden with pollen, probably attract them to a greater degree than any other food source. In the morning I would observe a maximum of eight birds feeding together. They appeared to meet to feed, for at other times it was usual to see either two or four. The groups of four may have been families, the two young already in adult plumage. When observing one such group, I noted how two were more playful, swinging on fronds and generally being more active, and I imagined that these were young ones. I saw only a single bird on one visit which I could identify as an immature. I watched one pair which spent long periods over three days investigating the fronds at the center of a palm tree, possibly looking for a nest site. Nesting had not then commenced.

On my second visit, I was accompanied by two friends. One morning we each took slightly different locations in the vicinity of Aitutaki Lodges, on the east coast, near the southern end of the island. We estimated that there were at least 12 birds there, and possibly 14 or 16—probably two groups of eights. We guessed that each eight occupied an area of about four acres, and that each pair spent much of the day resting in its own special territory. In the afternoon the blue lories would become active at about 4:30 P.M. or 5:00 P.M. and could be seen until dusk, at 7:00 P.M. They would be flying around after most birds had gone to roost. Extremely active when perched, and very fast on the wing, they must use up a lot of energy, and thus need to feed as early and as late as possible.

In July, 1993, Kerry-Jayne Wilson spent two weeks on Aitutaki, trying to determine why the Tahiti blue lory remains common there; she concluded that this was because of the absence of ship rats. Assuming that they are absent, ensuring that the island remains rat-free is of the highest conservation priority. If a deep-water wharf was built, ship rats would soon be introduced. At present, ships anchor offshore and small vessels ferry freight across the reef to the wharf. Wilson observed that villages and plantations, with their high diversity of plant species, were the most favored habitats. Fewer sightings were made in low diversity plantations, even when they contained favored plant species such as banana and coconut. Plant species used by this lory and the number of sightings at each plant were listed as follows: mango (*Mangifera indica*) 116; coconut (*Cocos nucifera*) 75; kapok (*Ceiba pentandra*) 69; "plum" (*Syzygium cuminii*) 56; banana (*Musa* sp.) 38; "ginger" (*Canna indica*) 25; breadfruit (*Artocarpus* sp.) 5; papaya (*Carica papaya*) 4; nano (*Morinda citrifolia*) 4; *Albizia* sp. 3; citrus trees 2; avocado (*Persea americana*) 1 ; unidentified trees 18; total observations: 432.

Of the most favored trees, all but *Syzygium cuminii* were flowering at the time (Wilson, report to Cook Islands Conservation Service).

In 1990, Roland and Julia Seitre spent four months in French Polynesia and observed four species of *Vini* lories. They commented on their vocalizations as follows: "Calls are relatively similar, a two-seconds 'wiiss' repeated a few times in flight or for minutes. The Bellingshausen population of *V. peruviana* often ended the call by a sort of goggle never heard from the Rangiroa birds. Calls can be easily imitated by humans and will often attract flying birds."

In 1986, from October to November, Curt Cappel spent three months filming in Rangiroa

and various coral islands in the Tuamotu Archipelago. There are 220 coral islands in the 240 km (150 mi.) long reef. Rats were present on islands involved in the copra industry, and on these the lory could not survive. But the more remote islands were free of these rodents. His observations there indicated that pollen and nectar of coconut palms were their main foods (Cappel 1988). On a previous expedition to Tikehau, from May to July, 1981, he saw small groups of blue lories high in the coconut palms. This was the dry season, and they were feeding young. However, he also saw young being fed during his visit to Rangiroa in October and November, 1986.

Nesting: Nests are difficult to find. W. Jankowski observed birds entering a hole where a branch had broken off a tree, about 11 m (35 ft.) above ground. The place was Aitutaki and the date October 17–24, 1993 (McCormack pers. comm. 1994). In 1993 an inhabitant of Aitutaki, whose information on this species agreed precisely with what I knew, said he had seen blue lories emerging from holes in pandanus trees along the coast. He believed that they nested there (Low 1994b). None of the students at Araura College, Aitutaki, who participated in a survey on the *Kuramo'o*, had seen a nest (McCormack 1993). Cappel (1988) stated that on Rangiroa and Tikehau the lories nest in holes in the tops of coconut palms, also in holes in other trees.

Status/Conservation: Endangered and declining. Its range has suffered a drastic retraction. Extirpation from certain islands probably occurred soon after the arrival of ship rats (*Rattus rattus*). Where it still survives these rats are either unknown or very rare. In July, 1993, Kerry-Jayne Wilson (1993) set about 90 traps on Aitutaki and caught only *Rattus exulans*. In March, 1994, Gerald McCormack and Eddie Saul set 526 traps in three horticultural areas. They caught only 27 *Rattus exulans*, plus mice and land crabs. Most people report only the "small rats" in the villages. However, some people claim to have seen larger rats and one *Rattus norvegicus* and three ship rats (*Rattus rattus*) were caught on Aitutaki in July, 1963 (McCormack pers. comm. 1994). The survival of the Tahiti blue lory is dependent upon keeping the few islands on which it still occurs free of ship rats. Dieter Rinke (pers. comm. 1986) suggested that Motu One and Fenua Ura should be established as reserves for this species. Cats are another threat; Seitre and Seitre (1991) state that on Scilly, the population of *peruviana* declined by as much as two-thirds after cats were introduced less than 15 years previously. Ship rats are not present on that island (or on Bellingshausen).

Population surveys have been carried out on Aitutaki. From June 6–10, 1992, Judith Kunzle and Gerald McCormack surveyed all the main roads and saw and/or heard 176 individuals. They conservatively estimated a population of more than 750 birds. During March, 1994, Gerald McCormack organized a survey for the students of Araura College which involved two counts along five transects. The first count was of 120 birds and the second of 123 birds. Students will monitor the transects annually to detect any fluctuation in the population. McCormack suggested that multiplying by ten the number of birds seen or heard provides a guide to the population total, in this case about 1,200 birds. He believed that this figure was not unrealistic (McCormack pers. comm. 1994). Less is known about Rangiroa. According to Bruner (1972), there were between 100 and 200 birds in the area of Tatiavoa. On Apataki, a team of biologists found a "good population" of Tahiti blue lories in April, 1989—but also evidence of a decline. Assessment of the status varied from "seriously endangered" (Forshaw 1989) to "they do not seem imminently threatened"—referring also to the ultramarine lory (Holyoak pers. comm. 1983). That the ultramarine was considered just ten years later to be critically endangered shows how rapidly the arrival of ship rats can decimate apparently heathy populations on small islands. Everything possible must be done to protect the Tahiti blue lory from these lethal invaders, and also to educate the local people regarding the importance of conserving this species. Seitre and Seitre state that in the Tuamotus, the situation is slowly improving, with the publication by the Ministry of Environment of a specific poster.

Aviculture

Status: Very rare

Clutch size: Two

Incubation period: 25 days; 24 days with parents, 24–25 days in an incubator (Rimlinger, San Diego Zoo, pers. comm. 1995).

Newly hatched chicks: Weight, 2 g; average weight of those hatched at San Diego Zoo, 1.84 g (Rimlinger pers. comm. 1995); covered in longish (not dense) white down; the beak is brownish.

Chick development: In hand-reared young (development not as rapid as parent-reared), the feather tracts of the second down appear under the skin at 10 days. The eyes start to slit at about two weeks, but may not be fully open until 18 to 21 days. At this time the dark gray second down is erupting, to become dense in due course. At five weeks the wing feathers are opening, those of the occiput following a few days later. As the feathers open, the chick assumes a distinctive appearance. The partly bare head has an elongated appearance (Low 1985b). At six weeks the wing and tail feathers are still partly encased in sheaths; by eight weeks the chick is fully covered with contour feathers except those on the rump, which appear soon after. A few whitish feathers appear on the cheeks at about eight weeks.

Young in nest: 61 days, 1 only (Low 1985b).

Chick growth

Shows weights of two chicks reared by the author, hatched in October, 1982. Their ages were estimated at seven and eight days when removed from the nest. (Parent-reared chicks would weigh more.)

Tahiti blue lory weight table

Age in days	Weight in grams	
	Chick No. 1	No. 2
7		5
8	6	
9		5
10	6	
12		6.5
13	7.5	
14		8.5
15	10	
16		10
17	11.5	
19		12.5
20	13.5	
21		14
22	16	
27		17
28	18	
29		19.5
30	19.5	
33		21
34	22	
39		25
40	27	
68 & 69	independent	

Average Weights of Young Hand-reared at San Diego Zoo	
Age in weeks	Weight in grams
1	2.44
2	4.0
3	6.47
4	11.0
5	20.25
6	26.55
7	25.65
8	25.25
9	25.1
10	26.0

Age at independence of hand-reared birds:
10–12 weeks, reared by author; 54 days at San Diego Zoo (Rimlinger pers. comm. 1995).

General: This lory has been represented in very few collections. It is a difficult avicultural subject. I have had more than 200 parrot species in my care and I consider this to be the most heart-breaking—yet in many ways the most exquisite. The first record of the Tahiti blue lory in aviculture would appear to relate to the 1930s. In 1936 an American resident on Tahiti, Eastham Guild, acquired a number of birds which were sent to the Duke of Bedford (Lord Tavistock). After several months most of them died. All but one pair of the eight survivors were then sent to Dr. Derscheid in Belgium. Lord Tavistock was successful in breeding from his remaining pair, an achievement which he described as "the greatest triumph of my avicultural career." Being very familiar with the problems involved in breeding this lory, I can share the sentiments of Lord Tavistock, considered the greatest parrot breeder of his era. The year before this success he had written: "These rare and interesting little birds are not too easy to manage as many cocks are apt to turn savagely on their hens with very little warning after living with them on apparently affectionate terms and feeding them. I had to separate them last winter on account of the cock's misbehavior, and in June put them into adjoining aviaries, the hen sharing her compartment with a hen blue-crowned hanging parrot whom she did not molest. Only when she seemed thoroughly anxious for his company and he was feeding her through the wire did I allow them together" (Tavistock 1937).

Lord Tavistock had special aviaries built for the lories. The outside flight measured 12 ft. (3.6 m) x 10 ft. (3 m) x 7 ft. (2.1 m) high and the shelter measured 3 m (10 ft.) x 1.5 m (5 ft.) x 2.1 m (7 ft.)

high. It was heated to maintain the temperature at 18–23°C (65–73°F). A hollow tree trunk about 1.8 m (6 ft.) high was provided for nesting. It was filled with peat to within about 30 cm (12 in.) of the entrance and a few inches of decayed wood were placed above the peat. The base of the trunk stood in a shallow container of water. The nest was provided in September, soon after male and female were reunited. The female laid two eggs. Tavistock noted that "they took turns at incubating, and when the baby was quite small, it was the mother who appeared to do most of the foraging for food, while the father stayed at home and looked after the nestling." (It was also my experience that males played an even more active role in rearing and at least an equal role in incubation.) While the young one was being reared the pair received four mealworms daily and would have eaten more. The young one was said to have left the nest "about eight weeks after the first egg had been laid." This is clearly an error, as young spend eight weeks in the nest. This pair reared two more chicks which fledged during March, 1938. A third clutch was infertile.

Their usual diet consisted of Allinson's food, diluted with its own volume of water, apple, pears and grapes. They occasionally chewed grass and would eat a little earth from the spaces in the tiled floor left for this purpose. In cold weather they seldom ventured into the outside flight but would sometimes rain bathe when the temperature was low (Tavistock 1938). Lord Tavistock described their flight as "strangely weak and labored, exactly like that of a very sick or young bird." I, too, gained this impression from captive birds—and they were not overweight. As birds in the wild are extremely swift in flight, the weak flight of captive birds is probably due to their inability to gain height or momentum in the space provided.

In the same era, Mrs. Gilbert Lee in California kept this species, but failed to rear young. Its avicultural history might have ended at this point but for the activities of bird smugglers in the 1970s. In October, 1977, two men, one from Tahiti and the other from New Caledonia, offered Tahiti blue lories to aviculturists in southern California (reputedly at $7,000 per pair). Customs agents were notified, the men were arrested, tried on October 25 and found guilty of illegally importing the birds. At that time, birds which had not been quarantined for the required 30 days were destroyed. As I recorded elsewhere (Low 1985b):

Because of their rarity, customs officials who legally had custody of the confiscated birds, sought an exemption to the destruction rule. While their fate was decided, the birds were in quarantine at San Diego Wild Animal Park. Their fortunes were followed by thousands of interested bird lovers and aroused national interest. I still have a file of newspaper cuttings and copies of letters to senators and various officials who literally pleaded for their lives.

I received a call from Dr. James Dolan, general curator of San Diego Wild Animal Park. One way to overcome the problem might be to quarantine them overseas. If this was permitted, would we take them, he asked? The Ministry of Agriculture responded to the emergency and issued a license to import them in record time. (Fortunately, in those days no other license was required.)

Thus it was on a frosty morning in early November that my husband and I anxiously awaited the arrival of a very precious consignment at London Airport. We carried the box out into the wintry sunshine and lifted the protective sacking to reveal the most exquisite birds imaginable. My first reaction was how tiny they were and my second wonder at the glossy depth of their midnight blue plumage. No picture can prepare one for their beauty and animation.

They were placed together in one flight in the quarantine room and were soon feeding and exploring. It occurred to me how like tits were their actions and call notes; they have a rasping contact note reminiscent of that of the Blue Tit *Parus caeruleus*. I can visualize groups of them working their way through coconut palms when feeding, rather like tits moving through trees in search of insects. Presumably they do not need to move far in search of food and have no need for rapid and sustained flight.

Forty years previously the Duke of Bedford had recorded sudden and unexplained losses in his group of Tahiti Blue Lories. Within the next few weeks we were to have the same problem. Nothing significant could be found on post mortem. In desperation I spoke to the late John Yealland, who was curator of the Duke of Bedford's birds when the Blue Lories (both species) were included in the collection. He told me

everything he could remember but we remained mystified concerning the losses. With hindsight and several years' experience of breeding these birds which, in temperament are totally unlike any other lory I have kept, I have little doubt that the losses were due to stress induced by other birds. Tahiti Blue Lories can be extremely aggressive. I do not mean that they inflict physical injuries on each other but that they can quite literally stress a companion to death.

A recent occurrence is a typical instance. Two year old birds, siblings, had lived together in a 6 ft. (1.8 m) flight all their lives. Late one evening, when I turned out the birdroom light and switched on the blue 15 w nightlight, one was chasing the other. At 6:45 A.M. next morning one bird was missing. When it appeared from behind the nest-box it was severely distressed. Guessing that its companion had been

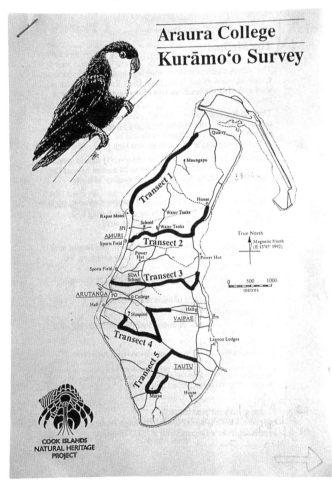

Known as Kuramo'o on Aitutaki, students at a college there participate in surveys to determine the size of the population.

Reproduced with the kind permission of Cook Island Natural Heritage

337

chasing it, I immediately caught up the aggressor and put a heat-lamp above the other. It sat on the nectar pot with eyes half closed and wings hanging down. I knew that the next half an hour would be crucial and that it would either lapse into a coma from which it would never recover, or it would gradually improve. Fortunately, the latter occurred. This bird had no injuries but had been frightened half to death by its companion. I never dared to reunite the 2 birds but after one week I introduced an unmated bird into the flight of the one which had been so near to death. I was surprised and delighted to find that they were compatible.

Fighting between siblings can occur before they are weaned at 12 weeks and separation has sometimes been necessary at this early age. On another occasion, when introducing 2 young birds, one hand-reared and the other parent-reared, in a flight which was neutral territory, the hand-reared bird behaved so aggressively that the 2 had to be separated after a few minutes.

At San Diego Zoo the same problem has been experienced. One morning keeper John Mitchell arrived just in time to rescue the female from being killed by the male. At the time the pair was rearing a chick. The female was removed and the male reared the chick on his own. During a visit to San Diego Wild Animal Park in 1992, where a pair was housed in a large aviary, I witnessed the continual harassment of the female by the male, and alerted staff. However, in 1996 this pair are still together and have produced young. When watching wild birds on Aitutaki in 1992, I saw a pair perched, and one suddenly turned aggressively on its companion. This is a naturally aggressive species, in my opinion, but in the wild, the harassed bird can simply fly away.

My experience was that these little lories can be greatly stressed by removal to new surroundings, even if only a few meters from the previous accommodation. On one occasion a pair was moved from a cage to a flight in the same birdroom. One turned on the other and started to attack it. On another occasion, four young birds were moved about 2 m (6 ft. 6 in.) from small holding cages to a small planted heated flight. One died within three hours. The birds were under constant observation and no aggressive behavior occurred. Another showed signs of stress and had to be moved back to its cage.

This aspect of keeping and breeding Tahiti blue lories was extremely discouraging. Nearly all the young were hand-reared and hours of work went into rearing each one. Because they are so small, newly hatched chicks needed a 3:00 A.M. feed (always my task!); after a few days I left them until 4:00 A.M., then 5:00 A.M., and when they were 10 or 11 days old, I could sleep until 6:00 A.M. As one pair was very prolific, and hatched about 20 young in a period of about three years, it was hard work. That I was used to—but the heartbreak of seeing these youngsters die from some incident which would not cause another lory species to bat an eyelid, was very hard to take. They were such exquisite birds and each loss affected me profoundly.

Some of the young went to Bronx Zoo in New York and ultimately to San Diego Zoo and a pair went to Loro Parque, Tenerife. The original intention had been to return the original birds to San Diego Zoo, but the attempt to do so resulted in tragedy. After the 90-day quarantine period in my home, they were boxed up waiting to be actually sealed up by a veterinary inspector at the moment of departure. Within two hours of placing them in the traveling boxes, two were dead and a third so severely stressed it never recovered. The attempt was abandoned and it was decided that only the young which had yet to be produced would travel. It is very easy to be wise after the event but I suspect now that these losses would have been averted if every bird had been in a separate compartment of the traveling box—but this was not how they had arrived.

However, in January, 1978, San Diego received a second group of confiscated Tahiti blue lories. (Possibly their source, and that of the previous group, was Rangiroa, as Seitre and Seitre state that birds were regularly taken from that island to Tahiti.) Precedent had been set for exemption from destruction and these birds were quarantined in Honolulu Zoo before going to San Diego. They were placed on exhibit in the zoo there in May of that year. Unfortunately, San Diego Zoo's experiences were no better than mine. One tragedy seemed to follow another. Just when the numbers had been augmented by breeding successes to a total of nearly thirty, about half that number was lost in an outbreak of sarcocystis. This protozoa was harbored in opossum droppings and carried into enclosures by cockroaches. Extensive control measures were then taken.

San Diego Zoo's records indicate, as I also

Three Tahiti blue lories hand-reared by the author, from left to right, 43 days, 8 1/2 months and 18 weeks.

Photo: R. H. Grantham

believe, that this species has the potential to be prolific. From the four males and four females the zoo received in 1978, and five more wild-caught birds in 1990 (a male and four females), and their offspring, 27 males and 21 females were reared. In addition, 19 more hatched; two were reared but not sexed and the others died at a few days old or when very young.

My experiences with this species extended from late 1977 until 1985 when I left London. However, two of the young which I had hand-reared came into my care again in 1987. They were on exhibit when I became curator at Loro Parque, Tenerife. They seemed very compatible and the female laid a clutch which was unfortunately infertile. Not long after, as I walked past their aviary early one morning I saw the female on the aviary floor, uninjured; the male had been chasing her. She was immediately taken to the clinic where she went into a coma from which

she did not recover. Sadly, it was not possible to obtain another companion for this bird. However, few of the thousands of parrots which have been exhibited at Loro Parque have given more pleasure to the public. I have watched on countless occasions as members of the public stood enthralled in front of his aviary. During my time at Loro Parque, the first thing I would do every morning, at 6:30 A.M., even when it was not yet light, would be to take him a little dish of tinned sweet corn, which he relished. At the time of writing, he is ten years old; ten of those at San Diego Zoo had lived for ten years or more by 1995 (see **Reproductive Span**).

His parents started to breed in 1979. The female had been in nest feather on arrival, i.e., she was less than six months old. In May, 1979 (just before she was two years old), she laid an egg from the perch. The next clutch consisted of two infertile eggs which, as always occurs in this species, were incubated by male and female. A third clutch that year produced two chicks, one of which died soon after hatching. Many more clutches followed. Infertile eggs and dead-in-shell were rare. However, I do recall an occasion when a chick died in the egg because, from the marks on the shell, it was obvious that male or female had tried to release the chick, which was overdue. Incubation usually commenced with the laying of the first egg. On a single occasion chicks hatched four days apart, instead of the usual interval of two days. Both chicks were removed for hand-rearing to ensure the survival of the second. On one occasion only, both hatched on the same day. The pair never reared young beyond the age of four weeks (Low 1985b). Each time (I cannot now recall on how many occasions, perhaps only two), a chick was killed in the fourth week and the surviving chick was then hand-reared. The risk in leaving them in the nest appeared too high, thus chicks were later removed at two weeks. I found that if chicks became noisy they were not adequately fed, and they were thus removed for hand-rearing.

It was interesting that although the wild-caught birds were prolific but not very good parents, a young sibling pair were superior parents. (However, they did have superior accommodation, so no direct comparison can be made.) I had no choice but to pair together brother and sister. They were hatched on February 26 and 28, 1982. While the other pairs were kept in indoor flights, this pair was maintained in a small planted house, thermostatically maintained at 18°C (65°F). This small house adjoined my kitchen, which had a large window looking directly into their house. This was a good site because it meant they were constantly under observation. Delightfully inquisitive, they would often fly on to the ledge beneath the window to observe what was happening in the kitchen. When nectar was being made, they would demand their share. The first egg from this female was laid on February 11, 1984 (again, two weeks before her second birthday). It was broken half way through the incubation period. The next egg was laid on about April 6 and hatched on May 1. The chick was exceptionally well fed and cared for. Because the parents were so tame, having been hand-reared, they did not object to nest inspection. The nest site was a small log with a natural spout and a wooden top which could be opened. The

nest litter was usually changed every week. The photograph which recalls my best memories of this species depicts me holding the chick in my hand while the litter was changed, with one of the parents just about to climb on to my hand. When I held it, the mother would perch on my hand and preen it or feed it. Interesting behavior was observed when the chick was 6 weeks old. They ceased to brood it at night and roosted elsewhere, in a nestbox. The young one left the nest on June 30, aged 61 days. Perhaps because it stayed with the parents until September 9, they did not make another attempt to nest that year. The chick was a male and a few weeks after removal from its parents' aviary, a hand-reared female, just weaned, was placed with it in a small flight. The two birds were immediately compatible. Young birds were very playful and could be seen every day rolling together on the floor. Even an adult pair would use the flat roof of the nestbox for this purpose. Almost any small item would be used in play. They delighted in swinging, especially from a twig suspended loosely from the roof of the flight. I often observed young birds swinging from the wing or the foot of another young one as it swung from beneath the perch by one foot.

The diet consisted of nectar, a small dish of sponge cake and nectar, a little tinned sweet corn, and fruit. The latter included apple and pear or, more favored, grapes. Their favorite food was pomegranates—but the season was short and they looked in vain for them when the supply ended. The nectar was given fresh two or three times daily.

Probably a few birds have always been kept on Tahiti and these are the source of those which have been exported. Indeed, in 1996 there were unconfirmed reports of several birds in France. It seems unlikely that this appealing little bird will ever be legally available to private aviculturists. However, in the event of this happening at some time, I would like to offer a suggestion to anyone who might be caring for them. Treat every male as a potential killer. For the enjoyment of birds and owner and for the protection of the female, keep pairs in planted aviaries. Each aviary should have a section to which the male can be confined at the first sign of aggression. Only when the female seems anxious for his company should the pair be reunited.

Ultramarine Lory *Vini ultramarina* (Kuhl 1820)

Synonyms: Goupil's lory, Marquesas lory

Description: One of the most unusually and exquisitely colored and patterned birds in the world. Its plumage is of contrasting shades of blue, set off with white. The forehead is turquoise; rump and uppertail coverts are bright sky blue. Forshaw (1989) described the upper parts, from skins, as "dull blue." Tavistock (1939), whose description was made from living birds, described the wings, back and tail as "a beautiful powdery blue of a rather unusual shade." The tail is pale blue tipped with white, with whitish blue on the underside. The narrow feathers of the crown and occiput are mauve blue, shaft-streaked with pale blue. The lores and part of the cheeks are white, the remainder of the face and upper breast are white and mauve in a unique pattern of almost triangular white marks on a mauve background; these markings are quite unlike any which exist in any other parrot species. There is a broad mauve band across the center of the breast and the feathers of the abdomen are blue at the base, white at the tip, their appearance being white. The underwing coverts are dull blue and undertail coverts are mauve, as are the thighs. The cere is orange, also the upper mandible which is tipped with brown; the lower mandible is black. The eyes are small; the iris is said to be yellow-orange. The feet are orange-pink.

Length: 18 cm (7½ in.). Tavistock (1939), who kept both species at the same time, stated that *peruviana* was smaller.

Weight: Assume about 40 g.

Immatures: Like *peruviana*, they lack white markings on cheeks, throat and abdomen, according to Salvadori (1891), who stated that the shade of blue is more greenish. Forshaw (1989) describes immatures as having the underparts dark blue with scattered white markings on ear coverts, breast and abdomen, and pale blue on the sides of the abdomen. This description would apply to young birds in the stage between nest feather and adult plumage. The young bird bred by the Duke of Bedford was described as having a bluish black breast and abdomen

with only a faint splash of grayish white, and a trace of the same color behind the eye. Bill and feet were black.

Mutation: An albino was collected on Ua Pou (Huapu Island) by the Whitney Expedition of 1921–23.

Natural History

Range: Marquesas Islands, French Polynesia (north of Tuamotu and Society island groups). It is known to survive only on the islands of Ua Huka, and on Fatu Hiva where a small number have been introduced. It may still occur on Ua Pou. Seitre and Seitre (1991) state: "In theory, a lory should only be considered extinct on an island when all motus have been explored, as large atolls can be 60 kms in diameter and count hundreds of motus." They point out that no scientist has ever had the time to explore all the motus. (An atoll can be fragmented into many motus, or islets, which are sometimes separated by large stretches of sea.) Some motus have never been visited by ornithologists.

Extinctions: Formerly more widely distributed throughout the Marquesas, it has probably recently become extinct on Ua Pou and certainly on Nuku Hiva. It was widely distributed on Ua Pou, an island of 104 km² (41 sq. mi.), in 1975, when its population there was estimated at approximately 300 pairs. The population on Nuku Hiva (337 km², or 130 sq. mi.) was believed to consist of about 70 birds at that time. However, the montane habitat made accurate figures difficult to obtain (Low 1994c). It formerly occurred on Hiva Oa (Kuehler and Lieberman 1992).

Habits: It is (or was) found in a variety of habitats—forest edge, coconut palms and banana plantations, from sea level and, on Nuku Hiva, at over 700 m (2,300 ft.). Seitre and Seitre (1991) point out that on Ua Pou the ultramarine lory had survived only at higher elevations, which was not primary habitat. Introduced plants invaded the native forest and, in time, eliminated even the larger trees. The undergrowth which replaced them was suitable habitat for rats, which then proliferated and gradually eliminated the lories. The population on Ua Huka is reportedly the result of multiple postwar introductions from Ua Pou. Seitre and Seitre state that it was introduced in the 1940s. The Vaipaee Valley is one location where they were readily observed in the 1960s, usually in banana plantations. Holyoak (1975) observed ultramarine lories on Nuku Hiva where they were concentrated on the mountain slopes and in the northwest valleys. They were usually seen singly or in pairs, more rarely in groups of up to six. They frequented the canopies of forest trees and occasionally the lower branches or understory bushes, where they clambered about in search of food. They were noisy and restless, always on the move and calling continuously. They allowed a close approach when feeding. Flight is swift and direct, with rapid wingbeats. (In all these details they do not differ from *Vini peruviana*.) When flying down mountain slopes they would glide for short distances with downcurved wings partially closed toward the body. Flight is generally at or below treetop level, but apparently on long distance flights they might spiral up to considerable heights. On Ua Pou, Holyoak documented their feeding habits. He made ten observations of ultramarine lories feeding on flowers, three of buds being consumed, two of birds eating fruits and two of them eating grubs. Crop contents of two birds collected there comprised nectar, pollen and the remains of buds and large grubs. They feed extensively on pollen and nectar but also have a liking for fruit. There are reports of them feeding on cultivated breadfruit and mangoes.

Nesting: Little is known. The nest is said to be a tree hollow or even a rotting coconut still attached to the tree. If the suggestion (Bruner 1972) is true that they use old nests constructed by finches and other birds, this is extremely unusual behavior for a lory.

Status/Conservation: Endangered (Collar et al. 1994). The introduction of ship rats (*Rattus rattus*) is probably solely responsible for this decline. Suggestions that avian malaria and the introduction of the common mynah and the great horned owl (*Bubo virginianus*) are partly responsible have never been substantiated. Ship rats are not known to be present on Ua Huka and this is probably the only island where the species has survived unaided.

Habitat destruction has not been involved in their disappearance; as pointed out by Seitre and Seitre, on Ua Huka the lories have benefitted from the transformation of native forest to gardens. A few centuries ago the human population there was probably much greater and areas which were then cultivated have reverted to forest. Bees and wasps are also suggested causes of the lory's decline, especially on Ua Pou, but they are widespread all over French Polynesia, so this seems unlikely.

Seitre and Seitre (1991) suggested that translocation to rat-free islands would be the best, and most urgent, solution to safeguard *ultramarina*, *stepheni* and *kuhlii*. In November, 1991, an expedition, co-sponsored by the Office of the Environment of French Polynesia and the Zoological Society of San Diego, was undertaken to the Marquesas. Although no lories were found on Nuku Hiva or on Ua Pou, expedition members suggested that between 1,000 and 1,500 ultramarine lories survived on Ua Huka. They commented: "Although this population is fiercely protected by the Ua Hukan islanders, its future is of much concern due to the prospect of the construction of a wharf to be built in 1993. Such development will allow the docking of large cargo ships which will lead to the potential invasion of exotic rat species and further anthropogenic activities, ie, industry, agriculture, and urban development" (Kuehler and Lieberman 1992). Fortunately, the wharf was not constructed.

In accordance with the draft recommendations of ICBP (now BirdLife International)/IUCN/CBSG, the Zoological Society of San Diego undertook the first step of an experimental translocation of ultramarine lories. They were moved to Fatu Hiva, the most southerly of the Marquesan islands.

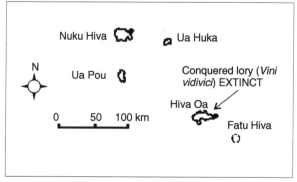

Range of ultramarine lory (*Vini ultramarina*). Not found on Hiva Oa.

This decision was based on the evidence of the prehistoric presence of this lory there, and on the pristine nature of the island. There is no wharf and no introduced predators or competitors. Plant species on which the lory feeds, such as coconut, banana, kava, mango and the coral tree (*Erythrina*), are thriving. In August, 1992, San Diego Zoo staff and personnel from the local Office of the Environment spent ten days on Ua Huka, attempting to mist net birds for translocation. Due to unseasonable rains caused by Hurricane Omar, only seven birds were caught. After being held for six days, they were transferred by boat to Fatu Hiva. There, the inhabitants of the village Omoa saw the birds before release and were told about the translocation program. The birds were released at first light in a foothill valley above Omoa, in an area rich in coconut and banana palms. Within minutes they were feeding on coconut flowers. Within an hour their foraging activities had taken them high into the hills, out of contact with human observers (Low 1994c). The intention was that the birds would be monitored by an employee of the Rural Economy Service. In November, 1993, seven more birds were netted and translocated to Fatu Hiva. By the end of 1995, the total number of ultramarine lories transferred to Fatu Hiva was 29. The good news was that young birds had been seen by employees of the Rural Economy Service, proving that breeding had occurred (Kuehler pers. comm. 1996).

This is one of the most vulnerable birds in existence. If rats reach Ua Huka the entire population could be gone in less than a decade, and some birds should at once be captured, in my opinion, in an attempt to establish them in captivity, unless it is known that a population is securely established on Fatu Hiva. Translocation and captive breeding are the only alternatives. However, recipients of any captive birds would need to be chosen with extreme care. Almost certainly, if captive breeding was deemed necessary, the birds would go to San Diego Zoo, which has long experience with the probably more difficult Tahiti blue lory.

Aviculture

Status: Currently unknown outside the Marquesas, except for a single bird in France. Traditionally, young birds were taken from nests as pets on Ua Pou, or adults were captured with bird lime. However, Seitre and Seitre (1991) reported that capture had ceased in the previous ten years, due to the rarity of the species.

Clutch size: Assume normally two, although Tavistock (1939) recorded two single egg clutches—but the female was in poor condition.

Incubation period: Unknown, but probably 25 days.

Newly hatched chick: Undescribed; assume similar to *Vini peruviana*.

Chick development: Unknown; assume identical to *Vini peruviana*.

Young in nest: Minimum of 54 days (one only), but hatch date unknown (Tavistock 1939). If the incubation period was 25 days, the young bird left the nest when 56 days old.

General: I have never seen the ultramarine lory alive. It is unknown in present-day aviculture, except for a single bird in France in 1990, the young bird depicted in the photograph. Its locality was never revealed to me but I believe that it was the survivor of two birds taken to France. In the 1930s this species was known in the collections of three eminent aviculturists, one of whom, Lord Tavistock, was successful in breeding it. He obtained a pair in 1936; at the same time he also kept the Tahiti blue lory. He described it as "better-tempered than its ally and we had no trouble with fighting at any time. The hen of my pair is very timid and rushes into her sleeping box on the slightest alarm, and as she cannot fly too well owing to cut wing feathers incompletely molted I never dreamed she would come into breeding condition. This, however, she did, not long after the other Blues, but twice had trouble with misshapen semi soft-shelled eggs. I therefore separated her from her mate and hope that by next year she will be in better fettle for breeding" (Tavistock 1937).

Lord Tavistock's hopes were fulfilled. His pair was successful in rearing a youngster. He described this event as "the apex of my achievements in rearing birds of the Parrot family, for the simple reason that for rarity and beauty, combined with need for very careful management, there is no species that I shall keep likely to be its equal" (Tavistock 1939).

He described their behavior and voice: "Ultramarine Lories have a gentle, squeaky, sibilant cry.

The flight is rather slow and heavy, though less so than that of *C. peruvianus*, but they climb actively and spend a lot of time running over the roof of the flight, upside down [a habit they share with *peruviana*]. They occasionally play together, but much less often than the other species. Now and again a small spider is eaten." He described *ultramarina* as "a much more gentle bird than the smaller *C. peruvianus*, a spiteful little creature, of which it is afraid." These comments of Lord Tavistock lead me to believe that captive-breeding of *ultramarina* would be easier than that of *peruviana* in that the problem of male aggression toward the female might not be encountered.

On the subject of diet, he recorded that the female "was only snatched back from the jaws of death by our discovery at the eleventh hour of the correct diet for this species new to aviculture—Allinson's Food, as for other Lories, but Allinson's Food diluted again with its own volume of water after it has been prepared for infants and sweetened. The mixture is given fresh twice daily with a little marmite and tomato juice added and lime-water is given when breeding is in prospect. Pear, grapes, and apple are supplied and,

when young are being reared, the insides of eight to twelve mealworms mixed with 2 to 3 teaspoons of honey and water."

The first two clutches which were incubated by his pair had consisted of a single egg—fertile but dented, probably due to a calcium deficiency. Lord Tavistock recorded that the male incubated—but not as much as a male *peruviana* during the early stages of incubation. When the chick hatched, however, he was very attentive and always spent the night in the nestbox. This was a tall box with the base standing in a container of water. During incubation more water was poured into the peat some way below the nest level, through a funnel, the base of which pierced the wooden side of the box. On top of the peat was a layer of decayed wood. In February, 1939, male and female ultramarine were re-united—without a nestbox, in order to discourage the female from laying. But she did lay, after being egg-bound. The egg was damaged. The female was returned to the aviary with a nestbox and a second egg was laid on February 28. A chick was heard on March 27 and the young bird left the nest on May 20, when it was well-grown.

Conquered Lory *Vini vidivici*

Extinct

Scientific name: A play on words, rooted in the Latin declaration "*Veni, vidi, vici*" (I came, I saw, I conquered) attributed to Julius Caesar. Steadman and Zarriello (1987) explain that this phrase may be projected into the prehistoric situation in the Marquesas and elsewhere in Polynesia, where people came to an island, saw the native parrots, and then conquered them, leaving behind only the bones.

Description: All that is known is that it was larger than any other known *Vini*, except *V. sinotoi*. In a photograph (see Steadman and Zarriello) showing the tarsometatarsi of selected Pacific parrots, including *Vini vidivici*, the length and width of that of *vidivici* is twice that of *Vini peruviana*, suggesting that its body size was very significantly larger.

Discovery: Its discovery was based on the tibio-

tarsus and tarsometatarsus, the most diagnostic postcranial bones in Old World parrots. In 1985, Dr. Sinoto and a colleague showed David Steadman a large number of unstudied bird bones from Ua Huka, Marquesas. They had been in the Bernice P. Bishop Museum for 20 years. They sorted these and others to produce 15,000 to 20,000 bird bones. These included about 200 parrot bones—of great interest because no parrots were known historically from Ua Huka. The prehistoric avifaunas of eastern Polynesia are composed mainly of species which did not survive into historic times.

Steadman and Zarriello (1987) stated that the tarsometatarsus of this species agrees with that of *Vini* and differs from that of all other lory species, plus that of 27 other parrot genera which they listed. They described the tarsometatarsus of the two extinct *Vini* species and how they differ from those of various other parrots, made detailed osteological comparisons and showed that the fossils clearly pertain to *Vini*.

Range: Known from two bones collected from Hiva Oa, with referred material from the islands of Ua Huka and Tahuata, Marquesas, also from Huahine in the Society Islands (Steadman 1989).

Historical: Steadman and Zarriello (1987) suggested that the two extinct species of *Vini* lived sympatrically on the Marquesas until after the arrival of humans. This was about 2,000 years ago—but before European contact. The data available suggests that the extinction of most land birds on Ua Huka, including the parrots, occurred within the first millenium of human settlement. The stratigraphic distribution of parrots at the excavation site suggested that *Vini ultramarina* survived longer on Tahuata than its larger, extinct, congeners, although all three coexisted at one time.

Reasons for extinction: Not known—but almost certainly attributable to predation by humans or predation by introduced mammals such as rats. According to Steadman (1989): "Among landbirds, rails, pigeons, and parrots were the main food items in Eastern Polynesia." As occurred on other islands, the lorikeet may also have been hunted for its plumage.

Sinoto's Lory *Vini sinotoi*

Extinct

Named for Dr. Yosihiko H. Sinoto, in recognition of his many accomplishments in the archaeology of eastern Polynesia, particularly in the Marquesas. The faunal material excavated by Dr. Sinoto has provided crucial evidence for understanding past relationships between animals and people on islands.

Description: The species is known only from archaeological fossils. It was the largest known member of the genus, perhaps with a length of 22 cm or 23 cm (9 in.); its tarsometatarsi averaged 22.8 mm.

(The tarsometatarsi of modern specimens of *solitarius* measure 15 mm, those of *australis* 13.3–13.9 mm and those of *kuhlii* 16 mm, for example.) (Steadman and Zarriello 1987.)

Range: Included Nuku Hiva, Hiva Oa, Ua Huka and Tahuata in the Marquesas Islands and Huahine in the Society Islands (Steadman 1989).

Historical: See *Vini vidivici*.

Reasons for extinction: See *Vini vidivici*.

Genus: *Charmosyna*

The largest genus, with 14 species, it consists mainly of small, elegant lorikeets of 50 g or under, with only two species above this weight. It is a genus of contrasts, containing some of the most beautiful of the lorikeets, and also some of the plainest; half the species are sexually dimorphic. The genus contains the smallest of all lorikeets (Wilhelmina's) and, if one includes the long tail, the longest of all, the Papuan (Stella's). There is a natural division between the mainly green species (ten) and the mainly red species (four); the green ones were formerly classified in the genus *Hypocharmosyna*. *Charmosyna* species are widely distributed throughout New Guinea and the Pacific islands.

They are closely related to members of the genus *Vini* and two species (*papou* and *josefinae*) share with some *Vini* species an unusual characteristic—the attenuated (slender) tips to the primaries of adult birds. But unlike *Vini* species, they have slender, pointed tails, thus are called lorikeets. Little has been recorded about most of the green species, some of which are unknown in aviculture. Those which are known share with the *Vini* lories the fact that male and female incubate and the male plays a larger part in rearing the young than in other lory genera. Another fact which points to the close evolutionary relationship is the similarity in the food-begging call of the chicks (see **Call, Juvenile Food-Begging**).

As avicultural subjects they are a little more demanding, with more specific dietary needs; a liquid nectar is the most important component of the diet. They tend to flick it, thus constant cleaning of cage, aviary and surroundings is essential. The small species have a high rate of metabolism; the faster rate of respiration is evident. They mature early and, perhaps, do not live as long as the larger lories.

Palm Lorikeet *Charmosyna palmarum* (Gmelin 1788)

Synonym: Vanuatu lorikeet

Native names: *Denga, Dedenga* or *Maramarei* on Santo; *Vini* on Tongoa Island (Shepherd Group).

Description: The adult male is almost entirely green, of a lighter and more yellowish shade on the underparts. The chin, the area around the base of the bill and some feathers on the cheeks are pale red. The mantle is washed with olive. Underwing coverts are grayish green. The tail is long and gradated, with the central feathers broadly tipped with yellow and the lateral feathers narrowly tipped with the same color. The bill is coral-red, the cere orange and the iris orange-red; legs and feet are orange-yellow.

Length: 15 cm to 16.5 cm (6 in. to 6½ in.), including the tail which measures about 8 cm (3 in.).

Weight: Estimated 35 g.

Sexual dimorphism: The female has few or no red feathers on the chin or around the bill's base.

Immatures: They are similar to the female but all the colors are duller. The iris is ocher and the bill yellow-brown (Bregulla 1992).

Natural history

Range: In the southwest Pacific, Vanuatu (see Gazetteer) and the Santa Cruz Islands (Treasurers, Taumoko in the Duff Group, Tinakula and Vanikoro—politically part of the Solomon Islands). By the 1960s, in Vanuatu it was confined to the following islands from Emae and Tongoa northward: Epi, Lopevi, Paama, Ambrym, Malakula, Pentecost, Ambae, Maewo, Santo and most of the Banks Islands. There are no records from the Tores Islands (Bregulla 1992). According to G. W. Stevens (pers. comm. 1976), this lorikeet also occurs in the Reef Islands, in the southern tip of the Solomons.

Natural history: Bregulla states that the palm

lorikeet perhaps exemplifies Vanuatu species that go through fluctuations in their distribution which could be related to cyclones or to colonization during periods of population growth. In 1870, observers reported that this lorikeet had appeared on Efate, in central Vanuatu, for the first time in 30 years. By the 1930s it was found on several islands of southern Vanuatu, including Aneityum. By the 1960s it occurred only on the islands listed above. It prefers uninhabited mountain forest and is not uncommon above 1,000 m (3,280 ft.) on the larger islands. It may be found wherever there is suitable food, in all types of montane forest. Predominantly nomadic, it travels widely between feeding areas. Its movements are largely determined by the flowering of certain trees, palms and vines. This lorikeet is usually seen in small flocks, singly or in pairs, but groups congregate to form larger flocks on preferred blossoms, often in the company of honeyeaters and white-eyes (*Zosterops*). Palm lorikeets are difficult to observe because of their small size and green plumage, but the constant chatter usually betrays their presence. Bregulla describes them as agile and acrobatic feeders, often hanging upside down to investigate blossoms. Flocks are usually seen climbing among branches covered with flowers, or flying high overhead. Flight is similar to that of Massena's lorikeet—swift, with rapid wingbeats which are clearly audible as they fly overhead. Long distance flight is high and direct but when flushed from a tree they weave their way through the treetops.

The palm lorikeet is an irregular and unpredictable visitor to the coastal lowlands; in some years it suddenly appears in large flocks. In other years it does not make an appearance. Sporadically, large numbers have been reported on the coasts of Santo and Malo when *Erythrina* trees are flowering. Generally, a few can be seen on flowering sago palms along the coasts of islands in the Banks group. The contact call, repeated at intervals during flight, is a short, high-pitched *tswit-tswit*, or a quickly delivered series *tswitswitswit*. A continuous shrill twittering is emitted while they are feeding; while resting they use softer notes.

The nectar and pollen of trees, palms, lianas and shrubs form their staple diet. Bregulla notes that flowering sago palms (*Metroxylon rumphii*) are a favored food source. They have been observed feeding on fruit of native figs and on berries, from which they extract the juice from the soft flesh. Small insects and larvae are probably ingested with nectar. G. W. Stevens, who formerly resided in the Solomon Islands, observed two pairs in the Reef Islands during the 1970s. This lorikeet had not previously been recorded from there. He commented that it was difficult to observe.

Status/Conservation: It is fairly common in mountain habitats. The species is protected in Vanuatu in as much as that country is a signatory to CITES, but Bregulla suggested that the designation of forest reserves may be necessary for its long term survival there. To date logging has been on a small scale, except on Efate where this species no longer occurs.

Nesting: Little is known; however, Bregulla described a nest found on Santo in December, 1961. It was located in cloud forest above the mountain village of Nokovula at approximately 1,600 m (5,200 ft.) in the hollow limb of a tree about 6 m (20 ft.) above the ground. Inside were two half-feathered chicks.

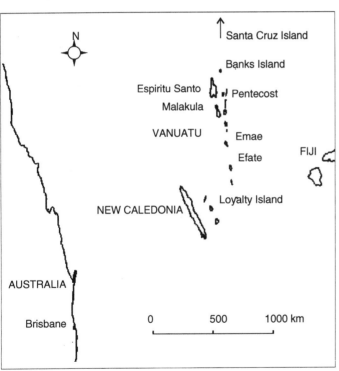

Species of Vanuatu and New Caledonia.
Vanuatu:

 Massena's lorikeet (*Trichoglossus h. massena*)
 Palm lorikeet (*Charmosyna palmarum*)
  also Santa Cruz and Banks Islands
New Caledomia:

 Trichoglossus h. deplanchii

Aviculture

Status: Extremely rare

Breeding: No records. One would expect the clutch size to be two and all other details to be similar to those of *placentis*.

General: This species is virtually unknown in aviculture. However, Homberger (1981), made observations on a captive bird in the 1970s. Received from Heinrich Bregulla in 1967, it was first in the collection of Dr. R. Burkard and later at the University of Zurich. Dominique Homberger had succeeded "in inducing feeding on seeds even in species that were previously thought not to eat seeds at all, such as the palm lory."

According to Homberger (pers. comm. 1996), a pair was kept for several years in the same aviary as estrildid finches (waxbills). They were fed on a liquid food containing various cereals with honey, and a mixture of grated raw carrots, hard-boiled egg and crumbled biscuit. Soaked millet sprays were relished. They did not breed but were "extremely gentle, friendly and vivacious." Burkard (1983) listed the palm lorikeet as among those which had bred (not necessarily successfully) in his collection. His only other mention of this lorikeet was with reference to diet, and contradicted other information: "Grains are taken by most species, especially if germinated"—but the palm lorikeet was an exception. Bregulla mentions captive birds, presumably on Vanuatu, and from their behavior, concluded that "pairs may spend the night in nesting hollows."

Red-chinned Lorikeet *Charmosyna rubrigularis* (Sclater 1881)

Synonym: Red-throated lorikeet

Description: The body plumage is green, except for the red chin and the red feathers at the side of the lower mandible. The shade of green is paler, more yellowish, on the underparts, and bluish green on the ear coverts which are faintly streaked with pale green. The underwing coverts are yellowish green and the underside of the flight feathers are marked with yellow. The tail is green, tipped with yellow, the inner webs of the feathers being red at the base. The bill is orange-red, the iris is orange or orange-red and the legs are orange in color.

The subspecies *krakari* described by Rothschild and Hartert in 1915 is not considered to be valid. It supposedly had an extensive area of red on the throat, bordered by yellow, and was slightly larger. It was named from a single specimen. The large series of skins from New Britain collected subsequently indicate there is much variation in the amount of red on the chin and in the presence or absence of yellow in birds from New Ireland, New Britain and Karkar. Diamond and Lecroy (1979) could see no consistent differences between the populations.

Length: 17 cm (6 3/4 in.)

Weight: Males, 34–40 g; females, 33–40 g (Diamond and Lecroy 1979); two birds, 31.5 g and 34 g (Beehler 1978).

Immatures: An assumed immature had a restricted area of red on the chin and a shorter tail, 83.5 mm in contrast to 94 mm in adult males and 92 mm in adult females (Diamond and Lecroy 1979). One would expect the iris to be grayish and the feet and beak to be dull colored.

Natural history

Range: New Britain and New Ireland (see Gazetteer) in the Bismarck Archipelago and Karkar Island off the northeastern coast of New Guinea.

Habits: Beehler (1978) described the red-chinned lorikeet as common in New Ireland, in small flocks above 1,500 m (4,900 ft.), where it fed on the inflorescences of a native palm of the high mountains. It was one of the most common birds on the summit of the Hans Meyer Range in February, 1976. Them (pers. comm. 1996), who visited New Ireland in 1996, also observed it frequently in small parties. On Karkar (the westernmost extent of its range), Diamond and Lecroy found that it occurred at all elevations to the summit but was very uncommon below 610 m (2,000 ft.). It flew over the canopy in flocks of up to ten. In flowering trees it fed with Massena's lorikeets, black sunbirds (*Nectarinia sericeus*) and Sclater's myzomelas (*Myzomela sclateri*). The diet is said to be comprised of pollen, nectar and blossoms, and perhaps soft fruits. Its voice was described as high and shrill, not as staccato or as sharp as some lories.

In southwestern New Britain, Gilliard found the red-chinned lorikeet abundant in the Whiteman Range between November, 1958, and February, 1959—but only above 500 m (1,640 ft.). Below this altitude it was replaced by the red-flanked lorikeet (*C. placentis*) (in Gilliard and Lecroy 1967).

Nesting: No information. The breeding biology is probably similar to that of *placentis*.

Status/Conservation: A common mountain species, probably with a stable population—but little information is available.

Aviculture: Unknown

Meek's Lorikeet *Charmosyna meeki* (Rothschild and Hartert 1901)

Description: Bright grass green, with olive brown on the mantle and bluish green on the crown. The shade of green is more yellowish on the underparts and underwing coverts and the lower part of the ear coverts and the sides of the neck are pale green streaked with darker green. There is a variable yellowish white band on the underside of the secondaries. The tail is dark green above, tipped with yellow, and bright yellow below. Bill and legs are bright red and the iris is yellow to orange.

Immatures: They have a brownish beak and shorter tail.

Length: 16 cm (about 6½ in.)

Weight: 23–32 g (Forshaw 1989)

Natural history

Range: The Solomon Islands (see Gazetteer)—major elevated islands only, such as Guadalcanal, New Georgia, Malaita, (Santa) Isabel and Kolombangara—and on Bougainville Island.

Habits: This species inhabits foothills and mountain forests. Jonathan Newman (pers. comm. 1994) recorded that it was abundant on Kolombangara; up to 50 a day were seen in August, 1990, especially between 900 m (2,950 ft.) and 1,500 m (4,900 ft.). They were rare in coconut palms, but occasionally seen in small numbers with *Trichoglossus* and *Chalcopsitta* lories. Groups of 20 to 25 visited a bottlebrush tree next to his camp, at 1,000 m (3,300 ft.) each morning. They arrived soon after dawn, when the flowers opened. After feeding for a few hours, all the Meek's flew down the mountains, presumably to feed at a lower altitude, although the *Chalcopsitta cardinalis* remained throughout the day. When feeding, Meek's lorikeets are very active, continually moving between blooms. They favor red flowers such as bottlebrush and a species with large white waxy umbels of flowers. Newman commented that the "golden" back of this species is very obvious when it is feeding acrobatically at flowers. His impression was that the area of skin surrounding the eye was dark reddish brown. Its voice is shrill and feeding pairs are very noisy; it is quieter in flight.

In southern Bougainville, between July and September, 1964, Schodde (1977) found this species in the canopy of inland hill forest or primary forest from about 300 m (980 ft.) up to 1,500 m (4,900 ft.), but it was rare above 1,200 m (3,900 ft.). Flocks of about 10 to 15 birds were observed. Elsewhere on this island it occurs in the canopy of stunted cloud forest at about 1,500 m. It has been observed in the company of duchess lorikeets (*C. margarethae*). Its food consists of pollen, nectar and blossoms.

Horace P. Webb was stationed on the island of Santa Isabel from September, 1986, to September, 1988. He made two trips up Mount Sasari but did not locate it there. However, he observed it on 4 occasions (July and August, 1988) near the summit of Mount Kubonitu at altitudes between 900 m and 1,000 m (2,950 ft. to 3,300 ft.). The lorikeets were in groups of two to five, feeding in the tops of flowering trees. They were identified by their "rather uniform green color, yellow underside of the tail, bright red bill and high pitched 'tweek tweek' call" (Webb 1992). He described this species as rare and local on Santa Isabel.

During several visits to Guadalcanal and Isabel during the mid-1990s (always in April), Jan van Oosten observed Meek's lorikeets; they were usually at forest edges or on the periphery of plantations, during mornings, or middle and late afternoons. Groups of five up to 30 were seen, occasionally with Massena's or duchess lorikeets, or with cardinal lories (van Oosten verbal comm. 1996).

Nesting:　　On Kolombangara Newman saw a pair at a clump of epiphytic moss, 6 m (20 ft.) up in a tree trunk. The location was stunted forest at 1,550 m (5,080 ft.). Chicks were heard when the pair entered the nest; the time was late August. This may be the first nest record for this species.

Aviculture:　　Unknown

Buru Lorikeet　*Charmosyna toxopei* (Siebers 1930)

Synonym:　　Blue-fronted lorikeet

Description:　　Mainly green, with the forepart of the crown blue. The shade of green is more yellowish on the underparts and underwing coverts; chin and throat are greenish yellow. There is a yellow band across the underside of the secondaries. The tail is green, narrowly tipped with dull yellow; the underside is dusky yellow with red markings at the base of the inner web. Bill and legs are orange and the iris is yellow-orange.

Length:　　16 cm (about 6½ in.)

Weight:　　Not recorded; assume about 28 g.

Sexual dimorphism:　　The blue on the forecrown is fainter and less extensive. The yellow markings on the secondaries are said to be more pronounced in the female.

Immatures:　　They are darker and duller, with the chin and throat more greenish. The yellow band on the secondaries is well defined.

Natural history

Range:　　Buru (see Gazetteer), Indonesia

Habits:　　Little is known about this rare species. While collecting on Buru from 1920 to 1922, the Dutch lepidopterist Toxopeus received seven live specimens which were caught on the west side of Danau Rana (in a valley of less than 1,000 m in altitude). They lived for only a week. No other specimens have ever been obtained. During 1980 Smiet (1985) visited Buru. He declared this species to be "quite common, in plantations, secondary and primary forest." Jepson (1993a) comments: "The ensuing literature has served to confuse the situation regarding Smiet's observations. White and Bruce (1986) state that Smiet only observed one individual and Forshaw (1989) questions Smiet's identification, arguing that the habitat is more in keeping with that of the red-flanked lorikeet *C. placentis* which Forshaw erroneously assumes occurs on Buru. Forshaw cites van Bemmel (1948) as his source for the occurrence of *placentis* on Buru, but this author misquoted an earlier paper by Vorderman who quoted a medical doctor who collected *placentis* on Amblan Island near Buru.

The Manchester Indonesian Islands Expedition of 1989 visited Danau Rana and the areas around Teluk Bara on the north coast, where Smiet found this species to be "quite common." Expedition members failed to find it around Danau Rana, but at Telek Bara brief views were obtained of four small lorikeets flying through coastal agricultural land. They reported: "Within the limitations of such views, we are confident that these were Blue-fronted Lorikeets."

Jepson commented: "The fact that only one out of the 24 collectors active on Buru recorded *C. toxopei* suggests that it is either very rare, nomadic or has specific habitat preferences. Smiet (pers. comm.) recorded the species in disturbed lowland forest between the band of coastal agriculture and the base of the hills. Our observations in the same area sheds little light on the situation, however we did find that little forest remained in this lowland belt, which is now dominated by grassland."

Multiagency teams of ornithologists, led by BirdLife International, visited Buru three times in 1995 and once in 1996. They observed the rare black-lored parrot (*Tanygnathus gramineus*) on one occasion but were not successful in recording the lorikeet. However, they were told by local residents it can be found "at inland sites and that in some years it appears close to the coast" (Poulsen and Purmiasa 1996).

Status/Conservation: If this species is confined to lowland forest on Buru, it must be seriously threatened. Jones et al. (unpubl. report 1990) state: "Very rare, only one probable sighting of two birds in an area of large but scattered trees, north west of Lake Rana at 800 m...Around Lake Rana (Lake Wakolo in earlier sources) there is some clearance through shifting agriculture but also some extensive areas of apparently untouched forest."

Jepson (1993a) states that the Department of Forestry has set aside 20.3 percent of the island for forest protection (although not all of this is primary forest), yet only two small areas have been designated as completely protected, and these cover only about 1 percent of the island. Poulsen and Purmiasa suggest that the proposed protected area of Mount Kepalat Mada probably holds all the forest species of conservation concern. They believe that the pressure should be removed from this area in the near future, as logging companies have taken all the timber required. But it is unknown whether the lorikeet occurs there.

Nesting: No information.

Aviculture: Unknown

New Caledonian Lorikeet *Charmosyna diadema* (Verreaux & Des Murs 1860)

Native name: *Kinkin-kunalu* (Mivart 1896)

Description: Male plumage undescribed. Female: the crown is violet blue and the throat and cheeks are greenish yellow. The forehead is emerald green; ear coverts are grass green streaked with emerald green. The thighs are tinged with violet blue. The area surrounding the vent is red. The rest of the plumage is green, paler on the underparts, and with a shade of brown on the mantle. The tail feathers are dark green above, washed with emerald green, contrasting with the lighter upper tail coverts. The three outer tail feathers on each side are reddish on the inner web, followed by a small area of black, and finally yellow, tinged with green; the underside of the tail is also green-tinged yellow. The bill and legs are orange. (This description is based on that of Mivart [1896], who examined the type specimen from the Paris museum.)

Length: 20 cm (7.9 in.)

Weight: Unknown; estimated 45 g.

Immature birds: Undescribed

Natural history

Range: New Caledonia

Habits: Unknown; possibly inhabits or inhabited scrub, rainforest and *Melaleuca* woodland (see below). This species is known only from two skin specimens, both females, collected in 1859. There were reported observations in 1913 of its presence in the northern forests near Oubatche. By the 1970s it was suggested that this lorikeet was extinct. However, Forshaw (1989) mentioned three unconfirmed sightings which were reported to Anthony Stokes who visited New Caledonia in 1976 on behalf of the Australian Museum in Sydney. When shown a colored plate of this lorikeet, one resident stated that he had seen such a bird in low scrub near Yate Lake, in the south, many years previously (perhaps in the 1920s). It was under observation for five minutes. Forshaw further stated that a senior forestry officer told Stokes that his first record of this species was in 1953 to 1954, on the road from La Foa to Canala. A pair was seen flying from rainforest to open *Melaleuca* woodland, where they alighted in a low

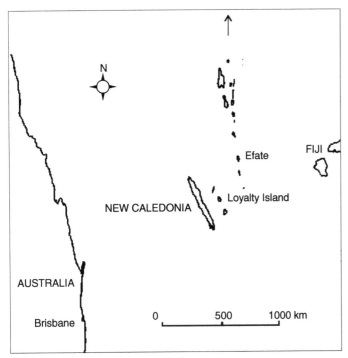

The New Caledonian lorikeet (*Charmosyna diadema*) occurs only in New Caledonia.

tree. He described them as being mainly greenish, with some yellow on the abdomen. The same observer saw this lorikeet again on June 3, 1976, to the west of Mount Panié, also in rainforest-*Melaleuca* ecotone. His attention was attracted by a call which differed from that of Deplanchi's lorikeet and, looking up, he saw two small green parrots dart from a tree and fly overhead. According to Heinrich Bregulla (1993), the New Caledonian lorikeet may still survive in the cloud forests of Mount Panié, Mount Humboldt and the Massif of Kouakoué.

Red-throated Lorikeet *Charmosyna amabilis* (Ramsay 1875)

Formerly *Hypocharmosyna aureocincta*

Synonyms: Golden-banded lorikeet, red-legged lorikeet

Native names: *Kula wai, Vuni as, Dre, Tikivili* (but unlikely to be in use now).

Description: Distinctively marked, it has red on the lores, forepart of the cheeks and on the throat, the latter bordered with a narrow band of yellow. The thighs are red. The rest of the plumage is bright green except for the tail which is tipped with yellow. According to Ramsay's original description, there is a small patch of crimson feathers around the vent, and a few feathers of bright yellow and of violet at the thighs. There is a large oblong patch of bright crimson on the inner webs of the three outer tail feathers; it is margined below with pale yellow in the male, but not in the female. The bill is orange; also the feet. The iris is orange-yellow.

Length: 17 cm (6.5–6.7 in.)

Weight: Estimated 34 g

Immature Birds: Overall duller, with the thighs grayish mauve or dull purple. Forshaw (1989) states that the yellow band on the throat is only faintly indicated. Ramsay (1875) attributes this characteristic to the female.

Natural history

Range: In Fiji (see Gazetteer), confined to the islands of Viti Levu, Taveuni and Ovalau. Clunie (1984) and Collar, Crosby and Stattersfield (1994) include Vanua Levu in its range; this island has been heavily logged and some or most former habitat has been destroyed.

Habits: It is a rare inhabitant of the upland forests of Viti Levu, usually above 500 m (1,640 ft.). According to Dick Watling there are recent unconfirmed reports from Taveuni, Ovalau and Vanua Levu. He states:

> It is less likely to survive on Ovalau now, than on the other two islands, but the chances that it is present on Taveuni and Vanua Levu are good. Its apparent restriction to mountain forests is probably artificial, a consequence of the location of good remaining habitat and fewer rats.

Predation by *Rattus rattus* is a likely explanation for its current scarcity; logging and lack of habitat may indirectly have contributed to its decline.

The call is a high-pitched squeaking, uttered frequently, not dissimilar to some calls of *Vini australis*. It is a quick and direct flier. I have seen only groups of less than five. I have recorded it feeding on the flowers of vuga (*Metrosideros collina*) and drala (*Erythrina*) species. This was in very disturbed country well outside of forest but at a high altitude (Watling pers. comm. 1996).

Thomas Arndt had a brief observation of this species in 1994. He was on Viti Levu, on the Kalambu road, in front of the entrance to the Coloi-Suva Forest Park, only 20 km (12 mi.) from Suva town. At about 11:00 A.M. Two *amabilis* came from a garden/agriculture area. They passed the road in a fast and somewhat undulating flight, disappearing into the forest (Arndt pers. comm. 1995). The possibility exists that it is less rare than generally suggested and that lack of sightings are due to its small size and mainly green coloration, making it difficult to observe in a forest setting.

Ramsay, who published his original description of this lorikeet in the *Sydney Morning Herald* on July 28, 1875, included an account of how the first specimens were obtained at Ovalau. The collector Pearce shot them from a large tree bearing yellow blossoms, on whose nectar they were feeding. The flock, of about 30 birds, did not appear until the tree was in flower. The stomachs of the birds collected contained nothing but nectar and a little pollen.

Since that time, few sightings of this species have been recorded in the literature. In 1926, Dr. Casey Wood, an authority on the birds of Fiji, believed that it was probably extinct on Ovalau but was still occasionally seen on Viti Levu and Taveuni, mainly in the high mountains of the interior.

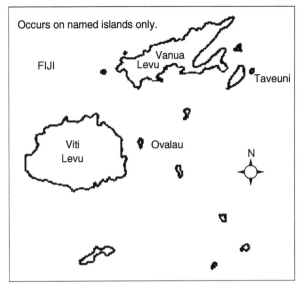

Distribution of red-throated lorikeet (*Charmoyna amabilis*).

Sydney Porter (1934) blamed the Whitney Expedition of the 1930s for its decline, stating that 47 birds were shot by expedition members, who were collecting skins for the American Museum of Natural History. Frank Chapman, curator of the bird department of that museum, denied this. "We secured twelve examples all taken during a short one-week trip, proof to those familiar with the difficulties of mountain forest collecting that the species is by no means rare; nor is it restricted to the mountains of Viti Levu, but is also found on the other large islands of the Fiji group. There are, for example, specimens in the British Museum from Taviuni" (Chapman 1934). No nests of this lorikeet have ever been recorded.

Aviculture

Status: Unknown, despite the claim by Silva (*A Monograph of Endangered Parrots*, 1989: 38) to have seen a specimen at Vogelpark Walsrode. It has never been held in that collection nor, probably, in any other.

Striated Lorikeet *Charmosyna multistriata* (Rothschild 1911)

Description: A most distinctive member of the genus, immediately recognizable by the unique bill coloration and by the brown patch on the nape. The plumage has various shades of green, brighter on forehead, throat and sides of head; the underparts are dark green, conspicuously streaked with yellow.

Around the neck the streaks are more greenish than yellow with a peculiar intensity which is almost luminous. The head coloration is interesting, being quite unlike that of any other lory. The hindcrown and nape are dark brown, followed by a small yellow patch, tipped with red in some birds. Forshaw (1989) states "variably spotted with orange-yellow on nape." Another unusual feature is the red feathers surrounding the vent. The tail is olive green above, tipped with dusky yellow and marked with red at the base; the underside is greenish yellow.

It is worth mentioning that one bird at Loro Parque, Tenerife, at least five years old at the time of observation, was more brightly colored on the head than some descriptions of this species lead one to expect. There is a broad yellow band above the eye and the area beneath the eye is more yellow than green. The depth and extent of the yellow around the eyes was not indicative of sex or maturity but apparently is a feature of individual variation in the eight birds examined by Schroeder. Most of these birds had some red tips on the yellow nape feathers.

The upper mandible is gray blue at the base, orange toward the tip, the colors divided in a straight line. According to Landolt (1981), the intensity of color varies in individuals; the lower mandible is orange. The iris is orange with an inner ring of gray-brown. The legs are gray.

Length: 20.5 cm (8 in.) This is the average of 8 captive birds (5 wild-caught and 3 captive-bred; 5 males, 1 female and 2 unsexed); length ranged from 19.5 cm to 22 cm but all wild-caught birds were 20.0 cm or 20.5 cm (Schroeder pers. comm. 1996).

Weight: 36–40 g (Landolt); 40–45 g (Michorius 1993).

Immatures: They have the head darker green with smaller orange-yellow spots on the nape; the streaks on the underparts are duller (Rand 1938). Landolt described a single newly fledged youngster as differing from the adults only in the smaller head and shorter tail. Paulik and Kruszona (1991) described a single newly fledged youngster as differing from the parents in the solid bluish gray color of the beak and less vibrant yellow streaks on the breast. They noted that "the beak changed to the two-tone color of the mature bird after approximately three months." Schroeder noted that "the immature birds have much less yellow around the eyes."

Sexual dimorphism: According to Michorius (1993), the male is larger with a longer beak and his colors are usually more vivid—but see Schroeder's remarks in previous column.

Natural history

Range: New Guinea, in the main mountain ranges from the Snow Mountains in Irian Jaya to the upper Fly River in Papua New Guinea. Beehler, Pratt and Zimmerman (1986) state that it occurs eastward to Mount Bosavi. More recent information gives the range as far east as Crater Mountains in the Chimbu Province, a range extension of 250 km (155 mi.) from the previous known limit (Mack and Wright 1996). It had previously been recorded in Papua New Guinea only from near the junction of the Fly and Palmer Rivers and in the Ok Tedi area near Tabubil.

Habits: This is a little known species, recorded from as low as 200 m (650 ft.) to as high as 1,800 m (5,900 ft.) in forested areas. Coates describes "scores" of striated lorikeets in the Ok Tedi Valley in mid-February, flocking in flowering trees with *placentis* and *pulchella*. Flight and feeding habits were similar to those of *placentis*, but the voice is quite different. In flight it usually gives a single drawn-out whistle but the call is sometimes composed of two or three notes. The striated lorikeet is described as "fairly common" in the Ok Tedi area "with a regular afternoon flyover at Dablin Creek, up to 125 birds (9 May 1993)." It has been recorded infrequently from the Ok Menga and Ok Ma areas. This is primarily a West Irian species and Tabubil is the easiest place to see it (Gregory 1995). The birds fly fast and straight, sometimes silently but often calling the distinctive whistled *Stree-eekt* note, sometimes repeated twice. Mr. Gregory (pers. comm. 1996) commented: "When I try to show them to visitors they need quick reactions as they literally streak over." At Dablin Creek, Tabubil, which lies at about 750 m (2,460 ft.), striated lorikeets fly over on most if not all afternoons, in numbers varying from single figures up to 140. They are seldom seen in the mornings, and then only in very small numbers. "I assume that they have a roosting site in the hills behind the town. I have recorded *multistriata* throughout the year, though the misty months of June to August see a drop in numbers which is, I suspect, related to feeding conditions and the lack of much flowering then. I have seen Striated Lorikeets feeding from flowering trees, sometimes in the company of *placentis*, which is very scarce at this site, the two species never mixing in flight in this particular locality, though they have fed together lower down in the town area. Flocks of Fairy Lorikeets sometimes have a few *multistriata* among them, and vice versa, though the former species is an erratic visitor here. Long-tailed Buzzards and Variable Goshawks are frequently seen and lorikeets seem to fly over regardless; quite why numbers vary from day to day is unknown. Ok Menga and Ok Ma are tributary valleys of the Ok Tedi, but the species is very sparse out there. I would not expect to find it on a day visit."

Nesting: Undescribed. Neff (pers. comm. 1996) describes eggs laid by captive birds as measuring 19.5 x 17.8 mm, 19.6 x 17.9 mm, 19.6 x 17.8 mm and 19.2 x 17.9 mm; two eggs weighed 3.43 and 3.27 g. Landolt (1981) measured one egg: 19.7 x 16.4 mm. Hubers (pers. comm. 1996) measured and weighed two eggs: 19 x 17 mm (3.15 g) and 21 x 17 mm (3.23 g).

Status/Conservation: Its status is unknown. As it has only rarely been captured for any purpose, the only threat to its existence would be deforestation. It has been recorded from only a few localities but whether this is indicative of rarity is not known.

Aviculture

Status: Rare

Clutch: Two

Incubation: "Generally lasts 25 days"; carried out by male and female but only the male incubated during the day (Michorius 1993). Paulik and Kruszona (1991) described a clutch, apparently incubated by the female, in which both eggs hatched 26 days after they were laid. Schroeder (pers. comm. 1996) found that the incubation period was always 26 days. He had no evidence that the male shared in incubation. Landolt stated that male and female incubated alternately but could not record the incubation period.

Newly hatched young: Have fine white down, about 1 cm long, on the back, neck and top of the head (Landolt); weight is unknown but probably about 3 g.

Chick development: At 3½ weeks old, the beak is nearly black; the quills of the tail feathers are about 1 mm long and yellow, and the longest wing feathers have grown 2 mm in length. No head feathers have erupted. At four weeks the first green feathers can be seen on the underparts; the iris of the eye is brown. At five weeks the green feathers on top of the head have opened and the wing quills are 1.5 cm long. The yellow tail feathers are about 3 mm long and the feathers on the abdomen are green with yellow shaft streaks. At 6½ weeks the head feathers have erupted fully; the middle of each feather is yellowish green. The brown neck feathers with orange points are fully developed. At 3½ months the young bird could be distinguished from the adults only by the color of the iris, which was yellow (Landolt 1981).

Young in nest: 53 days, one only (Landolt); 50 days, two young in one nest (Paulik and Kruszona 1991); "about 52 days" (Michorius); 46 to 54 days (Schroeder pers. comm. 1996).

Chick growth

No weights recorded except that of a parent-reared youngster which weighed 33 g at 8 weeks (Landolt).

General: The striated lorikeets received by Dr. Burkard in Zurich in 1978 may have been the first in aviculture. They originated from the region of Nabire in West Irian. Subsequently a few were exhibited at Vogelpark Walsrode, but none of these birds proved to be very long-lived, although a pair belonging to Dr. Burkard did rear at least one youngster. Losses were experienced with newly imported birds, which suffered pox, candidiasis and parasites. Dr. Burkard's birds were kept in cages 2 m (6 ft. 6 in.) long, 50 cm (20 in.) wide and 50 cm high (Burkard 1983). A partition could be inserted into the cage to keep the parents away from the nestbox so that the chick could be cleaned every week. In Germany, members of the avicultural society AZ reported breeeding eleven young from three pairs in 1992, one in 1993, five from one pair in 1994 and four from two pairs in 1995.

Paulik and Kruszona obtained two pairs in November, 1989. Initially they were kept in a small cage until one of the males was seen feeding a

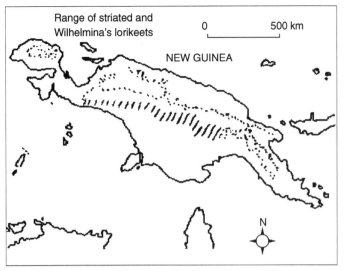

···· Striated lorikeet (*Charmosyna multistriata*)
//// Wilhelmina's lorikeet (*Charmosyna wilhelmina*)
N.B. both species—range imprecisely known.

female, after only a few days. The pairs were then housed in separate cages in the owners' apartment and provided with a budgerigar nestbox. One female laid two eggs about two months after purchase but the pair destroyed the eggs when they were disturbed. Both pairs then became ill with a severe *E. coli* infection. The female laid again, on June 15 and 18, and both the resulting chicks were parent-reared. The diet consisted of Avico Lory Life Nectar and fresh fruit. At the end of October (Paulik 1991) there were two more chicks, three weeks old.

H. J. Michorius of the Netherlands kept his striated lorikeets in cages measuring 80 cm (2 ft. 8 in.) square x 1.2 m (4ft.) high. Their nestbox was hung at an angle of 45° and measured 18 cm (7 in.) square and 30 cm (12 in.) high. The temperature at the nest site, which was very humid, was about 17°C (63°F). His chicks were ringed at 17 days with 4 mm rings. They were fully feathered at 40 days. He had the misfortune to lose the female of the pair when the chicks were three days old—but the male reared them unaided. His birds were fed on Aves Lory nectar (mixed with five parts of water) and small pieces of apple. Willow branches were regularly provided (Michorius 1993).

Landolt (1981) described the behavior of this species. The pair bond was very strong; especially during the breeding period male and female always moved together, when eating, stretching, preening,

etc. In display, they stretched themselves over the back of the partner, slowly moving back and stretching again. Presumably this means they stretched upward and leaned over the partner. The single bird on exhibit at Loro Parque was easily induced to display (probably being glad to receive some attention). It was very inquisitive and became quite excited, stretching upward to its full height, drawing the head in and out and from side to side, flicking the head and making a *huff* sound. It was fascinating to observe this display. The green streaks on the head then seemed very prominent.

In California, Dick Schroeder has kept this species since 1989. Eggs have been laid in most months of the year but more consistently in April, May and June. The male generally spends the morning in the nest, when there are eggs. Some of the early breeding attempts resulted in eggs with dead embryos, perhaps due to night disturbance by mice. The first successful rearing occurred when an egg was placed in the nest of a pair of Meyer's lorikeets; the chick was removed at ten days for hand-rearing. All subsequent chicks have been parent-reared and left with their parents for up to 90 days with no sign of aggression. The nest litter must be changed every three or four days or even more often during the colder months, when chicks are present, if the chicks are to survive. (Temperatures can fall as low as -4°C (mid 20°s F.) In 1996 Dick Schroeder had a breeding pair, four males (three wild-caught), and three unsexed young. He knew of only five others in the U.S.A.—in two collections in Florida (Schroeder pers. comm. 1996). In 1996, the Lory and Hanging Parrot Breeding Consortium requested permission from the U.S. government to import 40 birds—but permission was refused. The Consortium could have appealed if it had been able to show that removing 40 birds from the wild would not be detrimental to the wild population.

In South Africa, Dr. Chris Kingsley kept a female striated for eight years, but was aware of no other aviculturists in South Africa with this species. His bird liked soaked millet (Kingsley verbal comm. 1996). Worldwide, numbers in aviculture are so small that it seems unlikely that this extremely interesting lorikeet will be established, assuming that no more wild-caught birds are imported. In March, 1995, Indonesia banned the export of this species.

356

Wilhelmina's Lorikeet   *Charmosyna wilhelmina* (A. B. Meyer 1874)

Synonym: Pygmy streaked lorikeet

Description: Distinguished by the brown crown and nape and narrow blue streaking on the nape. The feathers of the upper breast are narrowly but not extensively streaked with yellow. The male has the lower back red and the rump dark purple-blue; his underwing coverts are red, and there is a broad band of red on the underside of the flight feathers. The tail is green, marked with red at the base. The plumage is otherwise green, paler on the face, forehead and underparts. The bill is orange, the iris is orange-red and the feet are gray. Astley (1908) described what was possibly the first to ever make the voyage to Europe as follows: "very dainty in its plumage of brilliant emerald green with frontal stripes of yellow, and a turquoise blue patch at the back of the head; a red bill and underwings, a patch of the same color on the rump, with violet just above the tail." This description of a male is of interest in suggesting the colors are more brilliant than those descriptions made from skins. The iris is orange-red (Neff pers. comm. 1995), and the bill is orange.

Length: 13 cm (5 in.) (Forshaw 1989); 11–13 cm (Coates 1985).

Weight: One female, 20 g (Neff).

Sexual dimorphism: The female lacks red in plumage, i.e., on lower back and wings. Neff described his female as follows: "The lesser underwing coverts are lime green and the greater underwing coverts are dark (more or less black), without any red. On the head (the crown) is a brown or more violet patch with bright stripes of blue on the nape. The lower back, rump and upper tail coverts are a wonderful violet." Neff commented on the lack of a wing stripe in this species (also lacking in *multistriata*).

Immatures: The males differ from adults in that they have the back dull purple, but they apparently have red in the wings. Immatures of both sexes have little or no streaking on the crown and breast.

Natural history

Range: In New Guinea, an apparently extensive range from the Arfak Mountains, Irian Jaya to the Huon Peninsula and the Mount Albert-Edward area in southeastern Papua New Guinea.

Habits: Coates (1985) states that it frequents forest and nearby areas which have been partly cleared, from near sea level up to about 1,800 m (5,900 ft.). It is generally scarce but sometimes locally common where there are flowering trees. Group size is generally up to 20 birds but sometimes large numbers may be seen with other species of nectar feeders. It is very active, and difficult to observe when feeding; its small size probably results in it being frequently overlooked. In the Port Moresby area, in the foothills and hills (Veomauri River and Sogeri Plateau) it is a scarce but possibly regular visitor between May and November. Coates describes its flight calls as similar to those of *placentis* but weaker, almost like those of a pygmy parrot (*Micropsitta*).

Clapp (1979) saw an estimated 100 or more Wilhelmina's lorikeets near Pongani, eastern Papua New Guinea, in April, 1979. They were in the company of other species, feeding at eucalypts bordering a road along a ridge. He had previously seen

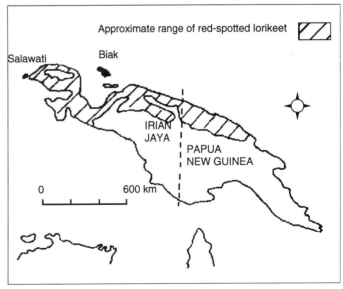

Approximate range of red-spotted lorikeet (*Charmosyna rubronotata*).

a smaller influx in June, 1978. On the second occasion, activity was most pronounced at 9:30 A.M., with groups of lorikeets darting from one tree to another along a 200 m (650 ft.) stretch of road. They clambered among the outer foliage to reach the blossoms. By 11:30 A.M. the few that remained were quite inactive. Diamond (1972) observed a flock at 1,500 m (5,000 ft.) on Mount Albert-Edward in southeastern Papua New Guinea. The birds were feeding at flowers of the oak *Castanopsis*. On Mount Missim, in the northern part of Papua New Guinea, Beehler (in Forshaw 1989) saw pairs sitting in the moss forest understory, only a few meters from the ground, probably resting.

Brown (1978) observed groups of five or six birds feeding on tall flowering eucalypts, among the other more numerous lorikeets. He was at Goroka, in the Central Highlands. "There were no more than 30 of this species in all which were incredibly agile, seemingly bundles of nervous energy as they dashed about." Newman (pers. comm. 1994) saw a single female on September 25 at 2,400 m (7,870 ft.) above Tari, Southern Highlands. She was very active, constantly flying between flowers, and was subordinate to all other species present in the red flowering tree. These included various honeyeaters, fairy lorikeets and, lower in the tree, Stella's lorikeets. Phil Gregory describes this species as very uncommon in the Ok Tedi area, Western Province, but easily overlooked. It has been recorded from the Ok Ma road, Ok Menga and at the mine site at 2,200 m (7,200 ft.). Maximum number of birds observed was five. Two and three were most often seen in flight (Gregory 1995; pers. comm. 1996).

Nesting: No information on nest sites. Two eggs in the British Museum (Natural History) measured 16.8 x 13.5 mm and 17.0 x 13.5 mm (Forshaw 1989). Eggs laid by a captive female measured 17 x 14 mm or 16 x 14 mm and weighed 1.68 g up to 1.79 g (Neff pers. comm. 1995).

Aviculture

Status: Extremely rare

Clutch size: Two (Neff)

General: Probably the first Wilhelmina's to reach Europe was collected by Walter Goodfellow in 1908. It endured the heat of the Red Sea but died as the ship reached the cold of Europe (Astley 1908). A second attempt was more successful; the well-known aviculturist E. J. Brook received a pair from Goodfellow in August 1909, along with three whiskered lorikeets, Stella's and fairies (Brook 1909). (They would have spent weeks en route, which indicates how skilled Goodfellow was at caring for small lories.) However, nothing else regarding this species appears to have been recorded in the avicultural literature until Robiller (1992) mentioned that C. Krause obtained several in 1978. They survived the long journey via Jakarta without problems, but sadly died of salmonellosis. In more recent years, only one bird is known to have survived in Europe—that which is illustrated, which belongs to Ruediger Neff in Germany. She was adult when obtained in 1988 and was still alive in 1995; every year she laid two or three clutches of two eggs, despite being without a male.

Red-spotted Lorikeet *Charmosyna rubronotata* (Wallace 1862)

Description: The male has the forehead and forepart of the crown red and the ear coverts are purple-blue streaked with brighter blue. The upper parts are dark green, and the lores, cheeks and underparts are light green. The sides of the breast are red and this color is also found on the underwing coverts. The tail is dark gray above; below it is pale orange, dark gray and, at the tip, yellow. The upper tail coverts are dark red, almost maroon. Bill and legs are red and the iris is orange.

Length: 17 cm (6 3/4 in.)

Weight: 30–35 g (Forshaw 1989)

Immature plumage: The sex is apparent from the color of the underwing coverts: red in the male and yellowish green in the female. Ruediger Neff

(1990) can sex the young by the time they are 22 days old. Carol Day (pers. comm. 1995) provided detailed plumage descriptions of her young. Both sexes have red on the forehead. In one particular nest pair, at 43 days the male had a few red feathers at the side of the breast. At the same time, the female's yellow streaks became apparent. At 12 weeks the red area on the female's head started to become smaller, with green feathers replacing red from the back of the head. (In a hand-reared female, the first green feathers appeared at the front of the head.) In hand-reared young, sex was recognizable at an earlier age (i.e., their development was more rapid); the red at the sides of the male's breast was visible at 31 days and the female's yellow cheek streaks at 38 days. In males, the red underwing coverts were apparent before the red at the side of the breast and some of the young males had red extending under their eyes. It is not possible to judge from the size and shape of the red on the forehead whether a bird is male or female. The first female reared from one pair had a very shallow "cap" and the male a full one; the next female from this pair had a full cap of red and the male a much smaller and shallower area of red.

Young in the nest have the beak dark brown; it starts to lighten soon after they leave the nest and by 11 to 13 weeks it is pink (Day pers. comm. 1995). The iris is brown.

Sexual dimorphism: The female has red in the plumage only on the uppertail coverts. The streaked feathers of her ear coverts are yellow and green streaked with yellow; the sides of the breast and the underwing coverts are pale green, with a yellow suffusion at the sides of the breast. There is a yellow band across the underside of the primaries and another across the secondaries (in adult and immature birds).

Subspecies: In *C. r. kordoana* the red on the crown of males is said to be paler and more extensive and the ear coverts more blue and less purple. Females do not differ from those of the nominate race.

Natural history

Range: The nominate race occurs in northwestern New Guinea, over much of Irian Jaya and eastward as far as the Ramu River, Madang Province (Coates 1985); it is also found on Salawati Island. *C. r. kordoana* is endemic to the island of Biak (see Gazetteer) in Geelvink Bay, Irian Jaya.

Habits: Coates states that it frequents forests, forest edges and coconut plantations. Occasional visits are made to flowering shrubs in open country. As it occurs from sea level up to about 900 m (2,950 ft.), its distribution overlaps that of *placentis*. According to Beehler, Pratt and Zimmerman (1986), in areas where its distribution is sympatric with that of *placentis* in the northern part of the range, it mainly inhabits hill country and red-flanked are mainly found in the lowlands. In Biak and other coastal areas, it frequents coconut palms and plantations. Red-spotted lorikeets sometimes associate with other small *Charmosyna* lorikeets but are unlikely to be the most numerous. They are not easy to observe as they are canopy feeders. Arndt (1992) described them as the most common of the four lories on Biak, but also the most difficult to observe, being perfectly camouflaged in foliage. They were numerous in secondary vegetation and in open areas with bushes. He would suddenly come across flocks of 15–30 birds which would take off, calling, when he disturbed them. Little is known about their natural history—but this is unlikely to differ from that of other small members of the genus.

Nesting: No information. Four eggs laid in the collection of Jos Hubers measured 16.8 x 14.0 mm, 16.9 x 14.1 mm, 17.0 x 13.9 mm and 16.9 x 14.0 mm; they each weighed 2.5 g (Hubers pers. comm. 1995). Carol Day, who has six laying females, commented that there is considerable variation in egg size and that the first eggs laid by a female are usually small (Day pers. comm. 1995).

Status/Conservation: Coates (1985) describes the species as locally common but usually scarce. It is rarely trapped and no threats to its survival are known.

Aviculture

Status: Rare, but numbers are increasing as a result of successes by specialist breeders.

Clutch size: Normally two; two females which had not laid before each laid four eggs over a period of ten days. One of these females laid three eggs on two subsequent occasions. It is of interest that the two females who laid large clutches were mother and daughter and that the mother was bred from Mrs. D. Cooke's female which laid three eggs (Day pers. comm. 1995).

Incubation period: Apparently only 21 days; when the period from laying to hatching is longer, it is because incubation did not commence immediately. It may be as long as one week before incubation commences. Usual period was 24 days after laying of second egg. Some males assume most of the incubation responsibilities (Day). Cooke (1991) described how the male of her pair incubated during the day, and the female at night.

Newly hatched chicks: Weigh about 2 g or just under. They are covered in fine white down (Day). Michi (1982) stated that newly hatched chicks resemble those of *placentis*.

Chick development: In parent-reared chicks, the eyes open between 11 and 14 days; at 23 days second down is thick and gray; green feathers on the breast are still in the quills; red feathers can be seen growing under the skin on the top of the head in both sexes; 40 days, wing feathering almost complete, with wisps of second down over their backs; 43 days, male showing red feathers at sides of breast and under the wings, also blue cheek patches; 43 days, female's yellow streaks on cheeks are apparent; tail feathers starting to erupt; back remains unfeathered (Day).

Young in nest: 50 to 62 days (Day).

Chick weights: Parent-reared male, 18 g at 23 days, 26 g at 47 days; parent-reared female, 18 g at 20 days, 30 g at 47 days (in same nest); a single male chick, 18 g (full crop) at 20 days (Day).

Chick growth

Shows the weights of two chicks removed from the nest aged 18 and 19 days when the female became sick. Both chicks were cold when removed (date August 21) by Carol Day. Weights shown are those before and after the first feed of the day. The chicks were fed during the night as well as during the day. Note that the scales weighed in increments of 2 g only but that odd numbers indicate where the scales fluctuated between the numbers above and below.

Red-spotted lorikeets weight table

| Age in days | Weight in grams | |
	Chick No. 1	No. 2
18		8/10
19	12/16	10/12
20	14/16	10/12
21	14/16	12
22	16	12/14
25		14/16
26	20/22	14/18
27	18/24	18/22
28	22/30	20/26
29	24/30	20/26
30	26/30	22
31	24/32	22/28
32	26/30	24/28
33	26/32	
35		26
36	26	26/34
37	28/34	
38		28
39	30	
40		30/36
41	32/36	
42		32
43	32	32/34
44	32/34	
45	32/34	32/36
46	30/36	
47		34/36
48	34/36	
49		30
50	31	
52		31
53	31	31
54	31	30
55	29	

Age at independence: The young of the above table were independent at 49 and 48 days. At 50 days they were placed in a small wire cage for exercise and one flew up to the perch. The cage was heated with a green light bulb, which they sat near for warmth. They voluntarily took a good bath after they were observed trying to bathe in a small food container so were given a dish of warm water. By the time they were 55 and 54 days, they were living permanently in the wire cage.

General: This appealing little lorikeet was unknown in aviculture until the late 1970s. Only small numbers reached Europe and probably even smaller numbers were imported into the U.S.A. C. Krause, a German collector in New Guinea, obtained a pair which were sent to Germany in 1980. They hatched two chicks on July 11, 1981; the chicks died on July 27 and 28 when the parents stopped feeding them (Michi 1982). The species was probably not imported again until 1988 when a few more reached

Europe. The first breeding in Britain apparently occurred in 1990 when the partnership of Blyth and Powell bred three and another breeder, Roger Bulloss, apparently bred one (Cooke 1991).

In 1990, Mr. and Mrs F. Cooke obtained a pair which were housed in an indoor birdroom in which the temperature was maintained at 23°C (74°F). In January, 1991, the female laid two eggs; one was fertile but the chick died before hatching. On February 24 there were two more eggs in the nest, the first of which hatched on March 16. By the 20th there were two chicks. Thenceforth the nest litter was replaced daily with warm shavings which had been rubbed to remove any spikes which might have injured the tiny chicks. Both proved to be males and left the nest on May 7. On May 23, one of the young was seen harassing his mother, so the two young were removed to another cage. The female had already laid again; there was a damaged egg in the nest. By early June there were two more eggs. These hatched on June 18 and 20. Again, nest litter was usually changed daily—but "at times when some of the nesting material was not removed and renewed, such as when egg-laying was taking place, the cock could be heard vainly trying to gnaw the wood of the box. He did, in fact, succeed in making a little heap of chips at one time when the box had become really sodden and the birds had endeavored to move their eggs into one corner near this little pile" (Cooke 1991). In 1992, the female laid three eggs; remarkably, three young were reared—two females and one male. At that time Mr. and Mrs. Cooke kept four pairs, two of which were of breeding age, and had reared ten young that year (Cooke 1993b).

Carol Day had 20 red-spotted lorikeets of three strains, most of which she had bred, in March, 1995. They proved prolific. Three pairs had hatched and fed chicks while the young from the previous nest were still with them. This was permitted because these females regularly laid again only one week after the young left the nest, and also because Mrs. Day rightly believes that it is beneficial to the young to stay with their parents longer. She recalled a fact which I found fascinating. On two occasions, two unrelated brother/sister pairs grew up together. The siblings kept together and roosted in their own nest-boxes until they were 11 months old, when they suddenly changed partners. Females usually start to lay when 13 to 14 months old and males are fertile at the same age. Therefore this change-over to un-related partners occurred at a most significant age in their maturity. If it had happened only once it would have been interesting but as it happened twice it almost suggests some involuntary mechanism at work!

Most of her pairs were kept indoors, maintained at temperatures of 21°C (70°F) or 18°C (65°F). Cage sizes are 2 m (6 ft. 6 in.) long x 61 cm (2 ft.) x 80 cm (30 in.) high or 1.5 m (5 ft.) x 47 cm (18 in.) x 80 cm (30 in.) high. Hazelnut branches and twigs are provided for the birds to chew and climb on. Nestboxes are the creche type, with the overall size of 38 cm (15 in.) x 18 cm (7 in.) x 18 cm. The inner compartment is 18 cm (7 in.) square. For 7 months several had lived outdoors in a shed heated to 15°C (60°F), with a 60 W infrared lamp for extra warmth. They had access to partly covered planted aviaries and spent much time outside, except when the temperature is below freezing. On most days they bathe in dishes in wire hanging baskets in the open flight. Carol Day believed that the captive-bred young would prove to be hardier than the wild-caught birds. Of interest is the use made of flowers. In 1994, two pairs hatched young in the spring; between them they reared four young on little else. They were offered mainly the flowers of horse chestnut candles (as it was easy to collect large amounts), rowan, elderflower, apple blossom and birdcherry. They fed part of willow twigs to their young. The only fruits they relished were apple, pear and pomegranates, also prickly pear when available. Carrot and papaya were liquidized in their food, because they would not eat it by choice. Occasionally, Chinese leaves were added.

No one should consider keeping red-spotted lorikeets unless they are prepared to pay special attention to their diet and to maintain them at a minimum temperature of 18°C (65°F). Successful breeders have noted that the best results are obtained at about 21°C (70°F). They are not a suitable species for beginners but in the hands of caring breeders they thrive and prove prolific, as the above descriptions indicate.

Red-flanked Lorikeet *Charmosyna placentis* (Temminck 1834)

Synonym: Pleasing lorikeet

Native name: New Guinea Daribi: *Abubage*

Description: The male is distinguished by his blue ear coverts and the scarlet down the sides of the body in line with the wings. The scarlet extends to the sides of the upper breast. The lores, cheeks, throat and lesser underwing coverts are also scarlet. The plumage is dark green above and lighter green below, except forehead and forecrown which are bright green. The ear coverts are mauve blue streaked with lighter mauve blue; the rump is dark blue. Greater underwing coverts are yellow and there is a yellow band across the underside of all the flight feathers, which are otherwise gray below. The tail is beautifully marked in fully adult birds: green above with the central feather terminating in a long narrow area of red that tapers into the shaft in the center of the feather and is tipped with yellow. The outer tail feathers are broadly tipped with yellow; the underside of the tail is also yellow. The bill and legs are coral red and the iris is orange.

Length: Usually 15.5 cm (6 in.) but larger birds occur; Diamond (1972) suggests an increase in size with altitude.

Weight: Males, 36–38 g; females, 36 g (weighed by author); Diamond gives an average weight of 35.5 g (+ or -4.3 g) for males and an average of 31.7 g (+ or - 3.5 g) for females collected at Karimui (1,100 m) and Soliabeda (600 m); 38–48 g for *subplacens* (Forshaw 1989)—but 48 g would surely be unusual.

Immature birds: Most can be sexed at about 35 days when the male's red flank feathers are apparent. There are at least two forms of immature plumage, probably dependent on the area from which they originate. In captive-bred young, the origin of whose parents was unknown, I have recorded the following forms. In one, the young of both sexes have red and yellow on the cheeks, but in females most of the red has gone by the age of five months. For example, a female hatched on December 24 showed the very last trace of red beneath the eye on June 19. Young males show the last traces of yellow

at about five months. In the second form, young females lack red on the cheeks (as described by Forshaw 1989) and males show very little yellow. (In one of the two pairs which produced such young I believe the male parent was *ornata*.) The beak is brown and the feet are gray, soon becoming pinkish gray, both becoming coral pink by four or five months. The iris and the cere are brown. On leaving the nest the young are very small—less than two-thirds the body size of an adult.

Sexual dimorphism: The female looks so different she could be mistaken for another species. There is no red in her plumage except on the tip of her tail. The ear coverts are broadly streaked with bright yellow on a greenish yellow background; the yellow streaks stand away from the head slightly. The forehead is the same shade of dark green as the upper parts and the underwing coverts and underparts are lighter green.

Mutation: A Belgian breeder produced a lutino or dilute female (mainly yellow, more greenish above, with red on the ear coverts); its photograph appeared in the April, 1992, issue of *Onze Vogels*.

Subspecies: The nominate race is described above. It is difficult to identify captive birds from written descriptions. Diamond (1972) states that in the Karimui and Soliabeda areas, the females are slightly darker and the males deeper red on the flanks than in the comparative material of nominate *placentis*. Forshaw (1989) describes the subspecies as follows: *C. p. intensior* differs in having a smaller blue patch on the rump, the color being blue violet; the forehead of the male is slightly greener (less yellowish); Arndt states that the shade of green is darker. Size is said to be slightly larger. *C. p. ornata* has the mantle slightly darker green, a large blue patch on the rump and slightly more extensive red on the throat. As pointed out by Arndt, the forehead is greenish yellow. A male which I believe to be of this subspecies had an irregular but almost unbroken band of scarlet across the breast—probably a very mature bird. A photograph in Arndt's *Lexicon of Parrots* shows to perfection the beauty of this subspecies. The other two forms are distinguished by lacking the blue rump. *C. p. subplacens* differs from the nominate race only in this respect. *C. p. pallidior* differs from *subplacens* in being paler throughout, especially on the ear coverts in males.

Natural history

Range: Indonesia, Solomons, New Guinea and adjacent islands. The nominate race is found in the southern Moluccas—Seram (see Gazetteer), Amboina, Ambelau and Pandjang—Aru Islands, Kai Islands and in southern New Guinea (eastward to the upper Purari River). *C. p. intensior* occurs on the northern Moluccan islands, including Obi, and in Gebe in the western Papuan Islands, Irian Jaya. *C. p. ornata* is found in northwestern New Guinea and in the western Papuan Islands, except Gebe. *C. p. subplacens* is found in eastern and northern New Guinea, east of Hall Sound in the south and in Sarmi, Irian Jaya, in the north. *C. p. pallidior* inhabits Woodlark Island (east of New Guinea), and is found through the Bismarck Archipelago (New Britain, New Ireland, New Hanover, Crown, Long, Tolokiwa, Umboi, Malai, Sakar, Witu, Lolobau, Watom, Duke of York, Tabar, Lihir, Tanga and Feni) to the easternmost Admiralty Islands (reported from Lou and Pak only), also on Bougainville and Nugaria (Fead). Next to *Trichoglossus haematodus*, this is the most extensive range of any lorikeet.

Habits: It occurs in a wide range of habitats, described by Coates (1985) as primary forest, forest edges, tall secondary growth, monsoon forest, partly cleared areas and coconut plantations, also savannah, sago swamps (when sago is flowering), gallery forest and, occasionally, mangroves. Lowlands and locally in lower mountains are its usual abode but in New Guinea it has been recorded from a number of locations between 1,300 m (4,260 ft.) and 1,600 m (5,250 ft.). However, where *rubrigularis* occurs, the latter usually replaces it above about 1,500 m (4,900 ft.). Diamond believed that its presence at Karimui, where it was abundant, might constitute an altitudinal record. It remained numerous on the lower slopes of the basin walls but was not present at 1,325 m (4,350 ft) on Mount Karimui. There are no other Eastern Highlands records for this species. Diamond recorded that *placentis* "congregated to feed in flowering trees,

where it was the most numerous species in the Karimui area and provided from 15% to as much as 65% of the bird usage. We had no record of it in the many fruiting trees kept under observation. In the flowering trees it distributed itself uniformly throughout the crown. At any given time a tree might hold up to 25 individuals of *C. placentis*, which came and went in groups of 2 to 10, kept up an incessant chatter, and often paused from feeding to chase each other."

It may be observed in pairs or small groups or, in flowering trees, in large flocks, often with other nectar-feeding species. Coates describes it as "active, noisy and quarrelsome when feeding though often very difficult to observe amid thick foliage. Feeds on pollen, nectar, flowers and seeds. Most often seen flying swiftly through or over the treetops in small, compact, noisy flocks."

In his classic *The Malay Archipelago*, first published in 1869, Alfred Russel Wallace described his visit to Batchian (Bacan) in the northern Moluccas, which was hitherto unexplored. It must have been intensior which he observed there, and described as follows: "When the jambu, or rose-apple (Eugenia sp.) was in flower in the village, flocks of the little lorikeet (*Charmosyna placentis*), already met with in Gilolo [Halmahera], came to feed upon the nectar, and I obtained as many specimens as I desired."

Jan van Oosten visited New Britain in 1993 and described *C. p. pallidior* as very common in Rabaul. He found it in another part of the city than the previous year; the trees which had been blooming then were not in flower. It was also very common in plantations, along with the purple-bellied lory (*L. h. devittatus*) and Massena's lorikeet (van Oosten unpubl. report 1993). During ten days in Rabaul in 1992, he spent at least one hour every morning and an hour each evening observing red-flanked lorikeets "as they covered every tree in town. These were not native trees, but had been introduced by the Germans, English and Japanese. They are very large trees that produced myriads of yellow flowers which the birds visited all day." They proved very easy to locate because of their "chattering." In 1993, few were seen in Rabaul but they were observed in a plantation, licking the sugary dew off casuarina needles. This species was not seen in the markets in Rabaul. He was told it was too nervous or did not live long in captivity (van Oosten 1993).

On Lihir, in the Bismarck Archipelago, Peter Odekerken made an interesting observation on August 7, 1995. He saw about 20 young *placentis* in a tree in which they were foraging—although no flowers were evident. No adults were present. Might this species have a creche system? he asked. It was very common on the island—the most abundant bird species seen during a three-week period (Odekerken pers. comm. 1996).

Nesting: Coates describes a variety "of actual or likely nesting sites." In New Guinea, near the Veimauri River, Central Province, he observed a pair excavating a hole in the base of a staghorn fern in mid-August. It was located 7.5 m (24 ft.) above ground in partly cleared forest. He saw a male enter 1 of 2 holes in a heavy moss "overburden" 21 m (69 ft.) above ground in the Ok Tedi Valley, Western Province; the tree was isolated in a cleared area and the month was February. Other observers have seen this species nesting in a "crow's nest" fern in late July, on Witu Island, Bismarck Archipelago (2 eggs); and climbing over a termites' nest on a savannah tree in February in the lower Trans-Fly region. In the Trans-Fly, a pair was seen excavating a tunnel in a large arboreal termitarium 6 m (20 ft.) above ground in a tea tree swamp, late in October. Another pair was visiting a termite nest high in a tree of mangrove association in April, in the Mappi region of the central south coast of Irian Jaya. Kenning (1994) found a nest of *pallidior* in Namatanai, New Ireland, situated at a height of 25 m (82 ft.) in an epiphyte. The nest was on the underside, protected from rain. Male and female were seen there. In New Britain, in Rabaul, van Oosten was shown a nest which he was told had been in use for eight years. It may be that nesting occurs in the wet season, as well as in the dry season. Three eggs collected from wild nests averaged 19.0 x 17.3 mm. Eggs laid by more than one female at Palmitos Park measured 19.2 x 16.8 mm, 19.5 x 16.9 mm, 20.1 x 17.0 mm, 20.4 x 17.5 mm and 20.5 x 17.5 mm.

Status/Conservation: Widespread and common, conservation measures aren't known to be needed.

Aviculture

Status: Uncommon but probably increasing due to breeding successes.

Clutch size: Two

Incubation period: 24 days, by male and female. Incubation rarely commences until after laying of second egg—as long as six or seven days in extreme cases (Low 1988). Eggs have hatched as long as 32 days after being laid, and 26 or 27 days is not uncommon.

Newly hatched chick: Weight, 2 g; the down is grayish white and thicker than in *C. josefinae*, for example, being longest and thickest on the back; the beak is pinkish brown and the feet are pink.

Chick development: In captive parent-reared young this varies, as evidenced by the fact that young are usually ringed at 15 or 16 days but have been ringed three days earlier or later; 4.5 mm rings are used. One chick was missed at 15 days and had to be ringed with a 5 mm ring at 17 days. By nine days much of the down has been lost, except on the back; the developing second down can be seen as black lines beneath the skin all over the head and body; the feet are becoming gray and the eyes are just starting to slit; the beak is brown. By 10 or 11 days the feet are dark gray and the ears are opening. By 17 days head and wings look dark with the quills forming; at 24 days the wing feathers are erupting and by 35 days the wings are fully feathered. The tail feathers start to erupt at about the same time as those of the head, i.e., at about 22 days, or later in some chicks. At this time the feathers of the underparts are just starting to erupt. At 26 days (where appropriate) there is a red blush on the cheeks from the feathers developing under the skin. At about 45 days young are fully feathered, except for the short tail, which does not grow significantly until after the young leave the nest.

Young in nest: 41–58 days recorded in birds in author's care, usually 46–49 days. In one case where the female died 23 days before the chick left nest, it was reared by the male and left nest at 56 days.

Chick growth

Shows the weights of the first three young reared by a captive-bred pair; the first two were single chicks. No. 1, a male, hatched on June 12, 1994, and No. 2, a female, on December 24, 1994. Chick No. 3 was the eldest of the next two chicks of this pair, hatched on March 29, 1995. It died of salmonellosis and soon after it was discovered that the male was carrying this disease. Note the difference in the weight gains of these three chicks. The lower weight of No. 1 can be explained by a lower protein diet. Nos. 2 and 3 were reared on nectar consisting of approximately equal parts of Nekton Lori and Milupa baby cereal, with honey added. There was no Nekton Lori in the food when the parents reared No. 1. In addition, fruit, mainly apple or pear, was offered once or twice daily. The high weights of No. 3 are difficult to explain; the weights of its sibling were also higher than normal but not as high as No. 3. The latter's weight gains remained higher than average until two days before it died. (Salmonellosis strikes quickly.)

Red-flanked lorikeet weight table

Age in days	Weight in grams		
	Chick No. 1	No. 2	No. 3
hatch	–	–	–
1	–	–	(March 30) 3
5	–	7 fc	–
6	–	4	8 fc
7	–	9 fc	13 vfc
8	–	10 fc	15 vfc
9	–	10 ne	17 vfc
10	8 nf	13 vfc	20 vfc
11	–	13 ne (ringed)	21 vfc
12	9 fic	14 fic	22 vfc
13	9 fc	17 vvfc	22 vfc
14	–	17 nf	23 fc
15	10 ne	18 nf (ringed)	25 vfc
16	–	19 nf	32 vfc
17	13 vfc	22 fc	31 vfc
18	–	22 nf	32 vfc
19	14 fc	25 vfc	35 vvfc
20	–	26 vfc	37 vfc
21		26 fc	33 fc
22	23 vvfc (ringed)		41 vvfc
23	20 vfc	29 fc	38 vfc
24		31 vfc	41 vfc
25		31 fc	39 fc
26	26 vfc	30 fc	36 ne
27		31 fc	44 vfc
28		32 fc	41 fc
29	30 fc	36 vfc	41 fc
30	29 nf	33 fc	42 fc
31	30 fc	34 fc	
32	31 fc	38 vfc	45 nf
33		37 fc	
34		39 fc	40 nf
35	35 fc	34 e	
36	34 fc	39 fc	42 nf
37	39 fc	39 nf	41 nf
38	37 fc	39 fc **	
39	35 fic	42 fc	35 e
40		43 fc	died
41	40 vfc	42 nf	
42	37 fc	43 fc	
43	34 fc *	39 fic	
44	34 ne	42 nf	
45	39 fc	45 fic	
46	left nest	39 ne	
48		40 ne	
49		38 e	
50		left nest	

* Sexable as a male: red feathers under wings; feathers not yet grown on flanks (or plucked by parents).
** Sex correctly noted.
Crop contents: e = empty; fic = some food in crop; nf = nearly full; fc = full crop; vfc = very full crop

General: This species was almost unknown in aviculture until about 1977. I acquired a pair then (privately imported) which remained so nervous I had no desire ever to keep red-flanked lorikeets again. Catchers and dealers generally had no interest in them because mortality rates were so high. However, by the mid-1980s, perhaps because some of the formerly common lories were becoming more difficult to obtain as the result of overtrapping, the first commercial importations reached Europe. Numbers were not large. The birds attracted the attention of exhibitors of foreign birds at shows (because of their beauty and small size) as well as lory breeders. In this way there were new recruits to the ranks of lory breeders. I bought, in December, 1986, a male from one of the first consignments which had arrived about nine months previously. Males had been in the majority. I could not resist his outstanding beauty, character and tameness. The latter was exceptional for a wild-caught bird. In the ensuing nine years, I have had many birds of this species hatch in my care, but there was never one which equalled him in any of these respects. He was by far the most colorful, with a wide broken band of scarlet on the upper breast, and probably belonged to the subspecies *ornata*.

At the time I obtained him, the chances of finding a female seemed very small. However, only two months later I moved to Tenerife in the Canary Islands, as curator at Loro Parque. I was able to exchange one of the lories I had taken with me for a female *placentis* (almost certainly of the nominate race). As no spare cage was available, I had to place the female in the male's cage instead of allowing her to take up residence in the cage on her own for a few days. I was unprepared for the aggressive reaction of the male. After watching anxiously for 20 minutes as he chased her about, I caught him and clipped the feathers of one wing. Being unable to fly had an instantly sobering effect on him and they soon settled down well. However, during their time together, the male was sometimes quite aggressive during courtship, something I have never observed with another male of this species. This pair was kept in an all-wire, suspended cage measuring 1.8 m (6 ft.) long, 95 cm (37 in.) wide and 1 m (40 in.) high. They were kept outdoors because in Tenerife the temperature does not normally fall below about 14°C (55°F) and the daytime temperature is usually much higher.

Their food consisted of a very liquid mix of honey, Milupa baby cereal and malt extract—about 80 percent of the diet. They consumed much fruit—apple, pear, grapes, soaked figs and soaked sultanas. Unlike almost all other *Charmosyna* lorikeets I have known, they would also eat spray millet.

The pair was introduced at the end of February. The female laid her first eggs on April 28 and 30. The chicks died in the egg nearly full term. The next two eggs were laid on May 20 and 22 and again the embryos died. I had then learned an important fact about this species: some birds do not keep the nest clean. Unlike most lorikeets which nest in holes in trees, they nest in clumps of moss or other vegetation; the nest walls are not solid and, in the wild, the liquid excreta would immediately drain away. In a wooden nestbox this cannot happen. During the incubation of the first clutch of eggs, I did not interfere—believing, wrongly, that perhaps this was their method of correcting the low humidity of the climate they were in, which would be the reverse of their natural climate. Thus the damp and dirty state of the nest caused the deaths of the embryos. During the incubation of the second clutch I did change the nest litter but apparently not frequently enough.

The eggs of the third clutch were laid on June 18 and 20; one egg pipped prematurely, then the chick died, and in the other egg the embryo died. Two more eggs were laid on August 9 and 11 and, again, the embryos died. In 1988, the female laid on February 19 and 22. I removed the eggs to a small Turn-X incubator on February 24. Neither egg had been incubated and both hatched on March 19. Therefore the incubation period was 24 days (it was a leap year). The next two eggs were laid on March 25 and 28 and were placed in the incubator on April 1, after incubation had commenced. They hatched on April 22 and 23. The chicks weighed 2 g on

hatching; the only problem they presented was the smallness of the beak! I used a small spoon to feed them and it was difficult to align this with the beak for the first couple of days. The first two young, both males, were reared to independence. Both chicks in the next nest died, one at four days, and the second at 33 days. Causes of death were not known. The next clutch, laid on April 22 and 24, resulted in two parent-reared young. The chicks thrived. They were plucked during the middle stage of the rearing period but were almost perfectly feathered when they emerged. The nest litter had to be changed every second day. The first youngster left the nest at 41 days. His father was so possessive and proud, he even drove away the female! I provided two containers of nectar and two containers of fruit to ensure that she was not kept away from the food. When the second youngster fledged the male was quieter and allowed the female to guard one. They made the most charming family group imaginable. The pride that the parents took in the young gave me intense pleasure. They were the first *Charmosyna* whose rearing I had observed. Since then I have often watched with interest the pride which members of this genus show towards their newly fledged young, which is much more noticeable than in other lory species. Often they also show great excitement when their young first leave the nest.

This species is very willing to nest when conditions are right. The temperature needs to be about 22°C (71°F) for breeding results to be consistently good. The first successful breeder was Leif Rasmussen of Denmark, whom I first visited in 1986. His birds were maintained at this temperature in an indoor birdroom. Interestingly, they were kept in glass-fronted cages 1 m square, each cage with its own lighting and, if necessary, also with a heat lamp. A unique aspect of management concerned the cage floors. In some cages the wood shavings on the floor had not been changed for many months and yet, reduced to the consistency of sand, appeared perfectly clean. Unseen workers that kept it so were dozens of mealworms! The fluorescent light in the cage below heated the floor but even in one place where this did not apply, the system worked. It supplied livefood for hanging parrots and marmosets! The lorikeets were seen grubbing on the floor; perhaps they ate the small mealworms. In the first four years of this system, 40 young were reared. At this time, red-flanked lorikeets were a rarity. Those who had bought Leif's young had, at that time, little success in breeding them. Yet his own young had proved fertile as early as nine months old. In 1985, for example, 17 were parent-reared from three pairs. The environment suited them perfectly—and this was the reason for his success. Their diet consisted of Milupa 7-Cereal, grape sugar (dextrose), malt extract and pollen meal.

Dulcie Cooke in England also kept red-flanked lorikeets in an indoor birdroom. She found that imported birds needed 18 months to two years to become acclimatized. During the first year they were kept at 23°C (74°F); the next year it was possible to lower the temperature to 20–21°C (68–70°F). She commented: "We know of one pair of these birds who reside in a gigantic aviary with water running through lawns planted with many small trees and shrubs where Red-flanked Lorikeets have scorned the heated house available and made their nest in a box in a dense shrub from which they produced healthy and extremely hardy young. It must be realized that these birds have access to enormous flying space, and can always return to a heated house if they wish. Yet other breeders seem to be successfully keeping them at rather lower temperatures than those which we have found suit our birds. I do not consider that Red-flanks would survive for long in unheated aviaries in the U.K. and with the restricted numbers there are in the country it would be most unwise to experiment with low temperatures where no heated house is available" (Cooke 1992). It was Dulcie Cooke's experience that even inspecting the nest to discover whether the female had laid was sufficient to prevent her from returning to the nest. However, her pair was wild-caught and she described them as very nervous. This is probably why their young also showed this trait.

When I became curator of the breeding center at Palmitos Park, Gran Canaria, in 1989, there were nine wild-caught birds which had been imported in the previous year. The males appeared to be 4 birds of the nominate race and one *intensior*. The females all looked alike, probably all of the nominate race. They were kept in an indoor cage together until May, 1989. They were then placed in pairs in suspended cages, each one of which measured 1.2

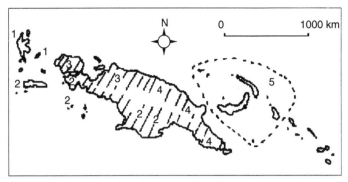

Range of red-flanked lorikeet (*Charmosyna placentis*).

1. *C. p. intensior*—north Moluccan Islands, including Obi; also Gebe.
2. *C. p. placentis*—Seram, Amboina, Aru Islands, Kai Islands and southern New Guinea.
3. *C. p. ornata*—northwestern New Guinea and western Papuan Islands.
4. *C. p. subplacens*—eastern and southern New Guinea.
5. *C. p. pallidior*—Bismarck Archipelago (including New Britain and New Ireland), Woodlark Island and Bougainville.

m (4 ft.) long, 75 cm (2 ft. 6 in.) wide and 1 m (3 ft. 4 in.) high, with a nestbox attached to the outside. The surplus male was paired with a female from my own pair which was hatched in March, 1988. (Her father was *ornata* and her mother *placentis*. One of her two female offspring had no color but green or yellow on the face, whereas she herself did show red in immature plumage. The head coloration of her other female youngster was unfortunately not noted.)

All five pairs nested and four pairs produced young. The failure of one pair was probably due to the fact that the female assumed the male's role during copulation. Her eggs were infertile. This female was the first to lay, on June 5. Other females laid on June 26, July 8, July 21 and September 3. In the following years I found that they would nest at almost any time of the year. October was the only month in which eggs were not laid and only one clutch was ever laid in November. Most eggs were laid from March to May. As recorded above, incubation (carried out by male and female) seldom commences until after the second egg is laid. Where both eggs hatched, this could occur on the same day, on consecutive days or at intervals of two days. It was interesting that there were no problems with captive-bred birds soiling the nest, necessitating changing the litter during incubation. However, after the chicks were about a week old, it was essential to change the nest litter daily or every other day, depending on the size of the chicks and whether there were one or two. If the nest litter is not changed often enough, the parents are likely to pluck the young. On leaving the nest, the young start to feed themselves within one or two days, although may be seen soliciting food from their parents for several days. On occasions young were left with their parents for many weeks after fledging. No aggression was ever observed, but if the female laid again the young had to be removed because the eggs were likely to be broken. Young mature very quickly and one male was seen mating when only six months old.

In England, D. and I. Bardgett's pair bred when the male was nine months old; one of the two eggs hatched. The first young reared by the male's parents left the nest at 50 days. The next two young were a male and a female and could be sexed at 30

days. In addition to nectar, they ate chopped apples and pears, grapes, soaked sultanas, defrosted peas and sponge cake soaked in nectar. Livefood was refused (Bardgett and Bardgett 1991).

This species should be provided with small twiggy branches in which they can sit—not a bare cage with two straight perches. They remind me of small softbills in some ways, such as the manner in which they move between closely spaced perches, appearing hardly to open their wings. Their movements are fast and jerky. In display, male and female erect the feathers of the crown and make a cracking sound with the wings. The male looks spectacular with his vivid colors, vivacious personality and rapid sideways movements of the head and tail. Its beauty, ease of sexing and willingness to breed, especially in a small cage, make this species ideal for the aviculturist who is prepared to provide the right degree of warmth.

Duchess Lorikeet *Charmosyna margarethae* (Tristram 1879)

Description: The yellow band, broad on breast, narrow on mantle, distinguishes this species from all other lorikeets. The band is bordered above by a narrow line of purplish black. Plumage is otherwise mainly red, except for the green wings, dull purple patch on the lower breast and purple-black patch on the crown and hindcrown. The tail is red, tipped with yellow. The male has the rump and upper tail coverts green washed with gold; in the female they are olive. Most striking, however, are the yellow sides of the female's lower back; this area is red in the male. The iris is yellow; the bill and feet are orange.

Length: 20 cm (8 in.)

Weight: 40–60 g (Forshaw 1989)

Immature plumage: Yellow band on breast and mantle less clearly defined and lacking the margin of purplish black. The red feathers of the body are variably margined with purplish black. The iris is gray; the bill is grayish orange and the feet are grayish.

Sexual dimorphism: Pronounced (see Description)

Natural history

Range: The Solomon Islands (including Guadalcanal and Kolombangara) and Bougainville (see Gazetteer).

Habits: This species inhabits hill forest, lowland forest and coastal coconut plantations. On Makira in the Solomon Islands Jonathan Newman found this lorikeet to be the most common parrot, in August, 1990. It was most numerous in hill forest and also occurred in mature trees around gardens. On Kolombangara it was seen mainly in coastal areas; the maximum altitude was 350 m (1,150 ft.). It panicked into noisy flight when a goshawk flew over. In coconut plantations, where it fed largely on nectar, it flocked with *Trichoglossus h. massena*. It was also seen in flowering trees where small groups of 15 or more birds fed together. Its voice was described as quite harsh for a *Charmosyna* (Newman pers. comm. 1994).

On the island of Isabel, Jan van Oosten searched for duchess lorikeets in 1993. He failed to find them in plantations so hiked up to 2,000 m (6,500 ft.). No duchess lorikeets were heard or seen until he had ascended to about 1,500 m (4,900 ft.), when a flock of about 20 birds was seen. "We observed them for about two hours as they fed and explored two flowering trees. They were indeed something to see with the red head and body with green wings and the very noticeable yellow band across their chest outlined in black" (van Oosten 1993). Further brief visits were made to Guadalcanal in 1994 and 1995. On one occasion, six birds were observed flying over the camp site early in the morning; the location was 2.5 km (1½ mi.) above Tenaru falls, in primary forest. In 1995, several flocks were seen further east from the falls trail in plantations close to primary forest. The lorikeets were feeding in coconut palms and in trees with yellow flowers. Massena's lorikeets and cardinal lories were also in the area and on occasions were feeding together (van Oosten pers. comm. 1996).

In 1975, Ken Stott recalled the birdlife of the Solomon Islands during the second world war. A good spot for birdwatching was the secondary road, west of Henderson Field, Guadalcanal, where it crossed the Tenaru River. "There the remnant of the foothill forest met the flat coastal plain's coconut and sugar-cane plantations. In this intermediate zone, clumps of high trees and lush secondary growth below proved sufficiently attractive to entice forest birds into more or less open circumstances to join riverine species that resided along the banks...A parrot that occurred rarely at Tenaru was the small (eight inches) but beautifully colored Duchess Lory. More typically a mountain bird on Guadalcanal and more abundant at low levels on other islands (Bougainville, for example) I saw it twice at the river crossing. Long-tailed and slim-bodied, it was mainly red with a yellow breast band" (Stott 1975).

On Bougainville it is said to be most numerous in the hills and lower mountains; it also occurs in the town of Kieta. Usually seen in small flocks of up to ten birds, larger congregations meet in flowering trees. Fruits form part of its diet, especially those of *Schefflera* (Coates 1985). Its call note is said to have a peculiar squeaky quality, unlike that of most other lories. In 1964, from July to Septem-

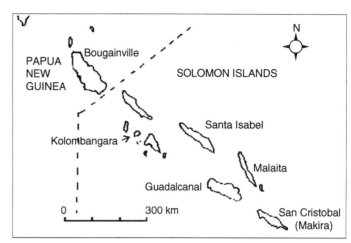

Range of duchess lorikeet (*Charmosyna margarethae*), named islands only.

ber, Schodde (1977) found them to be widespread in tall secondary and marginal primary forests in southern Bougainville. There, and at the edges of gardens, they occurred from about 5 km (3 mi.) inland up to about 750 m (2,460 ft.), but were most frequent between 100 m (330 ft.) and 600 m (2,000 ft.). He usually saw them in small, quiet flocks of between five and ten birds, either in swift, direct flight or in the canopy. They searched for fruits, preferably those of *Schefflera*.

There is no information on the breeding biology of this lorikeet. A male was seen displaying to a female in January, 1975. He fluffed out his head and breast feathers, dropped his wings and fanned his tail, then chased the female from branch to branch. The display ended when the female flew off, with the male following (Greensmith 1975).

Status/Conservation: Fairly common locally. Deforestation will be the main threat to its survival.

Aviculture: Unknown

General: After whom was this lorikeet named? Canon Tristram first described it, naming it after a recent guest, HRH the Duchess of Connaught. Mivart (1896) regretted the fact that it was not possible to illustrate the male in his *Monograph*. This was due to the fact that a sweep had stolen Canon Tristram's one and only skin!

Fairy Lorikeet *Charmosyna pulchella* (G. R. Gray 1859)

Synonym: Little red lory

Native name: Fore, *Súshuke*; Gimi, *áni* (Diamond 1972)

Description: The smallest of the red *Charmosynas*, it is identified by the band of narrow bright yellow streaks on the upper breast; the yellow streaks are broader than those on the thighs and/or lower flanks. This lorikeet is basically red, with the wings green. There is a variable blackish patch on the underparts. The flanks and sides of the rump are red in the male and green-tinged yellow in the female, and the rump is dull, dark blue. A patch on the crown is purple-black, as are the thighs. The underwing coverts are green and red. The tail feathers are green and red on the upperside, tipped with yellow; the undersides are yellow. Bill and iris are orange, and the feet are pink or orange.

Length: 18 cm (7 in.)

Weight: *C. p. pulchella*, one male 40.0 g, one female 37.5 g; *C. p. rothschildi*, four males and six females averaged 35.6 g, with the males averaging 37.5 g and the females 34.5 g (Bosch 1995).

Immature birds: The crown and nape are dull green or blackish green, and there are blackish markings above the eye. The breast is dull green without the yellow streaks, and the flanks also lack the streaks. The beak is dark brown, soon changing to dull coral. The feet are grayish pink and the iris is brown. An ancestral relic in young birds, commented on by Bosch, is the light band on the underwing coverts, that is normally lacking in adults. In this respect they resemble *papou* and *josefinae*. "The wing band remains with many females in a washed out form for years (or for life?). This instability in identifying characteristics is typical of changes that are not quite fixed."

Sexual dimorphism: Pronounced (see under Description)

Subspecies: *C. p. rothschildi* differs in that the yellow streaks of the breast are set in a broad band of green. Note that there could be confusion with immature birds of the nominate race which have the breast green; when they acquire the yellow streaks some green feathers have not yet been molted. In *rothschildi* the nape patch reaches to the back of the eye, whereas in the nominate race some black-margined feathers or black feathers may reach the eye but not in a solid band of black. The color of the abdomen is variable, being more black than red in many specimens. In the female, the undertail coverts are greenish yellow, not yellow as in the nominate race.

The subspecies *bella* is now considered synonymous with the nominate race. Bosch (1995), in an informative article on this species, suggests that *pulchella* and *rothschildi* "are so different in appearance and voice" that one might question whether they are in fact two species (Bosch 1995). When more is known about their distribution and whether they are sympatric, this question might be resolved.

Natural history

Range: Throughout the mountains of New Guinea, including the outlying ranges; it sometimes descends to the foothills but does not usually ascend above about 2,300 m (7,500 ft.). It is known from the Vogelkop, Irian Jaya, eastward to the Huon Peninsula and as far east as southeastern Papua New Guinea. *C. p. rothschildi* occurs in the Cyclops Mountains and on the northern slopes of mountains above the Idenburg River, Irian Jaya.

Habits: A nomadic species, it is generally scarce although may be fairly abundant in forest areas with flowering trees. It usually occurs in mid-montane regions, but has been recorded from sea level up to about 2,400 m (7,870 ft.). It rarely occurs in open country. In the Eastern Highlands, Diamond (1972) described it as sparsely distributed in hill forest between 2,000 ft. (600 m) and 5,800 ft. (1,760 m). At Mount Karimui all observations were between 4,400 ft. and 5,290 ft. (1,320 m and 1,760 m). Its altitudinal range is usually—but not always—below that of the Papuan lorikeet. During a week spent at Tari, in the Southern Highlands, Jonathan Newman saw fairy lorikeets in the same areas as *C. p.*

goliathina. The elevation was between 2,100 m and 2,400 m and the month was September. A group of five were seen feeding in a bottlebrush tree with *C. wilhelmina.* Smaller numbers, from one to three birds, were occasionally seen in trees with other lorikeets. They fed alongside *C. papou* and various honeyeaters (Newman pers. comm. 1994).

In the Ok Tedi area, Western Province, this species is fairly common. It has been recorded in flocks of up to 30, but more usually in groups of 5 or 6 birds, from Dablin Creek, the Ok Ma area and Ok Menga (Gregory 1995). It occurs up to 1,700 m (5,600 ft.) and was recorded at 2,180 m (7,150 ft.) on one occasion.

An interesting observation of the fairy lorikeet being the most abundant parrot in the area was made by Finch (1979). This was not a single occurrence but was noted on three visits to the Efogi district of the Owen Stanley Mountains, between early September, 1978, and mid-September, 1979.

Newman described a "whispering call note" of birds in flight and Diamond comments on "the short note similar to that of *C. placentis* but sweeter, less shrill or staccato." Beehler et al. (1986) comment on the geographical variation in the call note. "In the West, a nasal *ks* given two or threee times, and a weak *ss*. In the East, a short high-pitched note." Bosch (1995) described the voice of the nominate race as much quieter and less harsh. The pitch is higher. He states that the voice of *rothschildi* is unusually strong, also very deep and gruff. "Sometimes birds will produce a salvo of whistles, rasping and clucking loudly. The monosyllabic contact call, *dib-dib*, is followed often by a rougher pitch: *dib-dib-dib djadja.* Instead of the simple ritual hiss of the Josephine and Stella Lorikeets, they have usually in a duet a quieter, more frequent *ff-ffff-ff* that can be broken by a cracking noise."

Nesting: This species has been recorded nesting in the base of epiphytes—probably the normal nest site. Two eggs in the clutch of a female at Palmitos Park, Gran Canaria, measured 21.1 x 17.1 mm and 20.8 x 16.9 mm, and one egg in another clutch measured 19.9 x 16.3 mm; those of a second female measured 21.0 x 16.2 mm and 21.0 x ? mm (broken). Bosch recorded a weight range of from 3.05 g to 3.48 g (average 3.23 g) in six eggs of *rothschildi*, and average dimensions of eight eggs 21.7 x 16.7 mm.

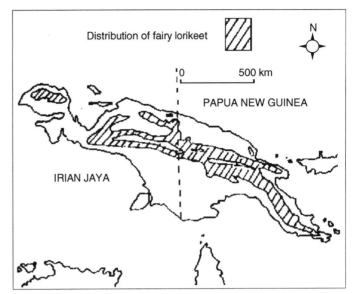

Range of fairy lorikeet (*Charmosyna pulchella*).

Aviculture

Status: Uncommon

Clutch size: Two, more rarely, one. Three eggs were laid once at Palmitos Park—and all hatched.

Incubation period: 24 days from start of incubation. In two pairs of *rothschildi* the incubation period was 25–27 days (probably indicating incubation did not commence when the first egg was laid) and in two pairs of *pulchella* it was 24 days (Bosch 1995). Schroeder (pers. comm. 1996) recorded 25 or 26 days in his pairs.

Newly hatched chick: Weight, about 2 g; two weighed by Bosch were 1.7 g and 1.9 g. They have long white down, sparse on crown; beak and feet, pink.

Chick development: At five to seven days feather tracts visible under skin of crop and upperparts; beak tinged with brown. At 11–15 days, only a few wisps of down remain and the dark gray second down is erupting; the feet are grayish and the beak is brown. By 19 days wings and body are covered in second down. Eyes open soon after. By 30 days chicks can be sexed by the yellow flank feathers of the female and red of the male. Some of the head and wing feathers have erupted. By about 34 days most of the body feathers have erupted and the tail feathers are just breaking through. By 42 days, young are fully feathered except for the short tail.

Young in nest: 50–58 days; 50–55 days (Schroeder 1995a).

Chick growth

The first two columns show weights of two parent-reared chicks in the same clutch at Palmitos Park. Column 3 shows weights of the third chick in a three-chick nest from the same pair, before and after the first feed of the day. It was removed for hand-rearing at 28 days as its weight was lower than that of its siblings.

Fairy lorikeet weight table

Age in days	Weight in grams		
	male	female	hand-reared
4		5 fc	
5	6 fc	6 fc	
6	6 fc	7 fc	
7	8 vfc	–	
11	–	9 e	
12	10 fic	14 vfc	
13	17 vfc	17 vfc	
14	20 vfc	–	
16	–	17 fic	
17	18 fc	–	
18	–	24 fc	
20	26 fc	–	
21	–	23 fc	
23	26 fc	–	
24	–	29	
25	33	29 fc	
26	32	–	
27	–	29 e	
28	31 fc	–	16/21
29	–	29	18/22
30	30 fic	33 fic	20/22
31	36 fic	33 fic	21/23
32	35 fic	33 fic	22/28
33	38 fic	–	24/29
35	–	42 fc	25/31
37	46 fc	40 fc	27/30
38	44 fc	32 e	28/32
39	36 e	–	29/35
41	–	–	32/39
43	–	–	33/42
47	–	–	33/39 perching
50	–	–	33/37
63	–	–	34

Crop contents: e = empty; fic = some food in crop; fc = full crop; vfc = very full crop.

General: This species has never been common or imported in large consignments. In the days of privately sponsored collectors and large private collections, in the early years of the twentieth century, a few pairs were kept by wealthy aviculturists. Thus the first recorded breeding occurred in 1914 in the collection of E. J. Brook of Hoddam Castle, Scotland. Five years after being imported they produced a single youngster, which left the nest after exactly two months. The parents were kept in an aviary with tanagers and sunbirds. Brook (1914) commented: "Always interesting with its vivacious impulsive movements, always on the move as if ready for any fun that may present itself, it is at the same time quite safe with other inmates of the aviary."

Not until 1973 did it become available commercially. In that year I looked after several birds of the subspecies *rothschildi* for the importer, Mrs. S. Belford. I kept four of them for a number of years, then wrote:

The appeal of this exquisite little bird is an instant one and not accounted for by color alone; its grace and daintiness are accentuated in its movements. This species is an absolute joy to watch; even caging it cannot diminish its fascination. It is seen to much greater advantage in an aviary, especially a planted one. I house one pair in a small enclosure which, in summer, is covered with the purple flowers of clematis and the white ones of that pretty weed convolvulus. It has an earth floor and contains a small cupressus tree which was soon partly destroyed by the Fairies but one of their chief delights is to bathe in it after it has been hosed down. This particular pair, unlike another, will bathe in a large plastic drinker hooked on to the front of the aviary.

These birds are easily induced to bathe by spraying. While wintering in my indoor birdroom they are sprayed regularly and nothing arouses greater excitement in them than the sight of the sprayer. As soon as they feel water on their plumage they flap their wings vigorously, whether they are perched or clinging to the wire netting, to receive the full benefit of the water. Those I have cared for have been very steady, almost fearless, and they do not retreat from the sprayer (Low 1977a).

Eggs were laid in both their winter birdroom and summer aviary. One August day when a pair had eggs, heavy rain fell for several hours. Afterward the birds were sitting outside their nestbox. I checked the nest to find it was flooded and the eggs were reposing in a puddle of water. I dried the box, placed a handful of peat inside with the eggs on top, and one of the pair returned immediately to incubate. Two weeks later a chick hatched. It died at a few days old and, sadly, that was the case with future clutches.

Soon after, breedings were reported in Europe. Vogelpark Walsrode had its first success in 1976. Dr. Burkard in Switzerland was also successful and, when his birds were placed with E. Zimmerli, third generation young had been produced by 1984. In Germany, Heiner Dahne produced young from a pair in a cage measuring only 50 x 40 x 30 cm (20 x 16 x 12 in.). The female laid on May 25 and 30 and did not start to incubate for another three days. They used a budgerigar nestbox (as did my own birds). The first chick hatched after 25 days and the second two days later. One chick died by the age of two weeks. The survivor, a male, left the nest at eight weeks. A second nest followed at once; this time a female was reared. The diet consisted of Milupa baby cereal and grape sugar, with apple and pear, and banana and orange on occasions (Dahne 1992). Mention is made above of the second egg being laid five days after the first. I once recorded an interval of six days between the first and second eggs (Low 1978a).

One of the most successful breeders was Jorgen Bosch in Germany. His pair, of the nominate race, produced 24 young in three years. Young were plucked in the nest, and some had to be removed for hand-rearing, thus up to five clutches were laid in a year. He stated that overbreeding should be prevented by keeping the birds in large planted aviaries (presumably for part of the year).

In the U.S.A., the fairy lorikeet (*rothschildi*) was first bred at San Diego Zoo. The nominate race was first hatched in 1989 by Dick Schroeder of California. They had two eggs in the nest by September 17, one of which hatched on October 10 or 11. When the chick was nine days old the parents started to eat more fruit—apple, pear, papaya and grapes; nectar was still the main item of the diet. By 30 days yellow feathers on the rump indicated the chick's sex. It left the nest on November 30, aged 50 days (Schroeder 1995a). This female was paired with a wild-caught male and produced fertile eggs at only nine months old. The first two young from this pair were retained for breeding; the male (paired with a wild-caught female) produced young in 1992. The wild-caught female who produced the first offspring died in 1994—but second and third generation young are still breeding. Most females will produce three clutches a year, although not all are viable (Schroeder pers. comm. 1996).

During my two years at Loro Parque there were imported fairy lorikeets of both subspecies in the collection. *C. placentis* was the only other small *Charmosyna* maintained. It was hardier than the fairies, which were the only parrots which needed to be taken indoors during the cooler months. From about December to February the temperature could drop as low as 10°C (50°F), which the fairies did not enjoy. On the neighboring island of Gran Canaria, in the south, where Palmitos Park was situated, the climate was much hotter and drier than the north of Tenerife. Here the fairies (wild-caught birds of the nominate race) thrived and bred—so at last I had a successful breeding pair in my care.

Success was not immediate. In the first year, 1989, two infertile clutches were laid. The nest was constantly wet and the eggs were dirty. On one occasion changing the nest litter caused them to desert their infertile eggs. Male and female incubate and, as in other *Charmosyna* species, incubation rarely commences when the first egg is laid. In 1990, the pair was incubating two eggs by March 27. As the nest was again wet and the eggs soiled, I gave the eggs to a pair of *placentis*. One hatched on April 18. Eleven days later I removed the chick for hand-rearing because one of its legs appeared slightly deformed. After a few days, during which calcium was added to its food daily, the leg was normal. The chick, which was independent by 52 days, grew into a perfect female. She was first offered the warm rearing food in a small container at 46 days and at once filled her crop. She flew for the first time at 60 days. The parents nested again. The eggs were again badly soiled and were exchanged with those of another pair of *placentis*. The fairy chicks hatched in the *placentis* nest on June 2 and 4 and were reared without incident. They were ringed on June 25 and 27 with 4.5 mm rings. These are standard rings whose width (as well as internal dimension) is 4.5 mm; however, they are really too wide for these small chicks and occupy the entire tarsus for some days. Ironically, the fairies did an excellent job of rearing the *placentis* and thereafter were allowed to rear their own young. They nested again at the beginning of December and the chicks hatched on January 2 and 3. They were fed by male and female from the start. Rearing food consisted of nectar (either Nekton-Lori or nectar made up from Milupa baby cereal and honey, or sometimes sugar), apple and pear. The chicks could be sexed by their flank feathers, as a male and female, at the ages of 30 and 29 days. They were ringed at 22 days old. They left the nest on March 1, after 58 and 57 days, and were seen feeding themselves on pear less than a week later. (Newly fledged young, as in *placentis*, are only about two-thirds the size of the parents, with very short tail feathers.) After a few days the parents took no interest in them. The two young would always be together at one end of the perch and the adults at the other end—rather unusual behavior for lorikeets. As the young grew and matured, they were placed together in a 2 m (6 ft. 6 in.) long cage with two nestboxes. All but two were males, but there was never any fighting in this small group. Indeed, they seemed to like being in a small flock. My hope was

that they could be housed in the butterfly house in the park—a wonderful, large, heavily planted and warm enclosure. But this never came about. Dr. Burkard kept a group of three pairs and three single birds. Perhaps this indicates that in an aviary of reasonable size with more nestboxes than pairs, this species might be kept in a colony. Whether more than one pair would breed successfully might depend on the size of the enclosure.

Fairy lorikeets are not well established in aviculture. They have little tolerance to low temperatures and females have always been in short supply. Captive numbers are low and it seems unlikely that they will survive over the long term. This is regrettable for the fairy is indeed one of the most delightful aviary birds in existence.

Josephine's Lorikeet *Charmosyna josefinae* (Finsch 1873)

Description: It can briefly be described as a smaller version of the Papuan lorikeet with the upper side of the tail red rather than green. The head and body are mainly red and the wings are green with red underwing coverts. Like *papou*, the narrow elongated feathers of the rear part of the crown are erectile; they are blue-gray in *josefinae*, gray in *sepikiana* and blue in *papou*; in *C. j. cyclopum* they are not present. There is a pointed black patch on the nape which extends to the rear part of the eye. The abdomen is green-tinged black; this area is extensive in *sepikiana*, usually reduced in the nominate race and almost nonexistent or absent in *cyclopum*. According to Hartert (1930), in *cyclopum* the feathers of the abdomen are blackish at the base but red on at least the distal half. The thighs are dull black with some (about eight) or very few yellow streaks. The rump is blue in the male, and the flanks and the upper part of the back are red. The female has the lower back and flanks yellow—but not such a bright yellow as in *papou*. According to Forshaw (1989), females of the nominate race have the lower back green instead of red, but after examining skins in various museums, Hubers (1993a) and van Dooren (pers. comm. 1995) stated that the nominate race cannot easily

be distinguished from *sepikiana*. Most females of the nominate race did not have a green rump; the rump was either greenish yellow or yellow.

The primaries have attenuated (narrow) tips. The uppertail coverts are green at the base, then gray-blue, and reddish in the center. Central tail feathers are gray-green at the base, then red, and tipped with yellow for about one-quarter of their length; lateral tail feathers are dark green on the outer web and tipped with yellow for half their length, and red on the inner web. The underside of the tail is yellow, with some gray at the base of the central feathers, except the long ones. Photographs of a *cyclopum* owned by P. Clear show the gray area to be darker and slightly more extensive, as compared with a *sepikiana* perched next to it. Also, the outer webs of the outer tail feathers are green in *sepikiana* and blackish tinged with green on the single *cyclopum*. (The latter also had a red bar on the upper median wing coverts which is probably just a feature of that individual.) In all subspecies the bill is deep coral and the iris and legs are orange.

Length: 24 cm (9½ in.); the tail accounts for half this length, but it is not elongated as in *papou*, thus the body size of *josefinae* is only slightly smaller.

Weight: Captive *sepikiana*, 70–75 g; information from labels on skins in an American museum, *cyclopum* 62 g and 67 g (Hubers pers. comm. 1995); *cyclopum*, seven birds weighed between 68 g and 75 g (Hartert 1930).

Immatures: They have a strong green tinge to the black on the abdomen and nape or, occasionally, the abdomen may be dark green. Shaft-streaking on nape and thighs is less well defined. The flight feathers do not have attenuated tips. Some females can be distinguished in nest feather—but this is not always the case, even in young from the same pair. In females which do not show yellow rump and flanks in nest feather, the yellow feathers usually start to appear at about four months; by the age of six months some females are in full adult plumage. In the experience of Anton Spenkelink (verbal comm. 1991), a few yellow feathers on the back or side of nestlings are not indicative of sex. In recently fledged young the beak is brown, gradually changing to orange; the cere is whitish (coral in adults), the iris is brown, the eye skin is paler gray and the feet are grayish pink.

Sexual dimorphism: Pronounced difference in color of lower back and flanks (see under Description).

Natural history

Range and subspecies: Confined to New Guinea. The nominate race occurs in the mountains of the Vogelkop, eastward to the Snow Mountains; *C. p. sepikiana* is from the Schrader Range in the Sepik region and from the Western Highlands, eastward to the Jimi River area and Mount Bosavi in western Papua New Guinea. *C. p. cyclopum* has been recorded only from the Cyclops Mountains in Irian Jaya. The distribution of this species in Papua New Guinea is not well known (Coates 1985). It has been recorded at least once from Lake Kopiago, in the Southern Highlands.

Habits: Not a lot is known about this lorikeet. Coates stated that it "seems to be rather scarce and local in Papua New Guinea though may perhaps eventually be shown to be locally numerous. Possibly nomadic." He described it as inconspicuous and unobtrusive, usually occurring from 760 m (2,500 ft.) to 1,770 m (5,800 ft.), occasionally lower. On the northern slopes of the Snow Mountains, Irian

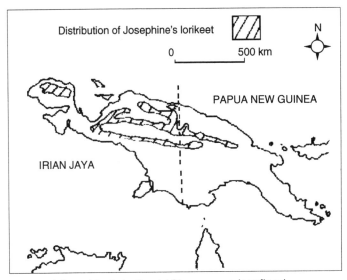

Range of Josephine's lorikeet (*Charmosyna josefinae*).

Jaya, it is common between 850 m (2,800 ft.) and 1,200 m (3,900 ft.) and has been recorded from near sea level at Jayapura. In the Western Province of Papua New Guinea, in the Ok Tedi area, it is rare. It has been recorded from Dablin Creek at 750 m (2,400 ft.), at Ok Menga at the same altitude and at Townsville drill site at 1,600 m (5,250 ft.) (Gregory 1995). Phil Gregory, who resides in Tabubil, augmented his published notes with the following: "I see Josephine's on just a handful of occasions each year, in groups of two or three only, and usually not with the other species. However, on September 9, 1992, a flowering tree in hill forest at 1,800 m on Mount Arik held 10 Papuan, 2 Josephine's, 3 pygmy and 15 fairy lorikeets. This probably reflected the general scarcity of food sources which concentrates birds in small areas, and causes them to be blossom nomads over wide areas of forest" (Gregory pers. comm. 1996).

In the Southern Highlands, above Tari, Jonathan Newman saw only three birds, at 2,500 m (8,200 ft.), in September, 1990. In December, 1994, when I spent five days in the same locality, I saw Josephine's with certainty on only one occasion. A pair was feeding in the crown of a tall tree, in which *C. papou goliathina* was also present. An unconfirmed sighting occurred about 5:00 P.M. on the main road, not far from Ambua Lodge; the birds were in the crown of a lower tree and were tentatively identified by their shorter tails and smaller body size; there is a possibility that they were young *papou*, yet no *papou* with short tails were seen during five days of observation. In the same area *goliathina* was very common; *josefinae* was evidently present in very low density. Whether it is more numerous in Irian Jaya is not known, but when it first became known to science, Dr. Meyer made this observation: "I found this bird on my voyage to New Guinea in June 1873, on the west coast of Geelvink Bay, where it was seen near the sea-shore in large flocks; but in no other spot, during my residence in the island, did I meet with it" (Gould 1875–88).

In July, 1994, Gert van Dooren saw small groups of Josephine's lorikeets at an altitude of 2,000 m to 2,500 m (6,500 ft. to 8,200 ft.), near Lluerainma (15 km northeast of Wamena). In the lowlands (Baliem Valley) most of the trees had been destroyed by local people for cooking purposes, but on the higher slopes the trees had survived. There the Josephine's were shy and did not appear when people were climbing the path to the salt mines. When it was quiet, they flew over, their flight fast, calling a high-pitched *kris*. (Vocalizations are similar to Stella's lorikeet, but quieter.) When they were perched it was difficult to see them, even when you knew where they had landed in the tree. They appeared to be quite common there. The Dani people kill Josephine's, also female Stella's and birds of paradise, for head decorations.

Nesting: Pratt (1982) observed Josephine's lorikeet frequently leaving an epiphytic clump from which chicks could be heard calling; the location was Mount Bosavi. These clumps are probably more numerous than suitable holes in trees. Eggs laid by two captive-bred *sepikiana* females owned by G. van Dooren measured 21.8 x 19.0 mm and 22.2 x 19.4 mm (pair No. 1) and 22.2 x 19.3 mm and 21.8 x 18.3 mm (pair No. 2); all eggs weighed 3.5 g when laid (van Dooren pers. comm. 1995). At Palmitos Park four eggs from one female measured 22.5 x 19.5 mm, 22.5 x 18.8 mm, 23.3 x 19.0 mm and 21.0 x 18.3 mm; the other female laid larger eggs, 25.5 x 19.8 mm and 25.5 x 19.5 mm.

Status/Conservation: No information but probably stable, although uncommon. No conservation measures are known to be necessary.

Aviculture

Status: Uncommon

Clutch size: Two

Incubation period: According to Hubers (1993a), artificially incubated eggs hatch after "about 25 days." Those left with the parents are not incubated until two or three days or more after the laying of the first egg. In the pair at Palmitos Park, the second egg usually hatched 25 to 27 days after being laid. In some of the instances when both eggs from this pair hatched, hatching occurred 27 and 25, 26 and 25, 29 and 27, 29 and 27, 28 and 26, and 29 and 28 days after they were laid; a single egg hatched after 29 days. Incubation is shared by male and female.

Newly hatched chick: Weight, 3 g or 4 g; covered in long, white down; beak pinkish, or with brown tip.

Chick development: Day 9, back darkening with down feathers growing under skin, down on back still intact, eyes slitting; day 11, feet darkening; day 18, eyes half open, feet gray; day 19, dark gray second down erupting; day 23, eyes fully open; day 24, red feathers erupting on the upper breast and flanks; day 25, wing and breast feathers erupting; day 29, tail feathers opening; day 32, tail feathers ½ cm long; day 37, blue-mauve head feathers erupting; 42 days, fully feathered except for short tail. Chicks are ringed at 17 to 21 days with 5.5 mm rings; G. van Dooren ringed two chicks at 12 days with 5.4 mm rings.

Young in nest: About 45–56 days; one of the two young in a nest fledged at 47 days and the other at 62!

Chick growth

Below are weights of three parent-reared and a hand-reared chick hatched at Palmitos Park. Chicks number 1 and 2 were nest-mates, number 3 was a single chick, reared when the parents consumed a dry food in addition to their usual diet, and chick number 4 was hand-reared from 11 days. Chicks 1 and 2 had captive-bred parents, the female being the daughter of the original breeding pair, and chicks 3 and 4 were from the original pair, the female of which was wild-caught.

Josephine's lorikeet weight table

Age in days	Weight in grams			
	Chick No. 1	No. 2	No. 3	No. 4
hatched	3	–	–	–
1	4 fc	–	–	–
3	–	7	–	6 fc
4	7 fc	9 fc	–	–
5	9 fc	10 fc	–	10 vfc
6	11 fc	–	–	10 vfc
7	–	11 fc	–	13 vfc
8	12 vfc	–	–	11 fc
10	18 vfc	–	–	14 vfc
11	–	15 fc	–	15 vfc
12	19 fc	–	–	13/14
13	–	22 vfc	–	15/17
14	24 vfc	–	–	17/20
16	–	23 vfc	–	21/24
19	–	27 vfc	–	25/29
20	32 vfc	33 fic	–	24/26
21	42 vfc	–	–	25/29
22	–	27 fic	–	27/29
23	36 fic	–	46 fc	30 *
25	–	34 fic	–	31/36
26	48 fic	–	–	31/35
29	–	–	57 fc	27/41
30	–	37 fic	–	36/46
31	53 vfc	–	–	38/46
32	–	41 fic	–	41/46
34	–	47 fc	–	44/50
35	72 vfc	46 fc	–	45/53
36	68 fc	–	55 e	45/51
39	64 fc	–	–	47/53
41	–	–	–	49/59
44	–	64 fc	–	–
45	90 fc	–	–	54
46	–	–	–	58
50	left nest	left nest	–	–
53	–	–	left nest	independent

* chick slightly unwell
Crop contents: e = empty; fic = some food in crop; fc = full crop; vfc = very full crop.

Age of independence of hand-reared young:
41 days to about eight weeks.

General: Josephine's could be described as a slightly less flamboyant and relatively little known version of Stella's. It did not appear in aviculture until 1977 when a single bird was obtained by San Diego Zoo. Soon after, a few birds were imported into Europe. The first breeding occurred in Germany in 1979 by Hans-Heiner Dahne. Only small numbers reached Europe and these numbers have only just been maintained. At the time of writing there are a mere handful of birds in South Africa and probably none in the U.S.A. Initially very few females were available. Clayton (1993) recorded that when visiting a dealer to buy two pairs, he saw 22 males and only three females.

I first had the pleasure of working with this species at Loro Parque, Tenerife—and I was immediately enchanted. With its graceful lines, gorgeous plumage and vivacious ways, it has everything to recommend it to the lory enthusiast. It is sexually dimorphic, matures early (about 15 months) and nests readily. In actions and behavior, it is much like Stella's, although the display is slightly less histrionic. These active birds are always a joy to observe. One day, in the breeding center at Palmitos Park, I recorded a few minutes of action as follows:

A young male of one year is kept with an adult male of seven years. They coexist amicably until something happens which causes the older male to exert his dominance. For example, on placing one soaked fig in the cage, the older bird at once acted as though the cage was his territory and no interloper would take a share of the fig. The older male became very excited and, as Josephine's do at such times, flicked his tongue around the beak (too fast for the human eye to follow). He then started to jump on the spot, jumping probably 4 cm off the floor, or jumping away short distances and returning to the fig. Then he would move off very fast, hopjumping sideways and returning immediately in the same fashion. I placed another fig in the cage, hoping that the young bird might get a bite while the older one was feeding. But immediately he took a bite, the older bird moved to drive off the young bird from his fig. At one point he even placed his foot firmly on the fig, as though claiming ownership! When I lodged a third fig in the welded mesh, the older bird did his round of the 3 figs at great speed. His companion was never able to take more than a single bite. At his most excited he swayed from side to side in an exaggerated manner, his feet firmly anchored to the welded mesh floor, then jumped at his companion, half turning on his back as though to take up the extreme defensive position of a lory—lying on the back with the feet in the air. The young bird never seemed afraid, and he no doubt had a fig to himself when the older bird had eaten his fill.

When excited, Josephine's flick their heads quickly; sometimes they wave their long tongue outside the beak. In the courtship display, without moving the feet, they jerk their body from side to side and keep the head low, almost in line with the body. Display is as pronounced in females as in males. Copulation is preceded by vigorous head bobbing in both sexes. At such times the female lifts her wings very slightly (making obvious the yellow patch above the thighs) and the feathers of the thighs stand out.

The older male referred to above was one of three hatched at Loro Parque destined for the surplus list. I bought them. When I moved to Palmitos Park, three wild-caught females were obtained. Sadly, two of them were light-beaked birds in poor health (see **Beak Color**); the third was always light-beaked yet she was a prolific breeder. The lories in the breeding center were housed in a spacious building to protect them from the strong sun and high winds in the mountain location. A section of the roof above each row of cages was covered in green shade cloth, which gave the building an airy, outdoor feel. However, after a flood (a freak of the climate which might occur once every 40 years) the shade cloth was replaced with solid windows which could be opened. The pairs of small lorikeets were housed in welded mesh cages on metal stands. The cages measured 2 m (6 ft. 6 in.) long, 76 cm (2 ft. 6 in.) wide and 1 m (3 ft. 3 in.) high.

A glance at the records of the Loro Parque hatched male and wild-caught female shows all their young hatched (1990–95) between early March and mid-July. In the first two nests, the young were removed for hand-rearing; the female laid again 31 and 19 days after the young were removed. Thereafter, most of the young were parent-reared. The pair was usually double-brooded or treble-brooded in the first year when the young from two nests were hand-reared. For example, in 1991 the female laid on March 28 and 31; the second egg hatched on April 27, the first egg was infertile. The chick was removed for hand-rearing at 11 days because the weather was not warm and the parents were spending too much time out of the nest. The chick was removed on May 8 but the female did not lay again until June 20 and June 23. These eggs hatched on July 18 and 19. As another example, in 1995 the two young of the first clutch hatched on March 10 and 12 and left the nest on May 10 and 11. They were separated from their parents on June 16 because the female had laid on

June 9. A chick hatched on July 7 and left the nest on August 28. The male was always mainly responsible for the care of the young after they fledged and, as in all the red *Charmosyna* species, played a large part in the incubation of the eggs and care of the chicks in the nest. He was always excited when his young left the nest for the first time, especially if one of them was a female. He was so excited that "false-mounting" behavior would occur; in other words, he made brief attempts to copulate with her. It was always a joy to watch the family together for the parents' pride in their young was plain to see.

The first youngster from this pair, a female, hatched on March 3, 1990, and was paired with another of the 1988-hatched Loro Parque males. She laid her first eggs on February 25 and 28, 1991, when she was 52 weeks old. The nest was very wet and the embryos died about two-thirds of the way through the incubation period. She laid the first egg of the second clutch on May 22 and the second probably on May 25. They hatched on June 20 and 21. The chicks were well cared for and left the nest on August 19, at 60 and 59 days of age (Low 1991c).

In the U.K., Josephine's was bred by G. Clayton in 1992 (Clayton 1993); whether this was a first U.K. success is not known. Of the two pairs in his care, one pair had produced a chick at the time of writing. It was ringed at 18 days with a size P ring. Because it was plucked by the parents, it was removed for hand-rearing. At five weeks old there was no yellow on the flanks but when it feathered up after being plucked, the flanks were yellow. The food provided consisted of nectar, apple, pear, grapes and pomegranates. When the young one was being reared, sponge cake soaked in honey was also taken. Josephine's readily sample new foods. Most fruits are relished, especially pomegranates and cactus fruits. They also like raw carrot, tinned sweet corn and chickweed. A dry food is consumed with enthusiasm when young are being reared.

This wonderful lorikeet is in danger of being lost to aviculture. All those fortunate enough to keep it should strive to maximize breeding successes. A studbook might help to secure its captive future.

Papuan Lorikeet   *Charmosyna papou* (Scopoli 1786)

Synonym: Stella's lorikeet (subspecies *stellae* and *goliathina* only)

Native names: *Divu*; Fore, *Waiya*; Gimi, *Waiya*; Daribi, *Hade*

Description: In full plumage this species is unmistakable, being the only lorikeet with a long trailing tail. Its plumage differs little from that of Josephine's but it is immediately distinguished by the green upperside of the basal half of the tail. The wings and mantle are dark green, the thighs are black and green and the abdomen is black. The head is red; the elongated, shaft-streaked feathers of the crown are blue-gray and fall over the black area which extends from behind the eye to the nape. The underparts and undertail coverts are red and the abdomen is black; the underwing coverts are red. An unusual feature is the elongated tips to the primaries. The tail is green, tipped with yellow or, in the two long central feathers, yellow for about half the length; the underside of the tail is orange-yellow. The rump is blue. In the nominate race there is a yellow mark on each side of the upper breast, partly hidden by the wing, and a yellow patch on the thighs. In the female, the red feathers on the sides of the rump are yellow at the base and there is a yellow patch above the yellow area on the thighs. The bill is deep orange to coral red; the upper mandible is slender and curved; at the widest point it was exactly 1 cm in a male and 85 mm in a female *goliathina*. The iris is orange, as is the cere; feet are coral.

Length: Body length is about 17 cm (7 in.); total length with full tail about 40 cm (16 in.) or up to 42 cm in *goliathina*. A long tail feather from a four-year-old male *goliathina* measured 27.4 cm (10 3/4 in.); it was green near the base, dark gray for the next 9 cm and yellow for 6 cm at the tip. At the base it measured 1.7 cm and at the tip about 0.01 cm. At the next molt, a year later, the long feather measured 29 cm (11¼ in.). It seems this increases in length slightly every year. One mature wild-caught melanistic male measured 48 cm (19 in.) and the long tail feather measured 32 cm (12½ in.).

Weight: *C. papou papou*, maximum of 60 g, according to Bosch (1994)—but this weight seems low; *goliathina*, captive, 94 g to 116 g; wild 85 g (Diamond 1972). Four birds at Palmitos Park weighed as follows on the same day: wild-caught red male 96 g; brother and sister melanistic, four years old, 94 g and 107 g; 20 week old male melanistic 72 g. Bosch (1994) weighed four females and five males aged 4–40 months; their weights ranged from 86 g to 116 g, the latter being a melanistic female. Females averaged 3 percent lighter than males.

Immatures: They have the red feathers of breast and neck faintly margined with black; the black feathers of the abdomen are tinged with green. There is a variable yellow band across the underside of the secondaries. The tips of the primaries are not elongated. The central tail feathers are short—about half the length of those of adult birds. The bill is brownish orange; the iris is brownish; the feet are grayish pink, and the skin surrounding the eye is darker than in adults.

Sexual dimorphism: The female of the nominate race has a yellow spot above the thighs (see above); in the other subspecies she has a conspicuous yellow patch on the flanks and rump.

Subspecies: *C. p. stellae* differs from the nominate race in having the black on the nape extending further down; the male lacks yellow on the thighs, and both sexes lack yellow on the side of the upper breast. The long central tail feathers are yellow-orange toward the tip. There is a melanistic phase—common in *stellae*, apparently rarer in *goliathina* and *wahnesi*—in which the red of the head and upper breast is replaced by black or brownish black. The long feathers of the nape are dull blue at the front and greenish gray at the rear, or green-gray throughout. In the male the flanks and the upper back are red; in the female they are green. The underside of the tail is dull yellowish green or dull green. The plumage is variable; rarely, in some melanistic birds it is nearer to maroon than black. According to Bosch, the melanistic birds from central Papua New Guinea have the red undertail coverts mixed with black or they are almost entirely black, whereas in melanistic *goliathina* the under tail coverts are red; the upper side of the tail is green at the base, gray-green in the center and yellow at the tip. *C. p. goliathina* differs from *stellae* in having the long tail feathers yellow ftoward the tip. In both forms and sexes the thighs are variably streaked with yellow; *C. p. wahnesi* has a distinctive yellow band across the breast.

Bosch suggests that the nominate race might be a separate species. It is geographically isolated and is the form which is most different in appearance.

Immature birds of *goliathina* have a brighter blue patch on the crown than adults. Red females have yellow on the rump and flanks but not as bright as in adults, some of the feathers being tinged with green or red. Some red males have some yellow feathers on the rump. Melanistic males have an area of red on the back which is not as large as in adults. Most melanistic birds show a pink tinge on the feathers surrounding the beak. The black areas are strongly tinged with blue and/or green and the patch on the crown is blue, not gray. The tail is colored as in adults, but is only half the length. I described one melanistic male at 62 days as follows: shaft-streaked feathers on crown brilliant blue of the same shade as the rump; forehead, neck and underparts purple-tinged black with some green on the crown; purple is most obvious on the forehead and forepart of the crown, and near the vent; lores and feathers around the beak, purplish pink; red sides and under tail coverts; blue thighs.

Natural history

Range: New Guinea in mountainous areas. The nominate race is confined to the Vogelkop, Irian Jaya. *C. p. stellae* occurs in southeastern New Guinea as far west as the Angabunga River and the Herzog Mountains. *C. p. goliathina* is found throughout the central mountain ranges and *C. p. wahnesi* is confined to the Huon Peninsula (Forshaw 1989). Range limits are not well known. Bosch (1994) saw two captive *stellae* which had been caught in the Wau area, far west of Angabunga.

Habits: This lorikeet inhabits montane forests; it does not occur in open country. Although it usually occurs between 1,500 m and 3,500 m (4,900 ft. and 11,500 ft.), it is found right up to the tree line which can be up to 3,900 m (12,800 ft.). It is generally

fairly common, especially where there are flowering trees. In at least two localities in Irian Jaya, melanistic birds predominated at midmontane altitudes between 1,800 m (5,900 ft.) and 2,800 m (9,100 ft.). Diamond (1972) reported that melanistic birds predominated in the Eastern Highlands. Black plumage is an advantage in a cool climate as a black body heats up faster and radiates heat more efficiently than a light-colored body. Bosch (1994) states: "It is possible that the *goliathina* population can segregate through active vertical migration with more blacks higher, while the reds tend to remain lower."

Below Tari Gap, Southern Highlands Province, Jonathan Newman made the following observations in mid-September, 1990:

Common in open roadside areas and stunted forest, at 2,200–2,500 m, single birds and pairs. When traveling between feeding sites, flies low and silently (compared with other *Charmosyna* which fly high in loose twittering flocks). It is beautiful in flight; pairs fly close together with streamers whipping behind. Red birds were more common than melanistic by 4:1. Most pairs were of the red phase, fewer of red/melanistic; no melanistic pairs were seen.

This species was closely tied to trees and shrubs of the genus *Schefflera*. Almost all plants of this genus contained *C. papou*. Birds fed on flowers (largely on pollen) and fruit. *Schefflera* fall into two categories: those with flat umbels and those with inflorescences resembling sprays of millet in form. *C. papou* favors the latter, feeding 1–3 m above the ground.

During six days in the same location, in December, 1994, I observed Stella's lorikeets daily. My first sighting occurred on December 3, at 8:00 A.M. A red male flashed across the Tari-Hagen road through a small gap in the trees. The flash of scarlet with long tail streaming behind was unforgettable. Later that morning I visited a Huli settlement lower down, at about 1,700 m (5,570 ft.). As I watched the male warriors don their paint, wigs and feathers in preparation for a sing-sing, a pair of *papou* flew over. These men prize skins and feathers of Stella's and Josephine's for their headdresses. The next day there were many sightings of *papou* at about 2,500 m (8,200 ft.) just below Tari Gap from 8:00 A.M. onward. At about 9:30 A.M. a lofty white-flowering tree was attracting four species of lorikeets like bees around a honey pot. The *papou*, red and melanistic, evoked gasps of admiration as the red birds contrasted with the foliage and moss high in the trees. The melanistic birds were more difficult to observe. As I watched, I realized why captive birds of this species, as well as *josefinae*, *pulchella*, *goldiei* and *musschenbroeki*, are so adept at running up and down welded mesh, rather like mice. In their natural habitat they run up and down moss-covered trunks and limbs! Feeding in the same tree, Goldie's and Musschenbroek's were much less conspicuous. In contrast even to the Stella's, the scarlet and black male red-collared myzomela (*M. rosenbergii*) honeyeaters positively glowed! Red Stella's were feeding in a nearby tree, lower down, where they could be observed at only 9 m (30 ft.) distance without the observer getting a crick in the neck! *C. p. goliathina* appeared to be the most common lorikeet, but it was probably merely the most conspicuous. They caught the eye as they fed and were spectacular in flight; furthermore, that unmistakable call note identified them instantly, even if they were not visible. No other lorikeet makes such a distinctive sound! Later in the morning, higher in the mountains, red and black Stella's were observed flying down to a bunch of greenish *Schefflera* fruits.

All observations of birds in flight, except around feeding trees, were of single birds or pairs. Unlike some lorikeets in the region, large flocks are not reported. They do not fly high, generally at about tree canopy height, and flight is not usually sustained over long distances. Their wings are quite loud in flight; whether the elongated primaries contribute to the sound is not known.

Nesting: No information. Two eggs laid by captive *goliathina* measured 24.0 x 22.5 mm and 25.0 x 22.0 mm and weighed 6.83 g and 6.78 g (Bosch 1994); 2 laid at Palmitos Park measured 26.0 x 20.5 mm and 25.2 x 20.3 mm.

Status/Conservation: Fairly common; no threats to it are known and no conservation measures seem necessary.

Aviculture

Status: Fairly common

Clutch: Two

Incubation: The actual incubation period is 25 days; this was determined by placing a newly laid egg under an iris lorikeet who was incubating; reports of 28 days (Bosch 1994) or longer originate due to the fact that incubation does not commence immediately, even though one member of the pair may be in the nest; periods as long as 32 days from laying to hatching have been recorded by the author, although chicks usually hatch after 28 or 29 days. Longo (1985) recorded that the incubation period is 24–25 days and that incubation usually starts one to five days after the second egg is laid. Incubation is carried out by male and female.

Newly hatched young: Weight, 4.5 g to 5 g; longish down, especially on back (longer than in josefinae), usually salmon-colored on back, gray on forehead, white elsewhere, or entirely white; feet pink. In one nest a red chick was pink-skinned with peach-colored down, and a melanistic chick hatched with gray skin and gray down; it is not certain whether this always applies, but it seems quite likely.

Chick development: Day 4, upper mandible turning brown; day 5, both mandibles brown, down feather tracts visible under skin; day 13 to 21, eyes starting to slit; day 15 to 23, eyes open; day 16 or 17, ringed with 6 mm rings; day 26, first feathers erupt on breast; day 30, covered in dark gray second down (denser than in *josefinae*), blue crown feathers opening, also tail feathers and coverts; day 37, flank feathers erupting; day 41, most feathers of the head, wings and underparts are free of the sheaths and the second down is visible mainly on the rump and sides; day 47, color of rump feathers may indicate sex; day 53, fully feathered except under wings, females have rump and back yellow except for a green line down middle.

Young in nest: 56–64 days (Bosch 1994); about 60 days (Longo 1985); in one nest of author's pair, eldest left at 56 days and youngest at 62 days; young in one nest, 71 and 72 days (Brockner 1991).

Chick growth

Shows weights of four young from the same pair. Chicks No. 1 (melanistic male) and No. 2 (red female) hatched at Palmitos Park in 1995; No. 1 was reared by Josephine's until removed for hand-rearing and No. 2 was parent-reared. Both were single chicks. Nos. 3 and 4, both red females, hatched in the author's collection in the U.K. in 1996 and were parent-reared in the same nest.

Stella's lorikeet weight table

Age in days	Weight in grams			
	Chick No. 1	No. 2	No. 3	No. 4
hatched	4*	5	–	–
1	4	–	–	–
2	6 fic	–	–	–
3	7 fic	8 fc	–	–
4	10 fc	9 fic	–	11 fc
5	11 fc	–	–	–
6	14 fc	13 fc	22 fc	–
7	16 fic	13 fc	–	19 fic
8	17 fic	–	–	–
9	22 fc	18 vfc	29 fc	–
10	21 nf	–	–	–
11	20 ne	23 vfc	–	–
12	23 fic	–	–	–
13	27 nf	26 vfc	–	–
14	30 fc	31 fc	–	–
15	35 vfc	29 fc	–	–
16	34 fc	36 fc	–	–
17	43 fc	38 fc	–	–
18	44 ne	–	–	–
20	50 nf	43 fic	–	53 nf
22	–	49 fc	71 fc	79 fc
24	–	–	–	78 vfc
26	78 fic	64 fic	–	81 vfc
28	77 fic	69 nf	80 vfc	79 vfc
30	83 nf	69 nf	84 vfc	85 vfc
32	91 nf	63 e	86 vfc	85 fc
33	93 fc **	71 fc	90 vfc	84 fic
34	–	83 fc	86 nf	94 nf
36	81/91	83 fc	92 fic	–
38	79/91	86 fc	–	95 nf
40	81/93	91 fc	103 fc	101 vfc
42	84/93 ***	90 fc	101 vfc	101 fc
43	93 independent	–	94 nf	97 nf
44	90/100 (self-fed)	87 fic	94 nf	96 nf
46	93	93 fc	95 nf	92 ne
48	96/102	–	–	–
50	102/112	91 nf	–	–
52	99/113	93 fic	–	–
54	100/109	–	–	–
55	99	–	–	left nest
56	99	–	left nest	–
60	93	–	–	–
64	90	left nest	–	–
68	89	–	–	–

vfc = very full crop; fc = full crop; nf = nearly full; fic = some food in crop; ne = nearly empty; e = empty.
* hatched in incubator, thus lower weight than normal
** removed for hand-rearing
*** offered warm food in dish for first time; filled crop.

Color inheritance: Inheritance is sex-linked and the melanistic form is dominant.

- Red male x melanistic female = melanistic males and red females.
- Melanistic male x red female = all melanistic young.
- Melanistic male x melanistic female = all melanistic young.
- Melanistic male bred from a red male x melanistic female = melanistic males and red and melanistic females.
- Melanistic male bred from a red male x red female = melanistic and red males and melanistic and red females.
- Red male x red female = all red young.
(From Bosch 1994.)

General: It was in the late 1970s when Stella's lorikeets (*goliathina*) were first imported commercially into Europe and the U.S.A. Numbers were small and the price was high. Inevitably, after a couple of years the price fell, although large consignments of this species were rare. Importation continued until about 1990. Many of the birds which came out of quarantine were in poor health, as their light beaks indicated. Most of these birds died after a few months with swollen livers. The reason for the condition is not clear. It has been suggested that overdosing of antibiotics during the quarantine period was the cause. Certainly I have never seen it in a captive-bred bird. (See **Beak Color.**) Aviculturists who obtained Stella's in good health were enchanted by them. They have all the attributes of the perfect aviary bird: exceptional beauty, grace, playfulness and readiness to breed. In addition, they are sexually dimorphic and neither noisy nor destructive. As well as these qualities, they are usually friendly and endearing. They are quite extraordinarily vivacious and charming. Like Josephine's, they are very quick in all their movements. They often move sideways by hopping very rapidly, or they may walk. An exaggerated walking gait is often used during display. Adult males are constantly showing off, as though well aware of their unique beauty. In display they will sway from side to side without moving the feet, or the male will stretch upward and lift first one foot and then the other, with a dancing motion. During display or when excited, the erectile feathers of the nape may be raised.

Bosch noted of captive birds: "When hopping on the ground they get easily into a display and go around with erect body and raised head in huge sideways leaps to the keeper or partner, a behavior which has little chance to be performed in the wild." Brook (in Page, 1910) commented that "the birds proceed by jumps or bounds, whether on the ground or on horizontal perch." In some respects their extreme activity and exaggerated behavior reminds me of birds of paradise; it is almost as though they wish to draw attention to their beauty. The male of my breeding pair is just as ready to display to me as he is to his female. I can never look at a Stella's in good condition without marvelling at its colors and form. Melanistic birds are equally as beautiful; their near-black plumage reminds me of velvet (as does the black plumage of some species of birds of paradise). Stella's lorikeet is undoubtedly one of the most beautiful birds in existence.

The first, and among the few, to specialize in breeding Stella's were Joe and Margie Longo, who at the time lived in Washington, U.S.A. They had a range of ten flights, each one of which contained a pair of Stella's. The flights measured 3.3 m (11 ft.) long, 91 cm (3 ft.) wide and 2.1 m (7 ft.) high. The roof was covered except for the last 1.2 m (4 ft.). Each flight contained a bird bath which the pairs used daily during the summer months, especially when incubating. The nestboxes were situated in the service passage and measured 23 cm (9 in.) square and 41 cm (16 in.) deep. There were no visual barriers between the pairs. It was believed that "the interaction between the flights seems to stimulate breeding activities in some of the younger pairs." Young fledged at about 60 days; they were removed from their parents by 90 days. Nesting could occur at any time and usually started in December and ended in July or August. Stella's will nest throughout the year, except perhaps in very hot weather, which they do not enjoy. As mountain birds, the often dismal climate of the U.K. and cooler parts of Europe suits them well. Indeed, Longo recorded that even in the coldest winters they did not seem uncomfortable in outdoor aviaries. They even appeared to enjoy the snow "getting down on the ground and rolling in it like kittens at play." However, if they are exposed to low temperatures, they should have an enclosed shelter and nestbox in which to retire at night. I have heard it suggested that melanistic birds breed more successfully than red birds in the U.K. because they brood their young for a longer period, thus aiding their survival in cold weather. All keepers of melanistics should be aware that they cannot tolerate high temperatures. Over about 40°C (104°F) they are liable to die from heat stress, although red Stella's in the same location will survive. At temperatures in the region of 37°C (100°F), they look uncomfortable and are breathing with the mouth open.

My breeding pair consists of a wild-caught red male and a melanistic female who was hatched at Palmitos Park in August, 1991. Her first chick hatched in May, 1993. In the breeding center at Palmitos Park most of the eggs from this pair were broken at varying stages during the incubation period if not fostered to another nest. It was a noisy environment, with more than 200 birds in that particular building, including Amazons nearby.

When I moved back to the U.K., this pair accompanied me. They were then housed in an outdoor birdroom containing only lories. It was a totally different atmosphere—smaller and quieter. In just over one year of residence, five clutches were laid. The birds arrived on October 26, 1995. The first eggs were laid on January 17 and 20, 1996, and incubation commenced on January 23. It lasted the full period but in these eggs the embryos died, perhaps because the heating system provided an environment which was too dry. The female laid again on March 10 and 13, and chicks hatched on April 7 and 9. They left the nest on June 2 and 3 and were removed on June 18—earlier than is my usual practice as I was to be absent for two weeks. The female laid again on July 1 and 4 and chicks hatched on August 1 and 2. They left the nest on September 26 and October 3 and were removed on October 26. The female laid again on November 17 and 19; the chicks died just before they were due to hatch; again, this occurred when the heating had to be kept on day and night. Three clutches in a year is sufficient. The intention is to remove the pair to an outdoor aviary in the spring. As aviary birds, Stella's are always moving rapidly in flight or by foot. The yellow rumps of females flash like beacons as they fly away from you. They are extremely active birds. A flight of a minimum length of 4.5 m (15 ft.) is essential to show them to advantage.

In the U.S.A., Jan van Oosten had an extremely prolific pair of wild-caught melanistic Stella's. The male was almost maroon in color, with some red markings on the breast. The female had a red spot behind each eye. Over a period of nine years, they reared 64 young, 32 red and 32 melanistic; each clutch consisted of one of each. After the female

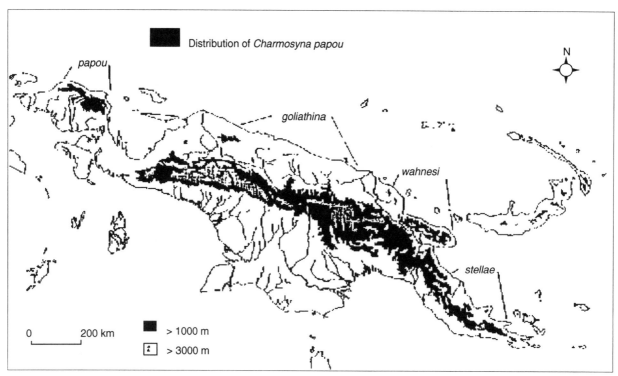

Range of *C. papou* (after Bosch,1994).

died, a red female was introduced to the male. Although they seemed compatible, copulation was not observed and no eggs were laid (van Oosten pers. comm. 1996). It is my belief that the pair bond is so strong in some pairs which have been together for years, that in many cases it takes a long time for the surviving bird to fully accept a new mate.

Stella's can be aggressive when breeding. My captive-bred female will try to bite my wrist daily (and often succeeds) when her chicks are small. The fault is probably mine, because the nestbox is too close to the feeders. On the other hand, the wild-caught male has not shown the slightest sign of aggression under any circumstances. Other breeders have also experienced nesting birds trying to bite them.

This species is very willing to try new foods and takes a wide range except, of course, seeds. All fruits are sampled; chickweed and raw carrot are among their favorite items. Bosch (1994) described the nectar which had proved successful for this species over a long period. Two heaped tablespoonfuls each of honey and pollen, 4 heaped tablespoonfuls of cereal-based baby food (without milk and sugar), half a tablespoonful of inactivated yeast powder (such as PYM), a very small pinch of mineral supplement and 2 drops of a multivitamin sup-

plement (high in Vitamin A) are stirred into half a liter of warm water, and the container is filled up to 1 liter with cold water. Fruit, pyracantha berries and chickweed are also eaten.

Stella's lorikeets (*goliathina*) are now well established in aviculture and are, of course, one of the most sought after lorikeet species. They are not unknown in Australia. They were first imported into the U.K. (subspecies *goliathina*, as depicted in H. Goodchild's exquisite plate in the March, 1910, issue of *Bird Notes*) as long ago as 1907. Walter Goodfellow collected a pair for Mrs. Johnstone, and 14 more birds in the following two years. In 1910, two pairs, which belonged to E. J. Brook in Scotland, reared young, although one of the females died in the process. The species was then unknown in aviculture for decades, perhaps until 1969 when San Diego Zoo obtained a single bird. In the summer of 1972, Mrs. S. Belford imported into the U.K. a single *C. papou papou* which she kindly lent to me for nine months. A young bird, still showing the dusky tinge on the feathers of the breast, he was engagingly tame. If I put my face close to his cage, he would push his slender head through the bars and stretch his tongue to its limit in an effort to reach my face. This species has an exceptionally long tongue.

He would wave it around his beak and could effortlessly reach his cere with it. He was easily induced to display by whistling to him, when he would stretch to his full height and bow and hiss and make a snapping sound with his beak. The yellow feathers on the thighs then became prominent and seemed to stand away from the body. I recorded at the time that the rump was red with a narrow blue line down the center. There was a small patch of yellow on each side of the breast which would have made a crescent shape if joined up (Low 1972). When the weather was warmer in the following year I kept this bird in an outdoor aviary. Looking at a slide taken at that time, when he was in his first full plumage, the tail is very long, as long as that of *goliathina*. He then went to Raymond Sawyer and lived in the huge planted aviary which is the show-piece of this well-known collection. Vogelpark Walsrode, Germany, has also kept this subspecies, which is perhaps now unknown in captivity. *C. p. wahnesi* is almost unknown; only a photograph (in Thomas Arndt's *Lexicon of Parrots*) indicates that it has been kept—but the location of this bird was unknown. I have seen *stellae* only in a private collection in New Guinea.

A final thought: no painting or photograph can "do justice to the glowing living beauty of this indescribable species" (Page 1910). You must see it for yourself.

Genus: *Oreopsittacus*

The single and diminutive species in this genus is very closely related to *Charmosyna*. It differs in two major respects: (1) the additional two tail feathers—14 in all, a total shared with no other parrot; and (2) the exaggerated length of the upper mandible which is apparent even in newly fledged young. Its voice is also weaker than that of a *Charmosyna* and has been described as "finch-like." Sexual dimorphism is pronounced in adults.

John Gould, writing in 1876 in *The Birds of New Guinea and the adjacent Papuan Islands*, commented thus on the number of tail feathers: "This peculiarity would almost be sufficient to place it in a separate genus; but this I cannot bring myself to do, in the face of its evident affinity to *P. placens* [*C. placentis*] and *P. wilhelminae*: but when we consider that certain snipes are still retained in the genus *Gallinago* which present similar variations in the number of tail feathers to that exhibited by these lorikeets, it is not unreasonable to keep the latter in one and the same genus."

This view is interesting, but Gould probably never observed *arfaki* alive. Had he done so, the differences in behavior between *placentis* and *arfaki* might have made him change his mind. Incidentally, Gould's first sight of a skin of *arfaki* caused him "no little astonishment." He exclaimed: "Fancy a little bird, scarcely bigger than a Bearded Reedling (*Calamophilus biarmicus* [*Panurus biarmicus*]) with a tail like that of a Minivet (*Pericrocotus*), and exhibiting a silvery tear-mark running down a cheek of smutty blue."

In the wild the whiskered lorikeet descends to the ground to drink and perhaps even to feed, which is unusual in lorikeets. In captivity it needs a plentiful amount of green food to maintain good condition. Do these two facts provide a clue to the reason for the unusual beak shape? If its diet was similar to that of *Charmosyna* species its beak would not differ. This might mean that it feeds extensively on, for example, seeds of a certain plant.

As a montane species, its chicks have thick down, similar to that of Stella's lorikeet (*Charmosyna papou*), which shares its habitat. Unfeathered chicks resemble those of Stella's, except for the beak shape. Neff (pers. comm. 1996) observed that chicks are similar to *Charmosyna* in their feeding behavior.

Whiskered Lorikeet *Oreopsittacus arfaki* (A. B. Meyer 1874)

Synonym: Plum-faced lorikeet

Native names: Fore, *Túshuke*; Gimi, *Gígi* (Diamond 1972)

Description: The unique feature of this species is the line of white feathers from the lores to the outer part of the cheeks—which provides the common name. The male has crown and forehead scarlet and the lores and cheeks mauve. As pointed out by Fred Barnicoat, "the white is vibrant and the streaks seem almost raised above the dark purple cheeks" (Barnicoat pers. comm. 1994). The abdomen is dull red, also the lower part of the flanks; underwing coverts and the sides of the breast are brighter red; there is a yellow band across the underside of the secondaries. The tail is distinctive—and unique for its 14 feathers. (All other parrots have 12 tail feathers.) The tail is beautifully marked and colored, being green above, tipped with rose red; on the underside it is rose red marked with dark gray transverse bands near the base. The beak is black and uniquely shaped, with a long, curved upper mandible, similar to most *Charmosyna* species but more pronounced. The iris is dark brown and the legs and skin surrounding the eye are dark gray.

Length: 15 cm (6 in.) (Forshaw 1989); 17–18 cm (7 in.), *major* (Neff 1992).

Weight: 16–23 g (Forshaw 1989); 20–21 g, *major* (Neff 1992); 2 males 22.3 g and 22.5 g; females 21.7 g, 21.8 g, 22.0 g and 22.3 g, *grandis* (Diamond 1972).

Immature plumage: Duller throughout, with red (or orange in females) only on the forehead; the crown is green. The white markings are duller and less clearly defined, especially in the female. The cheeks are otherwise greenish with a tinge of mauve in young males. Neff (1992) described the male (*major*) as having a broad red frontal band, and chin and cheeks dull reddish violet to rust red. The female's frontal band is narrower with an indistinct separation of the green of the head. The chin and cheeks are dusky gray-green. Females may be sexually mature before they attain full adult plumage. Neff (pers. comm. 1996) had two females which still had some reddish feathers at the sides of the forehead rearing young. The first red feather on the head of a young male hatched on September 11, 1995, was visible on January 12, 1996. To try to identify the sexes Neff plucked a few feathers from the head of some young birds a few days after they fledged. The new feathers were green but a few days later he examined the feathers more carefully to find that the tips of the feathers were green but the centers were red! They were males. One should therefore wait for three weeks after the young fledge before feathers are removed for the purpose of sexing.

Sexual dimorphism: The female lacks the scarlet forehead and crown.

Natural history

Range and subspecies: High mountains of New Guinea in the central ranges. In the west, the nominate race occurs in the Vogelkop, Irian Jaya; the subspecies *major*, which is distinguished by the scarlet-tipped tail (tipped rose red in nominate race), is confined to the Snow Mountains of Irian Jaya, and *grandis* (distinguished by the entirely green abdomen and lower flanks), occurs in the mountains of Papua New Guinea, eastward to the Huon Peninsula, and southward and eastward at least as far as Okapa (between Mount Hagen and Bulolo), where it was recorded by Diamond (1972). Coates (1985) states that it occurs as far west in PNG as the Hindenburg and Victor Emmanuel Mountains,

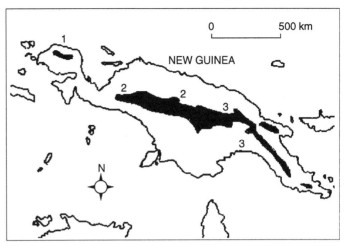

Range of the whiskered lorikeet (*Oreopsittacus arfaki*).

1. *Oreopsittacus a. arfaki*
2. *Oreopsittacus a. major*
3. *Oreopsittacus a. grandis*

where specimens were collected by Gilliard in 1954 (Gilliard and LeCroy 1961)—but there is little information on how far east it ranges. Ian Richardson, president of the Papua New Guinea Bird Society, has seen it in the mountains near Port Moresby, at Myola (Richardson pers. comm. 1996).

Habits: This is a mountain species, most at home in the moss forests. Diamond notes that "its vertical distribution on a given mountain seems to correlate with the distribution of this vegetational type. On Mount Karimui [near Okapa], where the cloud level is lower than in most other parts of the Eastern Highlands, *Oreopsittacus* descended to 5,600 ft. On Mount Michael, where the trees were heavily mossed only above 8,700 ft., *Oreopsittacus* was found only above this altitude and occurred up to about 10,500 ft. in the moss forest. On Okapa I collected up to 7,500 ft. and found neither heavy moss nor this parrot, but the Fore were familiar with the '*túshuke*' and said that it lived somewhere on top. It occurs in flocks of two to six and frequents flowering trees."

Forshaw (1989) observed whiskered lorikeets in the same region, the southern slopes of the Hagen Range. In disturbed *Nothofagus-Podocarpus* forest at 2,800 m (9,200 ft.), it occurred in small flocks, often with honeyeaters, flowerpeckers and other lorikeets, feeding in flowering trees.

Following a week spent in the area of Mount Wilhelm, at 4,720 m (15,500 ft.) the highest mountain of Papua New Guinea, Dave Coles' outstanding memory is of a pair of whiskered lorikeets. They were feeding on an umbrella tree (*Schefflera*) not more than 3 m (10 ft.) from where he was standing. "The pair was so engrossed with their feeding that they paid little attention to us and we were able to watch them for a good 5 minutes before they flew off, beaks covered in pollen. This must surely be one of the prettiest of all parrots" (Coles 1985).

I agree! And I had an almost identical experience just below Tari Gap, in the Southern Highlands. I set off early one morning with a couple of friends and Joseph Tano, the bird guide from Ambua, in search of this species. At about 6:30 A.M., we were not far below Tari Gap, at about 2,750 m (9,000 ft.). The sky was dull and the outlook was not promising, but very soon a pair of whiskered shot across our path. Having the benefit of a vehicle, we returned to Ambua Lodge for breakfast, watched many lorikeets feeding at a lower altitude, then went up again to the higher elevation to look for whiskered—that first glimpse having been so tantalizing. Here the vegetation was more stunted, more shrubby, than only a couple of kilometers below. We stopped where a couple of palms formed a backdrop for a low *Schefflera*. Suddenly, a pair of whiskered lorikeets, which had been feeding in dense shrubby vegetation not far from the road where we stood, flew toward us. They landed in the top of the *Schefflera*—which was only about 3 m (10 ft.) high and about 6 m from the road. Despite the excitement which their presence created, they fed unconcernedly for several minutes on the top of the *Schefflera*, where there were fruits. We had a wonderful view. They were exquisitely delicate and beautiful, the different plumages of male and female clearly visible through binoculars.

When they departed, we drove up to 2,745 m (9,000 ft.), where we ate our sandwiches; it was colder there and the sun burned our faces in the thin air. We drove down again, stopping near the same *Schefflera*. Suddenly a female Stella's (*goliathina*) alighted in the top of it; I was close enough to obtain a recognizable photograph with a 300 mm lens. When she departed, we sat on the ground in front of the same *Schefflera* in the hope of seeing more lorikeets. When a pair of whiskered—perhaps the same pair—arrived soon after, we were elated.

Two days later we returned to the same spot. It

was a lesson in not taking good fortune for granted! The weather was less sunny, more overcast. We returned to the same *Schefflera* and I set up a camera with a 500-mm lens on a tripod. But no birds came. Apparently this particular *Schefflera* was no longer attractive to the lorikeets. We walked slowly down the road as the sky darkened. A male ribbon-tailed bird of paradise (*Astrapia mayeri*) was feeding on *Schefflera* fruits high up in a tree; I watched him for ten minutes. Later we followed a deep red mountain firetail finch (*Oreostruthus fuliginosis*) carrying a long piece of grass to his nest. Away in front of me, on the side of the road, I saw a flock of small birds moving about. Assuming that they were hooded mannikins, I pointed them out to Joseph. "Plum-faced!" he exclaimed. He had already told me that they often alighted on the roadside to drink from the little streams and trickles of fresh water there. It had seemed an unlikely habit for a lorikeet, but as we moved quietly towards them, I knew that this was true. Although they allowed a fairly close approach when feeding in foliage, they took off long before we were near; no doubt they felt vulnerable on the ground. Later, a group of six whiskered lorikeets flew overhead, and later still, we crept close to a pair sitting quietly in a tree. They were not high up, only about 2.7 m (9 ft.), but the dense foliage shut out the light. They sat and preened. As I write, I look at a photograph that captured that moment. It shows two tiny objects blending with the leaves, only the red crown of one and the red tail-tip of the other making minute contrasts to the green foliage which fills the picture.

On another day at about 9:00 A.M., we saw these lorikeets at about 2,600 m (8,500 ft.) or lower, flying overhead. Later, we flushed three close by, from low in a grassy bank. What were they doing? Feeding on insects? Joseph told me that he had never seen them eating seeding grasses. The Kalam people have observed that they sometimes come down to the ground to salt pools (Majnep and Bulmer 1977)—but nowhere else have I seen mention of their habit of visiting the ground.

At the same locality, in August, 1990, Jonathan Newman saw whiskered lorikeets only once: three birds in flight over alpine grassland at 3,000 m (9840 ft.). He did not see them in flowering trees; neither did I. However, when Peter Them from Denmark went to the same location, with the same guide, Joseph Tano, in 1996, Joseph drew his attention to a small group of whiskered lorikeets feeding on blossoms. They were biting and pulling at the side of the corollas, and either detaching the corolla tube or tearing away about half of it. Then they fed on the nectar (and perhaps also the pollen) which was exposed between the ring of anthers and the long central style. This behavior would be unlikely to bring about pollination as the lorikeets did not come into contact with the flower stigmas.

In April, 1995, Jan van Oosten stayed at Ambua Lodge and went up to the Gap, but no whiskereds were observed. Carl McCulloch in California told me that a friend had videotaped this species eating moss [lichen?]. On the video they are apparently depicted feeding 60 cm (2 ft.) from the ground.

For me, the whiskered lorikeet epitomizes New Guinea—more than any bird of paradise. Yet this is not a typical lorikeet. So diminutive and slender is it! So active and so fast in its movements. It has as much substance as a finch. Its call has been described as a plaintive twittering on the wing or a soft squeaking while feeding. Diamond (1972) described it as a short, weak, repeated *ts* "weaker than that of other montane lories and more suggestive of a warbler than of a parrot." Despite its small size, it apparently thrives in a climate which is often cold and usually wet. It must be hardier than its porcelain appearance would suggest.

Nesting: Kalam people state that it roosts in epiphytic moss (Majnep and Bulmer 1977). These clumps of moss would make perfect nest sites for such a small bird. A near-confirmation of this comes from Australian ornithologist Len Robinson who has spent hundreds of hours observing parrots in the wild. On February 11, 1978, he made the following entry in his notebook:

> Tomba Gap (via Mt. Hagen) at about 9,000 ft. (2,700 m). While observing a male Ribbon-tailed Astrapia (*Astrapia mayeri*) a movement caught my eye. A pair of Plum-faced Mountain Lories was perched on a thin, dead hanging branch of a large tree. The lories were about 50 ft. away. The female flew to the under-surface end of a thick, deeply moss-covered limb. The limb angled down approximately 45 degrees from horizontal. The female paused, then rejoined the male. By their actions I suspected a possible nest. After a further 5 minutes the male

flew to an adjacent tree. The female became a little restless and appeared as if she would depart. However, she started to climb about on the dead twigs. Her behavior now certainly indicated a nest nearby. After a couple more minutes, she flew to the same spot on the thickly moss-covered limb and disappeared into the thick pile of moss. I am of the opinion that there was not a hollow in the actual wood of the limb but rather that the parrots had formed a cavity in the thick, closely-compacted moss.

Mr. Robinson could not see into the end of the limb—and the lorikeets could have excavated a hole in the decayed end—but he was totally convinced that they were nesting inside the thick moss.

There is no record of eggs being found in the wild. Two eggs laid by captive birds were round-oval in shape and weighed 2.48 g and 2.50 g (Neff 1992). Two others measured 17.0 x 15.3 mm and 17.0 x 15.1 mm (Neff pers. comm. 1996).

Status/Conservation: Common in moss forests, that is, usually above 2,000 m (6,500 ft.). No conservation measures are necessary. At the time of writing, there are no threats to this species.

Aviculture

Status: Rare

Clutch size: Two

Incubation period: 21 or 22 days (the first egg in a clutch had hatched by 5:30 A.M. on day 22 and the second egg hatched after 21 days at 8:30 P.M. (Neff 1992).

Newly hatched chicks: Weight, 1.67 g and 1.82 g; the down was gray-white and 1 cm long (Neff 1992).

Chick development: In parent-reared young, at four to five days the developing second down could be seen as dark spots under the skin; at ten days the eyes were slitting; at 10–12 days the chicks were ringed with 3.5 mm rings (4 mm was too large—too wide, and capable of being removed from adult birds!); at 13 days the second down was fully developed; at 15 days the yellow tip to the tail was apparent and the flight feathers were erupting; at 24 days the first red feathers on the forehead appeared and at 32 days they were nearly fully feathered, except on the back (Neff 1992).

Young in nest: The two young described above left the nest at 38 and 39 days, but could not fly until 44 days, when their tails measured 30–33 mm long.

Chick growth

Weights of the young described were as follows:

Whiskered lorikeet weight table

Age in days	Weight in grams	
	Chick No. 1	No. 2
12	10 g	10 g
24	19 g	19 g
37 (peak weights)	22 g	23 g
44	19 g	–
45	–	20 g

General: There is no record of this species in captivity until the late 1980s when whiskered lorikeets began to appear on the lists of exporters in Djakarta and Singapore. But either the trappers or the exporters failed to keep them alive. In 1990 or 1991 a few birds reached Germany. Fortunately for this species, some were acquired by Ruediger Neff of Fichtenberg—an exceptional aviculturist. By November, 1995, he had reared 44, including third generation young! All the wild-caught birds were still alive and producing three clutches a year! Lest this should be interpreted as this species being easy to breed, this is unlikely to be the case. Others have not been so successful, the usual problem being the death of the chicks with full crops when half-feathered. Undoubtedly Mr. Neff's success is due to the diet, which he described thus:

> The liquid food must include an essentially higher proportion of glucose (dextrose), similar to that used for Hummingbirds. The dry food, which contains bee pollen, oat cereal, glucose and brewer's yeast [see **Dry Diets**] is given on halved grapes daily. The other important point is to provide sufficient greenfood. Twice a day I attach fresh chickweed to the twigs of a branch. The birds are very eager for the green leaves and the seed capsules. The crops of young in the nest often appear quite green through the skin. Without green food, Whiskered Lorikeets do not have glossy, bright plumage and are less lively (Neff pers. comm. 1995).

Ruediger Neff told me: "This curious, inquisitive and playful species is a highlight for lorikeet enthusiasts." I agree! I have never kept them—only

had the pleasure of watching those at Loro Parque. What enchanting little birds they are! So quick and active and beautiful. Fred Barnicoat described the pair in his care in South Africa as follows:

> They are the most delightful of birds, being extremely lively and moving with a tit-like vivacity, constantly twittering and interacting with each other. They alight on a perch with a characteristic bobbing movement, at the same time flicking their tails to one side and immediately hop along the perch flicking their tails to the other side. Sometimes one will hop right over the back of the other bird. They often move right around when they are perching into an upside down position and up again in a sort of somersault movement.
>
> Pair bonding between them would appear to be exceptionally strong. When they were first put out into the aviary in January 1994 the hen could not fly at all. It was touching to see how the cock would not leave her alone for one moment but seemed anxious to teach her how to jump from one twig to another a little higher and so eventually to reach perch height in the shelter. He would be twittering in encouragement all the time, and she was replying to his calls. I doubt that she would have survived on her own (Barnicoat pers. comm. 1994).

Mr. Barnicoat commented that the birds in both consignments imported were in very poor condition, the feathers being stuck together in streaks. The liquid food was all over their feathers, possibly as a result of them bathing in it. Fortunately, because they were so fond of bathing, some of them cleaned themselves so that they could fly again. However, losses were extremely heavy. When a pair was put into a planted aviary, the male died suddenly within two hours, probably from the stress of the move.

Mr. Barnicoat recorded that initially the pair in his care roosted with their sides touching on a fine twig as high as they could get; in due course they took to sleeping in an L-shaped nestbox of the size and type provided for Gouldian finches. It had a thick layer of wood chips at the bottom, which they kept very clean. At night they were sensitive to disturbance and one night left the box in the dark because there was movement in the shelter—unlike other lorikeets. They were therefore placed in a small cage for the night and care was taken to ensure that they were not disturbed again at night.

It is very early in the morning, as dawn is breaking, that they come out of the shelter into the flight and jump excitedly about in the fine twiggy branches provided in the flight. At this time I have occasionally seen them mating. They become active ahead of all my other birds. They do not like strong sunlight and once the sun gets hot, they go into the shelter and eat, flit quietly about in their branch and sometimes doze on a branch, side by side. From midafternoon they come out into the flight for another active period with much twittering. They are the very last of all my birds to go into their shelter, and they prefer to do this at the very last moment when it is almost too dark to see and then they dive into their nestbox with a determined movement.

They are not tame, yet they never panic when I enter their aviary, provided that they can keep contact with each other, which they always do by their twittering calls. One of them occasionally gets through the door into the adjacent aviary, and then they panic until reunited. This is easily arranged by opening the door as they are so anxious to get back together again. I provide a dish of nectar as one would feed to sunbirds, and a dish of very thin lorikeet "porridge." They much prefer the former, and only when the nectar is finished might they be forced to take some of the thicker mixture. Though it was found in the quarantine station that they took more apple when it was grated, the pair in my aviary much prefer a half apple spiked on a twig.

The small size of these lorikeets is very striking. Though their length is almost that of a red-flanked or fairy, they are so slender, that they are really about half of that size. It is more like keeping some sort of finch than any parrot-like bird. Their high-pitched twittering is also very unparrotlike. Their extreme diminutiveness is something I find enchanting. I have watched these tiny birds with great concern on some very cold days when the temperature has even been below freezing. They do not seem to feel the cold, and in the frosty early dawn they are as keen to get out into the flight and they flit about as actively as ever, their water dish frozen solid! However, if they were not in good condition, any setback suffered from exposure to cold would be irreversible due to their delicate constitution and high metabolic rate.

Their passion for water is remarkable, even for a lorikeet. When a fine spray is turned on them they become more active than at any other time. Apart from hopping about in the greatest excitement, they love to swing into an upside down position on their perches and extend their wings ecstatically. At this time they also like to move along the wire roof of the flight in an upside down position or run up and down the wire sides in a somewhat rodent-like manner, and revel in becoming absolutely drenched to the extent that they cannot possibly fly. They adore fine rain, and on the very occasional cold, misty afternoon we have experienced on the Witwatersrand this winter, they have been the only birds to be out in their flight virtually revelling in the cold damp atmosphere (Barnicoat pers. comm. 1994)

Another aviculturist who is fascinated by this species is Hans Walser from Switzerland. First he obtained a male and later a pair, of the subspecies *major*. The three birds were kept in a 3 m (10 ft.) long aviary with small *Charmosyna* lorikeets. After 11 months, one male was feeding the female. In the evening the female became aggressive toward the other birds in the aviary. The female laid a single egg in a small horizontal nestbox. It hatched on December 20, 1991. A year later three young had been reared from this pair, plus three more from a pair bought in January, 1992. Hans Walser observed how clean the nest was while the chicks were being reared. Then he discovered why. He used horizontal nestboxes and the chicks would come to the entrance hole and squirt their feces outside!

The diet offered was as follows: at 6:00 A.M. Nekton Lori; at 8:00 A.M. a mixture which included Bimbosan baby food, wheat germ cereal, oats, millet and three-minute-maize; 3 dessertspoonfuls of this mixture were added to ½ liter of water and cooked for three minutes. Fruit sugar and cane sugar were then added. When the mixture cooled, CeDe Lori, meat extract and multivitamins were added. This food was fed warm (Walser 1993a).

Anyone fortunate enough to keep whiskered lorikeets should adhere closely to the observations of Fred Barnicoat and Ruediger Neff when working out the conditions which will suit them best. If special attention is not paid to the fact that they cannot be treated in the same way as other lorikeets, I fear the exquisite little whiskered will die out in aviculture.

Genus: *Neopsittacus*

The two members of this genus are among the most distinctive of lorikeets in appearance, behavior and dietary (captive) preferences. Indeed, Salvadori (1891) placed them with the fig parrots. Montane species, they occur only in New Guinea. The powerful bill of the Musschenbroek's indicates a diet which is less dependent on nectar and pollen than that of other lories. The brushes on the tongue are poorly developed. The two species are sympatric in many areas but no competition for food occurs as their diets differ. Little is known about their natural history and their nests have yet to be described. They are very active, with a jerky manner of moving, and are able to run very fast on any surface, up or down.

As avicultural subjects they are not well known and seldom easy to breed, but very attractive. They have striking red underparts and underwings; the tail is quite long. The voice is high-pitched.

Chicks are almost unique among the Loriinae in that in the second down stage they have a white patch on the nape, which contrasts with the otherwise dark gray down; such a white patch is faintly indicated in the dusky lory. A down nape spot also occurs in some Australian parrakeets, such as the blue-bonnet (*Psephotus haematogaster*).

Musschenbroek's Lorikeet *Neopsittacus musschenbroekii* (Schlegel 1871)

Synonym: Yellow-billed mountain lory

Native names: Fore, *Kása*; Gimi, *Kása*

Description: This species and *N. pullicauda* are notable for the striking scarlet coloration of the underparts, from throat and upper breast to vent. In mature *musschenbroekii* the entire underparts except a small area on the sides of the breast are scarlet. Underwing coverts are brilliant red, also the underside of the flight feathers for almost their entire length, except the green tip. The forehead is green; crown and nape are olive brown, or rich brown on the crown in some males, streaked with yellow; a narrow line above the cere and the lores is blackish; the cheeks are green streaked with dull yellow. Central tail feathers are green above tipped with yellow, and the outer feathers are green, broadly tipped with yellow and marked with red at the base; the underside is yellow or greenish yellow and pale red. The bill is dull yellow; the upper mandible is broad and strong. The iris is orange-brown and the feet and skin surrounding the eye are gray. The cere is yellowish.

Length: 21–23 cm (Coates 1985). According to Arndt (*Lexicon of Parrots*), the nominate race measures 23 cm in length; *medius*, 25 cm and *major*, 24 cm. Forshaw (1989) also gives the length of the nominate race as 23 cm. (Most birds in captivity measure about 21.5 cm.)

Weight: 43–55 g (Forshaw 1989); six males of subspecies *major*, 50–62 g, and seven females 49–53 g (Diamond 1972); of six wild-caught birds (five males and a female), probably *major*, on release from quarantine, three weighed 51 g each; one, 50 g; one, 48 g; and one, 46 g.

Immatures: According to Forshaw (1989), they are duller than adults, with less pronounced head markings and the underparts green, except for a variable red wash on throat and upper breast. However, young bred in my own collection and elsewhere (e.g., Thurlow 1989) are only slightly duller than adults with the scarlet on the breast slightly less extensive. The beak is brown and the cere is whitish. The iris is dark brown and, by the age of four months, brown red.

Subspecies: *N. m. medius* has the cheek streaks more yellowish, less green; perhaps not separable from *N. m. major* which has bright yellowish green streaking on the cheeks and generally paler plumage (particularly on the underparts which are more of a scarlet shade), according to Forshaw (1989).

Natural history

Range: New Guinea mountains, *N. m. musschenbroekii* in the Vogelkop (western) Irian Jaya, isolated from *major* (or *medius* if this race is valid); *medius* occurs in the Snow Mountains, Irian Jaya, and *major* is found throughout the highlands of Papua New Guinea, including the Sepik region and the Huon Peninsula.

Habits: A bird of montane forest, forest edges and even partly cleared areas, usually occurring between about 1,600 m and 2,700 m (5,000 ft. to 8,500 ft.) in the Eastern and Central Highlands. According to Diamond (1972), in the Eastern Highlands it is one of the few midmontane forest birds that have profited greatly from human activities—but he did not state how. (Most species in man-made habitats there were originally lowland birds.) It was numerous in partly cut forest around Okapa, Lufa and Mengino, and was the most common parrot at Miarosa village at 1,800 m (5,800 ft.). There, it congregated in casuarina stands along the gardens and roads in flocks of up to 50 birds. It apparently requires the proximity of forest and is absent in, for example, Goroka, where the forest has been destroyed for miles around. Martin (1982) observed it in the middle of Tari township at 1,650 m (5,400 ft.) in the Southern Highlands Province, feeding with green-naped lorikeets where there was some natural cover. Campbell (1979) was working in Tari at the Department of Primary Industry; he described this lorikeet as "very common on the station and surrounds, going in flocks of up to 30. Their feeding habits vary from the tops of Yar and Eucalyptus trees to feeding on annual weeds at ground level."

In December, 1994, I observed Musschenbroek's lorikeets in the Ambua region, on the Tari to Mount Hagen road, at about 2,500 to 2,650 m (8,200 to 8,480 ft.), to just below Tari Gap. It was common and was usually seen in flight, although it was difficult to distinguish by appearance from the smaller emerald lorikeet, which was also common there. Using binoculars to observe birds feeding in the crowns of trees about 30 m (98 ft.) high, I tried to observe the bill color and that of the underside of the tail, but usually much of the bird would be obscured by vegetation. Pairs were often seen flying high; then I would turn to Joseph Tano, a very knowledgeable ornithological guide, and ask: "Yellow-billed or orange-billed?" When he answered without even glancing upwards, I realized it was possible to distinguish the two species by their calls. That of Musschenbroek's was harsher. They almost invariably call in flight. One day a sudden shower of rain caused us to shelter close to a river. The view of the riverine forest was quite beautiful, in a hundred shades of green and brown. As I gazed at it, a Musschenbroek's landed on a small, dead branch over the river. It made me realize why so many lorikeets have red on the breast. It looked brilliant in the dead tree but in foliage it would scarcely be detectable because there are so many red or reddish leaves in the forest. Jonathan Newman, who observed this species in the same area in September, 1990, noted that they were "often seen perched along large branches, rather than across them" (Newman pers. comm. 1994).

The diet of this species is probably more varied than that of most small lorikeets. Stomach contents of birds collected by ornithologists have included dry and granular flower remains, fruits, berries, small hard seeds, small caterpillars and psyllid lerps. They feed on flowers (from high in the crowns of tall trees, right down to ground level) and *Schefflera* fruits. Thane Pratt saw Musschenbroek's taking the green seeds of certain trees. It is worth noting that of the eight lory species he observed at his study site in Mount Missim in 1977–80, this was the only one which consumed green seeds as well as nectar.

The crop of a *major* collected at Mafulu, at 1,250 m (4,100 ft.), was filled with seeds, according to Mayr and Rand (1937). At their campsite on the eastern slope of Mount Tafa, at about 2,000 m (6,500 ft.), this lorikeet was common. It frequented the tops of the lower trees in the "heavy forest," where *pullicauda* was absent. In a particular large flowering tree about 20 Musschenbroek's were feeding with *C. p. stellae*.

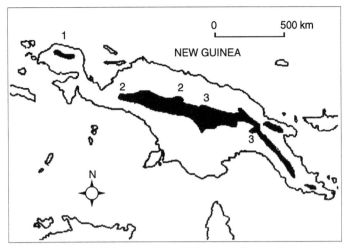

Range of Musschenbroek's lorikeet (*Neopsittacus musschenbroekii*)

1. *Neopsittacus m. musschenbroekii*
2. *Neopsittacus m. major*
3. *Neopsittacus m. medius*

Nesting: In the Eastern Highlands they breed in the rainy season, according to the local people (Diamond 1972), i.e., toward the year end. To the south of Wau, in northern PNG, Beehler (in Forshaw 1989) observed a male's precopulatory display in early December. In Irian Jaya, near Lake Habbema, a fledgling was taken in mid-November. Two eggs laid by a female in the author's collection (different clutches) measured 22.3 x 19.0 mm and 22.2 x 18.5 mm. Of six eggs laid by a female belonging to P. Clear, three measured 22 x 18 mm, two measured 21 x 18 mm and one was small, only 20 x 18 mm (Clear pers. comm. 1995). Two eggs laid at Chester Zoo weighed 5.05 g and 4.9 g (Wilkinson pers. comm. 1995).

Status/Conservation: Common and widely distributed, it is almost certainly found in many areas with little habitat disturbance. It is seldom trapped for trade although occasionally caught by native people who use the tail feathers for head decoration. No conservation measures are known to be necessary.

Aviculture

Status: Rare

Clutch size: Two

Incubation period: About 23 days; of six chicks hatched in author's collection, a single egg hatched after 23 days, and in three two-egg clutches a chick hatched 23 days after the second egg was laid, or 25 days after the first; in another clutch 23 days after the first and 22 days after the second; and in the third clutch 25 days after the first and 23 days after the second. In Germany, the first egg in a clutch hatched after 25 days and the second after 24 days (Volkmar 1995).

Newly hatched chick: Weight, 3 g or 4 g; it is sparsely covered in white down which is long on the back and flanks and shorter on the forehead and wings. The beak is brown with the egg tooth white.

Chick development: Second down can be seen under the skin as early as four days. By two weeks the chick is covered in the dense gray second down, except for a large startling white patch on the crown and nape. At nine days the eyes are just opening. At four weeks the wing and tail feathers are erupting through the thick covering of down (Low 1991b), and the red and green feathers of the breast are just emerging. At about 30 days the yellow and green tail feathers are just breaking through, also the feathers of the forehead and ear coverts; the tail is about 1 cm long. At about 32 days the feathers of the forehead and the ear coverts have erupted, the wings are three-quarters feathered and the red underwing coverts are erupting. The breast feathering is complete and the abdomen is half-feathered. Of two chicks being reared at the time of writing, at 35 days, one has the breast mainly green, while its younger sibling has the whole of the underparts red. (They proved to be a male and a female: the female parent also has much less red on the breast than the male.) By eight weeks young are fully feathered except for the nape; the tail is not yet full length.

Chick growth

Contrasts the weights of three chicks from the author's pair. No. 3, hatched in December, 1989, from an egg brooded by an iris lorikeet, was reared by and fledged with the iris. No. 1 and No. 2, hatched in January, 1991, were removed for hand-rearing at 27 and 28 days because the parents were plucking them.

Musschenbroek's lorikeet weight table

Age in days	Weight in grams		
	Chick No. 1	No. 2	No. 3
day hatched	–	–	–
1	4	–	–
3	–	7 fic	–
4	–	–	8
5	8 fic	10	10
6	–	11 fic	–
7	11	13 fic	12 fc
8	13 fic	15 fic	18
9	14 e	–	–
10	16 fic	17 fic	–
11	–	17 fic	24 fic
12	18 fic	18 e	–
13	19 fic	19 fic	28 e
14	19 e	20 e	–
15	22 fic	22 fic	34 e
16	22 e	23 e	–
17	25 fic	24 e	37 ne
18	25 e	26 fic	–
19	27 e	27 fic	40 fic
20	30 fic	29 fic	–
21	30 fic	31 fic	43 fc
22	31 fic	32 fic	–
23	33 fic	–	45 fic
24	34 fic	34 fic	–
25	38 fic	34 fic	48 fic
26	38 fic	35 ne	–
27	38 fic	35 ne	48 ne
28	39 fic	37/39	50 fic
29	38/39	37/41	–
33	39/43	40/44	53 fic
35	44/47	43/49	–
37	44/48	46/50	55 fic
41	48/50	49 *	55 ne
45	51/61 *	51	52 e
49	52/58	53/58	51 e
51	54/60	52/57 perching	–
54	49/54 **	50/55 biting!	46 e
56	49/53	49/51 flying	
67	48 weaned	47	–

* feeding alone, also spoon-fed ** occasionally spoon-fed
fc = full crop; fic = some food in crop; ne = nearly empty; e = empty

Young in nest: Two in a nest, 47 and 51 days (Volkmar 1995); 60 days, one only (Low 1991b); 50 and 51 days, two young in one nest (McDermid 1990); 50 or 51 days, one only (Martin 1982).

General: This species is little known in aviculture because it has not been available in large numbers and is difficult to breed. It was very rare until the late 1970s when it was imported into Germany and Denmark. The pair imported by Alfred Ezra in 1939 were believed to be the first to reach the U.K. A very few more were imported and then Musschenbroek's was unknown until 1986. Then about 30 birds were sent to a dealer who had not ordered them, in a consignment of other species. They were

therefore confiscated because they were not listed on the import license. Aviculturists hearing of their fate became concerned that they might pass into the hands of those unaware of their unusual dietary requirements. The birds were fortunate. The customs official, Mr. C. Miller, whose task it was to find buyers for confiscated birds, listened to the recommendations of zoo curators and private aviculturists. At that time confiscated birds were invariably placed in zoos. These birds were offered to several zoos, also to three private aviculturists, all with experience of keeping lories. They could purchase up to six each on condition that they retained them permanently in their collections.

I was fortunate to obtain six of these birds. I had them sexed surgically and they proved to be five males and a female. Not long after, the female was found dead in the nest. It was some months before I was able to locate a female and sought permission from Customs to obtain her in exchange for one of my males. Ten years later I am still breeding from these birds but, unfortunately, still with an excess of males.

After that initial consignment, a few more Musschenbroek's did reach the U.K., but then interest in the species seems to have waned. Perhaps it was because, despite their beauty, wild-caught birds remained aloof for a long time. And breeding them seemed beset with problems, especially egg-eating or egg-breaking, plucking of young, and parents attacking young after fledging. Or pairs might breed once or twice—and then cease. Or they made no attempt to breed for years. Probably the first breeding was that by L. Stokes; his birds apparently reared young in 1986, soon after being received. They bred when six were housed together! (S. Peacock verbal comm. 1996). Keeping them on the colony system is not recommended, but for the initial choice of partners it is the best method. Steve Peacock of Avon also had birds from the 1986 importation. One female did not breed for six years; then, up to October, 1996, in the final 12 months of her life, nine years after being imported, she produced the amazing number of eight young! They were parent-reared, the female laying again while young were still in the nest.

Another reason for a general lack of success is that this species is usually a winter breeder. Clear reported that even in the U.K. they nest early in the year. On two occasions his pair hatched young in January in an outdoor aviary: one chick died at 10 days after a night of frost and the other at seven days during a fall of snow (Clear pers. comm. 1995). Heated winter accommodation, or removal of chicks for hand-rearing at about four days, is essential for survival.

By 1996, total numbers in the U.K. were probably in the region of 30 birds. To me, Musschenbroek's is a delightful species. Its color, personality and tameness, are a constant delight. The way it moves about is fascinating. After seeing them in the wild, Forshaw described them as "running along stout branches in a rodent-like manner." This is also true of the way they move around on welded mesh or on a perch. They run rapidly for a couple of seconds, then stop abruptly. When they land on a perch, they often hold the wings open for a split second, displaying the red underwing coverts. They seem to jerk the head slightly upwards as they land, a sort of gesture of defiance toward the observer. The same behavior may be observed when they are running along the welded mesh floor of the cage, and suddenly stop.

The display of this species has been described by Thurlow (1989):

> I rarely saw the birds displaying, though the cocks would lean over their mates when excited and open their wings momentarily exposing the vivid red of their flight feathers. I also observed the cock of Pair 2 enact a display prior to mating in which he walked in a wide arc on a wooden platform positioned in the flight close to a perch on which the hen was sitting. His body was held at a slight angle away from his mate and his head was held outstretched. This was performed a number of times in quick succession, and on each occasion he would then fly to the hen and approach her before repeating the performance.

This is the only form of display I have ever seen in my own birds, so it appears that this species lacks the more exaggerated performance of, for example, a *Trichoglossus*.

The diet of captive birds requires little nectar if they are offered a variety of foods. Sunflower seed and the larger (Russian) pine nuts, are relished. They also like oats and spray millet and a few grains of hemp seed. I provide a mixture of small seeds in one container and sunflower seed in another. They seem to need the oil seeds and never put on weight. Favored fruits and vegetables include soaked dried figs, apple, guavas, pomegranates, thawed frozen peas, fresh corn, tinned sweet corn, carrot, chickweed and Swiss chard. But without any doubt, their favorite food is cracked walnuts! And they will play for hours with the shell after eating the contents. Thurlow (1989) described the diet of his Musschenbroek's as consisting of sunflower and safflower (rationed), canary seed, mixed millets, pine nuts, soaked millet sprays, nectar, diced apple, grapes and soaked dried figs, pomegranates, sowthistle, chickweed, lettuce, spinach and salad cress. Anyone who is fortunate enough to live in areas where *Casuarina* trees grow should offer branches containing the little "nuts." This will keep them amused for hours and is perhaps the only natural food available.

Most lories are nectar-feeding species and breed very readily in confinement. Musschenbroek's is more omnivorous and does not breed readily. Does the key to breeding this species lie in finding the correct diet? Or does the slightly different temperament of this species influence its behavior in the breeding aviary? These are difficult questions to answer, especially given the relatively small number of birds in the hands of aviculturists.

One of my pairs, until the untimely death of the female in 1992, were seasonal breeders. In the Canary Islands the female laid only between the end of November and March. The eggs of her first clutch were laid on March 24 and 26, 1988. They were infertile so I removed them on April 24. A single egg was found on March 28, 1989. (Perhaps one egg in the clutch was broken.) It hatched on April 17 but the chick died on the following day. Another single egg was found on November 21, and this was fostered to iris lorikeets. A chick hatched on December 14, left the nest briefly on February 5, and left the nest permanently on February 12. Removal of this egg induced the female to lay two more, on January 8 and 10, 1990. One hatched on February 2, but the chick died a few days later. The female laid again, on March 5 and 7, and chicks hatched on March 28 and 29. The second chick was very weak when it hatched and was removed at once for hand-rearing. After struggling initially to live, gaining only 1 g during the first week, it was successfully reared. The second chick was removed from the nest at the age of 11 days because it weighed only 13 g. (At the same age the iris-fed chick had weighed 24 g.) This chick seemed to thrive, but died suddenly at the age of 32 days. On autopsy it was found to have food in the trachea: it must have died of food aspiration.

This is the only time I have had this occur. Such accidents are extremely rare with spoon-fed chicks. The surviving chick was remarkable in that at 38 days it was offered the warm rearing food in a small container, consumed it at once and thereafter was completely independent. At the weaning stage its favorite foods were spray millet, apple and carrot. There was no need to handle it—and it did not welcome this. By 3½ months, it would draw blood if handled! Incidentally, Musschenbroek's has perhaps the worst bite for its size of any parrot, due to the power of its mandibles. Care is needed when handling this species!

When I returned to the U.K. to live in 1995, my Musschenbroek's were housed in a heated, outdoor birdroom during the winter. One female was seven years old and had never before produced young; I then discovered that this was because the male broke eggs at once. After the loss of the first clutch, the female laid again two weeks later. By checking the nest twice daily, I was able to remove the two eggs as soon as they were laid and place them in a Turn-X incubator; they were hand-turned. I replaced them in the nestbox with plastic eggs. Daily I could hear the male inside the nest literally kicking them about; they were incubated at night but usually left unattended during the day. During the early stage of the incubation period there was often a fight between male and female early in the morning. I suspected that the male was trying to make the female return to the nest. Both the eggs in the incubator were fertile; in the first, the embryo died just before it was due to pip internally. The second hatched after 24 days.

The other female caused some problems. Prior to removal to the U.K. she had been with her brother. When they came out of quarantine I tried to pair her with another male. He was afraid of her so, as I had two surplus males, I tried yet another. After they had been together for six weeks, and the female had shown little interest in the male, I found the male dead on the floor with blood coming from his mouth. Sadly, the female had killed him. I kept her alone for two weeks and then decided to try her again with her brother. The reunion was interesting.

Within a minute she was crouching for the male and they were mating. I placed them in the male's cage. For the next six days they were alternately mating and fighting. Finally, one fight was so vicious that both birds were locked together on the cage floor, and I had to separate them with a steel dish. I removed the male at once, fearing that one or the other would be killed. Four days later the female laid the first of two eggs, which she incubated assiduously, not leaving the box even when I was cleaning her cage. Ten months and a number of infertile eggs later, it has not been necessary to separate male and female again.

My young are ringed at 11 to 13 days with size 5 mm rings. Volkmar (1995) ringed his young at nine days with 4.5 mm rings. He bought two birds in 1988 and housed them in an aviary measuring 5 m (15 ft.) x 80 cm (2 ft. 8 in.) x 2 m (6 ft. 6 in.) high. They were fed on a mixture of sunflower, safflower and wheat (50 percent of diet), plus nectar and various fruits: egg food was offered when they were breeding and mealworms during the rearing period. Six months after they were obtained, the male's beak became a more intense yellow. The female laid on April 22 and 24 but the eggs disappeared. In the following year two eggs were laid with the same result. The pair was then moved to another aviary with a larger shelter and smaller inside flight and given an L-shaped nestbox. In 1991, no breeding attempt was made. In 1992 the two eggs laid in April were eaten. In the second clutch the first egg was eaten and the second was placed under a many-colored parrakeet (*Psephotus varius*). It hatched and the chick was fed by the foster parents along with their own chicks. However, their own young were perhaps more demanding, and by the time the Musschenbroek's was 26 days old, Mr. Volkmar fed it once daily to ensure it received enough food. At 29 days it was removed to a brooder and fed every four hours. By the age of ten weeks it was fed only twice daily. Not long after the young one was independent, the male was found dead. Another male was acquired and it soon became evident that the previous male had been responsible for eating the eggs. Again, the female laid in April, on April 7 and 9, 1993. The first egg hatched after 25 days and the second after 24 days. The young were reared without any problems and left the nest at 47 (first) and 51 days.

Few Musschenbroek's are kept in the U.S.A.;

in California there are three consistent breeders, Mike Jones, Frank Tromp and Dick Schroeder. One of the two pairs kept by Dick Schroeder initially mutilated or even killed chicks; but, with experience, this behavior ceased and the chicks were reared by the parents. The female of his second breeding pair plucks her own breast when laying or rearing, but plucks the young only minimally. Their diet consists of chopped fruit (papaya, apple, banana, grapes), frozen mixed vegetables (thawed), soaked Kaytee Exact mynah pellets and mealworms. When there are chicks in the nest, mealworm consumption increases greatly (Schroeder pers. comm. 1996).

This species was bred in Papua New Guinea, at the Baiyer River Sanctuary in the Western Highlands in 1978. A pair was housed in an aviary 6 m (20 ft.) square and 3 m (10 ft.) high, with two female crimson-winged parrakeets (*Aprosmictus erythropterus*). They chose the smaller of two nest-boxes—15 cm x 10 cm x 15 cm (6 in. x 4 in. x 6 in.). The diet consisted of bananas, milk and honey, split sugar cane, corn and sunflower seed, also a mixture of protein concentrate and biscuit meal (ratio of 2:1), with chopped papaya and crushed boiled eggs. On this nutritious diet they reared a young one. Although the male was a wary bird, the female and the young one were quite tame (Martin 1982).

Aggressive behavior toward newly fledged young has been reported from three different sources. The first two reared at Chester Zoo were attacked soon after they fledged, thus were removed from the aviary. Jos Hubers in the Netherlands reported the same problem, with males attacking and killing offspring soon after fledging (Hubers pers. comm. 1995). S. McDermid (1990) reported that the only two young his pair had reared were attacked by their parents three days after they left the nest. They were removed and fed three times daily with a mixture of Milupa baby cereal, glucose and Horlicks. At ten weeks old they were fully independent.

Only five zoos have reported breeding this species to the *International Zoo Yearbook*. Vogelpark Walsrode was the first, in 1987, when five were reared, followed by four more in 1990 and 1 in 1991. Chester Zoo in the U.K. reared two young in 1989. Four more were reared in 1990 and three in 1991. In New Zealand, Hamilton Zoo hatched four and reared three in 1990 and in the U.K., Penscynor Wildlife Park, near Neath, reared three in 1991. In the same year San Diego Zoo reared three. The Foreign Bird Federation's *Register of Birds Bred in the U.K. under Controlled Conditions* for the years 1988 to 1991, reported the following breedings by members of affiliated societies: 1987, none; 1988, 2; 1989, 15; 1990, 12 and 1991, 13.

In Germany, Karl and Michael Bruch produced a hybrid from a male Musschenbroek's and a female Rothschild's fairy lorikeet—two species which are very different behaviorally. A photograph depicted the hybrid as having a small beak like a fairy, but otherwise tending toward the Musschenbroek's in appearance, with red cheeks and throat (Bruch and Bruch 1983).

Emerald Lorikeet *Neopsittacus pullicauda* (Hartert 1896)

Synonyms: Alpine lorikeet, orange-billed mountain lory

Description: Briefly, a small version of Musschenbroek's. Entire underparts from throat to vent are scarlet, except the thighs and a small area on the sides (less extensive than in Musschenbroek's) are green. The forehead is yellowish green except (as in Musschenbroek's) for a very narrow brownish black line above the cere and along the lores. Just below the lores and on each side of the lower mandible is a small area of yellow; the cheeks are green, streaked with yellow and greenish yellow. Crown and nape are brownish green, lightly streaked with yellow and light green. Almost the entire underside of the wings is scarlet, except the tips of the feathers and a narrow area on the inner webs (which are brownish). Undertail coverts are green, the underside of the tail is dull olive green (much brighter in Musschenbroek's) and red. Upper parts are green, including the tail, which is not tipped with yellow as in Musschenbroek's. The upper mandible is orange and the lower mandible is yellow; the iris is orange-brown and the feet and the skin surrounding the eye are gray.

Length: 18 cm (7 in.)

Weight: 28–40 g (Forshaw 1989); 38 g (Diamond 1972); one captive female 40 g (Clear pers. comm. 1995).

Immatures: They are much duller than adults, with small areas of dull red on the underparts. The tips of the tail feathers are pointed (rounded in adults). The iris is dark. Clear (1993) described two newly fledged young as follows:

> Their heads were a medium shade of green suffused with faint striations. There was a distinctive yellow band across the brow and the cheeks were streaked with pale yellow, the back, rump and body were green, the upper breast was a light shade of red, extending across to the shoulders, the belly was a darker shade of red, separated from the breast by a band of green. The upperwing coverts were green, the primaries dark green edged with yellow, the underwing bright red and the flight feathers black. The short tail was green on the upper side and yellowish underneath. The beak was horn colored and the feet were light gray with dark nails. They were a little smaller in body size than their mother.

Subspecies: *N. p. alpinus* has the breast orange-red, in contrast to the darker red of the abdomen; upper parts are darker green. *N. p. socialis* differs from the nominate race in having the upper parts and the sides of the head darker green, with less olive brown on the nape.

Natural history

Range: The central, mountainous part of New Guinea. The nominate race is from southeastern New Guinea, west to the Sepik region; *alpinus* occurs in the Snow Mountains, Irian Jaya, eastward to the head of the Fly River and *socialis* has been recorded only from the Herzog Mountains and the highlands of the Huon Peninsula.

Habits: This is essentially a high mountain species, occurring mainly from 2,300 m (7,500 ft.) to 3,660 m (12,000 ft.) (Coates 1985), but not normally below 2,000 m (6,500 ft.), unlike Musschenbroek's. In some areas, both species are commonly found.

Then the less harsh call note of the emerald is usually the quickest way to identify it. Beehler, Pratt and Zimmerman (1986) describe its voice as "higher-pitched, more musical and quieter." They state that it feeds mainly on flowers, fruits, berries and seeds. It is found in moss forests, also nearby areas which have been partly cleared. Mayr and Rand (1937) described how "sometimes flocks of 28 to 30 gather together in the top of a pine tree, apparently to feed on the fruit; they came in twos and threes, with much twittering and squeaking as they moved about, and when they left they went in small parties, not as a flock...These parrots, as with *N. musschenbroekii*, are adaptable in feeding habits, eating either small fruits or flowers depending on their availability."

In December, 1994, I observed emerald lorikeets in the Southern Highlands. The first time I saw this species, a single bird was feeding in the crown of a tree about 3 m (10 ft.) high, at 7:30 A.M., along the main Tari to Hagen highway, close to Ambua Lodge. It was my first sight of either member of the genus in the wild and, until I could observe the bill color, though binoculars, I was unsure which species it was. We watched it for several minutes before it was joined by a second bird. As we walked back along the road, several more emeralds were briefly observed. These were my closest observations. Thenceforth, emeralds were mainly seen at a slightly higher altitude, over about 2,600 m (8,500 ft.). They were flying overhead or feeding in the crowns of a tall white flowering tree about 30 m (100 ft.) high. I was surprised to find *musschenbroekii* in the same locality. Diamond (1972) points out that there is an altitudinal range overlap of 500 ft. (150 m) to 1,000 ft. (300 m) and states that "this overlap may be made possible by the fact that *musschenbroekii* is about 40% heavier than *N. pullicauda.*" The former has a heavier bill so it probably exploits slightly different food sources. Diamond mentions catching a male in a mist net a few feet above ground in tall forest on level ground at about 2,400 m (7,870 ft.). He commented on this because he thought members of this genus usually stay in the canopy.

Nesting: No nests are known to have been described. Two eggs laid by a captive female measured 20 x 17 mm (Clear pers. comm. 1995). Hubers (pers. comm. 1995) weighed 4 new laid eggs; each was

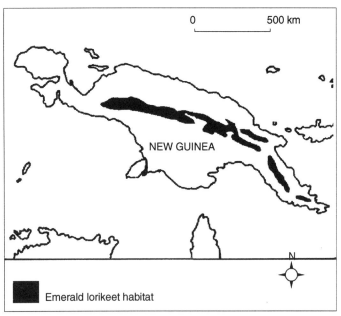

Range of the emerald lorikeet (*Neopsittacus pullicauda*).

4.4 g. One female laid large eggs, measuring 23 x 17 mm and 22 x 19 mm (Hubers pers. comm. 1996).

Status/Conservation: It is common and widely distributed. As an inhabitant of higher altitudes, much of its habitat is undisturbed. It is seldom trapped for any purpose. No conservation measures are known or likely.

Aviculture

Status: Very rare

Clutch size: Two (Baur 1992; Clear 1993)

Incubation period: "Approximately 24 days" (Clear 1993); 24 days, based on eight eggs (Walser 1993b); 24 days (Hubers pers. comm. 1996).

Newly hatched chick: Weight, 3.7 g (Hubers pers. comm. 1996); thin gray down, darkest on head and back.

Chick development: At 11 days, eyes opening, ringed with 4 mm rings; 15 days, covered in dark gray second down with a yellow patch on the nape; 22 days, feathers erupting; 30 days, 75 percent of body feathered, yellow patch still present on nape; 34 days, almost feathered, except under wings; 47 days, left nest (Hubers pers. comm. 1996). Baur (1992) mentioned the thick gray down and a photograph showing two chicks aged about two weeks depicts that they are similar to the young of *muss-chenbroekii*; a grayish white patch of down on crown and nape is clearly visible.

Chick growth

Shows weights of two young from a particular pair (Pair 1) and two from another (Pair 2) belonging to Jos Hubers in the Netherlands (Hubers pers. comm. 1996). The food consumed by the parents consisted of budgerigar mixture, sunflower seed and nectar.

Emerald lorikeet weight table

Age in days	Weight in grams			
	Pair 1		Pair 2	
hatched	3.7	3.7	3.5	3.5
1	4.2	4.0	3.9	3.9
2	4.9	4.7	4.6	4.3
3	5.8	5.5	5.3	4.9
4	6.9	6.6	6.5	5.9
6	9.0	8.2	8.6	8.0
8	11.5	10.8	10.8	10.0
9	14.5	13.8	13.9	13.6
11	19.6	19.5	19.5	19.1
14	24	24	24	23
17	26	25	26	26
19	28	26	29	28
22	31	28	32	30
26	36	30	36	33
30	37	32	40	36
34	39	34	42	36
38	37	33	41	35
42	36	34	38	34
46	fledged 33	33	34	32
48			fledged	

Young in nest: About seven weeks (Baur 1992); 47 and 48 days, two young in one nest (Clear 1993); about 50 to 60 days (L. Rasmussen pers. comm. 1994).

General: Apart from a single specimen in Australia, owned by Sir Edward Hallstrom in 1953, the emerald lorikeet was probably unknown in aviculture until the 1980s. Then, the odd bird or two appeared in consignments of Musschenbroek's lorikeets. However, the numbers exported have been very small—and females in the minority. The first recorded breeding appears to be that reported by Clear (1993). He obtained a female in 1986, the only one of its kind in an importation of Musschenbroek's. Not until 1989 was he able to locate a male. Unfortunately, it died before he could collect it. Despairing of ever finding another male, he then paired the female, which had laid eggs during the previous two years, with a male Musschenbroek's.

They produced two chicks which left the nest at 55 days but "for some unknown reason" died when aged 60 and 61 days. (They could have been killed by the male parent.)

A male emerald was then purchased from a small importation and, after a period caged side by side, the two birds were placed in a small aviary. It measured 2 m (6 ft. 6 in.) x 1.2 m (4 ft.) wide x 2 m high. There was a small shelter at one end; the floor was covered with washed beach pebbles. The two eggs laid in 1991 were infertile. The two eggs laid in March, 1992, hatched on April 11, after approximately 24 days. One young left the nest on May 28 and the other on May 29. They both proved to be females. During the weaning period the birds received about 20 mealworms daily. The usual diet included nectar made from fruit and vegetable flavors of Milupa baby food, sugars, soy flour, plus multivitamins. Malt extract was added twice weekly. They also received a dry mixture containing Readybrek (oat cereal), wheatgerm cereal, Vitafood (skimmed milk product, 21 percent protein, 3½ percent lactose, Boots plc), glucose, ground rice, Horlicks, demerara sugar, sunflower kernels and SA37 (a protein and vitamin supplement). The sugar and sunflower kernels were first reduced to a powder in a coffee grinder. Small Chinese pine nuts and mealworms were also offered. Other items in the diet included dried figs split open to expose the seeds, and soaked sultanas and raisins. A seed mixture was provided; about 40 percent of that consumed was safflower. The quantity of nectar usually consumed per bird per day was very small, about 0.5 fluid ounces—about one-quarter of that consumed by the Musschenbroek's. They took more when young were in the nest (Clear pers. comm. 1995). In contrast, Hubers found that his emerald lorikeets took more nectar than his Musschenbroek's.

In Denmark, Leif Rasmussen first bred this species about 1992. His emerald lorikeets have access to an outdoor flight even during the winter, when the indoor part of the aviary is heated to keep the temperature just above freezing. They appear to be very hardy and even breed during the winter. They do not take much nectar but consume a lot of seed—mostly sprouted—and many mealworms daily. Like Musschenbroek's, they tend to pluck the young in the nest (Rasmussen pers. comm. 1994.).

In Germany, the emerald lorikeet was bred by F. Baur. Originally, he tried to house several birds together in the hope that they would choose their own partners; but the more dominant birds were so aggressive that each time, a severely stressed bird had to be removed after a couple of hours. Eventually the birds were surgically sexed. They consumed a very small amount of nectar. They were offered the dry food Lori-Systemfutter B (Biotropic) and fruit, plus a handful of white sunflower seeds per pair daily and a few pine nut kernels, plus a millet spray and various fruits. Baur was convinced that oil seeds are essential to keep this species in good condition. When the temperature was higher than normal, their consumption of oil seeds decreased and their well-being was affected. Improvement was noticed after withholding other items of the diet to force them to eat the oil seeds. One pair reared two young. Two or three weeks later the female laid again (Bauer 1992).

In Switzerland, Hans Walser bought a pair of emerald lorikeets in the autumn of 1990. They were tame and friendly. He kept them in an aviary with an indoor section measuring 1.5 m (5 ft.) x 80 cm (2 ft. 8 in.) x 2 m (6 ft. 6 in.) high, with an outdoor flight of the same size. The horizontal nestbox was ignored. Three months later he placed another nestbox on another wall in the inside flight, and this was accepted. Thirty days after he first observed them mating, the female laid her first egg. By 1993, 8 young had been reared from four clutches! He was not expecting such good results because the birds are kept under conditions of high temperature and high humidity, which he believed would not suit mountain species. The temperature sometimes rose to over 30°C (86°F) during the day and was maintained at 18–22°C (65–72°F) during the night. The humidity was in the region of 60–80 percent. The diet consisted of approximately 50 percent lory nectar, 25 percent fruits, berries and green leaves and 25 percent seed mixture. During the summer, plenty of unripe seeds were offered. Favored items of fruit were golden delicious and Granny Smith apples, grapes and raisins. Green food was offered daily. Fresh cut branches were provided two or three times a week. Mr. Walser emphasized the great pleasure that he obtained from this species.

Jos Hubers obtained his first pair in 1990. The male died in an accident in 1993 and a new male was obtained. In 1994, the first youngster was produced. In 1995, two more were reared, and in 1996 nine young were reared from two pairs. All the young were delightfully tame from the time they left the nest and on occasions would even step on to the hand. Two other pairs laid only infertile eggs. By this time Mr. Hubers had six unrelated pairs and was working in conjunction with breeders of this species in four European countries (Hubers pers. comm. 1996). Only in this way can numbers be maintained. Whether the species can be established in aviculture from such a small number of birds remains to be seen but it appears increasingly hopeful.

While living in Papua New Guinea, Major Harry Bell kept an emerald and a Musschenbroek's lorikeet. He contrasted their different food preferences. The Musschenbroek's refused a honey and milk mixture but took, in order of preference, raisins, fruit cake, canary seed and flowers. In contrast, the emerald consumed, in order of preference: flowers, raisins, seed, cake, honey and milk mixture, and fruit. It favored, above all else, the flowers of *Clerodendron* species, tearing at the base of the inflorescences and eating the pith. Local people who knew both species confirmed that the Musschenbroek's favored fruit and seeds more than the emerald, which was the more nectivorous of the two. The emerald lifted food to its beak, even items as small as millet seeds. The only exceptions were flowers, which it ate directly with its beak, sometimes steadying the foliage with its foot. Raisins and cake were dipped in water before being eaten. Bell also commented on the voices of the two species, that of the emerald being "a low pleasant twittering squeaking note" in contrast with the "high-pitched disyllabic screech, almost unbearable to the human ear" of the Musschenbroek's. The emerald was very tame, and free-flying in the house. It had been acquired as a fledgling. It often displayed, bobbing its head about 15 times (Bell 1984).

Part 3

Lorikeets in Australian Gardens

Cities and Suburbs

Throughout Australia's cities and suburbs colorful native birds give enormous pleasure to householders. Among the most conspicuous and vivacious of garden visitors are the lorikeets, whose presence is actively encouraged by thousands. Easy to observe and brightly plumaged, they must awaken an interest in nature in many young children. In recent years the number of properties which are visited by lorikeets has greatly increased. As Jim Pearson explains: "Once barren farmland is being reclaimed to the benefit of native birds in some Australian suburbs." This occurs more by happy accident than careful planning; residents of newly developed areas naturally plant trees and shrubs to enhance the district—and lorikeets and honeyeaters are among the first to take advantage.

Adelaide

Mr. Pearson, who has lived in the outer northern suburbs since 1974, related what has happened in the Adelaide area of South Australia:

> Parts of this area are growing rapidly as the population expands, although other parts (the central area of Elizabeth, for example) have been established for over 30 years. Many native trees and shrubs have been planted both on private and public property (parks, nature strips), so there are now many food species for lorikeets (mainly *Eucalyptus*). Today, the total number of trees in the area must be in the tens, perhaps hundreds, of thousands. When I first moved to Elizabeth East (25 km north of Adelaide) in 1974 it was a new housing development with only a few mature trees which had been left standing. At this time there were not many lorikeets (or other nectar feeders) to be seen. I can recall only musks, in relatively small

numbers at irregular intervals. However, most people planted gum trees in their gardens and along road sides, so after a few years there were quite a few trees of flowering size. The lorikeet population has been on the increase ever since. The variety of gum trees, with different flowering times, seems to keep nectar feeders in the area almost continuously. Musks (in particular) and purple-crowned are seen flying or feeding in quite large numbers. I do not know where they breed since most of the trees here are too young to provide nesting hollows. The nearest areas with older trees in large numbers are over 10 km away. In the 1970s, if memory serves me correctly, most of the birds used to disappear during the winter. It is assumed that they moved to other locations, possibly in the Adelaide Hills, where food and breeding trees were more abundant.

> Although mainly a finch breeder, I kept purple-crowned lorikeets for about three years in the mid-1980s. There was a large sugar gum (*Eucalyptus cladocalyx*) in the back yard about 5 m from the lorikeet aviary, and two spotted gums (*E. maculata*) in the front. Both species flower quite prolifically for long periods and were regular feeding trees for musk and purple-crowned lorikeets and honeyeaters. These trees were visited daily by birds which fed from low levels (2 m—the flowers were no lower) to the tops of the trees. Mostly musks, they often flew quite low out of the trees and did not appear to be bothered by people. (I had to duck on more than one occasion to avoid being scalped by a low-flying lorikeet!) Some of the purple-crowneds would alight on my aviary, no doubt attracted by the captive birds.

> A neighbor at Elizabeth East had a red-flowering gum (*E. ficifolia*) which is a small to medium-sized tree with large clusters of red flowers that absolutely drip with nectar. This tree was frequented by musk lorikeets and bull

ants! The birds did not seem to like the ants (which are about 1 cm long) and while it did keep them from some parts of the tree, it did not prevent them from feeding in the tree altogether. Elizabeth has a large regional shopping center which is surrounded by acres of asphalt car park. Throughout the car park there are several hundred trees, presumably planted to shade parked cars. About half to two-thirds of these trees are various species of small to medium sized eucalpypts; some are red-capped gum, some possibly gungurru (*E. caesia*) with smallish red flowers and some are *E. torquata* with smallish pink flowers. Musk lorikeets can be seen in quite large numbers when the trees are flowering. Occasionally purple-crowned can be seen feeding at quite low levels, seemingly unperturbed by people and cars moving about within a meter or two of their feed tree. Most of the people in the car park never seem to notice the birds, either! Many of these trees flower during summer when temperatures can be up to or over 40°C (104°F). With radiation from the black asphalt, the temperature within a couple of meters of the ground must be considerably higher. However, this does not seem to bother the feeding lorikeets.

In 1990 my wife and I moved out to a newly developed area near Two Wells, 40 km north of Adelaide. The land was previously farmland—mostly cereal crops and hay, thus not too many trees. In the first couple of years I do not recall seeing a lorikeet in that area, although there were plenty of redrumps (*Psephotus haematonotus*). As this area became more populated, residents planted large numbers of trees. We planted about 300 trees and shrubs, mainly gum trees and *Melaleucas*. Unfortunately, we had to move back into suburbia but our son currently lives in the same street. Some of the trees (mostly planted from 1988 onward) are now growing well and are flowering. I have seen or heard musk lorikeets there. They seem to be spreading—a happy scenario—once barren lands being reclaimed to the benefit of native birds!

I now live at Munno Para, 10–15 minutes by car north of Elizabeth. This area was developed for housing about 16 years ago. There are plenty of gum trees and lots of lorikeets all year round. Many of these trees provide food for musks (quite large numbers), purple-crowned (not so many) and, on rare occasions, for rainbows. I have only ever seen three or four rainbows

feeding, on two or three occasions. Flocks of lorikeets often fly over my house in the mornings or towards dusk. This morning, for example, a flock of 70 to 80 musks flew overhead at about 8:00 A.M. A couple of smaller groups (eight to ten birds) were also seen. Usually there are more smaller flocks, not one large one. Groups of purple-crowns usually number up to eight or ten, but sometimes only two to five. Two other nectar feeding species are present here and at Elizabeth—white-plumed and New Holland honeyeaters.

Until a few years ago I used to work in Adelaide. The city is one square mile in extent and surrounded by a green belt of parks. The so-called parklands contain probably thousands of trees, many of which are eucalypts. On just about every occasion I went through the parklands I saw rainbow lorikeets. They seem to be resident in quite large numbers. Adelaide was established in the late 1830s so some of the gum trees are very old. No doubt they offer suitable nesting hollows for these birds (Pearson pers. comm. 1995).

Carl Clifton commented on the lory population in the Salisbury North area of Adelaide. "It has, I think, doubled in the past five years. Ten years ago one seldom saw any member of the parrot family in this area" (Clifton pers. comm. 1995).

Melbourne

In 1994, the Gould League of Victoria carried out a parrot count. When their February, 1995, newsletter *Greenprint* was published they came to the conclusion that "if Australia is the land of parrots, Melbourne must be the city of parrots." More than 4,500 record sheets had been returned, many with letters indicating that large numbers of parrot enthusiasts live in the suburbs. Maps produced from 3,700 records showed that habitat differences allow various species of parrots to live in the suburbs. The resulting map of rainbow lorikeet distribution in Melbourne is reproduced here, with kind permission of the Gould League.

Green corridors along creeks and river valleys are important pathways through which parrots enter Melbourne. Golf courses, with their mixture of grassy areas and tall, old trees are havens, also breeding areas for many birds. The record sheets returned showed that rainbow lorikeets are extremely common in Melbourne's suburbs; surpris-

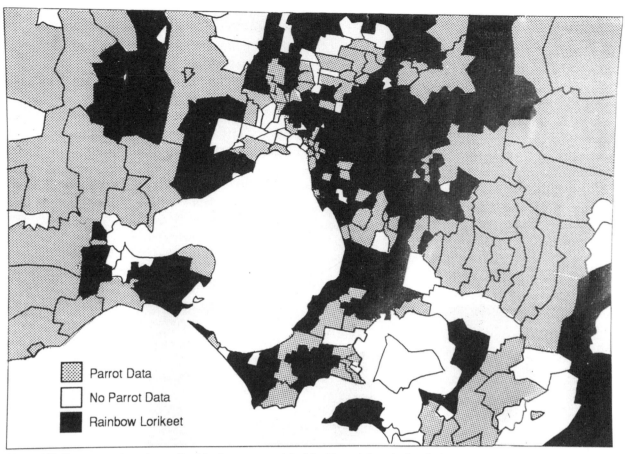

Map of Melbourne and environs; the black areas are inhabited by rainbow lorikeets

Reproduced by kind permission of the Gould League of Victoria

ingly, they are not so common on the outskirts of the suburban area. Apparently, they are not finding their way into the suburbs, as other parrots are doing, but are based in Melbourne, where they are breeding. Near Blackburn Lake, for example, in October, 1994, they were seen nesting in every available tree hollow, some of which were less than 2 m (6 ft. 6 in.) above ground. Two reasons were suggested for the rapid increase in numbers of this species: the planting of native flowering shrubs and the backyard feeding of lorikeets. Because they are very mobile, they can feed in any suburb and return to nest in reserves or golf courses. A major advantage of these lorikeets is that they can physically fight off introduced bird species which are competitors for nest hollows.

Queensland

At Mooloolah, in a rural/residential area 15 km (9 mi.) from the coast, lorikeets are present year-round in varying numbers, according to Phil Bender. The land is a combination of eucalypt forest, rainforest creek areas and planted gardens and orchards. The most important food trees for lorikeets on the Bender's property are bloodwood (*Eucalyptus intermedia*), rose gum (*E. grandis*), tallow-wood (*E. microcorys*), blackbutt (*E. pilulans*), brush box (*Lophostemon confertus*), euodia (*Euodia elleryana*) and the various species of *Banksia*, *Grevillea* and *Callistemon* that have been planted in the garden. Mr. Bender commented:

> When times are hard, lorikeets are found around the aviaries, eating spilled seed. In severe drought conditions, or in times of food shortage or constant heavy rain, we supply a wet-mix. At certain times of the year a fairly common tree in the area, the cheese tree (*Glochidion ferdinandi*) is nearly totally defoliated by insect larvae. Rainbow and a few scaly-breasted lorikeets congregate in these trees and I assume that they are feeding on the grubs, as there are no flowers or fruits. This often happens when other food is scarce (Bender pers. comm. 1995).

Backyard feeding

The harmful effects of backyard feeding of lorikeets are apparently not known by most Australians. They offer sugared water, with or without bread, and some also provide sunflower seed. In so doing they are causing the deaths of large numbers of lorikeets. Dr. Peter Wilson, veterinary surgeon at Currumbin Sanctuary on Queensland's Gold Coast, stated his concern. Many lorikeets, especially young ones, are brought to the surgery. They are unable to fly, 50 percent or more are suffering from leg paralysis, they are thin and suffering from a serious vitamin deficiency. The name Lorikeet Nutritional Myopathy was proposed for this syndrome (Wilson 1994).

Kevin Wilson reported the same, as a result of working at Taronga Zoo:

> Virtually every week, particularly in spring and summer there were young Rainbow Lorikeets being brought in that had either been kicked out of the nest or had been unable to fly adequately on fledging. If not brought in to the zoo there were scores of calls regarding the same. In just about all these cases the young birds were found to have stunted wings and tails, feet that were permanently crippled and or having fits, often leading to their death.
>
> As the years went by the numbers being brought to the Zoo seemed to increase. The problem was, it seems, that like myself, people were feeding the wild lorikeets on very poor (nutritionally) diets, mostly sugar or honey and water. The birds seemed to gorge on this and fed it to their young in the nest. The extremely liquid diet, already low in vital requirements, in many cases acted as a diuretic (increased excretion) and took with it what little valuable nutrition it was being fed (Wilson 1993).

One purpose of this book is to educate people to the harm that they are doing to the rainbow and, to a lesser extent, scaly-breasted lorikeets, which regularly consume artificial food at backyard feeders. I would ask them to discontinue this practice. This should not be done from one day to the next but gradually, so that the lorikeets have time to find natural food sources.

Lorikeets can be attracted to gardens and backyards by planting shrubs which bear nectar and pollen—their natural food. As an example, "Backyarder" (1994) described how, in 1988, he planted four red-flowering gums (*Eucalyptus ficifolia*) to commemorate Australia's bicentennial year. Six years later the trees were 5 m (15 ft.) high and had flowered after five years. He recorded: "The most exciting thing about our flowering gums is the native birds they attract, including a flock of approximately 20 Musk Lorikeets. They clamoured about the trees for two weeks and then disappeared as quickly as they had arrived."

During their visit they defended what they assumed to be their territory by chasing away any other bird in the vicinity, especially wattlebirds. The nectar from the flowers was their favorite food, too. A number of skirmishes were witnessed, but the lorikeets continually chased the wattlebirds away, "immediately resuming their place among the blossom and feeding throughout the day." The wattlebirds returned, along with crimson and eastern rosellas (*Platycercus elegans* and *P. eximius*), after the musks had moved on.

Other suitable plants are mentioned in the Garden Survey section.

Attempted rehabilitation

Veterinarian Dr. Adrian Gallagher sees many rainbow and scaly-breasted lorikeets suffering from malnutrition as a result of feeding in backyards. He stated that the majority of these birds test positive for Psittacine Beak and Feather Disease (PBFD) during active disease from feather tissue but can test negative in chronic cases. At his practice he sees a number of forms:

> In fledging or post fledging birds we see a range of feather loss from complete baldness to the more typical loss of tail and wing primary flights. The tail and primary wing flights can also be stunted and damaged, resulting in poor flight. Many of these birds are admitted as car, window and predator victims because of their lack of or poor flight. Some of these juveniles will spontaneously regrow the flights and tail within two months and appear grossly normal but these birds are all thought to be carriers for PBFD. The previous thinking was that these birds were immune and had eliminated the disease from their body. This has been disproved by testing these birds. They are a threat to our wild populations. Previously they would have been preyed on as fledglings. Now well-meaning carers raise them, then release them back into the wild. We are persevering to discourage this prac-

tice through writing to wildlife groups. Our present recommendation is to euthanize all suspect birds. The practice of rehabilitation and release should also be stopped (Gallagher pers. comm. 1996).

See also **Paralysis, Psittacine Beak and Feather Disease.**

Garden survey

In 1993, I gave out survey forms to aviculturists in Australia and the form was published in the March, 1994, issue of *Australian Aviculture*. Forty-two forms were returned. Over 50 percent were from Victoria, thus the findings of this survey do not give a true indication of the habits of lorikeets visiting gardens across Australia. There were 9 respondents from New South Wales, 5 from Queensland, 5 from South Australia and 24 from Victoria. (Their names are listed at the end of this section.) Only one species of lorikeet visited the gardens of 12 respondents, two species visited 16, three species visited the gardens of 11 and four species visited the gardens of four respondents. The rainbow was the most common visitor; it occurred in 31 of the 42 gardens. Next came the musk (27), followed by the scaly-breast and the little (13 each), and only nine gardens were visited by purple-crowned lorikeets. Survey results relating to each species are given below. Five and 10 years relate to those periods up to 1994.

Rainbow lorikeet

Present all year	Seasonal	Occasional
17	13	0

Numbers	increased	decreased	unaltered
5 yrs	8	1	5
10 yrs	11	1	3

Shrubs	increased	decreased	unaltered
5 yrs	4	0	10
10 yrs	10	0	6

At present address

1–5 yrs	6–10 yrs	11–15 yrs	16–25 yrs	over 25 yrs
7	6	5	5	4

Species on which they feed: Acacia; *Angophora* (rough-barked apple); apricot; box; *Banksia*; *Callistemon* (bottlebrush); eucalypts: blackbutt, ironbark, river red gum, spotted gum, white gum; hakea; firewheel tree; *Grevillea*, especially Robyn Gordon *grevillea*; *Melaleuca* (paperbark); ti-tree; wattle; also pear, plum and cherry, feeding on lerps.

Species which attract lorikeets most: Apricot; Callistemon; Grevillea; Eucalyptus: red box, red-capped gum, spotted gum, yellow box.

Has anything been done to attract lorikeets? No: 12; artificial food: 3; planting shrubs: 11.

Location of respondents: New South Wales, 6; Queensland, 4; South Australia, 1; Victoria, 15.

Scaly-breasted lorikeet

Present all year	Seasonal	Occasional
3	9	1

Numbers	increased	decreased	unaltered
5 yrs	1	0	4
10 yrs	6	1	2

Shrubs	increased	decreased	unaltered
5 yrs	1	0	3
10 yrs	4	1	2

At present address

1–5 yrs	6–10 yrs	11–15 yrs	16–25 yrs	over 25 yrs
2	4	1	4	2

Species which attract them most: See above, under rainbow lorikeet.

Has anything been done to attract lorikeets? No: 3; artificial food: 2; planting shrubs: 7.

Location of respondents: New South Wales, 4; Queensland, 5; Victoria, 4.

Musk lorikeet

Present all year	Seasonal	Occasional
8	11	1

Numbers	increased	decreased	unaltered
5 yrs	12	1	4
10 yrs	5	0	2

Shrubs	increased	decreased	unaltered
5 yrs	7	0	4
10 yrs	9	0	4

At present address

1–5 yrs	6–10 yrs	11–15 yrs	16–25 yrs	over 26 yrs
4	6	6	6	4

Species on which they feed: Acacia; apple; apricot; bottlebrush (*Callistemon*); eucalypts: flow-

407

ering red gum, ironbark, lemon-scented gum, red-capped gum, river red gum; Grevillea; hakea; ti-tree; wattle; also plum, pear and cherry, to feed on lerps. A food source in one garden was almonds before the nut dries (semi-mature or green kernel); the musk lorikeets fed only on the soft-shelled type called Californian papershell.

Species which attract them most:
Bottlebrush; eucalypts: flowering ironbark (*E. sideroxylon*?), yellow gum (*E. leucoxylon*), lemon-scented, red box, red-capped gum, spotted (*E. maculata*), bushy sugar gum and yellow box.

Has anything been done to attract lorikeets?
No: 4; planted shrubs and trees: 8.

Location of respondents: South Australia, 3; Victoria, 14.

Purple-crowned lorikeet

Present all year	Seasonal	Occasional
1	5	3

Only 2 replies referred solely to this species. Their numbers had increased during the 11 years and 16 years the respondents had lived at their present addresses.

Species on which they feed: Eucalypts: bushy sugar gum, ironbark, lemon-scented gum, red flowering gum, spotted gum, yellow gum; *Melaleuca*.

Location of respondents: South Australia, four; Victoria, five.

Little lorikeet

Present all year	Seasonal
2	10

Only two replies referred solely to this species. During the past five years, their numbers had increased in one case and were unaltered in the other. The former respondent had lived at his address for 23 years and the latter for five years. The lorikeets fed on yellow box and other eucalypts. One respondent resided in Victoria and the other in New South Wales.

Acknowledgments

Thanks are due to the following who took the time to respond to the survey: P. Bender, Mooloolah; J. Bennett, Blackburn; C. Boothman, Burwood; E. Bruton, Black Rock; J. Buchan, Glen Waverley; C. Clifton, Salisbury; L. Date, Port Stephens; G. Dosser, Benalla; D. Dunn, Ballarat; J. Flinn, Glen Waverley; P. Giddins, Lochinvar; N. Gunton, Melbourne; A. Harris, Newcastle; M. and B. Harvie, Nunawading; M. Heres, Yea; C. Hibbert, Eltham; G. Hyde, Elliminyt; S. Jones, Maitland; D. and E. Judd, Mildura; M. Kargulewicz, Melbourne; J. Kreuger, Townsville; T. Lang, Adelaide; I. Langdon, Mildura; D. Lee, Dewman; I. Lindsay, Portsea, Hacedon and Benalla; P. Mason, Mornington; G. Masters, Mulbring; G. and G. Matthews, Barmera; M. O'Brien, Wantirna; P. Odekerken, Buderim; M. Oxenbridge, Toowoomba; N. and N. Parker, Medowie; J. Pearson, Munno Para; P. Philp, Berri; J. and G. Redfern, Doncaster; W. Remington, Ballarat; A. Rotteveel, Toowoomba; M. Valk, Melbourne and S. Wilson, Coolah. The co-operation of Graeme Hyde, editor of *Australian Aviculture*, was greatly appreciated.

Part 4

Gazetteer

Members of the Loriinae range over a vast area, covering New Guinea and Indonesia, the center of their distribution, also Australia, the Philippines and Polynesia. In the west, lorikeets extend as far as Bali and Lombok (although probably now extinct on Bali) in the Lesser Sunda Islands. The lorikeet which ranges furthest north is the Mount Apo, from Mindanao in the southern Philippines; Tasmania, off the south coast of Australia, is the most southern extremity of the range of the family. In the tiny scattered islands of the South Pacific Ocean, lories reach their easternmost outpost. Half way between Australia and South America, in the Tuamotu and Marquesas Islands, the tiny exquisite *Vini* lories occur. But even further east, on Henderson Island, nearer to Peru than even to New Zealand, another member of the genus is found, on a single island, only 93 km^2 (36 sq. mi.) in extent. It has the smallest range of any lory. Some species are confined to one or two small islands; others range over an enormous area, are found on countless islands as well as in New Guinea and, as a result, have some or many subspecies. The green-naped lorikeet has the most extensive range of any lorikeet, also more subspecies than any other parrot. The red-flanked lorikeet has the second most extensive range.

Lories and lorikeets do not occur in swamps and grasslands but they are found throughout their region wherever there are flowering trees. From coastal coconut palms on tropical islands, to the umbrella trees (*Schefflera*) of the New Guinea highlands, lories will be found feeding. From palm-fringed beaches of coral sand, where the sun is always shining, to the damp, cloud-shrouded mountain heights of New Guinea and Indonesian islands, and the winter snows of Australia, lorikeets will be there. Many of the islands they inhabit are little known; their names will mean nothing to the aver-

age reader. But to judge how extensive are the ranges of the various species and subspecies, and the location of these islands, brief information is given below. The islands are listed alphabetically. In most cases they are islands on which more than one species of lory or lorikeets occur. (In many instances information is given in the text on the islands with a single species.) Lories found on these locations are listed, for general interest and for the sake of travelers who are also bird watchers.

Aitutaki

One of the Cook Islands in Polynesia, Central Pacific. Aitutaki is a 50-minute flight from the main island of Rarotonga, which has an international airport. An atoll, it has a land area of 18 km^2 (7 sq. mi.) and a length of 7 km (4 mi.). Species: *Vini peruviana*.

Admiralty Islands

A group in the Bismarck Archipelago with a land area of 2,070 km^2 (800 sq. mi.). Manus is the main island. Species: *Trichoglossus haematodus flavicans* and *Charmosyna placentis pallidior*.

Ambon (Amboina)

One of the Uliasser Islands, central Maluku, Indonensia, it has a total area of 777 km^2 (300 sq. mi.). It is the commercial and administrative center of the Moluccas. A mist-shrouded volcanic island, it is subject to earthquakes. Species: *Eos b. bornea*, *Trichoglossus h. haematodus* and *Charmosyna p. placentis*.

Aru Islands

Six larger, low-forested islands and more than 100 small ones in the Banda Sea, southern Moluccas, Indonesia. They have a land mass of 6,000 km^2 (2,300 sq. mi.). The islands are located about 640 km (400 mi.) southeast of Ambon and 240 km (150 mi.) west of New Guinea. Unlike other islands in the southeastern Moluccan chain, they are a sprawling mass of mangrove swamp and palm forest, harboring much interesting wildlife. Warmar is only 53 km^2 (20 sq. mi.) in extent, still with some undisturbed primary forest. Species: *Chalcopsitta scintillata rubrifrons, Trichoglossus haematodus nigrogularis* and *Charmosyna p. placentis.*

Australia

In the southwest Pacific, the smallest continent, about 7.7 million km^2 (nearly 3 million sq. mi.). This is the easiest place to see lorikeets in the wild. Species: *Trichoglossus haematodus moluccanus* and *T. h. rubritorquis, Trichoglossus chlorolepidotus, Trichoglossus versicolor, Glossopsitta concinna* and *Glossopsitta porphyrocephala* and *G. pulsilla.*

Bacan Islands

Bacan and its satellite islands are situated off the southwest tip of Halmahera. Bacan is mountainous, rising to over 2,000 m (6,500 ft.). The coastal plains have been cleared for cultivation; the remainder is forested but much of the forest is under concession for selective logging. Species: *Lorius garrulus flavopalliatus* and *Eos squamata.*

Bali and Lombok

East of Java, in the Lesser Sunda Islands, Indonesia. Bali has an area of 5,700 km^2 (2,200 sq. mi.) and Lombok 4,660 km^2 (1,800 sq. mi.). Bali is a mountainous island, 8 degrees south of the equator, extensively cultivated and with rice fields. Tourism is big business; it is the most visited island of Indonesia. In contrast Lombok is poor, subject to famine and uncommercialized. Species: *Trichoglossus haematodus mitchellii.*

Batanta

A small island north of Salawati, off the west coast of the Vogelkop Peninsula, Irian Jaya. Species: *Lorius lory lory, Chalcopsitta atra atra* and *Eos squamata.*

Biak (previously called Mysore)

An island of West Irian, about 60 km (37 mi.) in length. It is situated 200 km (124 mi.) east of the coast of the Vogelkop. It serves as an Indonesian naval base, has an international airport, increasing

tourism, a bird park and a nature reserve. The island of Supiori lies immediately to the north. Species on both islands: *Lorius lory cyanuchen, Eos cyanogenia, Trichoglossus haematodus rosenbergii* and *Charmosyna rubronotata kordoana.*

Bismarck Archipelago

Volcanic island group in the southwest Pacific; part of Territory of New Guinea. It includes New Britain, New Ireland and the Admiralty Islands. Total area is 49,000 km^2 (19,000 sq. mi.). See respective islands.

Bougainville

Largest of the Solomon Islands, west Pacific, with an area of 10,600 km^2 (4,100 sq. mi.). Volcanic mountains rise to 2,600 m (8,500 ft.). Species: *Chalcopsitta cardinalis, Charmosyna meeki* and *Charmosyna margarethae.*

Buru

A mountainous Indonesian island, in the province of Maluku, situated a few degrees south of the equator. It covers 8,000 km^2 (3,090 sq. mi.) and is notable for having ten endemic bird species. The west, central and southeastern parts of the island are mountainous, the highest mountains reaching almost 2,500 m (8,200 ft.). The higher mountains are covered in rainforest; nearly all forest under 1,000 m (3,300 ft.) has been logged; by 1990, 65 percent of the total land area had been cleared or designated for clear-felling. Only two very small areas are completely protected and much of the lowland is covered in rice paddies. Species: *Eos bornea cyanonothus, Trichoglossus h. haematodus* and *Charmosyna toxopei.*

Ceram, see Seram.

Cook Islands

Group in Polynesia, central Pacific; Rarotonga only sizeable island. (See also Aitutaki.) Species: *Vini australis* (Niue only) and *Vini peruviana* (Aitutaki only).

Fiji

Group of more than 300 islands in the Pacific; approximately 100 occupied. Land area covers 18,300 km^2 (7,000 sq. mi.), 86 percent of which is made up by Vitu Levu and Vanua Levu. Both are rugged, volcanic islands with mountainous interiors. Species: *Vini solitarius*, *Vini australis* and *Charmosyna amabilis*.

Halmahera (Djailolo)

An island of 19,400 km^2 (7,500 sq. mi.) in the north Moluccas, Indonesia. It is located between Sulawesi and New Guinea. Its avifauna is rich, with 26 endemic species. The rainforest is gradually being converted to secondary forest and cultivation since logging and development commenced in the mid-1980s. Species: *Lorius g. garrulus* and *Eos squamata*.

Henderson Island

An ancient coral reef in southeastern Polynesia, 200 km (125 mi.) east of Pitcairn Island and 2,300 km (1,430 mi.) southeast of Tahiti. It measures only 9.6 x 5.1 km (6 x 3 mi.), with an area of 36 km^2 (14 sq. mi.). It belongs to Britain. Species: *Vini stepheni*.

Indonesia

Indonesia consists of 13,000 islands. They extend along the equator in Southeast Asia, covering a distance almost as far as from New York to San Francisco. Many are still covered in tropical rainforest; this is one of the richest ecological zones on earth. One-sixth of the world's birds are found here. It has 75 parrot species, more than any other country, and more than 20 species of lories and lorikeets—far more than any other nation. In 1966, Indonesia was opened up to foreign investors and within ten years virtually all of the rainforests had been signed away as timber concessions.

Irian Jaya (West Irian)

Province of Indonesia; occupies western half of New Guinea (Netherlands New Guinea until 1963), 414,000 km^2 (160,000 sq. mi.) in extent and 2,500 km (1,550 mi.) from east to west. Nearly half the province is hilly or mountainous; there are ten peaks over 4,000 m (13,100 ft.). Mining and logging are encroaching on the rainforest but a system of reserves is being established.

Japen (Jobi or Yapen) Island

Off the northwest coast of Irian Jaya in Geelvink Bay, south of Biak, and approximately the same size. Species: *Lorius lory jobiensis* and *Pseudeos fuscata*.

Kai (Kei) Islands

Two larger islands (Kai Besar and Kai Kecil) and many small ones in Maluku Province, Indonesia. They are situated north of Tanimbar and 137 km (85 mi.) south of the Bomberai Peninsula in western New Guinea. Total area is about 900 km^2 (350 sq. mi.). Described as "as close as one will find to the perfect tropical islands" (Whitten and Whitten, 1992)—but they have an impoverished avifauna. Kai Kecil is heavily populated, Kai Besar has a sparse human population. Species: *Eos b. bornea* and *Charmosyna p. placentis*.

Lombok, see Bali

Louisiade Archipelago

Volcanic island group of the southwest Pacific, part of Papua New Guinea. Species: *Lorius h. hypoinochrous*.

Mios Num (Meos Num)

A small island in Geelvink Bay, south of Biak, about 40 km (25 mi.) long. Species: *Eos cyanogenia*.

Misool (Misol or Mysol)

One of the western Papuan Islands, east of Obi and north of Seram, about 80 km (50 mi.) long. Species: *Lorius lory*, *Chalcopsitta atra bernsteini*, *Eos squamata* and *Trichoglossus h. berauensis*.

Morotai

Island immediately north of Halmahera, Indonesia. Species: *Lorius garrulus morotaianus* and *Eos squamata*.

New Britain

Volcanic island in Bismarck Archipelago, southwest Pacific, 36,000 km^2 (14,000 sq. mi.) in area. The main town is Rabaul. It is part of the Territory of New Guinea. Species: *Lorius hypoinochrous devittatus*, *Trichoglossus h. massena*, *Charmosyna placentis pallidior* and *Charmosyna rubrigularis*.

New Caledonia

Volcanic island of west Pacific, a member of the French Community; area of 18,650 km^2 (7,200 sq. mi.). Species: *Trichoglossus h. deplanchii* and *Charmosyna diadema*.

New Guinea

The second largest island in the world—815,000

km^2 (315,000 sq. mi.), north of Australia. It is divided into Irian Jaya (West Irian) in the west (part of Indonesia) and Papua New Guinea in the east. It boasts the highest mountains in the Eastern Hemisphere after the Himalayas and is more than six times larger than England. More lories occur here than anywhere else.

New Hebrides, see Vanuatu

New Ireland

An island in the Bismarck Archipelago, 8,550 km^2 (3,300 sq. mi.) in extent. Species: *Lorius albidinuchus, Trichoglossus haematodus massena* and *Charmosyna placentis pallidior.*

Obi

An island about 64 km (40 mi.) long, south of Halmahera and northwest of Seram. Species: *Lorius garrulus flavopalliatus, Eos squamata obiensis* and *Charmosyna placentis intensior.*

Pohnpei (Ponapé)

One of the Federated States of Micronesia in the Caroline Islands about 7 degrees north of the equator in the western Pacific. Until 1914 it was a German colony; today it is greatly influenced by the U.S.A. Pohnpei Island is 334 km^2 (129 sq. mi.) in extent, the second largest island of the Carolines. Rainfall is high; in the inaccessible interior, where forest-covered mountains rise to 770 m (2,500 ft.) it is more than 10 m (33 ft.) annually.

Salawati

One of the western Papuan Islands, almost adjoining the westernmost point of Irian Jaya. It is approximately 50 km (19 mi.) square. Human settlement is mainly confined to the coast; only one road crosses the island. Species: *Lorius lory lory, Chalcopsitta atra atra, Pseudeos fuscata* and *Charmosyna rubronotata.*

Sangihe (Sangir)

The Sangihe-Talaud Islands are a volcanic island group in Indonesia, northeast of Sulawesi, covering approximately 780 km^2 (300 sq. mi.). The islands stretch almost to the Philippines, toward the southern tip of Mindanao, and have strong cultural links to that group. The avifauna is impoverished. Species: *Eos histrio* and *Trichoglossus ornatus.*

Seram

An island of the southern Moluccas, Indonesia, 18,650 km2 (7,200 sq. mi.) in extent and 340 km (211 mi.) from east to west. It lies to the west of the west coast of Irian Jaya and has a chain of high mountains, not volcanic. Species: *Lorius domicellus, Eos bornea rothschildi, E. semilarvata, Trichoglossus h. haematodus* and *Charmosyna p. placentis.*

Society Islands

Group in South Pacific, part of French Polynesia. Tahiti is the best known island. Species: *Vini peruviana* (extinct on Tahiti).

Solomon Islands

Volcanic archipelago in the southwest Pacific, covering 10,600 km^2 (4,100 sq. mi.). The islands include Bougainville and Buka which belong to the Territory of New Guinea. Most of the islands are a British protectorate, established in 1893. They include Kolombangara (heavily deforested) and San Cristobal (Makira) which is still largely primary forest. Logging concessions have been granted on most of the forest of the Solomons. Species: *Lorius chlorocercus, Chalcopsitta cardinalis, Trichoglossus h. massena, Charmosyna meeki, Charmosyna placentis* and *Charmosyna margarethae.*

Sulawesi (Celebes)

Indonesia's third-largest island, with an area of 173,000 km^2 (66,400 sq. mi.); it lies between Kalimantan and Maluku. Consisting of four long, narrow peninsulas, the main landmass is 1,300 km (806 mi.) long, although as narrow as 56 km (35 mi.) wide, in places. Large-scale conversion of forest to agriculture has occurred in most regions except the high volcanic mountains. The tropical scenery is spectacular. It has a flora and fauna of particular interest but this is jeopardized by a rapidly increasing human population. Species: *Trichoglossus ornatus* and *Trichoglossus flavoviridis meyeri.*

Sumba

Situated south of Flores and midway between Sumbawa and Timor, Indonesia, it is a dry, mostly barren island; 300 km (186 mi.) long by 80 km (50 mi.) wide. It has been extensively deforested, with a few relatively undisturbed small pockets of forest in the south. Species: *Trichoglossus haematodus fortis.*

Lobo River, Karakelong, in the Talaud Islands; habitat of the red and blue lory (*Eos histrio talauentis*).

Photo: Jan van Oosten

Tanimbar Islands (Tenimber Islands)

A group of more than 60 nonvolcanic islands in Indonesia, with a land area of 5,570 km^2 (2,150 sq. mi.). The most isolated group in southernmost Maluku, it is located 675 km (420 mi.) southeast of Ambon but only 480 km (300 mi.) north of Australia. The largest island is Yamdena which covers 5,082 km^2 (1,960 sq. mi.); most of it is still forested. These islands have empty beaches, pristine coral reefs, unique flora and fauna and superb scenery. Species: *Eos reticulata*.

Tasmania

Island state of Australia, off the southeast coast, 68,000 km2 (26,000 sq. mi.) in extent. Species: *Trichoglossus h. moluccanus* and *Glossopsitta concinna*.

Timor

A war-torn island, the largest of the Lesser Sunda group, it is 480 km (298 mi.) long and 80 km (50 mi.) wide. It is only 800 km (500 mi.) northwest of Darwin, Australia. Indonesia brutally invaded East Timor, which formerly belonged to the Portuguese,

in 1975. Thousands of people were killed. This is still a sensitive area, politically, and was reopened only in 1989. Because of the political situation, it is not well-known ornithologically. Little forest has survived in the eastern part of the island. Species: *Trichoglossus h. capistratus*, *Trichoglossus euteles* and *Trichoglossus iris*.

Tonga

Independent kingdom in the South Pacific, comprising about 150 islands, some with active volcanoes. Its total area is 675 km^2 (260 sq. mi.). Species: *Vini australis*.

Vanuatu (formerly New Hebrides)

An archipelago in the southwest Pacific; the volcanic islands span 15,700 km^2 (6,050 sq. mi.). The main group of islands cover 1,100 km (680 mi.). The 100 or so islands gained independence from joint British and French rule in 1980. The largest island is Espiritu Santo which forms one-third of the land mass. Species: *Trichoglossus haematodus massena* and *Charmosyna palmarum*.

Wetar

A small island 53 km (33 mi.) north of eastern Timor. Species: *Trichoglossus h. flavotectus* and *Trichoglossus iris wetterensis*.

List of Maps

References Cited

Amadon, D., 1942, Birds collected during the Whitney South Sea Expedition, *American Mus. Novitates*, July 24.

Anon, undated (produced in 1996), The Status and Conservation of Birds of the Sangihe and Talaud Islands, Sulawesi, Indonesia/University of York, England.

Anon, 1978, Recovery round–up, *Corella* 2,61.

Anon, 1985, Perfect Lorikeets (*Trichoglossus euteles*), *Magazine of the Parrot Soc.* XIX (1): 1–2.

Anon, 1989, Big eats—for small birds only, *Birds International* 1 (1):91.

Arndt, T., 1990–1996, *Lexicon of Parrots* Arndt-Verlag, Bretten. 1992, Die Papageien von Biak, *Papageien* 5 (2): 61–5.

Astley, H.A., 1908, A collection of rare birds from New Guinea, *Avicultural Magazine* New Series, VI (12): 334–5.

Attenborough, D., 1981, *Journeys to the Past* Lutterworth Press, Guildford.

'Backyarder', 1994, Bicentennial Gums, *Australian Aviculture* 48 (11): 253–4.

Bahr, P., 1911, Notes on some Fijian birds in captivity, *Avicultural Magazine* third series, iii (2): 49–56.

Baker, R. H., 1951, *The avifauna of Micronesia; its origin, evolution and distribution*, Univ. Kans. Publs. Mus. Nat Hist. 3: 1–359.

Bardgett, D. and I., 1991, The Pleasing or Red-flanked Lorikeet, *Foreign Birds* 7 (2): 46–48.

Barnicoat, F. C., 1976, Breeding of the Little Lorikeet, *Avicultural Magazine* 82 (2): 65–7.

——1983, A Lory Collection, *Avicultural Magazine* 89 (3): 181–5.

——1995, The first South African breeding of the Talaud Red and Blue Lory *Eos histrio talautensis*, *Lori Jnl Internationaal* 4: 42–45.

Bauer, F., 1995, Livefood for Lories, *Australian Birdkeeper* 8 (10): 500–502.

Baur, F., 1991, The Biak project—field research in Lories, *Proc. First Int'l Lory Conference* R1–5.

——1992, Der Smaragd-Gualori, *Papageien* 5 (2): 42–5.

Bean, N., 1996, Die Papageien der Sula-Inseln, *Papageien* 9 (3): 89–93.

Beehler, B. M., 1978, Notes on the mountain birds of New Ireland, *The Emu* 78 (2): 65–70.

——1991, *A Naturalist in New Guinea*, University of Texas Press, Austin, U.S.A.

Beehler, B. M., C. G. Burg, C. Filardi and K. Merg, 1994, Birds of the Lakekamu-Kunamaipa Basin, *Muruk* 6 (3): 1–8.

——, T. Pratt and D. Zimmerman, 1986, *Birds of New Guinea* Princeton University Press, Princeton, U.S.A.

Bell, H. L., 1966, Some feeding habits of the Rainbow Lorikeet, *Emu* 66 (6):71–72.

—— 1979, The effects on rain-forest birds of planting of teak *Tectona grandis* in Papua New Guinea, *Aust. Wildl. Res.* 6: 305–318.

—— 1984, New or Confirmatory Information on Some Species of New Guinean Birds, *Australian Bird Watcher* 10 (7): 209–228.

Bellingham, M. and A. Davis, 1988, Forest Bird Communities in Western Samoa, *Notornis* 35 117–128.

Bertagnolio, P., 1967, Breeding the Yellow-backed Lory, *Magazine of the Parrot Soc.* 1 (10): 7–8.

——1974, My birds in 1973, *Avicultural Magazine* 80 (4): 135–137.

——1978, Conservation through colony-breeding of the brush-tongued Parrots, *Magazine of the Parrot Soc.* XII (8): 197–20.

Blaber, S. J. M., 1990, A Checklist and Notes on the Current Status of the Birds of New Georgia, Western Province, Solomon Islands, *Emu* 90 205–214.

Blakers, M., S. Davies and P. Reilly, 1984, *The Atlas of Australian Birds* Melbourne University Press.

Boosey, E. J., 1956, *Foreign Bird Keeping* Cage Birds, London.

Bosch, J., 1987, Brutbiologie und Jugendentwicklung des Stellaloris (*Charmosyna papou goliathina* Rothschild und Hartert), *Trochilus* 8 (1): 7–12.

——1988, Der Grunschwanzlori, *Lorius chlorocercus*, *Trochilus* 9: 134–6.

——1993, The Goldie's Lorikeet, *Trichoglossus goldiei*, Sharpe, *Lori Jnl Internationaal* (2):39–42.

——1994, The Papuan Lory, *Charmosyna papou* (Scopoli), with particular details of the Mount Goliath or Stella Lory, *Charmosyna papou goliathina* (Rothschild and Hartert), *Lori Jnl Internationaal* (3): 51–70. (First published in 1988 in *Trochilus* 9: 35–50.)

——1995, The Fairy Lorikeet, *Charmosyna pulchella* (G. R. Gray), *Lori Jnl Internationaal* 4 (3): 57–72

Bowen, C., 1979, *A-Z of Health Foods*, Hamlyn, London.

Boyd, B., 1987, Little Lorikeets, a challenge to the dedicated, *Australian Aviculture* 41 (5): 114–7.

Brasseler, W., 1994, The Black-winged Lory, *Eos cyanogenia*, *Lori Jnl Internationaal* 3: 46–7.

Bregulla, H., 1992, *Birds of Vanuatu*, Anthony Nelson, Shropshire, UK.

——1993, Die Papageien Neukaledoniens, *Gefiederte Welt*, 117 (9): 310–3.

Brice, A., K. Dahl and C. R. Grau, 1989, Pollen digestibility by Hummingbirds and Psittacines, *The Condor* 91: 681–688.

——1991, Researcher's study pollen's role in bird nutrition, *Cage and Aviary Birds*, May 4, 5.

Brockner, A., 1989, Haltung und Zucht des Schimmerlori, *Papageien* (3): 74–6.

——1990, Haltung und Zucht des Erzloris *Lorius domicellus*, *Papageien* (1): 6–8.

——1991, Haltung und Zucht des Mount-Goliath Papualoris, *Papageien* (2): 38–40.

——1992a, Purple-naped Lory, *Bulletin ILS* 8 (1): 6–9. (Translation of 1990 reference.)

——1992b, Der Erzlori, *Papageien* (4): 131–2.

——1993a, Haltung und Zucht des Irisloris, *Papageien* 6 (2): 50–53.

——1993b, The Status of the Purple-naped Lory, *Australian Birdkeeper* 6 (7): 340–342.

——1994, Der Blauohrlori, *Papageien* 7 (1): 6–9.

Brook, E. J., 1909, A consignment of New Guinea rarities, *Avicultural Magazine* New Series VII (11): 336.

—— 1914, Breeding of the Fair Lorikeet, *Avicultural Magazine* Third Series V1. no 1, 29–30.

Brooker, Braithwaite and Estbergs, 1990, Foraging Ecology of Australian Wet-dry Tropics, *Emu* 90, 218.

Brown, P., 1978, The Parrots of Australasia, *Magazine of the Parrot Soc.* X11(6): 155–9.

—— 1996, Musk Lorikeets breeding in a Tasmanian winter, *Eclectus* 1 (1): 2.

Brown, W. H., 1966, Breeding Massena's Lorikeet, *Avicultural Magazine* 72 (6):155–6.

Bruch, K.and M. Bruch, 1983, Eine Mischlingszucht Gualori x Rothschild's Goldstrichellori, *Trochilus* 4: 44–46.

Bruce, M. D., 1973, Notes on aggressive and territorial behavior in nectar-feeding birds, *Emu* 73 (4): 186-7.

Brue, R. N., 1994, Concepts in adult and pediatric Psittacine nutrition, *Proc. Int'l Aviculturists Soc.*, Memphis.

Bruner, P. L., 1972, *Field Guide to the Birds of French Polynesia*, Honolulu: Pacific Scientific Information Center, Bernice P. Bishop Museum.

Buckell, T., 1992, Close Cousins; a look at three members of *Trichoglossus* genus, *Lori Jnl Internationaal* 1: 8–11.

—— 1996, Believed first UK breeding of the Yellow and Green Lorikeet, *Magazine of the Parrot Soc.* XXX (7): 217–8

Burkard, R., 1983, Practical experience with the hatch of lories and lorikeets, *Proc. Jean Delacour/Int'l Fdn for the Conservation of Birds Symposium:* 431–439.

Cain, A. J., 1955, A revision of *Trichoglossus haematodus* and the Australian Platycercine Parrots, *Ibis* 97: 432–479.

—— and I. C. J.Galbraith, 1956, Field notes on birds of the eastern Solomon Islands, *Ibis* 98: 100–134.

Cahaydin, Y., P. Jepson and B. Manoppo, 1994, The Status of *Cacatua goffini* and *Eos reticulata* on the Tanimbar Islands, Report PHPA/BirdLife International.

Campbell, G., 1993, Breeding the Blue-crowned Lory, *Avicultural Magazine* 99 (1): 27–29.

——1997, Solitary Lories, *Phigys solitarius*, in San Diego Zoo, *Lori Jnl Internationaal*, 6: 13–18..

Campbell, R., 1979, Birds of the Tari district, Southern Highlands Province, *New Guinea Bird Soc. Newsl* no.159: 9–14.

Cannon, C. E., 1984, Movements of Lorikeets with an Artificially Supplemented Diet, *Aust. Wildl. Res.* 11, 173–9.

Cappel, C., 1988, Saphirloris—Kostbarkeiten der Natur, *Geflugel-Borse* (13),July 8: 14–15.

Casmier, S. R., 1989, Boomer Came Back, *Bulletin of the International Loriidae Soc.* 5 (3): 8–9.

Casmier, S. R. and D. Wisti-Peterson, 1995, Notes on Lory Aggression, Diet and Housing, *A.F.A.Watchbird* XXII (5): 37–9.

Cayley, N. W., 1938, *Australian Parrots: Their Habits in Field and Aviary*, Angus & Robertson, Sydney.

Chapman, A. and S. Hazeldon, 1994, Rainbow Lorikeets breeding in Eastern Goldfields, *West. Aust. Nat.* 19 (4): 352.

Chapman, F. M., 1934, The American Museum's Whitney Expedition, *Avicultural Magazine*, Fourth series, XII (9): 247–8.

Christidis, L., R. Schodde, D. Shaw and S. Maynes, 1991, Relationships among the Australo-Papuan Parrots, Lorikeets, and Cockatoos (Aves: Psittaciformes): Protein Evidence, *Condor* 93: 302–317.

Churchill, D. M. and P. Christensen, 1970, Observations on pollen harvesting by brush-tongued lorikeets, *Aust. J. Zool.* 18: 427–437.

Clapp, E. G., 1979, Further sighting of the Pygmy Streaked Lory *Charmosyna wilhelminae* in the Northern Province, *New Guinea Bird Soc. Newsl.* no. 15, 5: 4–5.

Clayton, G., 1993, Breeding of Josephine's Lorikeet, *Magazine of the Parrot Soc.* XXV11 (6): 185–6.

Clear, P. S., 1993, Breeding the Emerald Lorikeet, *Avicultural Magazine* 99 (1): 21–3.

Clubb, S., 1996, Dry diets for Lories, *Just Parrots* Aug/Sept: 48–9.

Clunie, F., 1984, *Birds of the Fiji Bush*, Fiji Museum, Suva.

Coates, B. J., 1985, *The Birds of Papua New Guinea* vol. 1, Dove Publications, Alderley, Australia.

Coles, D., 1985, Papua New Guinea, *Magazine of the Parrot Soc.*, X1X (12):326–330.

Collar, N. J., M. J.Crosby and A. J. Stattersfield, 1994, *Birds to Watch 2*, Birdlife International, Cambridge.

Collard, W. H., 1965, Breeding of the Chattering Lory, *Avicultural Magazine* 71 (4): 111–112.

Collins, L. and D. Sefton, 1994, The Red Lory: A single species breeding program, *Bird Breeder* 66 (7): 34–45.

Cooke, D., 1990, The breeding and parent-rearing of *Eos squamata obiensis*, popularly known as Wallace's Lory, *Avicultural Magazine* 96 (1):24–9.

——1991, Breeding the Red-spotted Lorikeet, *Avicultural Magazine* 97 (3): 130–135.

——1992, Breeding the Red-flanked Lorikeet, *International Loriinae Soc.* (3): 5–9.

——1993a, Playful Obies are reasonably hardy, *Cage and Aviary Birds* Jan 23:6.

——1993b, Notes on the breeding and status of the Red Spotted Lorikeet, *Avicultural Magazine* 99 (3): 118–121.

Courtney, J., 1996, The Juvenile Food-begging Calls, Food-swallowing Vocalisation and Begging Postures in Australian Cockatoos, *Australian Bird Watcher* 16 (6): 236–249.

—— 1997, The Juvenile Food-begging Calls and Associated Behaviour in the Lorikeets, *Australian Bird Watcher* 17 (2):61-70.

Coyle, P., 1988, Rainbow Lorikeets (*Trichoglossus haematodus*) released on Rottnest Island in 1960, *West. Aust. Nat.* 17 (5): 109–110.

Cressman, W. and E. Cressman, 1993, A pair of Blue-streaked Lory *Bulletin Int'l Loriinae Soc.* 9 (3): 8–9.

Cristo, R., 1994, Nest Boxes, *American Lory Soc. Network* 1 (2): 9.

Dahne, H., 1992, The Fairy Lorikeet, *The Loriinae* 8 (2): 19–21.

Davidson, P. J., R. S. Lucking, A. J. Stones and N. J. Bean, 1994, Report on an ornithological survey of Taliabu, Indonesia, with notes on the Babirusa Pig, Jakarta.

Dawson, P., 1973, A yellow Scaly-breasted Lorikeet, *Sunbird* 4: 9–10.

Dear, A., 1986, The Red-collared Lorikeet, *Magazine of the Parrot Soc.* XX (10): 324–5.

Delacour, J., 1936, The Ruby Lory (*Vini kuhli*), *Avicultural Magazine* Fifth series vol. 1,5: 127-8.

Dhondt, A., 1976, Bird Observations in Western Samoa, *Notornis* 23, 29–43.

Diamond, J. M., 1972, *Avifauna of the Eastern Highlands of New Guinea*, Pbl Nuttall Orn.Club, no 12, Cambridge, Massachusetts, USA.

——, and M. LeCroy, 1979, Birds of Karkar and Bagabag Islands, New Guinea, *Bulletin of American Mus. Nat. Hist* 164 art.4: 467–531.

Dickinson, E. C., R. S.Kennedy and K. C.Parkes, 1991, *The Birds of the Philippines, An annotated check-list*, BOU, Tring, Herts.

Dios, M. A. R. de, and R. Sweeney, 1993, Keeping and breeding the Obi Lory, *Australian Birdkeeper* 6 (11): 543–5.

Donaghho, W. R., 1950, Observations of some birds of Guadalcanal and Tulagi, *Condor* 52: 127-132

Dooren, G. van, 1992, Experiences with the Perfect Lorikeet *Trichoglossus euteles*, *Lori Jnl Internationaal* 1: 4–7.

Dosser, G., 1990, Little Lorikeet, *Australian Aviculture* 44 (6): 133–134.

Engels, F. - J., and H. - D. Philippen, 1992, The Yellow-bibbed Lory, *Lorius chlorocercus* Gould (1865), *Lori Jnl Internationaal* 1:8–12.

Fergenbauer-Kimmel, A., 1984, Haltung und Zucht von Iris loris (*Psitteuteles iris*), *Trochilus* 5: 115-119.

——1992, Keeping and raising of *Psitteuteles iris* (Iris lorikeet), *The Loriinae* 8 (1): 25-29.

Filippich, L. J, and R. Domrow, 1985, Harpyryhnchid Mites in a Scaly-breasted Lorikeet, *Jnl. of Wildlife Diseases* 21 (4): 457–8.

Finch, B. W., 1979, An ornithological survey in the Efogi area, *New Guinea Bird Soc. Newsl.* no. 160: 2–14.

Finsch, O., 1900, Systematische ueberricht der Vogel der Sudwest-Inseln, *Notes Leyden Mus* 23: 237.

Forshaw, J. M., and W. T.Cooper, 1973, *Parrots of the World* Lansdowne Press, Sydney.

——1981, *Australian Parrots*, Lansdowne Press, Melbourne.

——1989, *Parrots of the World* 3rd (rev) edn., Blandford, London.

Franklin, D., 1996, A massive aggregation of the Varied Lorikeet, *Eclectus* 1 (1): 6–7.

Gammond, M., 1981, Breeding the Perfect Lorikeet, *Magazine of the Parrot Soc.* XV (11): 315.

Gerischer, B., 1990, Erfahrungen mit dem Veilchenlori, *Gefiederte Welt* 114 (9): 265-7.

——1991, Keeping, rearing and breeding the Red Lory, *Int'l Loriinae Soc.* 25, 5–8. (Translated from *Gefiederte Welt*, 1989, 113 (7)).

Gibb, J., P. Bull, W. McEwen and I. Sewell, 1989, The Bird Fauna of Niue compared with those of Tonga, Samoa, the Southern Cooks and Fiji, *Notornis* 36: 285–298.

Gibson, J., 1984, Breeding Blue-streaked Lories, *Magazine of the Parrot Soc., XV111 (12): 363–6.

Gilliard, E. T. and M. LeCroy, 1961 Birds of the Victor Emmanuel and Hindenburg Mountains, New Guinea. Results of the Amer. Mus. Nat. Hist.expedition to New Guinea in 1954. *Bulletin American Mus. Nat. History* 123: 1-86

——1967, Results of the 1958–1959 Gilliard New Britain Expedition: 4. Annotated list of birds of the Whiteman Mountains, New Britain, *Bulletin American Mus. Nat. History* 135: 173–216.

Goodfellow, W., 1906, Notes on Mrs. Johnstone's Lorikeet, *Avicultural Magazine* new series, 1V (3): 82–8.

—— 1933, Some reminiscences of a collector, *Avicultural Magazine* fourth series, Xl (7): 181–190.

Gould, J., 1875–88, *The Birds of New Guinea*, vol V, Sotheran, London.

Graves, G. R., 1992, The endemic land birds of Henderson Island, Polynesia: notes on natural history and conservation, *Wilson Bulletin* 104 (1): 32–43.

Greensmith, A., 1975, Some field notes on Melanesian Psittaciformes, *New Guinea Bird Soc. Newsl.* no 114: 7–10.

Gregory, P., 1995, *Birds of the Ok Tedi Area*, Ok Tedi Mining Ltd.

Haines, L. C., 1946, Lorikeets prefer different diets, *Emu* 46:76.

Hamley, T., 1977, Feeding behaviour of Scaly-breasted Lorikeets, *Sunbird* 8 (2): 37–40.

Harley, D., 1995, Conservation of Box-Ironbark Forests, *Field Nats News* June, p1.

Harrison, C. J. O., and D. T. Holyoak, 1970, Apparently undescribed parrot eggs in the collection of the British Museum (Natural History), *Bull. Brit. Orn. Club*, 90: 42–6.

Hartley, E., 1912, *Bird Notes* (September): 247.

Hartert, E., 1930, On a collection of birds made by Dr. Ernst Mayr in Northern Dutch New Guinea, *Novitates Zoologicae* vol. 36 (104–105).

Hayward, J., 1984, Blue-streaked Lory (*Eos reticulata*), *Parrot Breeder* no 3.

Hill, L. J., 1995, Lorikeet Feeding Exhibit—Marine World Africa USA, *A.F.A.Watchbird* XXII (5): 8–11.

Hiller, C. M., 1987, Notes on Dusky Lory (*Pseudeos fuscata*) Behaviors, *Loriidae Newsletter* 3 (1): 1–3.

Hesford, C., 1995, Yellow-backed Lory not a Lutino, *Cage and Aviary Birds*, March 18, p8.

Hodson, D. J., 1978, Ornate Lorikeets, *Magazine of the Parrot Soc.* XII (9): 231–2.

Holland, R., 1994, Fresh pollen improves the diet of lorikeets, *Cage and Aviary Birds* July 16.

Holmes, D. and K. Phillipps, 1966, *The Birds of Sulawesi* Oxford University Press, Oxford.

Holsheimer, J. P., 1996, Nutrition of Lories and

Lorikeets, *Parrot* (PS of NZ), vol 2, February: 39–42.

Holyoak, D. T., 1975, Les oiseaux des Iles Marquises, *Oiseau Revue fr. Orn.* 45: 341–366.

——1979, Notes on the birds of Viti Levu and Taveuni, Fiji, *Emu* 79:7–18.

Homberger, D. G., 1981, Functional morphology and evolution of the feeding apparatus in parrots, with special reference to the Pesquet's Parrot, in Pasquier, *Conservation of New World Parrots* ICBP Tech. Pub. No 1.

Hood, R., 1932—reprinted 1995, The story of our lories, *Avicultural Bulletin* 24 (12): 2–3.

Hopper, S. D., 1980, Pollen and nectar feeding by Purple-crowned Lorikeets on *Eucalyptus occidentalis*, *Emu* 80 (4): 239–40.

—— and A. A.Burbidge, 1979, Feeding behaviour of a Purple-crowned Lorikeet on flowers of *Eucalyptus buprestium*, *Emu* 79: 40–42.

Hubers, J., 1993a, A closer look at Josephine's Lory, *Lori Jnl Internationaal* 2: 21–23.

—— 1993b, The Yellow and Green Lorikeet *Trichoglossus flavoviridis flavoviridis*, *Lori Jnl Internationaal* 2:76–78.

1994, The Rosenberg Lory *Trichoglossus haematodus rosenbergii*, *Lori Jnl Internationaal* 3: 87–9.

—— 1995a, The Yellow-streaked Lory, *Chalcopsitta scintillata*, *Lori Jnl Internationaal* 4: 34–40.

——1995b, Flagellate infections with Lories, Lorikeets (Subfamily Loriinae) and Hanging Parrots (Genus Loriculus spp), *Lori Jnl Internationaal* 4: 46–47.

——1996, The Violet-Necked Lory, *Eos squamata*, and its sub-species, *Lori Jnl Internationaal* 5 (3): 64–69.

Hubers, J., and G. van Dooren, 1994, The Blue-Headed Lory, *Trichoglossus haematodus caeruleiceps*, a newcomer in aviculture, *Lori Jnl Internationaal* 3: 38–45.

Hutchins, B. R., and R. H.Lovell, 1985, *Australian Parrots, A Field and Aviary Study* Avicultural Society of Australia, Melbourne.

Hutchinson, L., 1978, Breeding the Violet-necked Lory, *Avicultural Magazine* 84 (2): 119.

Jepson, P., 1993a, Recent ornithological observations from Buru, *Kukila* 6 (2): 85–109.

——1993b, Secrets of Sumba, *World Birdwatch* 15 (4): 6–8.

Johnson, R., 1952, The breeding of the Purple-crowned Lorikeet, *Australian Aviculture* 6: 140–1.

Johnstone, Mrs., 1906, The nesting of *Trichoglossus johnstoniae*, *Avicultural Magazine* new series, V (1): 44–6.

Jones, B., 1995, My Little Dumplings, *American Lory Soc. Network* 2 (2): 4.

Jones, M., H. Banjaransaree and R. Grimmett, 1990, The Ecology and Conservation of the Birds of Sumba and

Buru, prelim. report.

Joshua, S., 1993a, The Goldie's Lorikeet (*Trichoglossus goldiei*), *Magazine of the Parrot Soc.* XXV11 (6): 200–204.

——1993b, *in* Greater range of foods now available to Lory breeders, *Cage and Aviary Birds* July 24, 12–13.

——1994a, Chromosomes solve puzzles of parrot taxonomy, *Proc. 3rd Int'l Loro Parque Parrot Convention* 69–78.

——1994b, The classification of the Lories, subfamily: Loriinae, *Lori Jnl Internationaal* 3: 19–23.

Kaal, G. T. F., 1991, Mycosis in Loriinae, *Proceedings 1st Int'n'l Loriinae Soc. Conference* 0–1 to 0–10.

Kapac, J., 1985, Adventures with Lories, *The Loriidae Newsletter* 1 (3): 11–12.

——1989, Experiences with Lories, *The Loriidae* 5 (3): 2-3.

Kenning, J., 1993–4, The Solomon Islands—A Pleasant Remembrance, *Australian Birdkeeper* 6 (12): 598–601.

——1994, Der Brillenkakadu auf Neubritannien im Bismarck-Archipel, *Papageien* (3): 89–92.

——1995, The Ponapé Lory (*Trichoglossus rubiginosus*), *Australian Birdkeeper* 8 (8): 390–92. See also Der Kirschlori auf Pohnpei, *Papageien* 9/95, 273–275.

King, H., 1940 (article title unknown), *Cage Birds* January 26.

Krupa, A. P., 1991, Hand-rearing of the Chattering Lory, *Magazine of the Parrot Soc.* XXV (9): 309–11.

Kuah, L., 1993, Breeding the Purple-naped Lory, *Avicultural Magazine* 99 (2): 72–5.

Kuehler, C. and A. Lieberman, 1992, French Polynesia, *PsittaScene* 4 (4): 11–12.

Kyme, R. T.,1975, Breeding the Iris Lorikeet, *Avicultural Magazine* 81 (1): 22–4.

——1979a, Breeding the Yellow-streaked Lory, *Avicultural Magazine* 85 (1): 2–4.

——1979b, Breeding of the Yellow-streaked Lory, *Magazine of the Parrot Soc.* XIII (7): 171–2.

Lambert, F. R., 1993a, Trade, status and management of three parrots in the North Moluccas, Indonesia: White Cockatoo *Cacatua alba*, Chattering Lory *Lorius garrulus* and Violet-eared Lory *Eos squamata*, *Bird Conservation International* 3 (2): 145–168.

——1993b, Some key sites and significant records of birds in the Philippines and Sabah, *Bird Conservation International* 3 (4): 281–297.

Lambert, F., R. Wirth, E. Seal. J. B. Thompsen and S. Ellis-Joseph, undated *Parrots, An Action Plan for their conservation*, 1993–1998 draft report BirdLife International, IUCN and others.

Landolt, R., 1981, Zucht des Vielstrichelloris, *Charmo-syna multistriata* (Rothschild), und des Schonloris, *Charmosyna placentis subplacens* (Sclater), *Gefiederte Welt* 105: 181–184.

Lavery, H. J. and J. G. Blackman, 1970, Sorghum damage by lorikeets, *Queensl, Agric. J.*, 2–4, Nov.

LeCroy, M., W. S. Peckover and K. Kisokau, 1992, A Population of Rainbow Lorikeets *Trichoglossus haematodus flavicans* Roosting and Nesting on the Ground, *Emu* 92: 187–90.

Leggett, R., and P. F. Woodall, 1987, Hybrid Scaly-breasted x Rainbow Lorikeets, *Australian Bird Watcher* 12: 122–126.

Leighton, M., 1986,Hornbill social dispersion: variations on a monogamous Theme, *In Ecological aspects of social evolution* 108–30, Princeton University Press, Princeton, U.S.A.

Lendon, A., 1946, Memories of the Moluccas, *Avicultural Magazine* 52 (6): 206–213.

——1973, *Australian Parrots in Field and Aviary*, Angus & Robertson, Sydney.

——1979, *Australian Parrots in Field and Aviary* (rev.edn), Angus and Robertson, London and Sydney.

Lepschi, B. J.,1993, Food of Some Birds in Eastern New South Wales: Additions to Barker and Vestjens, *Emu* 93, 195–199.

Lewins, E., 1996, Blue-crowned Lories, *Vini australis*, at the San Diego Zoo, *Lori Jnl Internationaal* 5: 31–33.

Lindholm. J. H.,1995, Lories may be hazardous... *A.F.A.Watchbird* XX11 (5): 22–27.

Loman, B. and H. Loman, 1992, The Cardinal Lory, *Chalcopsitta cardinalis*, *Lori Jnl Internationaal* 1: 4–6.

Longo, J., 1985, Breeding the Stella's Lorikeet, *The Loriidae Newsletter* 1 (1): 3–5.

Low, Rosemary, 1972, Long-tailed Papuan rarity, *Cage and Aviary Birds*, Dec.14,1 and 4.

——1974, Identification, care and breeding of *Trichoglossus haematodus*, *Avicultural Magazine* 80 (6): 203–4.

——1975, Talking of Parrots, *Cage and Aviary Birds*, April 24: 2.

——1976, The 1975 Breeding Season, *Avicultural Magazine* 82 (1): 11–16.

——1977a, *Lories and Lorikeets*, Pauk Elek, London.

——1977b, Hand-rearing Meyer's and Iris Lorikeets, *Avicultural Magazine* 83 (1): 12–17.

——1977c, Breeding the Black Lory, *Cage and Aviary Birds*, Dec. 29: 5–6.

——1978a, The 1977 Breeding Season, *Avicultural Magazine* 84 (2): 88–94.

——1978b, Breeding the Black Lory, *Avicultural Magazine* 84 (3): 121–4.

——1979a, Josephine's Lorikeets rear young in Germany, *Cage and Aviary Birds*, Nov.17: 6.

——1979b, These Lorikeets are perfect parents, *Cage and Aviary Birds* Aug. 25.

——1980a, Breeding results for 1978–9, *Avicultural Magazine* 86 (4):220–221.

——1980b, Ideal temperament of Goldie's Lorikeet, *Cage and Aviary Birds*, Nov.15: 6.

——1984, Breeding Duivenbode's Lory, *Avicultural Magazine* 90 (1): 18–26.

——1985a, A decade with Dusky Lories, *Cage and Aviary Birds,* Mar.30:3.

——1985b, Breeding the Tahiti Blue Lory, *Avicultural Magazine* 91 (1 & 2): 1–14.

——1986a, Nothing left to chance with the rarer Lories, *Cage and Aviary Birds* Jan. 22.

——1986b, *Parrots, their care and breeding* (revd edn), Blandford Press, Dorset.

——1988, Placentis: eight chicks in six months, *Int'l Loriidae Soc* 4 (4): 6–10.

——1989, News and Views, *Avicultural Magazine* 95 (1): 45.

——1990, Highlights of a visit to Australia, Part 2, *Australian Birdkeeper* 3 (1): 18–21.

——1991a, Breeding Lories, *Proc. First Int'l Loriinae Soc. Conference* F1–18.

——1991b, Musschenbroek's—not the easiest species to breed? *International Loriinae Soc.* 7 (2): 5–7.

——1991c, Breeding Josephine's Lorikeet at Palmitos Park, *Avicultural Magazine* 97 (4): 167–173.

——1992a, *Parrots, their care and breeding*, third (revd) edn, Blandford Press, London.

——1992b, Lories—captivating and colourful, *Cage and Aviary Birds*, Aug. 22 6–7.

——1992c, The Tahiti Blue Lory on Aitutaki, *Lori Jnl Internationaal* 1: 14–18.

——1992d, Breeding four species of *Charmosyna* at Palmitos Park, Gran Canaria, *Lori Jnl Internationaal* 1: 11–19.

——1993, Breeding the Rajah Lory, *Chalcopsitta atra insignis*, at Palmitos Park, *Lori Jnl Internationaal* 2: 67–71.

——1994a, Fiji's most exquisite bird, the Collared Lory, *Lori Jnl Internationaal* 3: 91–94.

——1994b, Observing the Tahiti Blue Lory on Aitutaki, *The Loriinae: Bulletin of the ILS* 10 (4): 5–7.

——1994c, *Endangered Parrots* (revd edn), Blandford, London.

——1995, Haltung und Zucht des Braunloris, *Papageien* 8 (3): 78–80.

Mack, A. L., and D. Wright, 1996, Notes on occurrence and feeding of birds at Crater Mountain Biological Research Station, Papua New Guinea, *Emu* 96 (2): 89–101.

Mackay, R.D., 1971, Observations for September, *New Guinea Bird Soc. Newsl.,* no 71: 3.

MacKinnon, J., 1990, *Field Guide to the Birds of Java and Bali*, Gadjah Mada University Press, Indonesia.

MacWhirter, P., 1994, Malnutrition in *Avian Medicine, Principles and Application* Eds Ritchie, Harrison and Harrison, p850.

Majnep, I .S. and R. Bulmer, 1977, *Birds of My Kalam Country* Auckland, Oxford University Press.

Manning, A., 1994/5, Dusky Lory Hybrid, *Parrot Magazine* no 6: 18–19.

Marchant, S., and P. J.Higgins (Eds), 1993 *Handbook of Australian, New Zealand and Antarctic Birds* vol 11, Oxford University Press, Melbourne.

Marsden, S., 1995, The Ecology and Conservation of the Parrots of Sumba, Buru and Seram, PhD thesis, Manchester Metropolitan University, UK.

Marsden, S. and Jones, in prep.

Martin, M., 1982, Musschenbroek's Lorikeet, *Australian Aviculture* 36 (3): 52–4, 57–8.

Mason, V. and F. Jarvis, 1989, *Birds of Bali*, Berkely: Periplus.

Masters, B., 1989, *The Passion of John Aspinall*, Coronet Books, p. 134.

Maxwell, P. H., 1952, My Rosenberg's Lorikeet, *Avicultural Magazine* 58 (3): 116.

Mayr, E. and A .L .Rand, 1937, Results of the Archbold Expeditions, X1V. Birds of the 1933–1934 Papuan Expedition, *Bull. Am. Mus. Nat. Hist.* 73, art.1: 1–248.

——and R. de Schauensee, 1939, Zoological results of the Denison-Crockett Expedition to the South Pacific for the Academy of Natural Sciences of Philadelphia, 1937-1939, Part IV. Birds from northwestern New Guinea, *Proc. Acad. Nat. Sci. Philad.* 91: 97-144

McCormack, G., 1993, The Kuramo'o of Aitutaki, *Cook Islands News* Dec.11, p10.

McCormack, G., and J. Kunzle, 1993, The 'Ura or Rimatara Lorikeet (*Vini kuhlii*): its former range, its present status, and conservation priorities. Unpubl report, presented at Seminaire Manu-Connaissance et Protection des Oiseaux (9–12 Novembre 1993), Cook Islands Natural Heritage Project.

McCullough, C., 1985, News and Views, *The Loriidae Newsletter* 1 (3): 12.

McDermid, S., 1990, Breeding of Musschenbroeks Lorikeet (*Neopsittacus musschenbroekii major*), *Magazine of the Parrot Soc.* XX1V (9): 289–90.

McGregor, K., 1991, Breeding Blue-streaked Lories, *American Cage-Bird Magazine*, November 1991: 41–44.

McWatters, A., 1996, Candida prevention—an alternative cure, *Bird Breeder*, 68 (5): 16–21.

Mees, G. F., 1965, The avifauna of Misool, *Nova Guinea Zool.* series 31: 139–203.

Merck, W., 1983, Keeping and breeding Goldie's Lorikeet (*Trichoglossus (Psitteuteles) goldiei*), *Avicultural Magazine* 89 (2): 69–70.

Meyer, A. B. and L. W. Wiglesworth, 1898, *The Birds of the Celebes and the Neighbouring Islands* 1, Berlin: Friedlander.

Michi, H., 1982, Rotstirn—und Schonloris—kleine Kostbarkeiten aus Neuguinea, *Trochilis* 3: 37–40.

Michi, M. and H. Michi, 1983, Der Meyerslori (*Trichoglossus flavoviridis meyeri*), *Trochilus* 4: 85–8.

Michorius, H. J., 1993, The Multistriated Lorikeet, *Charmosyna multistriata*, *Lori Jnl Internationaal* 2: 18–19.

Michorius, H. J. and J. Wierda, 1994, The Meyer's Lorikeet, *Trichoglossus flavoviridis meyeri*, *Lori Jnl Internationaal* 3: 85–6.

Milton, G. R. and A. Marhadi, 1987, An investigation of parrots and their trade on Pulau Bacan (North Moluccas) and Pulau Warmar, Aru Islands,. Report to WWF/IUCN.

Mitchell, P., 1979, Scaly-breasted Lorikeets at Mount Eliza, Victoria, *Australian Bird Watcher* 8 (3): 99–100.

Mivart, St.G., 1896, *A Monograph of the Lories or Brush-tongued Parrots*, Porter, London.

——1992, ibid, Fundacef-Editions, Seibersbach, Germany.

Morford, T., 1980, Breeding of the Cardinal Lory, *Magazine of the Parrot Soc.* XIV (4):84–5.

Morrison, D., 1987, Musk Lorikeets at Wattle Park, Melbourne, *The Bird Observer* 668: 106.

Muller, B. and H. Muller, 1992, The Ornate Lory, *Lori Jnl Internationaal* 1: 21–24.

Muller, K., 1991, *Spice Islands, The Moluccas*, Periplus Editions, Ca.

Müller, M. and N. Neumann, 1996, Die Haltung und Zucht des Kardinalloris (*Chalcopsitta cardinalis*), *Papageien* 9 (2): 38–41.

Nash, S. V., 1993, Concern about Trade in Red-and-Blue Lories, *TRAFFIC Bulletin* 13 (3):93–96.

Nauschutz, W., 1979, Handrearing a Chattering Lory, *Magazine of the Parrot Soc.* XIII (8): 195–6.

Neff, R., 1990, Neue Erkenntnisse zum Geschlechtsdimorphismus juveniler Rotstirnloris, *Gefiederte Welt* 114 (11): 330–332.

——1992, Der Grosse Bergzierlori, *Oreopsittacus arfaki major*, *Gefiederte Welt* 116 (2): 42–45.

——1994, Uber den Schuppenlori, *Gefiederte Welt* 118 (1): 6–8.

Neumann, N. and S. Patzwahl, 1992, Deutsche Erstzucht des Apoloris, *Papageien* (6): 178–80.

North, A. J., 1911, Nests and eggs of birds found breeding in Australia and Tasmania, vol 3, *Australian Mus. Spec. Cat.* no 1: 1–176.

O'Brian, B., (sic, O'Brien), 1985, Breeding Australian Lorikeets, *Magazine of the Parrot Soc.* XIX (10): 262–4.

Odekerken, P. 1988, undated, The Varied Lorikeet (*Psitteuteles versicolor*) *Australian Birdkeeper* 2 (7): 248–249.

——1992, Musk Lorikeet, *Parrot* 1 (1): 14–17.

——1993a, The Swainson's (or Blue Mountain) Lorikeet, *Trichoglossus h.moluccanus*, *Lori Jnl Internationaal* 2: 52–7

——1993b, Lories and Lorikeets, part 2, *Australian Birdkeeper* 6 (7): 349–352.

——1994, The Varied Lorikeet, *Psitteuteles versicolor*, *Lori Jnl Internationaal* 3: 10–14.

——1995a, Red-Collared Lorikeet, *Lori Jnl Internationaal* 4: 51–54.

——1995b, A Guide to Lories and Lorikeets, *Australian Birdkeeper*, New South Wales.

Oliver, M., 1976, Hand-reared Swainson's an 'incredible mimic', *Cage and Aviary Birds*, Mar.25.

Oosten, J. R. van, 1993, Lories/Lorikeets in the wild, *Bulletin Int'l Loriinae Soc.* 9 (1): 3–6 and 9 (3) 13–14.

——1995, Breeding the Johnstone's Lorikeet, *A.F.A.Watchbird* XXII (5): 28.

Orenstein, R., 1976, Birds of the Plesyumi area, central New Britain *Condor* 78: 370-74.

Pace, D., 1996, A Guide to the Status of Australian Parrots and Cockatoos in Victorian Aviaries, *Australian Aviculture* 50 (10): 236–241.

Page, W. T., 1910, The Stella Lory, *Bird Notes* New series, vol 1, No. 3, 65–68.

Pagel, T., 1985, *Loris*, Ulmer, Stuttgart.

Pagel, T. and H. Greven, 1990, On the development of the tongue papillae in lorikeets (*Loriinae, Psittaciformes*), *Acta Biol. Benrodis* 2: 253–254.

Paquet, M., 1994, Loriquet magnifique, *La revue des Oiseaux Exotiques* No 189, 39–40.

Patten, R. A., 1941, Observations on the Solitary Lory in captivity, *Avicultural Magazine* Fifth Ser. VI (3): 72–5.

——1947, Observations on Kuhl's Ruffed Lory (*Vini kuhli*) (sic) in captivity, *Avicultural Magazine* 53 (2): 40–43.

Patzwahl, S., 1991, Keeping and breeding the Loriinae at Vogelpark Walsrode, *Proc. First Int'l Loriinae Soc. Conference*.

Paulik, M., 1991, Short note, *International Loriinae Soc.* 7 (1): 11.

Paulik, M. and D. Kruszona, 1991, Breeding Striated Lorikeets, *A.F.A.Watchbird* 18 (2): 4.

Pearson, D. L., 1975, Survey of the birds of a lowland forest plot, in the East Sepik District, Papau New Guinea, *Emu* 75: 175-7.

Penny, A., 1996, Mad Max, *Just Parrots and Parakeets* 10 (June/July): 32.

Peters, J. L., 1937, *Check-list of Birds of the World* III, Cambridge, Mass.

Phillipps, R., 1903, The Varied Lorikeet, *Avicultural Magazine* new ser, I (9): 287–91.

Pittman, Tony, 1996, Lori-Suche im Paradies, *WP-Magazin* 2 (2): 22–24.

Porter, S., 1934, The American Whitney Expedition, *Avicultural Magazine* Fourth series, X11 (12): 327–9.

——1935, Notes on Birds of Fiji, *Avicultural Magazine* Fourth series, X111 (5): 126–139.

——1938, Notes from Australia, *Avicultural Magazine* Fifth series, 3 (11): 309–320

Poulsen, M. K.,1996, White Cockatoo, *World Birdwatch* 18 (4): 20–21.

—— and D.Purmiasa, 1996. Buru, a Forgotten Island, *World Birdwatch* 18 (3): 16–19.

Pratt, H. D., P. L.Bruner and D. G.Berrett, 1987, *A Field Guide to the Birds of Hawaii and the Tropical Pacific*, Princeton, Princeton University Press.

Pratt, T. K., 1982, New Guinea Lories in the Wild, *Newsletter of American Lory Soc.* 1 (4): 1–4.

Prestwich, A. A., 1963, *I Name This Parrot* author, London.

Pszkit, S., 1991, My Experience in Handraising a Red Lory, *Australian Birdkeeper* 4 (8): 378.

Ramsay, E. P., 1875, Linnean Society, *Sydney Morning Herald*, July 28.

Rand, A. L., 1938, Results of the Archbold Expeditions, No XIX. On some non-passerine New Guinea birds, *Am. Mus. Novit.* no 990: 1–15.

—— and E. T.Gilliard, 1967, *Handbook of New Guinea Birds*, Weidenfeld & Nicholson, London.

Regler, B., 1994, Der Veilchenlori, *Die Voliere* 17 (7): 196–9.

Rensch, B., 1931, Die Vogelvelt von Lombok, Sumbawa und Flores, *Mitt. Zool. Mus. Berlin* 17: 451-637.

Riffel, M. and S. Dwi Bekti, 1991, A further contribution to the knowledge of the birds of Sumba, Indonesia, unpublished report.

Riley, J., 1995, Preliminary assessment of the status and utilization of the Red and Blue Lory (*Eos histrio*) on Talaud and Sangihe, unpubl. report, University of York.

Rinke, D., 1985, Zur Biologie des Blaukappchens (*Vini australis*), mit Bemerkungen zum Status der Maidloris (Gattung *Vini*), *Trochilus* (6) 29–38.

—— 1986, Notes on the avifauna of Niuafo'ou Island, Kingdom of Tonga, *Emu* 86: 82–6.

——1991, Birds of 'Ata and Late, and additional notes on the avifauna of Niuafo'ou, Kingdom of Tonga, *Notornis* 38: 131–151.

——1993, The status and conservation of parrots in the Kingdom of Tonga, *PsittaScene* 5 (3): 4–5.

——1994, Haltung und Zucht des Blaukappenloris, *Papageien* 7 (3): 78–82.

Riplay (sic), S. D., 1938, Round about Dutch New Guinea, *Avicultural Magazine* Fifth series III: 267-274.

Ripley, D., 1947, *Trail of the Money Bird*, Longmans, Green, London.

Ripley, S. D., 1964, A systematic and ecological survey of birds of New Guinea, *Bull. Peabody Mus. Nat. Hist.* no 19:1-85.

Ritchie, B. W., G. J. Harrison and L. R. Harrison, 1994, *Avian Medicine: Principles and Application*, Wingers Publ., Lake Worth.

Robiller, F., 1992, *Papageien*, Band 1, Deutscher Landwirtschaftsverlag, Berlin.

Roth, H. and I., 1976 (article title unknown), *Stuefuglene* (51): 162.

Rucker, 1980, Blue-streaked Lory (*Eos reticulata*) a successful breeding, *Magazine of the Parrot Soc.* XIV (11):265–6.

Ruggles, A., 1991, What nobody told me about keeping lories and lorikeets, *Proc. First Int'l Loriinae Soc. Conference* I1–11.

Rutgers, A., 1965, *The Handbook of Foreign Birds in Colour* vol 2, Blandford, London.

Ryan, P., 1988, *Fiji's Natural Heritage*, Southwestern Publishing Co. Ltd, Auckland.

Salisbury, C. A., 1985, The successful breeding of four Australian lorikeet species and one hybrid at the Currumbin sanctuary, Queensland, Australia, *The Loriidae Newsletter* 1(3): 2-10.

Sallien, G., 1994, Lori a nuque verte, *La revue des Oiseaux Exotiques* no. 190, Dec.: 19.

Salvadori, T., 1891, *Catalogue of birds in the British Museum* vol 20, Psittaci, British Museum (Natural History), London.

St. A. S., Olive, 1938, Alfred, *Avicultural Magazine* 5th series, lll (7): 200–202.

Schodde. R., 1977, Contributions to Papuasian ornithology, VI. Survey of the birds of southern Bougainville Island, Papua New Guinea, C.S.I.R.O. Div. *Wildl. Res. Tech. Pap.* 34: 1–103.

Schonwetter, M., 1964, *Handbuch der oologie* vol 1 (9), Akademie, Berlin.

Schroeder, D., 1991, Lory housing and diets in the USA, *Proc First Int'l Loriidae Soc. Conference* N1–5.

——1994, Designing nest boxes for lories, *American Lory Soc. Network* 1 (2): 10.

——1995a, Breeding the Fairy Lorikeet, *Bird Talk* 13 (2): 92–3.

——1995b, Lorikeets in Planted Aviaries, *A.F.A.Watchbird* XX11 (5): 4–7.

Segal, H. G., 1994, Der Kirschlori *Trichoglossus rubigi-*

nosus, Papageien (6): 182–3.

Seitre, R., 1991, Der Rubinlori, *Papageien* 4 (3): 81–84.

——1994, Endangered Parrots of the French Pacific, *Proc. 3rd Int'l Loro Parque Parrot Convention*, 54–59, Loro Parque, Tenerife.

Seitre, R. and J. Seitre, 1991, Current situation of Loriinae in French Polynesia, *Proc. First Int'l Loriidae Soc. Conference* J1–7.

Serpell, J. A., 1979, Comparative ethology and evolution of communicatory behaviour in the loriine parrot genus *Trichoglossus*. Ph D thesis, University of Liverpool.

—— 1981, Duets, greetings and triumph ceremonies: analogous displays in parrot genus *Trichoglossus*, *Z.Tierpsychol.* 55: 268–283.

—— 1982, Factors influencing fighting and threat in the parrot genus *Trichoglossus*. *Animal Behaviour* 30: 1244–1251.

Shephard, M. and C. Welford, 1987, The Status of Australian Parrots in South Australian Aviaries, *Magazine of the Parrot Soc.* XX1 (5): 149–160.

——, J. Pyle and S. Fairlie, 1991,The Status of Australian Parrots in South Australian Aviaries—a Five Year Comparison 1986–1991), *Bird Keeping in Australia* Nov. 162–169.

Simmons, M. L., 1994, Lories as pets, *The Loriinae* 10 (2): 8–9.

Sindel, S., 1987, *Australian Lorikeets*, Singil Press, New South Wales.

Smiet, F., 1985, Notes on the field status and trade of Moluccan parrots, *Biol. Conserv.* 34: 181–194.

Smith, G. A., 1975a, Observations on *Trichoglossus haematodus*, *Amazona* and *Forpus* species, *Avicultural Magazine* 81 (4): 237–8.

——1975b, Systematics of parrots, *Ibis* 117, 18–68.

——1985, Mystery of the Rainbow and the King, *Cage and Aviary Birds*, October 26: 7.

——1988, News and Views, *The Loriidae Bulletin* 4 (4): 10.

Sorgenfrei, B. Von, 1978, Letters to the Editors, *Magazine of the Parrot Soc.* XII (2): 41.

Spence, T., 1955, Breeding of the Purple-capped Lory, *Avicultural Magazine* 61 (1): 14–17.

Steadman, D. W, 1985, Fossil birds from Mangaia, Southern Cook Islands, *Bull. British Orn. Club* 105, 58–66

——1989, Extinction of Birds in Eastern Polynesia: A Review of the Record, and Comparisons with Other Pacific Island Groups, *Jnl of Archaeological Science* 16: 177–205.

——1991, Extinct and Extirpated Birds from Aitutaki and Atiu, Southern Cook Islands, *Pacific Science* 45 (4): 325–47.

——1993, Biogeography of Tongan birds before and after human impact, *Proc. Natl. Acad, Sci.* 90:818–822.

—— and M. C. Zarriello, 1987, Two new species of parrots (Aves: Psittacidae) from archaeological sites in the Marquesas Islands, *Proc. Biol. Soc.Wash* 100 (3): 518–528.

Stevens, G. W., 1969, *Cage and Aviary Birds*, Studying Psittacenes in the South Pacific, July 10: 21.

Stott, K., 1975, Parrot Fever in the Solomons, *Zoonooz* XLVIII (12):11–13.

Stresemann, E., 1914, Die Vogel von Seram (Ceram), *Novitates Zoologicae* 21: 25-153

——1940, Die Vogel von Celebes, Teil III, Systematik und biologie, *J. Orn.* 88: 389-347

Stuckey, P., 1994, Australian Parrots—breeding with purpose, *Jnl of Parrot Soc. of Aust* (NSW) VI (3): 3–12 and (4): 7–11.

Sweeney, R., 1992a, Artificial incubation and hand rearing of a Black winged Lory (*Eos cyanogenia*) at Birdworld, *Magazine of the Parrot Soc.* XXVI (6): 202–7.

——1992b, Artificial incubation and hand-rearing of Blue-streaked Lories at Birdworld, *Magazine of the Parrot Soc.* XXVI (12):403–8.

——1993a, Perfect Lorikeet first breeding in the Philippines, *Cage and Aviary Birds*, November 13:9.

——1993b, Incubation artificielle et elevage a la main du Lori a Raies Bleues a Birdworld, *Les Oiseaux* (9): 37–39.

——1994a, A strong personality makes the Cardinal Lory a favourite, *Cage and Aviary Birds* Oct.8: 15.

——1994b, The Mount Apo Lorikeet, *Trichoglossus johnstoniae*, now looks set to become established in aviculture, *Lori Jnl Internationaal* 3:32–5.

——1994c, Lory care, *Bird Keeper*, February: 50-51

——1995, Importante reussite d'elevage avec le lori du mont Apo a Loro Parque, *Les Oiseaux du Monde* no. 132, Decembre, 4–5.

——1996, Colours of the rainbow, *Cage and Aviary Birds* ,September 7: 17.

Tarr, H. E., 1963, The Varied Lorikeet in Northern Australia, *Australian Bird Watcher* 2 (part 1), June: 18–19.

Tavistock, Lord, 1937, Breeding results for 1937, *Avicultural Magazine* Fifth Series, II (12): 349–350.

—— 1938, The breeding of the Tahiti Blue Lory, *Avicultural Magazine* Fifth Series, III (2): 34–38.

—— 1939, The breeding of the Ultramarine Lory, *Avicultural Magazine* Fifth Series, IV (9): 292–294.

Taylor, E. J., 1986, Breeding the Perfect Lorikeet, *Magazine of the Parrot Society* XX (3): 84.

Taylor, J., 1991a, Report from Indonesia, *PsittaScene* 3 (1): 6–8.

——1991b, Die Papageien von Ceram, *Papageien* (5): 149–55.

——1992, Die Papageien von Sangir-Talaud, *Papageien*

(5): 156–9.

Taylor, P. and J., 1995, Breeding Obi Lories, *Eos squamata obiensis*, *Lori Jnl Internationaal* 4: 27–31.

Thurlow, J. G., 1989, Breeding Musschenbroek's Lorikeet, *Avicultural Magazine* 95 (1): 4–8.

Tiskens, A. and P., 1993, The Jobi Lory, *Lorius lory jobiensis*, *Lori Jnl Internationaal* 2: 28–34.

Tiskens, P., 1995, The Mitchell's Lorikeet, *Trichoglossus haematodus mitchellii*, *Lori Jnl Internationaal* 4: 11–16.

Toft, A., C. H. Langley and D. M. Brown, 1994, Studies of Parrot evolution at University of California, Davis, *The Amazona Quarterly* 13 (2): 3–6.

Trevelyan, R., 1995, The feeding ecology of Stephen's Lory and nectar availability in its food plants, *Biological Jnl of the Linnean Society* 56: 185–197.

Tucker, S., 1994, Construction of an aviary: adventures and misadventures, *The Loriinae* 10 (1): 6–9.

Utschick, H. and R. Brandl, 1989, Roosting activities of the Rainbow Lory (*Trichoglossus haematodus*) at Wau, Papua New Guinea, *Spixiana* 11 (3): 303–310.

Vanderhoof,1994, Handfeeding Loridae, An Alternative Method, *American Lory Society Network* 1 (3): 2.

Volkmar, A., 1995, Erfahrungen mit Gualoris, *Papageien* 8 (1): 10–12.

Wallace, A. R., 1986 (reprinted from 1896), *The Malay Archipelago* Oxford University Press, Singapore.

Walser, H., 1993a, Grosser Bergzierlori (*Oreopsittacus arfaki major*), *Gefiederter Freund* (1): 14–17.

——1993b, Der Smaragd-Gualori, *Neopsittacus pullicauda*, *Gefiederter Freund* (5): 135–137.

Warham, J., 1958 [article title unknown], *Cage Birds*, July 3.

Watling, D., 1982, *The Birds of Fiji, Tonga and Samoa*, Millwood Press, Wellington

——1983, Ornithological Notes from Sulawesi, *Emu* 83: 247-261

——1995, Notes on the status of Kuhl's lorikeet *Vini kuhlii* in the Northern Line Islands, Kiribati *Bird Conservation International* 5: 481-489.

Webb, H. P., 1992, Field Observations of the Birds of Santa Isabel, *Emu* 92: 52–57.

Wenner, M-L., 1979, *Domicella Garrula Garrula* author, Zurich.

White, C. M. N. and M. D. Bruce, 1986, *The Birds of Wallacea* Brit. Orn. Union, London.

Whitten,T. and J. Whitten, 1992, *Wild Indonesia*, New Holland/WWF, London.

Wilkinson, R., 1994/5, EEP Parrot TAG Annual Report 1995, 323–328, in *EEP Yearbook 1994/5* Eds. F. Rietkerk, K. Brouwer and E. Smits, EAZA/EEP Executive Office, Amsterdam.

——, A.Woolham, P. Morris and A. Morris, 1993, Notes on the husbandry and breeding of Musk Lorikeets at Chester Zoo, *Avicultural Magazine* 99 (4): 188–92.

Willis,R., 1991, Portrait of a pet: an Australian First, *Australian Natural History* 23 (9): 680–1.

Wilson, K., published in 1988, but undated, Red-Collars...Rubies of the sky, *Australian Birdkeeper* 2 (8): 271–4.

——1991-2, The Littlest Lorikeets, *Australian Birdkeeper* 4 (12): 560–3.

——1992, Musks, the lorikeets of levity, *Australian Birdkeeper* 5 (2): 71–4.

——1993, Rakish Rainbows, *Australian Birdkeeper* 6 (7): 321–326.

Wilson, K. - J.,1993, Observations on the Kuramo'o (*Vini peruviana*) on Aitutaki Island, Cook Islands, *Notornis* 40: 71–75.

Wilson, P., 1994, Nutritional myopathy in free-living lorikeets, *Annual Conference 1994 Proc. ACAAV*: 229–234.

Wood, K. A., 1992, Cockatoos, parrots and lorikeets eating food from introduced trees and shrubs in the Illawarra region, *Aust. Birds* 25 (3): 79–81.

Wright, A. J., 1981, Hand-rearing and fostering lories, *Magazine of the Parrot Soc.* XV (12): 340–1.

Wright, C. K., 1977, The first breeding of two Lory species? *Avicultural Magazine* 83 (4): 189–90.

Zawila. A. R., (1992), The real cost of owning a parrot *Bird Talk* 10 (4): 10–12.

Zompa, T., 1995, Bacterial Culture and Antibiotic Therapy in Lories and Lorikeets, *American Lory Soc. Network* 2 (1): 3–4.

*Note: The publications entitled *The Loriidae Newsletter*, *The Loriidae*, *The Loriinae* and *The Loriinae: Bulletin of the International Loriinae Society* are all publications of the latter society which changed its name from the International Loriidae Society.

Index

Scientific names (main text reference only)

Zimmerli, E. 373
Zompa, Ted 20
Zurich, University of 348

Zoos mentioned in text

Assiniboine Park Zoo 325
Barcelona Zoo 262
Birdland, Bourton-on-the-Water 194
Birdworld, Farnham 209, 211, 213
Busch Gardens, Tampa 66
Bronx Zoo 174, 338
Chester Zoo 26, 174, 250, 283, 310, 311, 312, 394, 398
Cologne Zoo 266
Currumbin Sanctuary 65, 161, 259
Hamilton Zoo 398
Honolulu Zoo 338
Jurong BirdPark 56, 154, 182, 220
Kelling Park Aviaries 261, 271, 282
London Zoo 66, 249, 262, 321, 329
Loro Parque, Tenerife 9, 51, 56, 57, 60, 82, 162, 198, 204, 205, 214, 220, 229, 231, 247, 249, 251, 256, 257, 265, 266, 268, 273, 274, 338, 339, 354, 365, 373, 377, 378
Los Angeles Zoo 271
Marine World Africa, U.S.A. 161

Munich Zoo 232
Natal Lion Park 265
Palmitos Park 9, 27, 31, 36, 57, 148, 162, 169, 170, 171, 190, 193, 194, 196, 198, 222, 224, 228, 235, 239, 240, 243, 244, 245, 267, 297, 364, 367, 371, 373, 376, 377, 378, 379, 381, 383
Pearl Coast Zoo, Broome, Australia 223
Penscynor Wildlife Park 398
Rainbow Jungle 315
Rotterdam Zoo 56, 311, 312
San Diego Wild Animal Park 162 ,337, 338
San Diego Zoo 18, 56, 82, 83, 148, 149, 161, 177, 178, 182, 191–192, 198, 207, 208, 210, 261, 282, 312, 318, 320, 322, 323, 324, 325, 326, 332, 335, 336, 338, 339, 342, 343, 356, 373, 377, 384, 398
Stuttgart Zoo 255
Taronga Zoo 88, 232, 318, 321, 330, 406
Umgeni River Bird Park 18
Vogelpark Walsrode 16, 78, 182, 196, 205, 215, 249, 250, 275, 283, 322, 353, 355, 373, 398
Wassenaar Zoo 250, 264
West Berlin Zoo 201
Wildlife World 161
Woburn Safari Park 162

432